Warm Climates in Earth History

The geologic record contains evidence of greenhouse climates in the earth's past, and by studying these we can gain greater understanding of the forcing mechanisms and feedbacks that influence today's climate. By examining several warm intervals of geologic time from the same perspectives (oceanic and terrestrial; theoretical and observational), the chapters in this book illuminate the differences and commonalities among various greenhouse climates.

Leading experts in paleoclimatology combine in one integrated volume new and state-of-the-art paleontological, geological, and theoretical studies of intervals of global warmth. The book reviews what is known about the causes and consequences of globally warm climates, demonstrates current directions of research, and outlines the central problems that remain unresolved. The chapters present new research on a number of different warm climate intervals from the early Paleozoic to the early Cenozoic. They also integrate a range of approaches, including paleoclimate simulations using coupled GCMs, paleoclimate reconstructions using paleontological and geochemical data, refinement of paleogeography, and the study of the effects of climate change on marine and terrestrial organisms.

The book will be of interest to researchers in paleoclimatology, and will also be useful as a supplementary text for advanced undergraduate or graduate level courses in paleoclimatology and earth science.

Brian T. Huber is Research Curator in the Department of Paleobiology, National Museum of Natural History, Smithsonian Institution, Washington, DC

Kenneth G. MacLeod is Assistant Professor at the Department of Geology, University of Missouri

Scott L. Wing is Research Curator, Department of Paleobiology, National Museum of Natural History, Smithsonian Institution, Washington, DC

WARM CLIMATES IN EARTH HISTORY

EDITED BY

BRIAN T. HUBER, *Smithsonian Institution*
KENNETH G. MACLEOD, *University of Missouri*
SCOTT L. WING, *Smithsonian Institution*

PUBLISHED BY THE PRESS SYNDICATE OF THE UNIVERSITY OF CAMBRIDGE
The Pitt Building, Trumpington Street, Cambridge, United Kingdom

CAMBRIDGE UNIVERSITY PRESS
The Edinburgh Building, Cambridge CB2 2RU, UK   www.cup.cam.ac.uk
40 West 20th Street, New York, NY 10011-4211, USA   www.cup.org
10 Stamford Road, Oakleigh, Melbourne 3166, Australia
Ruiz de Alarcón 13, 28014 Madrid, Spain

First published 2000

Printed in the United Kingdom at the University Press, Cambridge

Typeset in Times

A catalogue record for this book is available from the British Library

Library of Congress Cataloguing in Publication data

Warm climates in earth history / edited by Brian T. Huber, Kenneth G.
  MacLeod, and Scott L. Wing.
      p.   cm.
  Includes index
  ISBN 0 521 64142 X hardback
  1. Paleoclimatology. 2. Earth sciences. I. Huber, Brian T.
II. MacLeod, Kenneth G., 1964– . III. Wing, Scott L.
QC884.W37   1999
551.6′09′01–dc21   98-51724 CIP

ISBN 0 521 64142 X hardback

Contents

List of contributors   vii

Preface   xi

Part I   Approaches to the study of paleoclimates   **1**

1   Warm climate forcing mechanisms
Paul J. Valdes   3

2   Recent advances in paleoclimate modeling: toward better simulations of warm paleoclimates
Robert M. DeConto, Starley L. Thompson, and Dave Pollard   21

3   Comparison of zonal temperature profiles for past warm time periods
Thomas J. Crowley and James C. Zachos   50

Part II   Case studies: latest Paleocene–early Eocene   **77**

4   Comparison of early Eocene isotopic paleotemperatures and the three-dimensional OGCM temperature field: the potential for use of model-derived surface water $\delta^{18}O$
Karen L. Bice, Lisa C. Sloan, and Eric J. Barron   79

5   Deep-sea environments on a warm earth: latest Paleocene–early Eocene
Ellen Thomas, James C. Zachos, and Timothy J. Bralower   132

6   Mountains and Eocene climate
Richard D. Norris, Richard M. Corfield, and Karen Hayes-Baker   161

7   An early Eocene cool period? Evidence for continental cooling during the warmest part of the Cenozoic
Scott L. Wing, Huiming Bao, and Paul L. Koch   197

Part III   Case studies: Mesozoic   **239**

8   Paleontological and geochemical constraints on the deep ocean during the Cretaceous greenhouse interval
Kenneth G. MacLeod, Brian T. Huber, and My Le Ducharme   241

9   Late Cretaceous climate, vegetation, and ocean interactions
     *Robert M. DeConto, Esther C. Brady, Jon Bergengren, and
     William W. Hay*   275
10   Jurassic phytogeography and climates: new data and model
     comparisons
     Peter McA. Rees, Alfred M. Ziegler, and Paul J. Valdes   297

Part IV   Case studies: Paleozoic   **319**
11   Permian and Triassic high latitude paleoclimates: evidence from
     fossil biotas
     Edith L. Taylor, Thomas N. Taylor, and N. Rubén Cúneo   321
12   Organic carbon burial and faunal dynamics in the Appalachian
     Basin during the Devonian (Givetian–Famennian) greenhouse:
     an integrated paleoecological and biogeochemical approach
     *Adam E. Murphy, Bradley B. Sageman, Charles A. Ver Straeten,
     and David J. Hollander*   351
13   Glaciation in the early Paleozoic 'greenhouse': the roles of
     paleogeography and atmospheric CO_2
     Mark T. Gibbs, Karen L. Bice, Eric J. Barron, and Lee R. Kump   386

Part V   Overview: climate across tectonic timescales   **423**
14   Carbon dioxide and Phanerozoic climate
     Thomas J. Crowley   425

Index   445

Contributors

Huiming Bao
Department of Geosciences, Princeton University, Princeton, NJ 08544, USA

Eric J. Barron
Department of Geosciences and Earth System Science Center, Pennsylvania State University, 248 Deike Building, University Park, PA 16802, USA

Jon Bergengren
National Center for Atmospheric Research, PO Box 3000, Boulder, CO 80307, USA

Karen L. Bice
Department of Geosciences and Earth System Science Center, Pennsylvania State University, 248 Deike Building, University Park, PA 16802, USA

Esther C. Brady
National Center for Atmospheric Research, PO Box 3000, Boulder, CO 80307, USA

Timothy J. Bralower
Department of Geology, University of North Carolina, Chapel Hill, NC 27599-3315, USA

Richard M. Corfield
Department of Earth Sciences, Oxford University, Parks Road, Oxford, OX1 3PR, UK

Thomas J. Crowley
Department of Oceanography, Texas A&M University, College Station, TX 77843, USA

N. Rubén Cúneo
Museo Paleontológico Egidio Feruglio, 9 de Julio 655, (9100) Trelew, Chubut, Argentina

Robert M. DeConto
Department of Geosciences, 233 Morrill Science Center, University of Massachusetts, Amherst, MA 01003, USA

My Le Ducharme
Department of Paleobiology, National Museum of Natural History, Smithsonian Institution, MRC: NHB 121, Washington, DC 20560, USA

Mark T. Gibbs
Center for Climatic Research, University of Wisconsin at Madison, 1225 West Dayton Street, Madison, WI 53706, USA

William W. Hay
GEOMAR, Wischhofstr. 1–3, D-24148 Kiel, Germany

Karen Hayes-Baker
Department of Earth Sciences, Oxford University, Parks Road, Oxford, OX1 3PR, UK

David J. Hollander
Department of Geological Sciences, Northwestern University, Evanston, IL 60208, USA

Brian T. Huber
Department of Paleobiology, National Museum of Natural History, Smithsonian Institution, MRC: NHB 121, Washington, DC 20560, USA

Paul L. Koch
Department of Earth Sciences, University of California, Santa Cruz, CA 95064, USA

Lee R. Kump
Department of Geosciences and Earth System Science Center, Pennsylvania State University, 248 Deike Building, University Park, PA 16802, USA

Kenneth G. MacLeod
Department of Geology, University of Missouri-Columbia, Columbia, MO 65211, USA

Adam E. Murphy
Department of Geological Sciences, Northwestern University, Evanston, IL 60208, USA

Richard D. Norris
Woods Hole Oceanographic Institution, MS-23, Woods Hole, MA 02543-1541, USA

Dave Pollard
Earth System Science Center, Pennsylvania State University, 248 Deike Building, University Park, PA 16802, USA

Peter McA. Rees
Department of Earth Sciences, The Open University, Walton Hill, Milton Keynes, MK7 6AA, UK, and Paleographic Atlas Project, Department of Geophysical Sciences, University of Chicago, 5734 S. Ellis Avenue, Chicago, IL 60637, USA

Bradley B. Sageman
Department of Geological Sciences, Northwestern University, Evanston, IL 60208, USA

Lisa C. Sloan
Department of Earth Sciences, University of California, Santa Cruz, CA 95064, USA

Edith L. Taylor
Department of Botany and Natural History Museum and Biodiversity Center, University of Kansas, Lawrence, KS 66045, USA

Thomas N. Taylor
Department of Botany and Natural History Museum and Biodiversity Center, University of Kansas, Lawrence, KS 66045, USA

Ellen Thomas
Center for the Study of Global Change, Yale University, PO Box 208109, New Haven, CT 06520-8109, and Department of Earth and Environmental Sciences, Wesleyan University, Middletown, CT 06459-0139, USA

Starley L. Thompson
National Center for Atmospheric Research, PO Box 3000, Boulder, CO 80307, USA

Paul J. Valdes
Department of Meteorology, University of Reading, PO Box 243, Earley Gate, Reading, RG6 6BB, UK

Charles A. Ver Straeten
Department of Geological Sciences, Northwestern University, Evanston, IL 60208, USA

Scott L. Wing
Department of Paleobiology, Smithsonian Institution, Washington, DC 20560, USA

James C. Zachos
Department of Earth Sciences, University of California at Santa Cruz, Santa Cruz, CA 95064, USA

Alfred M. Ziegler
Paleographic Atlas Project, Department of Geophysical Sciences, University of Chicago, 5734 S. Ellis Avenue, Chicago, IL 60637, USA

Preface

The first-order climate history of the Phanerozoic eon is now known, and it reveals that globally warm intervals without major icecaps represent approximately 75% of the last 540 million years (Frakes, 1979; Fischer, 1982; Crowley, Chapter 14). Study of these greenhouse climates is taking on increased importance in light of climatic warming predicted to result from the doubling of pre-industrial CO_2 concentrations over the next century (Broecker, 1997). In spite of the prevalence of warm climates in earth history and the potential practical significance of understanding them, the fundamental causes, nature, and mechanics of warm climates are still poorly understood. We have assembled this book as a way of reviewing what is known about the causes and consequences of globally warm climates, of demonstrating current directions of research on warm climates, and of outlining the central problems that remain unresolved. In serving these goals the chapters present new research on a number of different warm climate intervals from the early Paleozoic to the early Cenozoic. The chapters also integrate a range of approaches, from paleoclimate simulation with atmospheric and oceanic general circulation models, to paleoclimate reconstruction from paleontological and geochemical data, to refinement of paleogeography, to the study of the effects of climate change on marine and terrestrial organisms.

Much of the research effort on climate history has been devoted to studies of the Pleistocene–Holocene, a period for which climate change can be studied with a temporal resolution of years to centuries and for which excellent geographic coverage is available across many environments. Paleoclimate studies of the more distant past are generally confined to temporal resolutions of millennia or greater and suffer from less complete spatial coverage, but they greatly expand the range of earth climates that can be studied. Climate and climatic variables exhibit relatively small fluctuations during the glacial/interglacial cycles of the last 2–3 million years. For example, during the peak of the last glaciation atmospheric CO_2 concentrations were ~30% less than pre-industrial levels, and anthropogenic emissions over the last 125 years have increased the pre-industrial level by ~30%. Predicted doubling of pre-industrial pCO_2 for the coming century (Broecker, 1997) greatly exceeds the maximum CO_2 concentrations reached during the recent glacial/interglacial cycles. Estimates of pCO_2 for ancient periods of global warmth, on the other hand, are as high as 2× pre-industrial values in the early Eocene (Part II), 5× pre-industrial levels in the late Mesozoic (Part III), and 12–15× pre-industrial values in the early Paleozoic (Part IV). It is difficult to rigorously document these ancient warm climates, but larger and better-refined proxy data sets (fossil and geochemical) and advances in atmospheric and oceanic models are allowing examination of climate–ocean dynamics at increasingly fine temporal and spatial scales. Proxy data and model simulations have often led to conflicting conclusions, and failure of these two approaches to converge suggests gaps in our understanding of greenhouse climates. Study of warm intervals in the distant past is thus an important testing ground for establishing a general understanding of climate.

Improvements in computer performance and increased sophistication of climate simulations are narrowing some of the discrepancies between geological observations and climate model outputs by enabling more realistic portrayals of modern and ancient climatic boundary conditions and interactions (DeConto *et al.*, Chapter 2). For example, early general circulation models for the Eocene predicted seasonal freezing in the North American continental interior (Sloan and Barron, 1990). These results directly contradicted paleontological inferences of mild winters based on the presence of palms, crocodilians, and other organisms that do not tolerate severe frosts (MacGinitie, 1969, 1974; Hickey, 1977; Wing, 1991; Markwick, 1994). A more recent modeling study with more realistic paleogeography, including the presence of a large lake, predicted mean annual temperatures in better agreement with the paleontological data (Sloan, 1994). However, significant disagreement remains between simulations of cold monthly mean values and geological proxy data (Greenwood and Wing, 1995; Bice *et al.*, Chapter 4).

Controversy is not limited to comparisons between geological data and modeling results. Modern tropical sea surface temperatures (SSTs) are generally between ~24 and 28 °C, but estimates of ancient tropical temperatures vary much more. For the Eocene, estimates of tropical SSTs based on $\delta^{18}O$ measurements range from ~18 °C (Shackleton *et al.*, 1984) to 18–24 °C (Bralower *et al.*, 1995), whereas the distribution of Eocene reefs and other tropical organisms suggests tropical SSTs that were nearly the same as the modern (Adams *et al.*, 1990). Similarly, many Late Cretaceous isotopic studies conclude that tropical SSTs were ~10 °C below modern values (e.g., Douglas and Savin, 1978; D'Hondt and Arthur, 1995), whereas recent analyses of exceptionally preserved Late Cretaceous rudistid bivalves and early marine cements indicate tropical temperatures at least as warm as the modern (Wilson and Opdyke, 1996). Diagenetic artifacts, salinity effects on $\delta^{18}O_{seawater}$, and secular variations in $\delta^{18}O_{seawater}$ are all sources of error in paleotemperature estimates and need to be more carefully considered in paleoceanographic studies (Zachos *et al.*, 1994; Bralower *et al.*, 1995; D'Hondt and Arthur, 1995; Huber *et al.*, 1995; Crowley and Zachos, Chapter 3; Bice *et al.*, Chapter 4; MacLeod *et al.*, Chapter 8).

Terrestrial paleoclimate estimates suffer from similar levels of uncertainty. Perhaps the most widely used technique for inferring continental temperatures is Leaf Margin Analysis, which is based on the strong positive correlation observed in living vegetation between the proportion of species in a local flora that have entire (smooth) margins and the mean annual temperature under which the flora grows (Wolfe, 1979). In spite of fairly wide use, the underlying ecophysiological basis of the correlation is not well understood (Wing et al., Chapter 7), and proper estimates of the statistical error in the method have only recently been developed (Wilf, 1997). Other sets of correlations between leaf morphology and climate are also being used to estimate paleoclimate features (Wolfe, 1993), again with little knowledge of the functional causes of the correlations. Correlations between leaf morphology and precipitation are even less precise than those between leaf morphology and temperature (Wilf et al., 1998). Other methods of continental paleoclimate reconstruction generally yield semiquantitative information (e.g., presence of palms inconsistent with ground frost), which makes them difficult to compare with simulation results. Furthermore, all methods of paleoclimate reconstruction based on organisms (terrestrial or marine) assume to some degree that there are consistent relationships between climate and the distribution of lineages or morphologies; this assumption becomes progressively more tenuous with increasing geologic age. Clearly, a better understanding of the physiological basis for the correlations we use would improve confidence in paleontologically based estimates of paleoclimate.

Advances in climate modeling and reconstruction of paleoclimates from proxy data have resolved some controversies but have underscored areas where understanding of greenhouse climates remains imperfect. Latitudinal temperature gradients are particularly problematic. Simulations of greenhouse climates that seem to reproduce accurate globally averaged temperatures predict steeper latitudinal temperature gradients (both warmer tropics and colder poles) than proxy data seem to allow. Possible resolutions of this disagreement include examining climate-dependent, systematic variations in the extent and type of cloud cover with latitude; increased oceanic heat transport, perhaps through a reorganization of ocean circulation; unrecognized climatic interactions such as positive feedback between high latitude forest and polar warmth; differences in plate configurations and/or sea level; artifacts of poor resolution and processing time of climate simulations compared with the complexity of climate on geologic time scales; and flaws in paleotemperature estimates introduced by salinity and diagenetic influences on the $\delta^{18}O$ record. Chapters in this book examine the problem of understanding greenhouse climates and provide current discussion of these issues. Despite a wide range of approaches among the authors, our focus is on the differences between data and theory, with an emphasis on the use of multiple independent lines of evidence as a means of testing results from numerical climate simulations.

The book is divided into five parts. Part I discusses important general factors in paleoclimate studies. The first chapter, by Valdes, addresses the influences on global climate along a spectrum from purely external forcing mechanisms (e.g., changes in solar luminosity across geologic time) to internal feedbacks (e.g., ice albedo) from the perspective of simple equations describing the earth's heat budget. Chapter 2, by DeConto *et al.*, reviews approaches, advances, and limitations of paleoclimate modeling studies. In Chapter 3, Crowley and Zachos compare estimates (from oxygen isotopes and transfer functions) of tropical sea surface temperatures for five warm intervals. Especially for oxygen isotope data, they examine sources of random and systematic error in an attempt to determine uncertainties in making estimates of ancient tropical sea surface temperatures. In short, the three chapters in this part summarize the basic mechanics of climate, basic methods and approaches to climate modeling, and the reliability of one of the most widely used proxies (oxygen isotopes) for estimating ancient paleotemperatures.

Most of the remaining chapters are case studies that present modeling results and paleontological and/or geochemical data for specific Phanerozoic intervals of global warmth. Contributions come from climate and ocean modelers, paleontologists, and stable isotope geochemists often working collaboratively on given chapters. It is impossible to provide uniform coverage of greenhouse intervals across the Phanerozoic while including in-depth discussion of the latest research results, but chapters in Parts II–IV are organized to collectively discuss geologic data (continental and oceanic) and modeling results for progressively older time windows. By including chapters that examine the nature and behavior of greenhouse climates for different times from different perspectives, it is hoped that the book as a whole will allow the reader to begin to evaluate whether there are general greenhouse climatic rules or if the grouping of intervals of climate history by average temperature is more convenient than informative (Crowley, Chapter 14).

The best-studied interval of global warmth, the late Paleocene–early Eocene, is the focus of Part II. The part begins with a comparison by Bice *et al.* (Chapter 4) of the fit between global ocean temperatures predicted from computer climate models and oxygen isotopic paleotemperatures estimated from benthic and shallow-dwelling planktic foraminifera for the early Eocene. These authors found generally good agreement between model-predicted and isotopically based estimates of surface water paleotemperatures but found the modeled deep water temperatures are significantly less than paleotemperature estimates. In the following chapter, Thomas *et al.* (Chapter 5) interpret the global paleoceanography of the latest Paleocene–earliest Eocene time based on new and published paleontological and stable isotopic evidence, with an emphasis on description and likely causation of the Latest Paleocene Thermal Maximum event. Norris *et al.* (Chapter 6) discuss different approaches for reconstructing continental paleotopography and propose a way to estimate paleoaltitudes using stable isotopic data. They apply this method to the Eocene Green River Formation of the western United States and conclude that the Laramide mountains were as high or higher than the modern Rockies. Part II ends with a chapter by Wing *et al.* (Chapter 7) who find that paleobotanical and geochemical records from the northern Rocky Mountains are consistent with 0.5–1.0 million years of cooling during the otherwise globally warm early Eocene. This cooling implies unanticipated variability in the strongest greenhouse interval of the last 65 million years, and may have had major effects on the timing and rate of change in terrestrial faunas and floras.

Part III includes case studies for periods of extreme global warmth during the Mesozoic. The part opens with a chapter by MacLeod *et al.* (Chapter 8), who document the global distribution of deep sea inoceramid bivalves across the Late Cretaceous. They compare this distribution to parallel studies of stable isotope data and discuss the results relative to possible changes in the pattern of deep water circulation. DeConto *et al.* (Chapter 9) attempt to construct a realistic simulation of Campanian climate. Their simulation uses elevated pCO$_2$ and oceanic heat transport similar to earlier studies, but also includes a predictive vegetation model that allows climate and biology to interact. They conclude that vegetation may have played a key role in maintaining low meridional thermal gradients during the Late Cretaceous. The final chapter in this section is a study by Rees *et al.* (Chapter 10) who develop a new method for treating floristic data quantitatively, then use a global database of Jurassic plant assemblages and climatically sensitive sediment types to show that a warm, stable climate persisted throughout this geologic period.

Part IV includes three chapters addressing primarily Paleozoic intervals. Despite varied spatial, temporal, and topical focus, the three Paleozoic case studies employ the same multidisciplinary approach as the earlier chapters, and feature discussion of discrepancies between climate model simulations and geological data. According to Taylor *et al.* (Chapter 11), exquisitely preserved cellular growth structures in Permian and Triassic fossil wood from Antarctica indicate that the south polar climate was much warmer and less seasonally extreme than climate models predict. They suggest that inaccurate portrayal of land surface albedo and topography may account for some of the mismatch between the models and data. An earlier Paleozoic greenhouse phase is investigated by Murphy *et al.* in Chapter 12. For Middle to Late Devonian time (an interval marked by a prolonged extinction of shallow marine invertebrates), these authors propose a linkage between marine biogeochemical cycles, benthic community dynamics, and climate. The final chapter of Part IV, by Gibbs *et al.* (Chapter 13), presents parallel climate simulations of the Late Ordovician (glacial) and Early Silurian (non-glacial/greenhouse) to address the relative importance of paleogeography and pCO$_2$ in forcing the transition from a greenhouse to icehouse global climate state. In addition to the paired comparison between adjacent, climatically dissimilar time intervals, this chapter presents an interesting test of greenhouse simulations. Unlike Mesozoic and Cenozoic climate models, these early Paleozoic model experiments assume significantly lower solar luminosity, much higher (10–18× present day) pCO$_2$, and negligible terrestrial vegetation cover.

In the closing chapter of the book, Crowley (Chapter 14) steps back and addresses Phanerozoic climates as a whole and discusses a possible hierarchical ordering of climatic forcing mechanisms. He demonstrates a strong correlation between pCO$_2$ and climate state during the Phanerozoic but also considers paleogeography and solar luminosity as critical factors in the global paleoclimate equation. He suggests that orography, ocean heat transport, vegetation, and instabilities in the climate system act as secondary feedback mechanisms that become most important when global climate is at a bifurcation point. He concludes that climate forcing mechanisms will have to be integrated in a more systematic and parsimonious way in order to improve agreement between paleoclimate model results and proxy data.

As expected, the contributions in this book make clear that there is still a lot left to understand about warm climate intervals. However, recent paleoclimate research is already paying off. Broad patterns in the history of climate are generally accepted and the paleontological and geochemical evidence for shallow latitudinal temperature gradients and much warmer winter temperature in mid–high latitude continental interiors now seems solid. Perhaps most importantly, though, integrated studies have revealed consistent areas of disagreement between empirical and theoretical analyses. Increasing temporal and geographic resolution in models and proxy data will only improve the ability to use deep time as a testing ground for our understanding of the causes, nature, and mechanics of globally warm climates.

ACKNOWLEDGEMENTS

This book would not have been possible without the enthusiastic cooperation of chapter authors from the time of the North American Paleontological Convention symposium on this topic (Washington, DC; 1996) through the receipt of final revised manuscripts. We would also like to thank the following manuscript reviewers, as they provided many insightful suggestions and helpful editorial comments: Eric Barron, Lisa Boucher, Timothy Bralower, Peter Brenchley, Carleton Brett, Daniel Bryant, Andrew Bush, Will Clyde, Tom Crowley, Peter Fawcett, Larry Frakes, Kate Freeman, Katherine Gregory-Wodzicki, Mitch Lyle, Bette Otto-Bliesner, Dorothy Pak, Judy Parrish, Chris Poulsen, G.D. Price, Peter Schultz, Robert Spicer, Ellen Thomas, Paul Valdes, Joen Widmark, and Fred Ziegler. We are especially indebted to Karen Bice for the extraordinary amount of time and effort she put in as a reviewer and as a sounding board for the entire book. Thanks are also extended to Mary Parrish for producing the cover illustration and My Le Ducharme for her editorial assistance. Funds for production of color inserts were received from Eric Barron (Earth System Science Center, Pennsylvania State University), Robert DeConto (NSF grant no. NSF-EAR 9405737 and National Center for Atmospheric Research funds), Ellen Thomas (Grants in Support of Scholarship, Wesleyan University), and the Smithsonian Institution's National Museum of Natural History.

REFERENCES

Adams, C. G., Lee, D. E. and Rosen, B. (1990). Conflicting evidence for tropical sea-surface temperatures during the Tertiary. *Palaeogeography, Palaeoclimatology, Palaeoecology*, **77**, 289–313.

Bralower, T. J., Zachos, J. C., Thomas, E. *et al.* (1995). Late Paleocene to Eocene paleoceanography of the equatorial Pacific Ocean: Stable isotopes recorded at Ocean Drilling Program Site 865, Allison Guyot. *Paleoceanography*, **10**, 841–65.

Broecker, W. (1997). Will our ride into the greenhouse future be a smooth one? *GSA Today*, **7**, no. 5, 1–7.

D'Hondt, S. and Arthur, M. A. (1995). Interspecific variation in stable isotopic signals of Maastrichtian planktic foraminifera. *Paleoceanography*, **10**, 123–35.

Douglas, R. G. and Savin, S. M. (1978). Oxygen isotopic evidence for the depth stratification of Tertiary and Cretaceous foraminifera *Palaeogeography, Palaeoclimatology, Palaeoecology*, **3**, 175–96.

Fischer, A. G. (1982). Long-term climatic oscillations recorded in stratigraphy. In *Climate in Earth History*; Washington, DC: National Academy Press. eds. W. Berger and H. Crowell, pp. 97–104.

Frakes, L. A. (1979). *Climates throughout Geologic Time*. Netherlands: Elsevier.

Greenwood, D. R. and Wing, S. L. (1995). Eocene continental climates and latitudinal temperature gradients. *Geology*, **23**, 1044–8.

Hickey, L. J. (1977). Stratigraphy and paleobotany of the Golden Valley Formation (early Tertiary) of western North Dakota. *Geological Society of America Memoir*, **150**, 1–181.

Huber, B. T., Hodell, D. A. and Hamilton, C. P. (1995). Mid- to Late Cretaceous climate of the southern high latitudes: Stable isotopic evidence for minimal equator-to-pole thermal gradients. *Geological Society of America Bulletin*, **107**, 1164–91.

MacGinitie, H. D. (1969). The Eocene Green River flora of northwestern Colorado and northeastern Utah. *University of California Publications in Geological Sciences*, **83**, 1–202.

MacGinitie, H. D. (1974). An early middle Eocene flora from the Yellowstone–Absaroka volcanic province, northwestern Wind River Basin, Wyoming. *University of California Publications in Geological Sciences*, **108**, 1–103.

Markwick, P. J. (1994). "Equability", continentality, and Tertiary "climate": The crocodilian perspective. *Geology*, **22**, 613–16.

Shackleton, N. J., Hall, M. A. and Boersma, A. (1984). Oxygen and carbon isotope data from Leg 74 foraminifers. In *Initial Reports of the Deep Sea Drilling Project*, vol. 74, eds. T. C. Moore, Jr., P. D. Rabinowitz *et al.*, pp. 599–612. Washington, DC: US Government Printing Office.

Sloan, L. C. (1994). Equable climates during the early Eocene: Significance of regional paleogeography for north American climate. *Geology*, **22**, 881–4.

Sloan, L. C. and Barron, E. J. (1990). "Equable" climates during Earth history? *Geology*, **18**, 489–92.

Wilf, P. (1997). When are leaves good thermometers? A new case for Leaf Margin Analysis. *Paleobiology*, **23**, 373–90.

Wilf, P., Wing, S. L., Greenwood, D. R. and Greenwood, C. L. (1998). Using fossil leaves as paleo-rain gauges – an Eocene example. *Geology*, **26**, 203–6.

Wilson, P. A. and Opdyke, B. N. (1996). Equatorial sea-surface temperatures for the Maastrichtian revealed through remarkable preservation of metastable carbonate. *Geology*, **24**, 555–8.

Wing, S. L. (1991). Comment and reply on '"Equable" climates during Earth history'. *Geology*, **19**, 539–40.

Wolfe, J. A. (1979). Temperature parameters of humid to mesic forests of eastern Asia and relation to forests of other regions of the northern hemisphere and Australasia. *US Geological Survey Professional Paper*, **1106**, 1–37.

Wolfe, J. A. (1993). A method of obtaining climatic parameters from leaf assemblages: *US Geological Survey Bulletin*, **2040**, 1–71.

Zachos, J. C., Stott, L. D. and Lohmann, K. C. (1994). Evolution of early Cenozoic marine temperatures. *Paleoceanography*, **9**, 353–87.

I

Approaches to the study of paleoclimates

1

Warm climate forcing mechanisms

PAUL J. VALDES

ABSTRACT

Warm climates present a particularly challenging test of our understanding of climate system processes. This paper reviews the possible mechanisms that can affect climate on long timescales. Increased radiatively active gases can act to warm climate but cannot by themselves simulate correctly the temperature gradient between equator and pole. Ocean and atmospheric heat transport, large bodies of water, a modified cryosphere, clouds, and surface vegetation are all thought to be of importance, but our limited ability to model some of these processes means that at present we cannot fully explain the mechanisms behind warm periods in earth history.

INTRODUCTION

Warm climates in the distant past provide one of the most challenging tests of our understanding of climate system processes and our ability to predict them. The differences between these climates and our present one are dramatic. It is hard to imagine a world in which, for instance, there were no ice caps at either pole. Why were these periods so warm? What processes could lead to a climate regime that was so radically different from our present one? What are the parallels with possible warmer worlds that are predicted for the future? These questions provide the motivation for paleoclimate studies of these periods, and elucidating them is of fundamental importance for our understanding of climate dynamics in warmer climate regimes.

The evidence for warm climates in the past is addressed in other chapters in this volume (e.g., Crowley and Zachos, Chapter 3; Bice *et al.*, Chapter 4; Thomas *et al.*, Chapter 5; MacLeod *et al.*, Chapter 8). Data from a wide variety of ocean and terrestrial sources all agree on some basic aspects. The Mesozoic and early Tertiary were undoubtedly warmer than the present, especially at mid- and high latitudes. Ice cover was either absent or very restricted. Low latitude temperatures seem to have been similar to the present, at least within the bounds of observational error (*c.* 2 °C; (Crowley and Zachos, Chapter 3), and thus latitudinal temperature gradients appear to have been very much reduced. This chapter discusses the possible forcing

3

mechanisms that could produce such a warm, low gradient climate and examines the role of climate modeling in explaining past warm periods.

DEFINITION OF THE PROBLEM

The increasing amount of knowledge that we are gaining of past climates enables us to ask ever more subtle questions, but there are probably three main issues. Firstly, why was the mean global climate warmer for some periods in the past? Secondly, why was the temperature gradient between equator and pole reduced? Geological evidence indicates that the greatest warming was at high latitudes, and that temperatures in the tropics were similar to, or even slightly cooler than, those at present (Crowley and Zachos, Chapter 3, this volume). Early climate model simulations for the Cretaceous suggested that the amount of CO_2 (or other radiatively active gases) required to produce sufficient warmth at high latitudes would result in overheating in the tropics (Schneider et al., 1985). Thus warm periods in the past cannot be explained solely by increased concentrations of CO_2 and other radiatively active gases. Furthermore, a reduced temperature gradient between equator and pole suggests relatively sluggish atmospheric zonal winds (although it should be noted that the geological record measures only the surface temperature gradient; a vertically averaged temperature gradient would be more useful but is geologically unattainable).

The third issue of importance, not least because of the contradictions between model results and data, is continental interior temperatures. The geological data suggest that these were more equable than today: for example winter temperatures are thought to have been above freezing in the Eocene and Cretaceous (e.g., Wing and Greenwood, 1993; Markwick, 1994; Herman and Spicer, 1997; Norris et al., Chapter 6, this volume; Wing et al., Chapter 7, this volume). However, most model simulations to date have produced too low winter temperatures (e.g., Barron and Washington, 1984; Barron et al., 1993). This result arises from the simple fact that land has a relatively low effective heat capacity compared with sea ice or ocean (a ratio of 1:5:60), and thus cools rapidly in winter. The contradiction between the data and model results has led to claims that either the data are incorrect or the models and our understanding of climate are fundamentally flawed (Sloan and Barron, 1990; Wing, 1991).

A SIMPLE CLIMATE MODEL

Before discussing possible mechanisms it will be useful to develop a conceptual model of how our climate system works. To a first approximation, and for time periods in excess of decadal, the climate system must reach an equilibrium in which the global annual mean of incoming solar energy is balanced by the outgoing long-wave radiation (Crowley and North, 1991). If this were not true then the earth would warm or cool until a new equilibrium was reached. This 'energy balance' can be written as

$$S_0 \times (1 - \alpha_p) = \varepsilon_p \sigma T_s^4, \tag{1.1}$$

where S_0 is the solar constant (currently about 1370 W m^{-2} but 4.5% less in the late Ordovician); α_p is the planetary albedo (proportion of solar radiation reflected back to space, currently approximately 0.30); ε_p is proportional to the infrared emissivity, and is currently approximately 0.61; σ is the Stefan–Boltzmann constant (5.67 × 10^{-8} W m^{-2} K^{-4}), and T_s is the surface temperature (in kelvin). The flux of energy released from the earth's core is small compared with the other terms and can probably be ignored (but see Roach, 1998).

For the earth at present, this corresponds to a global annual mean surface temperature of 15 °C, which is approximately correct. The model is so simple that it has to apply in all climate regimes, whether past, present, or future. Unfortunately, it is also so simple that it is not really a predictive model but rather a diagnostic one. There are many parameters that have to be specified but which can and do vary depending on the climate system. For instance, the planetary albedo depends on the surface conditions (especially ice cover, land extent, and vegetation type) and also on the amount, type, and microphysical properties of clouds. The infrared emissivity depends on atmospheric composition and cloud cover. In addition, these parameters strongly depend on latitude, so a refinement to the simple energy balance model is to include latitude and time dependence:

$$C \times (\text{rate of temperature change}) = S_0 \times (1 - \alpha) - \varepsilon\sigma T_s^4$$
$$- \text{divergence of latitudinal heat transport}, \tag{1.2}$$

where C is the heat capacity of the climate system, and C, S, α, ε, and T are all now functions of latitude. S depends on the orbital parameters, and α generally varies with temperature, including a marked change in albedo when the temperature decreases and ice forms (typical values are from 75% for ice-covered conditions to 25% for ice-free conditions). If the climate is in equilibrium, the rate of change of temperature will be zero and we return to pure energy balance. The most important new term, however, is related to the total transport of heat by the atmosphere and ocean. This transport is a necessary part of the climate system because the solar input is greater in the tropics than at high latitudes but the long-wave cooling is more uniform with latitude. Thus the tropics receive more energy than they emit, and the polar regions emit more energy than they absorb, and hence there must be transport of heat from the equator toward the poles.

Again, such a simple model has to apply to all periods, but there are still many parameters that need to be specified rather than predicted. Indeed, there are now more parameters since we have to know the latitudinal transport of heat (which is largely unknown for periods other than the present). For the present climate, the transport is divided approximately equally between the ocean and the atmosphere. For the atmosphere, observations suggest that three main processes accomplish the heat transport. Firstly, the zonal mean flow (the Hadley and Ferrel cell) and the zonal variations in the mean flow induced by orography and land/sea temperature contrasts transport heat toward the pole. Secondly, mid-latitude depressions transport a substantial amount of heat toward the pole. Indeed, it can be argued that the only reason we have these transient systems is because of their efficiency in transporting heat from the subtropics to polar regions. Thirdly, observations suggest that a substantial amount of heat transport is associated with movement of water vapor (Peixoto and Oort, 1992).

A common approach in these simple models is to make transport of heat proportional to the latitudinal temperature gradient: the transport is weaker if the equator-to-pole temperature gradient is weaker. This leads to a potential paradox when considering a climate that is warmer than the present at high latitudes (Bice *et al.*, Chapter 4, this volume; DeConto *et al.*, Chapter 9, this volume). Equation (1.2) would suggest that a warmer pole would require more heat being transported from tropical regions. However, the reduction in the temperature gradient implies a reduction in poleward heat transport.

In part, this paradox arises from a too simplistic treatment of the equation. Firstly, in a warmer climate there would be less ice and hence more solar radiation would be absorbed at higher latitudes (Rind and Chandler, 1991). This would be only partially offset by increased emission of long-wave radiation (and both processes could be influenced by changes in cloud cover). Thus there would be less need for transport of heat to the poles. Secondly, the processes that transport heat from equator to poles are more complicated than can be modeled based on temperature gradient alone. Changes in the land/sea contrast, orography, and ocean seaways could all be important, as could the vertical ocean temperature gradient. Thus the paradox can be addressed but the challenge is to determine if we can answer it and do so in a quantifiable way.

An additional point to note is that complex climate models (called general circulation models, GCMs) are just an extension of Equation (1.2) in an attempt to remove some of the arbitrariness of the parameters. Rather than specifying albedo, emissivity, and heat transport, they calculate them from 'first principles' by considering Newton's laws of motion and the first and second laws of thermodynamics. However, the models cannot solve these equations exactly and the approximations (called parameterization schemes) effectively have parameters which are the equivalents of those in Equation (1.2) but are generally less understandable. The parameterizations are derived from present-day observations of the physical processes but may in reality depend heavily on the climate state. It is this which is the main cause of uncertainties in climate model predictions of the future and is why it is vital to test the models on past climate regimes. Further, this testing must be for climate regimes warmer than present so that the models can be properly evaluated in an appropriate parameter range.

CLIMATE PROCESSES

The earth's climate is one of the most complex physical systems studied by science. It encompasses a huge range of different processes and time and spatial scales. It is potentially a highly non-linear system, and thus to separate 'cause and effect', or forcing from response, is often extremely difficult. However, for the purposes of this paper it is useful to separate processes that force climate and climate change from the response of the climate system. It is also useful to further subdivide forcing mechanisms into external processes, which are completely unaffected by the state of the earth's climate, and internal processes, which can be modified and feed-back with climate.

Wigley (1981) suggested the following processes:

- changes in the position of the solar system relative to the galactic center
- evolution of the sun and solar variability
- changes in the orbit of the earth
- continental drift and orography
- volcanic activity
- evolution of the atmosphere
- albedo feedbacks and the land surface
- ocean and atmospheric circulation.

The first four processes are external to the climate system, while the subsequent processes become more and more internal. Ocean and atmospheric circulation should probably be viewed as part of the response of the climate system, rather than a forcing process.

The effect of the first process is difficult to quantify. It may be of importance (McCrea, 1975), but there is insufficient knowledge, data, and understanding to assess this and thus it is not discussed further here. The other processes are discussed below.

Evolution of the sun and solar variability

Solar evolution models (e.g., Endal and Sofia, 1981) generally suggest that solar output has increased by about 5% over the last 500 Ma, and by no more than 1% over the last 100 Ma. We can use our simple energy balance model to estimate the importance of these changes. If we assume that both albedo and emissivity are unchanged from present-day values, then a 1% decrease in solar constant would imply a temperature decrease of approximately 1 °C. We will see later that if we include feedback processes that modify the albedo and emissivity this could double the result. However, the key point is that solar variability seems to be in the wrong direction to explain past warm periods, and we need to find other processes that will more than compensate for these solar changes (Gibbs *et al.*, Chapter 13, this volume).

It should also be noted that some solar physics models (e.g., Gough, 1977) have suggested that Kelvin–Helmholtz instabilities could cause short-term solar output variability of up to 5% (generally a reduction). Such changes would last for the order of 10 Ma. The corresponding temperature changes would be substantial and, in general, there is little evidence of short-term cold spells during the last few hundred million years.

Changes in the orbit of the earth

Orbital changes are thought to be the main cause of the climate variations over the last 2 Ma. These orbital changes are caused by interactions between the gravitational forces of the earth, moon, and sun, and other planets. The changes modify the seasonal distribution of incoming solar radiation, but have very little effect on the annual average radiation. Feedback processes in the climate system, especially in the cryosphere and the ocean, convert the seasonal changes into annual mean changes. For instance, at 115 000 years ago the orbit was such that there was less insolation during the northern hemisphere summer. This resulted in cooler summers, winter snow not completely melting, and a gradual build-up of snow and ice. This increased the planetary albedo and cooled climate, in terms of both the global and the annual average. This is a very strong positive feedback process and greatly amplifies the direct orbital forcing.

Orbital changes have certainly been occurring in the distant past, but it is not possible to calculate the precise orbital configuration beyond approximately 5 Ma ago (Berger et al., 1989). As a result, any study of the effects of orbital changes during the Mesozoic and early Tertiary can be based only on taking both typical and extreme values for the eccentricity, obliquity, and precession of the earth's orbit (e.g., Crowley and Baum, 1995). Such studies have shown that orbital changes can strongly influence climate, especially at high latitudes and on a seasonal basis. Whether this gets converted into an annual mean change will depend largely on the state of the cryosphere: if there is no ice cover at either pole the impact of orbital changes is likely to be very subdued (compared with the present); if ice is present, then it becomes likely that the orbital variations will manifest themselves in large-amplitude oscillations of temperature at mid- and high latitudes.

It has also been suggested that the obliquity of the earth was substantially smaller during the Eocene (Wolfe, 1978), based on the presence of light-limited floras. However, if this were true it would act to increase the equator-to-pole temperature gradient (Barron, 1984), and thus would not help to explain past warm periods.

There are a number of further caveats to the above discussion. Firstly, only in some cases can the geological data resolve the relatively rapid orbital oscillations. In such cases the amplitude appears to be large, even if we are not sure of the response mechanism. There is a small possibility that there could be a preservational bias to one particular phase of the orbital variations. If this were the case, the data would give a misleading view of the warm periods. Although this idea cannot be completely discounted, it is probably unlikely because of the large range of proxy climate indicators. Further, Oglesby and Park (1989), Valdes et al. (1995), and others have shown that, at least for the Cretaceous and the Jurassic, even using extreme orbital parameters cannot reproduce mid-latitude continental warmth.

Secondly, there is a problem with evaluating the effects of orbital variations. Many GCMs have failed to simulate glacial inception, at 115 000 years ago (e.g., Rind et al., 1989; Phillips and Held, 1994). There is currently only one atmospheric GCM that has successfully predicted snow to last over the summer season (Dong and Valdes, 1995). Other GCMs have simulated glacial inception when ocean processes (Sytkus et al., 1994) or vegetation feedbacks have been included (Gallimore and Kutzbach, 1996). Thus there is some uncertainty about the ability of models to correctly simulate the climate change associated with orbital parameter changes.

Continental drift and orography

Changes in continental configuration and orographic relief have been widely discussed as possible mechanisms for climate change (e.g., Barron and Washington, 1984). They can operate in several ways. Firstly, the distribution of land can profoundly alter the climate regime and circulation. Large continents will generally have large seasonal variations of temperature because the thermal capacity of land is lower than that of ocean or ice. If the continents are at tropical latitudes, summers will be characterized by a monsoon-type circulation with heavy rainfall (Kutzbach and Gallimore, 1989). At higher latitudes, continents near the pole can also strongly influence snow accumulation (e.g., Crowley *et al.*, 1987; Gibbs *et al.*, Chapter 13, this volume). Barron and Washington (1984) found that paleogeography could explain a substantial fraction of the Cretaceous warmth. However, Barron *et al.* (1993) found that the inclusion of a seasonal cycle resulted in a major change in these conclusions. In these simulations, the paleogeography was found to be of less importance than CO_2 and sea-ice distribution.

This issue of continentality has often been raised as central to discussions of mid- and high latitude warmth in the middle Cretaceous and Eocene. The effects of lakes (Sloan, 1994) and interior seaways (Valdes *et al.*, 1996) have both been proposed as possible mechanisms of amelioration of low winter temperatures. Such processes might explain the warmth of some regions, but results from Eurasia (Herman and Spicer, 1997) show that high continental temperatures, especially in winter, are widespread and therefore more difficult to explain by the introduction of lakes or seaways.

Secondly, the distribution of continents can profoundly influence ocean circulation. The opening up of seaways has been argued to be important for the formation of the Antarctic ice cap (Kennett, 1977), and for late Cenozoic climate change when the Panama isthmus closed (Maier-Reimer *et al.*, 1990). Also, during periods with a Pangean supercontinent there was effectively only one ocean and thus only one western boundary current (compared with the two at present: the Kuroshio Current and the Gulf Stream). This might have resulted in considerably reduced transport of heat by the ocean (Valdes, 1993).

Orography can also play a major part in altering regional climate (Norris *et al.*, Chapter 6, this volume), although the direct effects on global mean temperatures are less pronounced (e.g., Barron, 1985). Obvious local responses include reduced surface temperature in the vicinity of the mountains and rain shadows on the downwind side but larger-scale effects of orography can be more important. Much recent attention has focused on the effects of the growth of Tibet on climate (see the review by Molnar *et al.*, 1993). Tibet acts as a heat source that intensifies the Asian monsoon and hence strongly influences the climate regime of a large part of the tropics. Further, the increased weathering due to Tibetan uplift has been suggested to be important for atmospheric CO_2 concentrations and it has been claimed that the growth of Tibet resulted in global cooling during the Cenozoic (Raymo *et al.*, 1988; Raymo and Ruddiman, 1992).

A further effect of orography is that the cold elevated surfaces may be where ice sheets start to form, first as small mountain glaciers and then spreading to form a larger area. The ice sheets have a high albedo, and hence reflect more solar radiation back to space, resulting in a positive feedback.

Volcanic activity

Volcanic eruptions can influence climate on a short timescale by the emission of sulfate aerosols which cool the climate system. If these aerosols remain only in the troposphere they are relatively rapidly (in a few weeks) removed by precipitation processes and do not have a strong effect on climate. However, if the eruption is sufficiently intense to emit the aerosols into the stratosphere the climatic effect of the eruption will last for considerably longer. From experience of modern eruptions (Robock and Free, 1995), the timescale is a decade and the climatic effect is a cooling of a few degrees Celsius.

Most geological indicators of climate have very crude temporal resolution, so that individual volcanic activity would be climatically recordable in the rock record only if there were periods of sustained activity well beyond any bounds based on the recent past. Bralower *et al.* (1997) have suggested that massive volcanic activity played a role in enhancing low latitude intermediate water production, leading to a methane clathrate disassociation (see below). Thus this indirect effect of volcanism can result in substantial warming.

Volcanic activity also emits CO_2 and this could have a more important effect on long-term climate. Long-term accumulation of CO_2 from outgassing from oceanic ridge systems as well as from intra-plate volcanism can result in substantial variations of atmospheric CO_2. Thus periods with above-average total areas of mid-ocean spreading are also likely to have enhanced atmospheric and oceanic CO_2 levels (see next section).

Evolution of the atmosphere

The most common explanation for the global warmth of distant past climates is elevated concentrations of atmospheric CO_2 (e.g., Barron *et al.*, 1993; Sloan and Rea, 1996). There is much evidence, from sources such as the isotopic carbon ratio in paleosols (e.g., Cerling, 1991) and geochemical modeling (e.g., Berner, 1994), that suggests that this was true for most of the Mesozoic and early Tertiary. Geochemical modeling can estimate CO_2 concentrations throughout the last 500 Ma, although the uncertainties become large further back in time. Even for the late Mesozoic, estimates are from two to eight times present-day, pre-industrial values. The radiative effect of CO_2 is proportional to the logarithm of the concentration, and thus some of the proposed large changes in CO_2 are less dramatic than they at first appear.

An increase in CO_2 concentration results in a reduction in the emissivity and hence warming. In the absence of other feedback processes, Equation (1.1) would suggest that for a doubling of CO_2 the global annual mean warming would be about 1.1 °C. An eight-fold increase would suggests a warming of about 3.3 °C. The geological data imply a 4–6 °C global warming during the middle Cretaceous, and thus an eightfold change appears to be insufficient. However, the starting assumption of no feedbacks is almost certainly incorrect.

Less is known about other radiatively important gases, including methane. Methane is known to vary naturally in glacial and interglacial times. However, there is no direct or indirect estimate of methane concentrations during warm periods in the past. There are also no model estimates for methane. In part this is because we do not have a full understanding of the sources and sinks, even for the present-day climate. Further, the residence timescale for methane can be of the order of 10–20 years and thus methane concentrations are closely linked to short-timescale processes.

Recently there has been increased interest in the effect of catastrophic releases of methane on both future climate change (Harvey and Huang, 1995) and past climates (e.g., Dickens *et al.*, 1997). It is suggested that large oceanic reservoirs of natural gas hydrates can occasionally release large amounts of methane, which is then oxidized to CO_2. These dramatic releases have been suggested to be important for short-lived intervals of extreme warmth, such as that seen at the Paleocene/Eocene boundary. It has also been proposed that increased levels of methane could lead to enhanced polar stratospheric clouds and substantial polar warming (Sloan *et al.*, 1992).

Oxygen itself is not an especially important radiatively active gas, but ozone is. The concentration of ozone is proportional to the concentration of oxygen. Thus if oxygen concentrations varied, stratospheric ozone concentrations would vary proportionately. This would result in large changes in temperature in the stratosphere but more modest changes near the surface. Hansen *et al.* (1997) showed that the climate response to changing ozone concentrations can vary from a 1–3 °C regional warming (if ozone is removed from above 10 mb) to a similar cooling if ozone is removed from between 70 and 250 mb.

Another important radiatively active gas is water vapor. Water vapor has a very short residence time in the atmosphere and is heavily influenced by the state of the climate. The amount of water vapor in the atmosphere strongly depends on temperature. Typical mixing concentrations for the surface in the tropics are $20\,g$ kg^{-1} whereas near the pole they can be less than $1\,g\,kg^{-1}$. Thus water vapor is normally considered to be a feedback of the climate system rather than an explicit forcing. GCMs predict the concentration of water vapor based on estimates of evaporation, precipitation, and moisture transport. The simpler models incorporate the effect of water vapor through modifications to the emissivity in Equations (1.1) and (1.2).

Water vapor is a very strong absorber of long-wave radiation, and since its concentration increases with temperature it generally represents a positive feedback process. For example, if an increase in CO_2 caused some warming this would result in more water vapor in the atmosphere, which would enhance (or amplify) the initial CO_2-induced climate change. Estimates suggest that the water vapor feedback can almost double the effect on the climate system of changes in CO_2 (e.g., Cess, 1989), so that for a doubling of CO_2 the total warming is expected to be 1.8 °C (cf. 1.1 °C for no feedback).

However, this argument has been challenged by some researchers (e.g., Lindzen, 1990). Near the surface there is already so much water vapor that increasing it has no effect on the radiative properties of the atmosphere. Higher in the atmosphere, water vapor is sparser and therefore here it is more effective as a climate forcing process. Lindzen (1990) suggests that in a warmer atmosphere convection is likely to be more vigorous and this would result in a drying of the upper troposphere. The reduced water vapor concentrations would act as a cooling mechanism and thus water vapor feedback would be a damping (or negative feedback) process, with a doubling of CO_2 producing a less than 1 °C warming. Such arguments are at the heart of the debate on future climate change. From a paleoclimate perspective this presents a further challenge: almost all attempts at explaining warm climate periods invoke CO_2, at least as part of the forcing mechanism; if the sensitivity of the climate system to increases in CO_2 is indeed so small then a different mechanism will have to be found.

A further problem arises when considering CO_2-induced warming. In simple and complex models an increase in CO_2 leads to warming everywhere, even in the tropics. Barron and Washington (1985) found that for a CO_2 concentration four times that of the present day the global temperature increase was reasonable but the tropics were too warm. One possible explanation for this can be deduced from Equation (1.2). Increased transport of heat between equator and poles would cool the tropics and warm high latitudes. Sloan *et al.* (1995) proposed this for the Eocene. In simple models the total heat transport is specified (or modeled in a very simple way) whereas atmospheric GCMs predict the atmospheric component but specify the ocean transport. The resulting simulation cannot be viewed as a complete prediction because the results are strongly dependent on the choice of ocean transport. However, the results show that a combination of increased radiatively active gases plus changes to total heat transport could produce a world that was warmer and had a reduced equator-to-pole temperature gradient. However, uncertainty in the ocean heat transport element has resulted in many authors electing to specify sea surface temperature, which allows a fuller investigation of other parameter space.

Albedo feedbacks and the land surface

Equations (1.1) and (1.2) show that the energy balance of the earth is strongly influenced by planetary albedo. The earth's albedo is controlled by surface conditions and by cloud cover. At the earth's surface the most marked effects occur for ice and snow. Fresh, clean snow can have an albedo of more than 90%, older snow and ice will have somewhat lower albedos and sea ice may be nearer 55%, but all of these are much larger than other surface types. In addition, sea ice plays a crucial role in insulating the atmosphere from the ocean, and has a smaller heat capacity than open ocean. Barron *et al.* (1993) found that changes in sea-ice distribution were of fundamental importance in explaining Cretaceous warmth in GCM simulations.

Snow and ice act to cool the system by reflecting solar radiation back to space before it can warm the earth, but this should not be viewed as an external forcing factor. Like water vapor, snow and ice are strongly dependent on climate. If the earth warms, ice will retreat and the amount of solar radiation absorbed by the surface will increase. This will further warm climate and thus snow/ice albedo is a positive feedback mechanism. This could enhance the warming caused by a doubling of CO_2 from 1.8 °C to 2.3 °C.

The effect of ice albedo can dramatically modify the need for transport of heat from equator to poles. In warmer periods when ice caps were not present at either pole the albedo was reduced and more solar energy absorbed. Thus it is possible to have a warmer high latitude climate without necessarily increasing the transport of heat between equator and poles (Rind and Chandler, 1991).

There are again some caveats to the above arguments. In a recent set of GCM simulations (Randall et al., 1994) it was found that a few models did not show snow albedo feedback as positive. In these models, when snow melted it was replaced by cloudy conditions so that the planetary albedo was either unaltered or even slightly increased. Even for those models that showed a positive feedback there was considerable variation in the magnitude of the response. This again shows that we lack a full understanding of the climate system.

These results emphasize that it is not just ice but also cloud cover that controls the high latitude albedo. Unfortunately the prediction of clouds is one of the most unreliable parts of any climate model yet clouds have profound effects on the energy balance of the earth. They reflect solar radiation and therefore act to cool climate. They also prevent long-wave radiation escaping to space and in this way act to warm climate. Observations of the present climate system suggest that currently the net effect of clouds is to cool the climate system (e.g., Ramanathan et al., 1989). However, the key issue is to understand how that will change in different climate regimes. Further, at high latitudes in the present climate the insulating effect of clouds appears to dominate over the albedo effect and hence clouds act to warm the high latitudes. Some GCM simulations have suggested this to be important (Sloan et al., 1992; Sellwood and Valdes, 1997) but these results should be treated with caution because our current ability to predict clouds is limited. Thus we cannot prove or disprove the idea that warm periods were significantly cloudier at mid- to high latitudes in winter.

Other albedo effects are related to vegetation. A tropical rainforest has an albedo of 13%, grassland an albedo of 20%, and desert an albedo as high as 40%. In addition, high latitude forests can modify the snow albedo feedback because snow falls through the branches of the trees and hence the albedo remains low, even when the snow is deep (Bonan et al., 1992). Vegetation also alters the transfer of heat, momentum, and moisture between the surface and the atmosphere and thus affects atmospheric circulation. Further, Charney et al. (1977) have suggested that in semi-arid regions there are important vegetation/climate feedbacks. If vegetation cover decreases, the albedo increases and this results in a decrease in precipitation that would be liable to result in further vegetation decreases. Further, Clausen (1994) showed that in semi-arid Africa there was the possibility of multiple equilibria. If the region was initially vegetated, there continued to be enough precipitation to maintain the vegetation. However, if the region was initially unvegetated, then the precipitation remained weak and vegetation was not able to grow. Thus the distribution of vegetation has the potential to affect climate. On a global mean basis, GCM simulations suggest that vegetation warms climate by 1–2 °C (Dutton and Barron, 1996). However, regionally this effect can be much larger and vegetation may be an important component in explaining the problems of warm continental interiors (Dutton and Barron, 1996; Otto-Bliesner and Upchurch, 1997; DeConto *et al.*, Chapter 9, this volume). It should be noted that changing the albedo of high latitude forests is most important in springtime; during winter there is only limited sunlight.

Ocean and atmospheric circulation

Circulation changes in the atmosphere and ocean depend on all of the factors discussed above and these cannot really be considered as a forcing process. They are part of the response of the climate system to the other forcing mechanisms. The atmosphere is the fastest-changing component of the whole climate system and any attempt to understand climate change must consider the role of the atmosphere. In particular, a GCM dynamically simulates the transport of heat and moisture by the atmosphere, given suitable boundary conditions such as the solar constant, orbital variations, atmospheric composition, land/sea contrast, orography, and surface type. An atmospheric GCM also requires specification of the sea surface temperatures, or a simple ocean model may be used which requires specification of the heat transport in the ocean. The models are then run until some form of dynamic equilibrium is achieved. Typically this requires the model to simulate a decade or more. The resulting 'prediction' of climate can be compared with the geological record and successes or failures can be noted. The models are becoming increasingly sophisticated and fewer factors have to be specified as boundary conditions (see DeConto *et al.*, Chapter 2, this volume), but this means that there are many more parameterizations that may, in part, depend on present-day climate.

As discussed earlier, GCMs are not perfect and one of the most important areas where they appear to fail is with regard to continental interiors. There continues to be some debate about whether the models and the data are truly in disagreement because the data are not as far into the continents as would be desirable. However, it is useful to address the question of what processes could warm continents in winter. Our simple energy balance model acts as a useful guide. One way of warming the interior would be to increase the transport of heat into the interior. Heat is currently being transported from the relatively warm oceans on to the continents. The mid-latitude flow will always be from the west (unless the poles were warmer than the tropics) and a key question is where is the nearest water upstream to the continental interior? Sloan (1994) and Valdes *et al.* (1996) have shown that if a water mass is added near to the region in question, this will act to warm the climate. Thus this solution to cold continental interiors is not a 'general' solution but depends on finding evidence for large water masses upstream of the data points.

Another aspect of the problem is the role of orography (Norris *et al.*, Chapter 6, this volume). Continental sedimentary deposits tend to record the climate best at lower elevations, where the sediment will accumulate. Most GCMs have a relatively coarse resolution which has a tendency to smooth orography. Peaks are smaller but valleys are higher. The crudest method for correcting this error is to correct the temperature, assuming a uniform lapse rate of approximately $6 \, \text{K} \, \text{km}^{-1}$

(e.g., Rees *et al.*, Chapter 10, this volume). Further sophistication can be achieved through the use of downscaling, or limited area models.

Other methods for increasing heat transport are less successful. Heat transport can be enhanced if the winds strengthen or if the nearest water masses are warmed. Climate model simulations (e.g., Schneider *et al.*, 1985; Valdes and Sellwood, 1992) have found that there is a canceling effect in that warmer high latitude sea surface temperatures reduce the windspeed so that the net effect of the warming is diminished. However, this is a zonally symmetric viewpoint and it is possible to imagine a situation whereby, for instance, the introduction of a mountain could lead to a local enhancement in the transport of heat from the south. This emphasizes the importance of correct reconstructions of orography (see Norris *et al.*, Chapter 6, this volume).

Finally, we need to consider the role of the oceans. Until recently all climate model simulations had used either specified sea surface temperature or a specified transport of heat in a simplistic ocean model. The resulting simulations could generally not be tested against ocean data since this was part of the input to the model. The ideal would be to run an ocean model as well as an atmospheric model, and this is beginning to happen (e.g., Barron and Peterson, 1990; Schmidt and Mysak, 1996; Bush and Philander, 1997; Bice *et al.*, Chapter 4, this volume). The work has two benefits over atmosphere-only models. Firstly, it is now possible to validate the models over the entire globe. The simulations are independent of ocean data and hence we have a much richer source of data for comparisons (MacLeod *et al.*, Chapter 8, this volume). This is especially important because coupled ocean–atmosphere models are now the main tools for predicting future climate change and these models must be validated against past climate data. Secondly, by predicting the ocean as well as the atmosphere, we can come much nearer to a complete answer to the question posed at the start of this paper: why were there warm periods in the past, with no polar ice caps?

However, there is one very important limitation on ocean models being driven by atmospheric models. Our current understanding of the present-day coupled system is very poor, and we cannot model it well. Many models for the present and future have to resort to arbitrary 'flux' corrections in order to get the present-day climate anywhere near correct. It has been suggested that this may be related to problems with the representation of clouds in the atmospheric component of these models (Gleckler *et al.*, 1995). Thus the use of coupled models for paleoclimate simulations should currently be viewed as an approximate method more suited for examining processes than for its truly predictive capability.

Further, it is worth pointing out that the problem of continental interiors can still be addressed without the need for ocean–atmosphere models. Sensitivity experiments using specified sea surface temperatures allow an examination of the parameter space to determine if there are any combinations of sea surface temperature that could produce warm winters. If such a distribution is found, then an ocean–atmosphere model will be required in order to examine if the sea surface temperatures are in any sense consistent with the rest of the climate system.

SUMMARY AND FINAL REMARKS

In order to answer the three questions posed at the start of this paper, we must seek physical processes which

1. decrease the global average planetary albedo and/or emissivity (e.g., changes in CO_2, changes in CH_4, large decreases in high clouds)
2. increase the poleward heat transport, and/or decrease the high latitude albedo (e.g., stronger ocean currents, or removal of polar ice caps)
3. cause changes in cloud cover, land surface type, decreased continentality and/or increased heat transport from ocean to land (e.g., boreal forests).

The first two factors suggest that there are possible processes that change the mean global temperature and the equator-to-pole temperature gradient. More problematic is the third issue relating to continental interior temperatures. Some success in the matching between data and models has been achieved only through the consideration of 'special cases', such as lakes or interior seaways, coupled to other feedback processes such as vegetation. This raises another issue. Normal scientific method should always include an estimate of the uncertainties and error bars due to the various processes. For our purpose we need to define both the geological proxy data errors and the model errors. Gates *et al.* (1996) showed that there is large intermodel variability in the prediction of present-day surface temperature for atmospheric GCMs. At high latitudes the differences can exceed 10 °C. Simulations with coupled ocean–atmosphere models will almost certainly have an even wider spread of results. Simulations of the past will have a similar range of scatter. Thus it could be said that the models and data agree to within the error bars. However, this interpretation of modeling results is controversial since a similar argument applied to future climate predictions would suggest that the predicted change in future climates in mid- and high latitudes does not exceed the modeling errors!

Finally it is worth considering a further complicating factor, that of chaos. Chaos theory has been applied to weather forecasting, and has sometimes been referred to as 'the butterfly effect.' The basic idea is that we will never be able to observe perfectly the state of the atmosphere (perhaps because a butterfly flaps its wings). The uncertainty will grow rapidly and result in an inability to deterministically forecast the weather beyond a timescale of a few weeks. The extent to which chaos theory applies to climate is still open to debate. There are some indications that climate is not as deterministically predictable as we would like to think. Several studies have shown that the climate system can exhibit multiple equilibria. Examples include the ice albedo feedback (as discussed above; see also Gibbs *et al.*, Chapter 13, this volume) and the thermohaline circulation in the North Atlantic (Manabe and Stouffer, 1988). For a given forcing, the climate can be in two (or more) different states and thus the present climate state depends both on the current forcing processes and on its history. For modeling work this would correspond to the simulation being sensitive to the initial conditions. Further, in the case of the thermohaline circulation, it has been suggested that if the climate system is close to a transition point then relatively small differences could result in major changes to climate (Rahmstorf, 1996).

If such considerations are truly important then we cannot ask the question 'Why were some periods in the past warm?' without asking an additional question about the state of the system beforehand. For instance, the real reason for the Cretaceous being ice free may not be high CO_2 levels or changes in ocean circulation but because the previous period(s) were also ice free. The key aspect to be explained would then be what causes the changes from one climate regime to another.

In practice, our knowledge of the climate system and especially particular climates is still sufficiently elementary that it is premature to pursue the issue of chaos too far. However, such richness of behavior of the climate system will insure that there will continue to be important challenges for scientists studying past and future climate change.

A corollary to the above discussion is that the climate system is non-linear and the climate dynamics of 'warm' periods are likely to be very different from those of 'cold' periods, such as the Last Glacial Maximum. This highlights the importance of studying warm periods. It is essential that we use our knowledge of these past periods in combination with the climate modeling tools that are being used to predict the future. These models must be thoroughly tested against past climates before we can have any real confidence in their ability to predict future climate.

ACKNOWLEDGEMENTS
Reviews of this manuscript by Karen Bice and Andrew Bush were greatly appreciated.

REFERENCES
Barron, E. J. (1984). Climatic implications of the variable obliquity explanation of Cretaceous–Paleogene high latitude floras. *Geology*, **12**, 595–8.

Barron, E. J. (1985). Explanations of the Tertiary global cooling trend. *Palaeogeography, Palaeoclimatology, Palaeoecology*, **50**, 45–61.

Barron, E. J. and Peterson, W. H. (1990). Model simulation of the Cretaceous ocean circulation. *Science*, **244**, 684–6.

Barron, E. J. and Washington, W. M. (1984). The role of geographic variables in explaining paleoclimates: results from Cretaceous climate model sensitivity studies. *Journal of Geophysical Research*, **89**, 1267–79.

Barron, E. J. and Washington, W. M. (1985). Warm Cretaceous climates: high atmospheric CO_2 as a plausible mechanism. In *The Carbon Cycle and Atmospheric CO_2: Natural Variations Archean to Present*, eds. E. T. Sundquist and W. S. Broecker, pp. 546–53. Geophysical Monograph No. 32. American Geophysical Union.

Barron, E. J., Fawcett, P. J., Pollard, D. and Thompson, S. (1993). Model simulations of Cretaceous climates: the role of geography and carbon dioxide. *Philosophical Transactions of the Royal Society*, **B341**, 307–16.

Berger, A., Loutre, M. F. and Dehant, V. (1989). Influence of the changing lunar orbit on the astronomical frequencies of pre-Quaternary insolation patterns. *Paleoceanography*, **4**, 555–64.

Berner, R. A. (1994). GEOCARB II: A revised model of atmospheric CO_2 over Phanerozoic time. *American Journal of Science*, **294**, 56–91.

Bonan, G. B., Pollard, D. and Thompson, S. L. (1992). Effects of boreal forest vegetation on global climate. *Nature*, **359**, 716–18.

Bralower, T. J., Thomas, D. J., Zachos, J. C. *et al.* (1997). High-resolution records of the late Paleocene thermal maximum and circum-Caribbean volcanism: Is there a causal link? *Geology*, **25**, 963–6.

Bush, A. B. G. and Philander, S. G. H. (1997). The late Cretaceous simulation with a coupled atmosphere–ocean general circulation model. *Paleoceanography*, **12**, 495–516.

Cerling, T. E. (1991). Carbon dioxide in the atmosphere: Evidence from Cenozoic and Mesozoic paleosols. *American Journal of Science*, **291**, 377–400.

Cess, R. D. (1989). Gauging water vapour feedback. *Nature*, **342**, 736–7.

Charney, J., Quirk, W. J., Chow, S.-H. and Kornfield, J. (1977). A comparative study of the effects on albedo change of drought in semi-arid regions. *Journal of Atmospheric Science*, **34**, 1366–85.

Clausen, M. (1994). On coupling global biome models with climate models. *Climate Research*, **4**, 203–21.

Crowley, T. J. and Baum, S. K. (1995). Reconciling Late Ordovician (440Ma) glaciation with very high (14X) CO levels. *Journal of Geophysical Research*, **100**, 1093–101.

Crowley, T. J. and North, G. R. (1991). *Paleoclimatology*. New York: Oxford University Press.

Crowley, T. J., Mengel, J. G. and Short, D. A. (1987). Gondwanaland's seasonal cycle. *Nature*, **329**, 803–7.

Dickens, G. R., Castillo, M. M. and Walker, J. C. G. (1997). A blast of gas in the latest Paleocene: simulating first-order effects of massive dissociation of oceanic methane hydrate. *Geology*, **25**, 259–62.

Dong, B. and Valdes, P. J. (1995). Sensitivity studies of Northern Hemisphere glaciation using an atmospheric general circulation model. *Journal of Climate*, **8**, 2471–96.

Dutton, J. F. and Barron, E. J. (1996). GENESIS sensitivity to changes in past vegetation. *Palaeoclimates*, **1**, 325–54.

Endal, A. S. and Sofia, S. (1981). Rotation in solar-type stars, I, Evolutionary models for the spin-down of the Sun. *Astrophysical Journal*, **243**, 625–40.

Gallimore, R. G. and Kutzbach, J. E. (1996). Role of orbitally induced changes in tundra area in the onset of glaciation. *Nature*, **381**, 503–5.

Gates, W. L., Henderson-Sellers, A., Boer, G. J. *et al.* (1996). Climate Models – Evaluation. In *Climate Change 1995, The Science of Climate Change*, eds. J. T. Houghton, L. G. Meira Filho, B. A. Callender, N. Harris, A. Kattenberg and K. Maskell, pp. 229–84. Cambridge: Cambridge University Press.

Gleckler, P. J., Randall, D. A., Boer, G. *et al.* (1995). Interpretation of ocean energy transports implied by atmospheric general circulation models. *Geophysical Research Letters*, **22**, 791–4.

Gough, D. (1977). Theoretical predictions of variations in solar output. In *The Solar Output and Its Variation*, ed. O. White, pp. 451–74. Boulder, CO: Colorado Assoc. Univ. Press.

Hansen, J., Sato, M. and Ruedy, R. (1997). Radiative forcing and climate response. *Journal of Geophysical Research*, **102**, 6831–64.

Harvey, L. D. D. and Huang, Z. (1995). Evaluation of the potential impact of methane clathrate destabilisation on future global warming. *Journal of Geophysical Research*, **100**, 2905–26.

Herman, A. B. and Spicer, R. A. (1997). New quantitative palaeoclimate data for the Late Cretaceous Arctic: evidence for a warm polar ocean. *Palaeogeography, Palaeoclimatology, Palaeoecology*, **128**, 227–51.

Kennett, J. P. (1977). Cenozoic evolution of Antarctic glaciation, the circum-Antarctic ocean, and their impact on global paleoceanography. *Journal of Geophysical Research*, **82**, 3843–60.

Kutzbach, J. E. and Gallimore, R. G. (1989). Pangean climates: Megamonsoons on the megacontinent. *Journal of Geophysical Research*, **90**, 2167–90.

Lindzen, R. S. (1990). Some coolness concerning global warming. *Bulletin of the American Meterological Society*, **71**, 288–99.

Maier-Reimer, E. K., Mikolajewicz, U. and Crowley, T. J. (1990). Ocean GCM sensitivity experiments with an open central American isthmus. *Paleoceanography*, **5**, 349–66.

Manabe, S. and Stouffer, R. (1988). Two stable equilibria of a coupled ocean–atmosphere model. *Journal of Climate*, **1**, 841–66.

Markwick, P. J. (1994). 'Equability', continentality, and Tertiary 'climate': the crocodilian perspective. *Geology*, **22**, 613–16.

McCrea, W. (1975). Ice ages and the galaxy. *Nature*, **255**, 607–9.

Molnar, P., England, P. and Martinod, J. (1993). Mantle dynamics, uplift of the Tibetan Plateau, and the Indian monsoon. *Reviews of Geophysics*, **31**, 357–96.

Oglesby, R. J. and Park, J. (1989). The effect of precessional insolation changes on Cretaceous climate and cyclic sedimentation. *Journal of Geophysical Research*, **94**, 14793–816.

Otto-Bliesner, B. L. and Upchurch, G. R. (1997). Vegetation-induced warming of high latitude regions during the Late Cretaceous period. *Nature*, **385**, 804–7.

Peixoto, J. P. and Oort, A. H. (1992). *Physics of Climate*. New York: American Institute of Physics.

Phillips, P. J. and Held, I. M. (1994). The response to orbital perturbations in an atmospheric model coupled to a slab ocean model. *Journal of Climate*, **7**, 767–82.

Rahmstorf, S. (1996). On the freshwater forcing and transport of the Atlantic thermohaline circulation. *Climate Dynamics*, **12**, 799–811.

Ramanathan, V., Cess, R. D., Harrison, E. F., Minnis, P., Barkstrom, B. R., Ahmad, E. and Hartmann, D. (1989). Cloud-radiative forcing and climate: results from the Earth Radiation Budget Experiment. *Science*, **243**, 57–63.

Randall, D. A., Cess, R. D., Blanchet, J. P. *et al.* (1994). Analysis of snow feedbacks in 14 general-circulation models. *Journal of Geophysical Research*, **99**, 20757–71.

Raymo, M. E. and Ruddiman, W. F. (1992). Tectonic forcing of late Cenozoic climate. *Nature*, **359**, 117–22.

Raymo, M. E., Ruddiman, W. F. and Froelich, P. N. (1988). Influence of late Cenozoic mountain building on ocean geochemical cycles. *Geology*, **16**, 649–53.

Rind, D. and Chandler, M. (1991). Increased ocean heat transport and warmer climate. *Journal of Geophysical Research*, **96**, 7437–61.

Rind, D., Kukla, G. and Peteet, D. (1989). Can Milankovitch orbital variations initiate the growth of ice sheets in a general circulation model? *Journal of Geophysical Research*, **94**, 12851–71.

Roach, W. T. (1998). Can geothermal heat perturb climate? *Weather*, **53**, 11–19.

Robock, A. and Free, M. P. (1995). Ice cores as an index of global volcanism from 1850 to the present. *Journal of Geophysical Research*, **100**, 11549–68.

Schmidt, G. A. and Mysak, L. A. (1996). Can increased poleward oceanic heat flux explain the warm Cretaceous climate? *Paleoceanography*, **11**, 579–93.

Schneider, S.H., Thompson, S.L. and Barron, E.J. (1985). Mid-Cretaceous continental surface temperatures: Are high CO_2 concentrations needed to simulate above freezing winter conditions? In *The Carbon Cycle and Atmospheric CO_2: Natural Variations Archean to Present*, eds. E. T. Sundquist and W. S. Broecker, pp. 554–9. Geophysical Monograph No. 32, American Geophysical Union.

Sellwood, B. W. and Valdes, P. J. (1997). Geological evaluation of climate general circulation models and model implications for Mesozoic cloud cover. *Terra Nova*, **9**, 75–8.

Sloan, L. C. (1994). Equable climates during the early Eocene: significance of regional paleo-geography for North American climate. *Geology*, **22**, 881–4.

Sloan, L. C. and Barron, E. J. (1990). Equable climates during Earth history? *Geology*, **18**, 489–92.

Sloan, L. C. and Rea, D. K. (1996). Atmospheric carbon dioxide and early Eocene climate: A general circulation modelling sensitivity study. *Palaeogeography, Palaeoclimatology, Palaeoecology*, **119**, 275–92.

Sloan, L. C., Walker, J. C. G. and Moore, T. C. (1995). Possible role of oceanic heat-transport in early Eocene climate. *Paleoceanography*, **10**, 347–56.

Sloan, L. C., Walker, J. C. G., Moore, T. C., Rea, D. K. and Zachos, J. C. (1992). Possible methane induced polar warming in the early Eocene. *Nature*, **357**, 320–2.

Sytkus, J., Gordon, H. and Chappell, J. (1994). Sensitivity of a coupled atmosphere dynamic upper ocean GCM to variations of CO_2, solar constant, and orbital forcing. *Geophysical Research Letters*, **21**, 1599–602.

Valdes, P. J. (1993). Atmospheric general circulation models of the Jurassic. *Philosophical Transactions of the Royal Society*, **B341**, 317–26.

Valdes, P. J. and Sellwood, B. W. (1992). A palaeoclimate model of the Kimmeridgian. *Palaeogeography, Palaeoclimatology, Palaeoecology*, **95**, 47–72.

Valdes, P. J., Sellwood, B. W. and Price, G. D. (1995). Modelling Late Jurassic Milankovitch climate variations. In *Orbital Forcing Timescales and Cyclostratigraphy*, eds. M. R. House and A. S. Gale, pp. 115–32. Geological Society Special Publication No. 85.

Valdes, P. J., Sellwood, B. W. and Price, G. D. (1996). The concept of Cretaceous climate equability. *Palaeoclimates*, **1**, 139–58.

Wigley, T. M. L. (1981). Climate and palaeoclimate: what can we learn about solar luminosity variations? *Solar Physics*, **74**, 435–71.

Wing, S. L. (1991). Equable climates during Earth history?: Comment. *Geology*, **19**, 539–40.

Wing, S. L. and Greenwood, D. R. (1993). Fossils and fossil climate: the case for equable continental interiors in the Eocene. *Philosophical Transactions of the Royal Society*, **B341**, 243–52.

Wolfe, J. A. (1978). A paleobotanical interpretation of Tertiary climates in the Northern Hemisphere. *American Scientist*, **66**, 691–703.

2

Recent advances in paleoclimate modeling: toward better simulations of warm paleoclimates

ROBERT M. DECONTO, STARLEY L. THOMPSON, AND DAVE POLLARD

ABSTRACT

Numerical climate models have been applied to problems of paleoclimatology for several decades. During that time, climate models have progressed in complexity from relatively simplistic Energy Balance Models (EBMs) to sophisticated three-dimensional General Circulation Models (GCMs) of the atmosphere and oceans. Recent advances in climate modeling, mainly the incorporation of climate system processes once generalized or ignored and the interactive 'coupling' of climate system components, are leading toward more realistic paleoclimate simulations and a new understanding of warm paleoclimate dynamics. The future promises even more comprehensive models running at higher resolutions, allowing paleoclimate modeling studies to focus on subcontinental to regional-scale problems. However, as the models gain sophistication, paleoclimate modeling strategies become more complex. The set-up and initialization of GCMs using Atmospheric General Circulation Models (AGCMs) coupled to full-depth, dynamical Ocean General Circulation Models (OGCMs) will provide one of the primary challenges for the paleoclimate modeling community in coming years.

INTRODUCTION

Climate, defined as the mean state of the atmosphere along with some statistical description of variability, has undergone constant change throughout earth history. Paleoclimatology is the study of those ancient climates. Perhaps no subject within the earth sciences is as interdisciplinary, bringing together atmospheric science, oceanography, biogeochemistry, biogeography, and paleontology with the solid-earth sciences of geology, marine geology, and geophysics. In order to understand the possible results of climatic forcing, including anthropogenic forcing, we must first learn about the first principles of physics and chemistry related to earth system processes. The past may contain no direct analogs for the climatic future. However, the geologic record provides many clues to the factors controlling climate (Hay *et al.*, 1997). A goal of paleoclimatology and paleoceanography is to determine the relative importance of those factors, which may be crucial to our understanding of global warming scenarios and our biological future.

Climate is the net result of external forcing, such as changes in insolation caused by changing orbital parameters and solar luminosity, and complex, sometimes non-linear interactions between the atmosphere, hydrosphere, cryosphere, biosphere, and the solid earth. Some of the major forces of climate change are summarized by Valdes (Chapter 1, this volume). Despite the complex nature of the climate system, attempts to model climate as a numerical representation have been made for more than a century (e.g., Arrhenius, 1896). Climate models test the sensitivity of the climate system to forcing. They also provide a means for interpreting geological observations and identifying the physical mechanisms responsible for climates reconstructed from the geologic record. In turn, paleoclimate simulations provide an important evaluation of climate model performance by testing their ability to reproduce the varied climates of the geologic past.

The simplest numerical climate models, zero-dimensional Energy Balance Models (EBMs), represent the earth as a single, non-dimensional entity, without the explicit representation of the dynamics of the general circulation (North, 1975; North *et al.*, 1981). Three-dimensional General Circulation Models (GCMs) are at the complex end of the climate model hierarchy (Schneider and Dickenson, 1974), with many models of intermediate complexity suitable for specific applications (see Crowley and North, 1991). The advantage of EBMs lies in their ability to examine the long-term effects of specific climatic forcing scenarios, using relatively little computational power. Computationally expensive three-dimensional GCMs provide a more detailed simulation of climate phenomena with the spatial and temporal resolution required to account for both the global and regional transient responses to forcing. GCMs have been applied to studies of warm paleoclimates for nearly two decades (e.g., Barron and Washington, 1984; Oglesby and Park, 1989; Barron *et al.*, 1993*a,b*, 1995; Otto-Bliesner, 1993; Sloan and Rea, 1995; Sloan *et al.*, 1996; Valdes *et al.*, 1996; Otto-Bliesner and Upchurch, 1997), including several in this volume (Bice *et al.*, Chapter 4; DeConto *et al.*, Chapter 9; Gibbs *et al.*, Chapter 13). GCMs are at the leading edge of our ability to represent the physical climate system numerically and are the focus of this brief overview of recent advances in modeling techniques as they relate to past warm climate periods.

GENERAL CIRCULATION MODELS (GCMS) – AN OVERVIEW

The term GCM as used here refers to a General Circulation Model of either the atmosphere (AGCM) or the ocean (OGCM). Although a complete description of the formulation of GCMs is beyond the scope of this brief introduction, some general, non-technical background is useful in understanding the terminology used in the case-studies that follow. More complete descriptions can be found in Washington and Parkinson (1986), Henderson-Sellers and McGuffie (1987), Schlesinger (1988), Trenberth (1992), and McGuffie and Henderson-Sellers (1997).

AGCMs describe the dynamical, three-dimensional character of the atmosphere by solving the fundamental physical equations for conservation of mass, energy, momentum, and moisture, and the equation of state relating atmospheric pressure, density, and moisture. However, even the most sophisticated AGCMs are limited in spatial detail, requiring small-scale phenomena to be described collectively

(a) Cartesian Grid GCM   (b) Spectral GCM

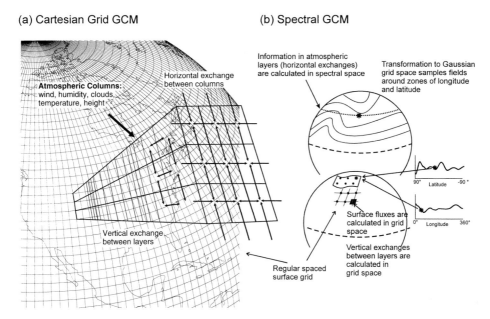

Figure 2.1. Schematic representation of (a) Cartesian grid and (b) spectral GCMs (General Circulation Models). In Cartesian grid GCMs, information is passed between adjacent layers and columns and model resolution is determined by the spacing of grid points. In spectral GCMs, vertical exchanges are computed in grid space, as in the Cartesian grid GCMs, while horizontal exchanges are handled in spectral space. The resolution of spectral GCMs is determined by the number of zonal waves used to describe the atmosphere. The surface grid of spectral GCMs often follows a straight-forward regular-spaced Cartesian format. The surface grid shown in (a) has a 2° × 2° resolution, the same as GENESIS v. 2.0. (Redrawn from McGuffie and Henderson-Sellers, 1997.)

by parametric representation (parameterization), sometimes determined empirically rather than from first principles.

AGCMs divide the earth's surface into a horizontal grid, with the grid extending up into the atmosphere in a series of adjacent columns divided into vertical layers (Fig. 2.1; McGuffie and Henderson-Sellers, 1997), or down into the ocean in the case of OGCMs. Typical resolutions are between 2° and 7.5° of latitude and longitude. Most AGCMs have 10–20 vertical layers with vertical spacing depending on the importance of a particular layer. AGCMs based on a regular horizontal and vertical grid, with the finite difference exchange of information between adjacent grid cells and layers, are called gridded GCMs. Most GCMs currently used in paleoclimate research are *spectral* GCMs. In spectral GCMs the atmospheric fields are manipulated in spectral space in the form of waves defined by the Fourier transforms of the original functions. The advantage of using Fourier transforms is that fewer numbers are required to represent global atmospheric fields, reducing computational requirements. Although horizontal atmospheric fluxes are represented spectrally, vertical fluxes rely on regular grid passing techniques as in the gridded GCMs

(Fig. 2.1; McGuffie and Henderson-Sellers, 1997). Usually between 15 and 106 waves represent each variable in each latitude zone and within each vertical layer. Terminology describing spectral GCM resolution is often written as 'R15' or 'T31' with the 'R' and 'T' referring to the method of truncation (rhomboidal or triangular, respectively) and the number referring to the truncation wave number. Larger truncation wave numbers represent higher Gaussian resolutions, corresponding to approximations of evenly spaced latitude–longitude grids of 4.75° × 7.5°, 3.75°, and 2.8° for R15, T31, and T42 spectral resolutions, respectively. Coarse resolutions have been shown to bias results (Boville, 1991). In particular, coarse resolution spectral GCMs poorly represent surface orography (via spectral truncation), precluding the accurate simulation of some regional climate patterns, especially precipitation in mountainous regions. However, higher resolutions come at the expense of computational efficiency. In general, modeling strategies use the highest resolution permitted by available computer resources.

Limitations of GCMs applied to paleoclimate studies

Despite the widespread use of modern climate models (e.g., Houghton *et al*., 1990), parameterizations derived empirically from present-day observations rather than from first principles may not be valid for all climates of the geologic past. In addition, boundary conditions based on paleogeographic reconstructions and estimates of, for example, atmospheric chemistry, solar luminosity, and orbital parameters, all become less well constrained with increasing age.

Because computational requirements generally limit GCM simulations to time intervals of decades to centuries, the evolution of climate over geologic time cannot be simulated directly. The results of paleoclimate simulations represent an equilibrium climate state, valid for the applied boundary conditions only, providing a snapshot of climate in the geologic past. Ultimately the validity of numerical simulations may depend on the existence of a single equilibrium state for the boundary conditions applied, a much debated topic for the earth's climate system (e.g., Schneider and Gal-Chen, 1973; Hay *et al*., 1997).

The earth system approach to climate modeling

Until recently, GCMs have been developed to model the physical climate system, with the land and ocean surface boundary conditions specified and fixed. The ultimate goal of earth system models is to include all aspects of the climate system: the atmosphere, ocean, cryosphere, biosphere, and terrestrial and marine biogeochemical cycles, with all the components interacting in realistic ways to allow feedback mechanisms that may affect the resulting simulated climate. Accounting for these interactions provides one of the greatest challenges for climate system modelers. Earth system models have yet to achieve such ultimate comprehension. However, the inclusion of more detailed treatment of climate system components is allowing the examination of climate system interactions under different forcing scenarios.

AGCMs with slab ocean parameterizations

The oceans play an important role in regulating the earth's energy budget through their storage and meridional transfer of energy. Model integrations of past, present, or future climate longer than a few months should include a representation of at least the surface ocean. GCM modeling schemes developed to take into account the climatic role of the oceans include: (1) AGCMs coupled to 'swamp' oceans, in which sea surface temperatures (SSTs) are computed from surface energy balance only, without considering the thermal storage capacity of the ocean surface layer or the transport of heat by ocean circulation; (2) AGCMs coupled to 'slab' ocean models, in which SSTs are calculated from a consideration of both surface energy balance and seasonal heat storage in a representation of the ocean surface mixed layer (usually 50 m deep) with no explicit ocean circulation; and (3) AGCMs coupled to OGCMs in which SSTs are calculated from the surface energy balance considering the full three-dimensional circulation of the oceans (see Meehl, 1992). It is the second type of GCM, the AGCM coupled to a slab ocean, that has become the most commonly applied tool to problems of paleoclimatology. One such model, the GENESIS (Global ENvironmEntal Simulation of Interactive Systems) GCM (Fig. 2.2) has been designed specifically for paleoclimate research following the earth system modeling philosophy (Thompson and Pollard, 1995; Pollard and Thompson, 1997; Thompson and Pollard, 1997). GENESIS has been used in several warm paleoclimate modeling studies, including the Paleozoic (see Gibbs *et al.*, Chapter 13, this volume), Eocene (see Bice *et al.*, Chapter 4, this volume), and Cretaceous (see DeConto *et al.*, Chapter 9, this volume).

Limitations of AGCMs with slab ocean parameterizations

At present the oceans contribute approximately as much of total poleward heat transport as the atmosphere (Carissimo *et al.*, 1985; Peixoto and Oort, 1992; Trenberth and Solomon, 1994), with the oceans contributing more heat transport at low latitudes (equatorward of about 30°) and the atmosphere contributing more at high latitudes. Increasing the ocean heat transport in AGCMs with slab ocean models effectively cools the equator and warms the high latitudes. This explains why enhanced ocean transport has been proposed as a mechanism for maintaining the low meridional thermal gradients of the Cretaceous and other warm climate periods (Barron, 1983; Schneider *et al.*, 1985; Covey and Barron, 1988; Covey and Thompson, 1989; Rind and Chandler, 1991; Barron *et al.*, 1993*b*, 1995; Johnson *et al.*, 1996). However, slab ocean models require a prescription of the heat transport term in the ocean model scheme and do not allow the explicit investigation of the role of ocean circulation in maintaining warm paleoclimates. To date, mechanisms accounting for large increases in ocean transport have not been adequately explained (Sloan *et al.*, 1995). To account for the role of the oceans during warm climate modes, a representation that describes the circulation of the oceans in response to both the overlying atmospheric forcing and the solid-earth boundary conditions (the distribution of the continents, the presence of ocean gateways, and bathymetry) specific to the geologic age in question must be included.

(a)   **The Global Climate System**

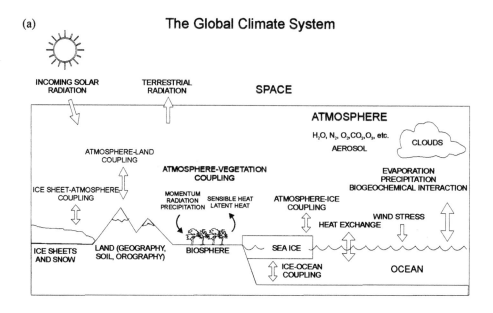

(b)   **GENESIS Version 2.0 Earth System Model**

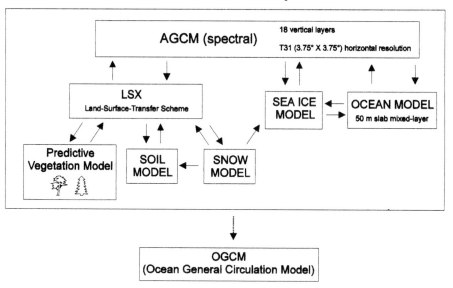

Figure 2.2. (a) Schematic representation of the primary components of the climate system (after Houghton, 1984) and the interactions between them and (b) an example of an earth system model used in paleoclimate research (GENESIS). Bidirectional arrows indicate interactive communication between climate model components, allowing feedbacks. The incorporation of predictive vegetation models as interactive components of paleoclimate simulations is an important new advance in paleoclimate modeling. The 'slab' ocean model component of GENESIS is a simplistic representation of the ocean surface mixed layer, accounting for the seasonal thermal capacity of

Ocean general circulation models (OGCMs)

OGCMs simulate the three-dimensional circulation of the oceans from surface to bottom, and several have been applied to paleoclimate problems (e.g., Modular Ocean Model: Cox, 1989; Large-Scale Geostrophic Model: Maier-Reimer *et al.*, 1993; Parallel Ocean Climate Model: Semtner and Chervin, 1992; Washington *et al.*, 1994). Summaries of OGCM formulation can be found in Robinson (1983), O'Brian (1986), and Haidvogel and Bryan (1992). OGCMs have much in common with their atmospheric counterparts (AGCMs). In fact, many of the relevant equations of motion and numerical methods used in AGCMs are found in OGCMs. However, boundary layers exist on the sides of the ocean as well as on the bottom, and ocean basins have very complex geometries affecting the mean flow. A wide range of temporal and spatial scales exists in the oceans, ranging from tens to hundreds of kilometers and days to thousands of years. The mesoscale eddies of the ocean, the ocean analog of synoptic scale atmospheric circulation, provide most of the ocean's kinetic energy and may contribute a significant fraction of the total poleward heat transport, but are sub-grid scale to most global ocean models. OGCMs are not as computationally demanding as AGCMs, but much higher resolution is required to resolve features at the eddy scale and the slow reaction of the deep ocean to perturbations at the surface requires very long integrations (10^3 years).

In general, OGCMs solve the primitive equations of motion (see Veronis, 1973) to calculate ocean currents, temperature, and salinity in three dimensions, from the surface to the ocean bottom as a result of surface forcing and geographic boundary conditions. These models not only account for the energy balance and seasonal heat storage of the surface, but also include the three-dimensional circulation of the oceans, including the gyre circulation in the horizontal plane, thermohaline circulation in the vertical plane, and parameterizations of sub-grid scale processes. Modern OGCMs applied to paleoclimate problems typically have resolutions of $2°$–$3°$ in the horizontal and 10–20 vertical layers – too coarse to explicitly resolve mesoscale eddies, but capable of representing the major wind-driven and thermohaline current systems.

Typically, OGCM simulations of present-day circulation produce systematic errors in regions where the surface freshwater balance is influenced by seasonal sea-ice formation. However, the application of OGCMs to warm paleoclimates without extensive sea ice precludes sea-ice related biases. Ironically, OGCMs designed to simulate the modern ocean circulation may perform better in ice-free paleoclimate applications than in late Cenozoic scenarios.

Caption for Fig. 2.2 (*cont.*)
the surface ocean and providing a simple scheme to account for the meridional transport of heat, without explicitly accounting for ocean circulation. OGCMs (Ocean General Circulation Models) can be forced by climate model output as stand-alone model components, without allowing the ocean to affect the atmosphere. This obvious limitation can be precluded by using GCMs (General Circulation Models) that include OGCMs as interactive climate model components. Computational limitations and caveats associated with coupled GCMs are discussed in the text.

OGCMs can be driven by atmospheric forcing provided by prior AGCM simulations using a simpler swamp or slab ocean scheme as *stand-alone* model components (Fig. 2.2). This uncoupled method has been applied to paleoclimate modeling studies of the Cretaceous (Barron and Peterson, 1990; Brady *et al.*, 1998) and the Eocene (Bice *et al.*, Chapter 4, this volume). However, this approach has obvious limitations owing to the lack of interaction between the atmosphere and the ocean. The obvious solution is to replace the simplistic non-dynamical ocean model in the initial simulation with an OGCM, interactively coupled with the atmosphere by way of surface heat balance, wind stress, and freshwater flux. Such coupled Atmosphere–Ocean General Circulation Models (AOGCMs) have recently begun to be applied to pre-Quaternary paleoclimate applications (e.g., Bush and Philander, 1997), providing an important new tool for describing ancient oceans.

Coupled GCMs

OGCMs have been developed with a wide range of complexity, from non-eddy resolving OGCMs using simplistic boundary conditions, to high resolution, eddy resolving OGCMs requiring detailed geographic boundaries and daily–monthly atmospheric forcing. The latter are now being successfully coupled to AGCMs as the ocean model component of climate system models. Perhaps the single greatest difficulty in coupling OGCMs to AGCMs (Fig. 2.3) lies in the range of response times to perturbations. The atmosphere typically responds to forcing in weeks or months, whereas the deep ocean reacts in hundreds to thousands of years. In addition, even small systematic errors can lead to slow drifts away from the equilibrium state (Bryan, 1998). Another limitation in the application of this class of GCM is computational expense. Coupled atmosphere–ocean models can respond only as fast as the slowest responding component, the deep ocean. Thus, AOGCMs must be run for hundreds of years to reach equilibrium in response to climatic perturbations, whereas GCMs using simpler slab layer ocean models require only a few decades of integration to reach equilibrium. More computationally efficient coupled atmosphere–dynamical ocean models, using two-dimensional EBMs instead of AGCMs as the atmospheric component of the coupled system, have been applied to paleoclimate simulations (Mikolajewicz and Crowley, 1997). Other coupling schemes have applied energy and moisture balance models to simplified dynamical ocean models (Schmidt and Mysak, 1996), ignoring winds and the wind-driven surface circulation in order to economically assess changes in ocean heat transport by the thermohaline circulation. However, the more computationally demanding AOGCMs allow the study of processes and sensitivities not explicitly accounted for by the simpler modeling schemes.

Fully *synchronous* AOGCM coupling schemes allow communication between atmospheric and ocean model components at least once a day, and more often when the AGCM component incorporates a diurnal cycle in its radiative scheme. Fully coupled AOGCMs are the most computationally demanding class of GCMs, but provide the most realistic representation of the natural system. Other coupling schemes have been developed, providing more economical alternatives. *Asynchronous* and *periodically synchronous* model schemes (e.g., Sausen and

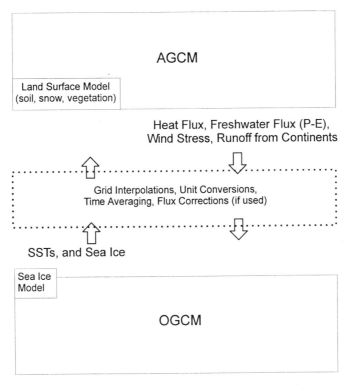

Figure 2.3. Schematic representation of a coupled atmosphere-ocean GCM (General Circulation Model), showing the primary model components commonly represented in modern GCMs. In order for an AGCM (Atmosphere General Circulation Model) to communicate with an OGCM (Ocean General Circulation Model) and vice versa, fields (two-dimensional arrays of predicted variables) must be made compatible and the frequency of communication must be structured. (After Meehl, 1992.)

Voss, 1996; Kutzbach and Liu, 1997; Voss *et al.*, 1998), in which the atmosphere and ocean are coupled but communicate less frequently than fully coupled models, require less computer time. These methods take advantage of the fact that the atmosphere takes more computer time per model year of integration, but responds to forcing more rapidly than the ocean. The asynchronous approach has been popular because of its relative computational efficiency, but may not be the ideal modeling scheme if computational requirements were not a consideration.

Limitations of coupled GCMs and additional considerations

The ideal simulation of paleoclimates would include complete interaction between the atmosphere and the ocean, with the surface and deep ocean temperature, salinity, and circulation responding to the atmosphere above, and the atmosphere allowed to 'feel' the simulated sea surface temperature distributions predicted by the ocean model. However, the complex nature of coupled atmosphere–ocean models presents a number of problems specific to paleoclimate applications.

The sensitivity of ocean circulation to changing solid-earth boundary conditions (the distribution of land and sea, the location of oceanic gateways, and bathymetry) has been speculated on by geologists for decades (e.g., Berggren and Hollister, 1974; Kennett, 1977; Berggren, 1982). The sensitivity of ocean circulation to paleogeography has been explored in several sensitivity experiments using OGCMs (e.g., Gill and Bryan, 1971; Cox, 1989; England, 1992; Mikolajewicz *et al.*, 1993; Toggweiler and Samuels, 1995; Mikolajewicz and Crowley, 1997), suggesting that changing paleogeography has indeed played an important role in ocean circulation, ocean heat transport, and perhaps global climate. Thus, paleogeographic reconstructions for paleoclimate simulations using OGCMs should include detailed reconstructions of the ocean basins at the resolution of the ocean model. Recently, quantitative methods of paleobathymetric reconstruction have been applied to pre-Quaternary ocean modeling studies (e.g., Bice *et al.*, 1998; Brady *et al.*, 1998). These methods are based on plate tectonic and shoreline reconstructions, published magnetic lineation data (e.g., Cande *et al.*, 1989), age–depth relationships for ocean crust (e.g., Stein and Stein, 1992), and corrections to account for the accumulation of sediment (e.g., Crough, 1983). Unfortunately, paleobathymetric reconstruction becomes more difficult with increased age, as the amount of surviving ocean crust diminishes.

Stand-alone OGCMs and coupled atmosphere–ocean GCMs have been shown to be sensitive to the treatment of continental runoff, reflecting the sensitivity of the thermohaline circulation to surface salinity (Bice *et al.*, Chapter 4, this volume). Most AGCMs do not include explicit schemes for the channeling of water based on continental topography, precipitation balance, and regional runoff. Instead they use a common approach involving the instantaneous distribution of excess continental water flux distributed evenly over the oceans and sea ice, with some adjustment to maintain a freshwater balance. Other schemes have been devised that divide runoff into drainage basins according to paleotopography, with the net runoff distributed among the ocean grid cells adjacent to each basin (Bice *et al.*, 1997). Explicit river routing will need to be included in future schemes of land surface hydrology (e.g., Miller *et al.*, 1994), within the coupled atmosphere–land surface–ocean framework. This may be especially important to simulations of warm paleoclimates, when precipitation was augmented by the overall warmth of the atmosphere, enhancing the hydrological cycle and the net poleward transport of freshwater.

In coupled modeling schemes, systematic errors resulting from deficiencies in the individual component models, errors in the component models as a product of coupling, or errors caused by coupled interactions between deficient model components can cause 'climate drift', a slow (10^2 years) trend toward an unrealistic climate state (Meehl, 1992). If the systematic errors in a coupled integration are large, they can be 'fixed', by a process termed flux correction (Sausen *et al.*, 1988). Flux correction involves the addition of corrective values of surface heat flux and/or freshwater flux at the ocean–atmosphere interface at each model timestep. However, in the context of pre-Quaternary paleoclimate studies, the application of corrective fluxes becomes problematic because sea surface temperature and salinity are the desired variables and should be allowed to respond freely to the overlying atmosphere and ocean circulation as the net result of external forcing and paleogeography. In addition, pre-Quaternary observations suitable to providing a means of flux adjustment do not exist. The alternative approach to flux correction is to allow the propagation of systematic errors in the coupled system in an internally consistent approach that may allow more meaningful interpretations of sensitivity to perturbations from specific forcings, even if the resulting climate is less accurate. As individual model components improve and systematic errors diminish, so will the need for flux correction. Several AOGCM modeling efforts do not use flux correction, allowing more straightforward applications to paleoclimatic scenarios. The Climate System Model (CSM; Boville and Gent, 1998), developed at the National Center for Atmospheric Research, is an example of a non-flux corrected AOGCM designed to consider interactions among additional climate system components. CSM is similar to other coupled atmosphere–ocean models (see Gates *et al.*, 1996; Kattenberg *et al.*, 1996), but is unique in its model component coupling strategy. CSM is designed to allow the inclusion of different component models (atmosphere, ocean, land surface, sea ice, etc.), communicating through a driver program called a flux coupler. The use of a flux coupler allows users to 'plug in' their own sub-models as interactive components, enabling a great deal of flexibility when designing experiments.

In order to perform meaningful paleoclimate simulations using coupled GCMs, compatible, stable, and relatively drift-free initial states must be reached within the individual component models (atmosphere, ocean, and sea ice) before the actual coupled integration can begin. This procedure is called 'spin-up'. Initial conditions for the OGCM component of coupled integrations prescribed at the start of the spin-up process include three-dimensional global fields of temperature, salinity, and sea ice. This becomes problematic because three-dimensional 'observational' data are poorly constrained for the pre-Quaternary oceans, further complicating the application of coupled GCMs to paleoclimatic problems.

The sensitivity of equilibrium climate states to different initial conditions, but using the same forcing, is still largely unknown for coupled GCMs. Much more experimentation will be needed as spin-up procedures for paleoclimate simulations are developed. The spin-up of synchronously coupled GCMs will provide one of the primary challenges in the application of these models to pre-Quaternary paleoclimates in coming years.

SEA ICE, ICE SHEETS, AND POLAR PALEOCLIMATES

As more evidence of significant high latitude ice is uncovered from the geologic record (e.g., Frakes and Francis, 1988; Stoll and Schrag, 1996), our concept of warm paleoclimates becomes more tolerant of subfreezing seasonal temperatures at polar latitudes. The accurate simulation of polar climate in the context of any climate study, past, present, or future, is critical, because of the strong amplifying affect of ice–albedo feedback on global climate (Budyko, 1969), and the sensitivity of deep ocean circulation to the seasonal changes in sea-ice distributions (Hay, 1993). Unfortunately, high latitude climates are difficult to simulate using GCMs. Spherical coordinates converge at the poles producing numerical difficulties, and water transport becomes poorly represented at high latitudes, especially in spectral GCMs, as do clouds, due to microphysical processes not explicitly represented. The sum of model biases at polar latitudes are typically manifested in excessive precipitation, too cold temperatures in the upper troposphere, and poor seasonal representations of cloudiness, surface pressure, and long-wave radiation (Tzeng and Bromwich, 1994; Briegleb and Bromwich, 1998; Weatherly *et al.*, 1998). Several modeling studies have shown changes in the sensitivity of the coupled atmosphere–ocean system when sea-ice dynamics and thickness are explicitly represented (e.g., Pollard and Thompson, 1994; Rind *et al.*, 1995; Weatherly *et al.*, 1988). The thermohaline circulation in OGCMs is very sensitive to the distribution of seasonal sea ice. Sea-ice model development will be an important area of focus in future coupled GCM model development.

To understand all warm climate modes thoroughly we must consider the dynamics of alpine and continental ice. Land-based ice provides thermal inertia and influences land-surface albedo. Ice-sheet models have been developed to simulate the initiation and development of ice sheets as a function of net snow accumulation and ice-sheet dynamics (e.g., Huybrechts, 1993). However, until recently most ice-sheet modeling studies have used either empirical parameterizations or simple energy balance climate models for their surface mass balance forcing (e.g., Oerlemans, 1982; Huybrechts and Oerlemans, 1990; Huybrechts, 1993; Peltier and Marshall, 1995). The use of climatological input (precipitation, surface heat flux, windspeed, surface pressure, cloudiness, etc. from a GCM with a three-dimensional atmosphere responding to realistic boundary conditions is a significant advance in the potential for predicting past ice-sheet dynamics. GCM–ice sheet coupling strategies are being developed (e.g., Verbitsky and Oglesby, 1995; Verbitsky and Saltzman, 1995; Pollard and Thompson, 1997), providing valuable tools for determining the likelihood of high latitude ice accumulations during warm paleoclimates such as the Cretaceous and Eocene and the mechanisms contributing to glacial-interglacial climate transitions.

Vegetation–atmosphere interactions

The Devonian expansion of land plants marked the development of an important new component in the global climate system. The physical characteristics of terrestrial ecosystems (e.g., canopy and understorey structure, total leaf area, and seasonality of leaf display) affect albedo, land surface roughness, surface hydrology, and the partitioning of sensible and latent heat. In turn, these factors affect vertical fluxes of momentum, radiation, heat, and water vapor near the ground. Terrestrial ecosystems also play an important role in the budgets of atmospheric carbon dioxide and methane, important 'greenhouse' gases, and affect the global biogeochemical cycle by providing an additional reservoir for carbon and aiding mineral weathering (Knoll and James, 1987). Dickinson *et al.* (1986) and Sellers *et al.*, (1986) recognized the importance of these factors and formulated explicit physical models of vegetation for use with AGCMs. These models and their derivatives (i.e., the Land-Surface-Transfer Scheme (LSX) in GENESIS; Fig. 2.2) have demonstrated that biosphere–atmosphere interactions can influence global paleoclimate dynamics via surface albedo-radiative and atmospheric moisture forcing associated with changing distributions of evergreen, deciduous, and single canopy (tundra) ecosystems (Bonan *et al.*, 1992, 1994; Dutton and Barron, 1996; Gallimore and Kutzback, 1996; Otto-Bliesner and Upchurch, 1997).

The structural and physiological characteristics of different terrestrial eco-systems provide important boundary conditions for paleoclimate simulations. However, reconstructing the distribution of vegetation from a fragmented geologic record is difficult, especially at the resolution required by most modern climate models. Vegetation distributions used in pre-Quaternary climate simulations are often prescribed and fixed using one average vegetation type as the global default (e.g., Barron *et al.*, 1993*a*, 1995) or a 'best guess' vegetation distribution based on point data from the geologic record (Otto-Bliesner and Upchurch, 1997). The application of *predictive* vegetation models (DeConto *et al.*, Chapter 9, this volume), interacting with the overlying atmosphere, provides an alternative to prescribed vegetation distributions.

One such model, IBIS (Integrated BIosphere Simulator; Foley *et al.*, 1996), developed specifically for coupling with an AGCM, provides a framework for simulating vegetation dynamics, terrestrial carbon balance, and land surface physics at a global scale. The predictive Equilibrium Vegetation Ecology (EVE) model (Bergengren and Thompson, 1999) describes the physical characteristics of vegetation as a function of climate (monthly mean values of temperature, precipitation, and relative humidity) and fundamental ecological principles (including algorithms for competition and succession after disturbance). IBIS uses nine *plant functional types* to represent vegetation canopy characteristics. EVE divides the present-day global vegetation into 110 lifeforms. The lifeforms define the individual components of a plant community, such as the individual and species levels of vegetation, but are based on their physiognomic and ecologic characteristics at the biome level (Box, 1981). Because the functional types and lifeforms are defined by physiognomy and not taxonomy, these models can be applied to paleoclimate simulations by including only those plant types and lifeforms with known physiognomic analogs from a particular time in the geologic past (DeConto *et al.*, Chapter 9, this volume).

Exchanges of water vapor and carbon dioxide between the atmosphere and vegetation canopies play an important role in continental climate through the partitioning of fluxes of sensible and latent heat (Pollard and Thompson, 1995). These exchanges are largely controlled by the physiological processes governing photosynthesis and stomatal conductance, which are likely to have varied in the geologic past (Beerling, 1994). However, direct, quantitative estimates of stomatal resistance based on fossil cuticle anatomy are difficult to derive (Upchurch, personal communication).

Models describing photosynthesis in terms of physiological processes (e.g., Farquhar *et al.*, 1980), coupled to mechanistic models of stomatal functioning (e.g., Farquhar and Sharkey, 1982; Ball, 1986; Collatz *et al.*, 1991, 1992) can be used to predict the stomatal conductance of ancient vegetation under varying environmental conditions (i.e., temperature, atmospheric CO_2, and O_2). The inclusion of these interactions (IBIS; Foley *et al.*, 1996) may improve the overall quality of continental paleoclimate simulations for periods when temperatures and atmospheric chemistry were very different from today.

We feel that the application of truly interactive climate–vegetation schemes is a significant advance in paleoclimate modeling. This is likely to become a more common approach, leading toward a better understanding of the role of the biosphere in paleoclimate dynamics, and providing a testable hypothesis as to the distribution of vegetation in the geologic past.

CLOUDS IN GCMS

Clouds, like water vapor, may provide important feedbacks, associated with changing levels of atmospheric trace gas concentrations (e.g., Ramanathan *et al.*, 1985, 1989; Ramanathan and Collins, 1991). Clouds can either warm or cool climate, depending on their type, altitude, water vapor content, ice content, and latitudinal distribution (Sloan *et al.*, 1992). In addition, clouds play an indirect role in climate via their influence on tropospheric chemical processes. However, most cloud model schemes in GCMs are crude, relating cloudiness to a simple dependence on relative humidity, and ignoring the complex microphysical processes controlling the three-dimensional distribution of cloud water, ice, and crystal size. Although more complex cloud schemes are now being incorporated in GCMs (e.g., Smith, 1990; Slingo and Slingo, 1991; Hack, 1994), they are still relatively simple compared with nature. The global sensitivity of temperature to forcing from clouds differs greatly among GCMs, ranging from a small cooling feedback to a large amplification of warming. However, the overall combination of water vapor and cloud feedbacks in response to increases in atmospheric CO_2 is generally accepted as positive, leading to enhanced warming (Cess *et al.*, 1989).

In general, the deep convection and associated cumulus formation at tropical latitudes have both a large long-wave (LW) and a large short-wave (SW) forcing which tend to cancel each other out. However, Ramanathan and Collins (1991) showed that clouds with higher water content have greater optical thickness and SW forcing potential. With very warm SSTs (>32 °C), cumulus convection can lead to cloud decks with very high albedos, resulting in more negative radiative forcing (cooling). This low-latitude 'tropical-thermostat' cooling feedback could contribute to apparently cool tropical SSTs suggested by tropical isotopic data (e.g., Sellwood *et al.*, 1994; D'Hondt and Arthur, 1996; Crowley and Zachos, Chapter 3, this volume). However, GCMs may not capture this phenomenon, as most modeling studies yield increased tropical SSTs in response to increased atmospheric carbon dioxide (CO_2). At high latitudes, increased stratospheric moisture from increased methane (CH_4) concentrations can increase the occurrence of high latitude stratospheric and noctilucent clouds, composed of ice crystals at the cold season mesopause (Sloan *et al.*, 1992). Enhanced polar stratospheric cloud development would have a positive warming effect over high latitudes, owing to the clouds' opacity to outgoing long-wave radiation, possibly contributing to high latitude warmth during the Eocene and other warm paleoclimates (Sloan *et al.*, 1992).

Examination of the role of clouds during warm paleoclimates will be facilitated by the inclusion of more explicit cloud schemes allowing cloud formation to be affected by atmospheric chemistry (Hauglustaine *et al.*, 1994), sea surface temperature, paleogeography, the distribution of terrestrial vegetation, land surface hydrology, and marine biogeochemistry. Atmospheric chemistry models are under development to predict spatial and temporal distributions of climatically important chemical species (i.e., CO_2, CH_4, O_3, and N_2O) and their direct and indirect effects on climate (e.g., Taylor and Penner, 1994; Roelofs and Lelieveld, 1995; Moxim *et al.*, 1996; Wang *et al.*, 1998). Representations of photochemistry and heterogeneous chemistry of statospheric aerosols and polar stratospheric clouds have already been developed (e.g., Brasseur and Granier, 1992; Lefèvre *et al.*, 1994; Tie *et al.*, 1994). Comprehensive treatment of these chemical processes has begun to be incorporated into GCMs as interactive model components, allowing chemistry to feedback on the simulated climate dynamics (e.g., Rasch *et al.*, 1995), and facilitating interpretations of the forcing potential of these important feedbacks.

The application of an implicit scheme for CH_4 flux (sources minus sinks), within the context of a land surface modeling package, will be an important improvement, given the direct and indirect climate forcing potential of large increases in atmospheric CH_4 (Lelievield and Crutzen, 1992; Sloan *et al.*, 1992). However, not until surface–atmosphere chemical fluxes and atmospheric chemistry are treated as fully interactive components of GCMs, allowing direct two-way interaction with clouds, can these issues be resolved.

Marine biogeochemistry

In addition to physical processes contributing to the redistribution of heat on the earth's surface, the oceans can play a fundamental role in the global climate system via the effect of marine biogeochemical processes on the concentration of climatically important trace gases (Sarmiento and Bender, 1994). The most important of these is CO_2 (the largest contributor to the greenhouse effect after water vapor), but nitrous oxide (N_2O) and dimethyl sulfide gas (DMS; produced by phytoplankton in ocean surface waters) are also important (Najjar, 1992). The long recognized correlation between atmospheric CO_2 concentration, global mean temperature, and ice volume has led to the examination of processes that affect the distribution of total dissolved carbon and alkalinity in the oceans, and the exchange of CO_2 between the atmosphere and the oceans as a function of marine productivity (Sarmiento and Toggweiler, 1984) and calcite preservation (Archer and Maier-Reimer, 1994). However, the explicit representation of marine biogeochemistry in a modeling context is difficult; most biogeochemical modeling efforts have focused on the better-known physical and chemical processes that determine the distribution of CO_2 in the atmosphere and oceans, without considering the roles of N_2O and DMS.

Geochemical models that consider weathering rates, crustal degassing, and sedimentation (e.g., Berner, 1994), consider the fundamental processes that control atmospheric CO_2 on geologic timescales. Box models of ocean and atmosphere CO_2 exchange have also been explored for decades (e.g., Revelle and Suess, 1957; Oeschger, 1975; Sarmiento and Toggweiler, 1984). The box models represent the ocean as a small number of well-mixed reservoirs, with transports between the reservoirs represented by simple parameterizations. Biology is represented by considering the export of nutrients from the surface ocean to the deep ocean as sinking organic material, and the replenishing of nutrients into the surface ocean by mixing from below. These models have been useful in studying the cycling of nutrients and carbon and the mechanisms controlling atmospheric CO_2. However, most of the biogeochemical ocean models developed thus far have been designed to simulate the modern ocean's uptake of anthropogenic CO_2, and use parameterization schemes based on modern (late Cenozoic), cold ocean conditions. These models may not be applicable to warm paleoclimates when deep ocean temperatures (solubilities) and modes of thermohaline circulation (mass transports) were very different from today.

To consider lateral advection, mixing, and the role of the deep oceans in biogeochemical cycling, we must consider the full three-dimensional ocean circulation, along with nutrient cycling and productivity. Biological models of upper ocean productivity and the cycling of nutrients and carbon have been coupled to OGCMs (Bacastow and Maier-Reimer, 1990; Najjar *et al.*, 1992; Sarmiento and LeQuere, 1996; Sarmiento *et al.*, 1998). Models of this type have been used to account for changes in atmospheric CO_2 after perturbation, considering interactions between the atmosphere, oceans, and biogeochemical cycles. Physical–biogeochemical models will eventually be applied to warm paleoclimate simulations, perhaps contributing to our understanding of the dramatic changes in the global carbon cycle, long recognized in the isotopic record of deep-sea sediments.

REGIONAL PALEOCLIMATE MODELING

Most applications of GCMs in paleoclimatic modeling require several decades of model integration to average out the effects of interannual variability, and to spin up the predicted sea surface temperatures and sea ice if the model includes an explicit ocean component. Given current practical limits of computing power, this means that the horizontal resolution of atmospheric GCMs is limited to about 200–400 km. However, many paleoclimatic problems are regional in nature and therefore involve smaller scales. A number of down-scaling approaches have been developed to increase the resolution of climate models over a limited region, using either varying resolution within a GCM (Deque and Piedelievre, 1995; Leslie and Fraedrich, 1997) or a separate high-resolution regional model (Giorgi and Mearns, 1991; Giorgi *et al.*, 1993a,b). The first technique is promising, but is relatively new with few applications to date; the second technique using regional climate models has been used extensively since the early 1990s, applied mainly to present-day climate. These regional models contain physics similar to GCMs, but since their domain spans a limited area (typically continental or smaller), horizontal resolutions in the order of 50 km or less are affordable. At the boundaries of the domain, meteorological variables such as air temperatures, humidities, and wind are prescribed at intervals of a few hours or less, obtained from a previous GCM simulation or observed climatologic data, and interpolated to the regional model grid. Nested regional models have been run in this way for timespans from months to several years over various regions of the USA (Giorgi, 1991), western Europe (Giorgi and Marinucci, 1996), central Africa (Semazzi *et al.*, 1993), Australia (Walsh and McGregor, 1995), the Arctic (Lynch *et al.*, 1995), and the Antarctic (Hines *et al.*, 1995; Walsh and McGregor, 1996), and generally have been successful in simulating small-scale features of the climate not resolved by coarser GCMs. In Europe and the USA, many of these smaller-scale features are associated with topography below the GCM resolution. In one of the few paleoclimatic applications to date, Hostetler *et al.* (1994) used a regional climate model to investigate the climate over the paleolakes of southwestern USA around the Last Glacial Maximum.

Although many aspects of paleoclimate research are continental in scale and can adequately be addressed by current GCMs, others involve smaller-scale features and progress will require the use of high-resolution regional models. Present-day applications of regional models have produced promising results. However, they have demonstrated that their validity depends on the quality of the GCM simulations providing the boundary data. That will be an important concern for paleoclimate applications where the accuracy of the GCM simulations is harder to assess.

Boundary conditions for paleoclimate simulations

Modern GCMs adapted to paleoclimate research require a comprehensive and detailed description of external forcing factors and the physical characteristics of the earth's surface as boundary conditions. Typically, this includes descriptions of orbital parameters, the solar constant, atmospheric chemistry, the distribution of land–sea–ice (surface type), elevation, sub-grid scale orographic roughness, soil texture, vegetation, and bathymetry of the ocean basins if an OGCM is being used. Ocean heat flux is sometimes prescribed when slab ocean models are used as the ocean representation. Atmosphere and ocean climate simulations are sensitive to paleogeography and uncertainties can bias results (Hay, 1996; Crowley, 1998).

Data available for pre-Quaternary paleogeographic reconstructions are often sparse. In the past, boundary conditions applied to pre-Quaternary climate model simulations have been assigned average global values, reconstructed at relatively coarse resolutions from qualitative interpretations, or held at present-day values. However, as model resolution is increased so must the resolution and realism of the boundary conditions being applied (see Crowley, 1998). Both increased resolution, including the fine-scale grid of regional models discussed above, and the use of full-depth ocean models will increase the importance of better paleogeographic reconstructions, perhaps necessitating the development of new techniques for high resolution paleoenvironmental reconstructions of the distant geologic past.

Computational requirements

Variability within the time averaged state of the modeled climate system, especially in GCMs using AGCMs coupled to slab oceans or OGCMs that exert thermal inertia on the coupled system, require very long numerical integrations to achieve equilibrium. A 10-year global simulation of climate using a state of the art GCM requires tens of hours of central processing unit time on modern supercomputer systems capable of more than one GFLOP (10^9 floating point calculations per second). Sensitivity tests using GCMs running at different resolutions show a strong correlation between resolution and the quality of the simulation (Simmons, 1990;

Boville, 1991). In paleoclimate applications, high resolution simulations are required to resolve regional paleoclimate problems and to discern the role of paleogeography on the climate of a specific location. However, even modest increases in resolution, added atmospheric layers in the vertical, interactive coupling with ocean models, and the inclusion of more climate system component models (snow, soil, biogeochemistry, atmospheric chemistry, vegetation, etc.) can dramatically increase computational requirements. In the near future, the comprehensive earth system models at the high end of the GCM hierarchy will require computer performance exceeding 1000 GFLOP, a level of performance that is likely to come from large-scale parallel processing (Hack, 1992).

VERIFICATION OF PALEOCLIMATE SIMULATIONS

One of the fundamental challenges of pre-Quaternary paleoclimate modeling lies in the validation of the results, which becomes more difficult with increasing age. The most common approach is to compare proxy climate indicators from the geologic record with the climate simulated by the model at those same locations. Proxy climate indicators can include information on temperature, precipitation, seasonality, relative humidity, storminess, and paleowind direction and intensity (see Allen, 1994; Sellwood and Price, 1994; Rees *et al.*, Chapter 10, this volume). Terrestrial climate indicators often applied to paleoclimate reconstruction include the distribution of plants and animals (e.g., Adams *et al.*, 1990; Horrell, 1991), vegetation physiognomy (e.g., Wolfe, 1971, 1979; Upchurch and Wolfe, 1987; Spicer, 1989; Herman and Spicer, 1997), sedimentary facies such as evaporites, storm deposits, glacial sediments, glendonites, coals, aeolianites (e.g., Blackett, 1961; Parrish *et al.*, 1982), and paleosols (Retallack, 1990). In the marine realm, oxygen isotopic ratios in ancient carbonates provide valuable information regarding paleotemperature (see Crowley and Zachos, Chapter 3; Bice *et al.*, Chapter 4). Warm- and cool-water carbonate facies can be used to reconstruct the location of warm and temperate marine environments (see Sellwood and Price, 1994). Climate-induced fluctuations in productivity may be reflected in the depth of the CCD (Calcite Compensation Depth) and the accumulation rate of silica-rich sediments caused by changes in upwelling intensity. Analysis of particle size, composition, and mass accumulation rate of deep-sea eolian sediments has shown that these record the intensity of atmospheric circulation (Rea *et al.*, 1985). Winter storms and summer–autumn hurricanes have the ability to affect coastal sedimentation, recognized in hummocky cross-stratification and megaripples, thought to be indicative of inner-shelf storm deposits. Biogeography can be used to infer ocean circulation patterns (Gordon, 1973).

Proxy formation models

Much of the information listed above provides only qualitative and indirect evidence of ancient climates. The need for a quantitative comparison between GCM output and the geologic record encouraged the recent development of Proxy Formation Models (PFMs; Pollard and Schulz, 1994). PFMs simulate the physical, chemical, and/or biological conditions necessary for the formation of a specific climatically sensitive geologic deposit, such as evaporites. Evaporites form where evaporation exceeds total precipitation plus inflow. Because evaporites can accumulate very quickly (up to 100 m per 1000 years), short episodes of deposition can lead to the accumulation and subsequent preservation of huge thicknesses. Owing to the solubility of most evaporite minerals, preservation requires special circumstances related to the post depositional climate and burial history. Thus, not all of the arid regions (high E − P) that existed in the ancient world will be documented by evaporite deposits. However, those locations where evaporites do occur must have had a high E − P value.

The PFM for evaporites (Pollard and Schulz, 1994) predicts E − P over a 50-m column of perfectly mixed water. The PFM can be initialized with different salt contents sufficient to precipitate halite (385 psu) or gypsum (175 psu). PFM simulations of halite and gypsum-anhydrite, driven by GENESIS GCM simulations of the Triassic (Pollard and Schulz, 1994) and the late Cretaceous are in good agreement with the geologic record, suggesting that the spatial patterns of precipitation are being realistically simulated by the models. Eventually, PFMs will be applied to the prediction of other climate sensitive sediments and soils such as coals and laterites (Wold, personal communication). Algorithms are under development to quantitatively evaluate the predicted occurrence of proxies with observations (Wold, unpublished data).

Oxygen isotopes and AOGCMs

Oxygen isotope analysis of monospecific planktonic and benthic foraminifera and bulk carbonate (nannofosssils) has long been used to estimate ancient surface and deep-water temperatures for the reconstruction of past climates (Savin, 1977), and for comparison with GCM output as a means of model–data comparison (see Bice *et al.*, Chapter 4, this volume). The method is based on the observation that the ratio of the two stable isotopes of oxygen (^{18}O and ^{16}O) is temperature dependent when calcite is precipitated from a surrounding solution. Decreasing temperatures result in an enrichment of the heavier isotope (^{18}O) in the calcite. However, interpretation of the $^{18}O/^{16}O$ in the shells of foraminifera remains equivocal because it is a function of both the $^{18}O/^{16}O$ in the seawater and the temperature. The $^{18}O/^{16}O$ in seawater is a function of the volume and isotopic composition of glacial ice, which changes the ratio by the same amount throughout the ocean, and the hydrological cycle (the balance between evaporation and precipitation and runoff, as a result of atmospheric forcing) which induces regional differences in the isotopic ratio that are sometimes referred to as the 'salinity effect.' The combined effects of temperature, the hydrological cycle, and ice volume do not allow a unique interpretation of measured $^{18}O/^{16}O$ values.

A test of paleoclimate model performance could be achieved by predicting the isotopic composition of precipitation, runoff (e.g., Joussaume *et al.*, 1984; Jouzel *et al.*, 1987, 1996; Charles *et al.*, 1994), and ocean water, by accounting for the processes that fractionate the stable isotopes of oxygen and hydrogen in the atmosphere. The incorporation of such a model into an AOGCM could predict the three-dimensional distribution of $\delta^{18}O$ in the ocean. Because the ocean model component will also predict temperatures, it is possible to develop a proxy model that will predict the $\delta^{18}O$ in the tests of planktonic and benthic foraminifera. The predicted distributions could be tested against the distribution of isotopic values obtained from the geologic record of specific periods, removing the ambiguity inherent in the interpretation of the data and providing more meaningful model–data comparisons.

Predictive vegetation models as validation tools

Modern vegetation can be described, mapped, and modeled without consideration of taxonomy (e.g., Holdridge, 1947; Box, 1981; Mathews, 1983; Prentice *et al.*, 1992), with patterns of plant distributions displaying an observable relationship with climate (Spicer and Corfield, 1992; Rees *et al.*, Chapter 10, this volume). Similar physiognomic characteristics and adaptations have been observed in unrelated plants growing in similar environments (Spicer, 1989), suggesting that in addition to plant physiognomy, vegetational units up to the biome level are directly reflective of climate. Thus, an important advantage is provided by paleoclimate simulations using predictive vegetation models, over those using fixed land surface schemes. Because vegetation distributions provided by predictive vegetation models (i.e., EVE and IBIS) reflect the seasonal climatology of specific locations, they can be directly compared with fossil vegetation assemblages, providing a more comprehensive evaluation of paleoclimate model accuracy (DeConto *et al.*, Chapter 9, this volume).

SUMMARY

Recent advances in paleoclimate modeling are providing new tools for the study of the earth's climatic history. A new generation of earth system models, allowing realistic feedbacks between the atmosphere, oceans, cryosphere, solid-earth, terrestrial ecosystems, and biogeochemical cycles, is enabling a more comprehensive investigation of the sensitivity of warm paleoclimate dynamics to changing boundary conditions than ever before. General circulation models of the coupled atmosphere–ocean system are now being applied to the pre-Quaternary. However, the paleoclimate application of these models including full-depth, dynamical oceans, will necessitate careful considerations of model set-up and initialization, as the sensitivity of coupled models to initial conditions becomes more clear. Thus, the immediate challenge for the paleoclimate modeling community will be to resolve caveats related to the application of ancient boundary conditions to coupled atmosphere–ocean GCMs, before these models can become a readily applicable tool for the investigation of pre-Quaternary paleoclimates that were fundamentally different from today. Ultimately, this may require more paleoenvironmental data (especially from the oceans), at higher temporal and spatial resolution than presently available. Nonetheless, the next decade promises exciting advances in paleoclimate modeling as coupled climate models become a reality and progress is made toward the inclusion of more climate system processes, resulting in a new order in the climate modeling hierarchy and converging model–data comparisons.

ACKNOWLEDGEMENTS

We thank Esther Brady, Bette Otto-Bliesner, and Phil Rasch at NCAR for their insightful suggestions. We also thank Peter Schultz and Karen Bice for their thorough and thoughtful reviews and valuable comments.

REFERENCES

Adams, C. G., Lee, D. E. and Rosen, B. R. (1990). Conflicting isotopic and biotic evidence for tropical sea-surface temperatures during the Tertiary. *Palaeogeography, Palaeoclimatology, Palaeoecology*, **77**, 289–313.

Allen, J. R. L. (1994). Paleowind: geological criteria for direction and strength. In *Paleoclimates and their Modelling: With Special Reference to the Mesozoic Era*, eds. J. R. L. Allen, B. J. Hoskins, B. W. Sellwood, R. A. Spicer and P. J. Valdes, pp. 27–33. London: Chapman & Hall.

Archer, D. E. and Maier-Reimer, E. (1994). Effect of deep-sea sedimentary calcite preservation on atmospheric CO_2 concentration. *Nature*, **367**, 260–3.

Arrhenius, S. (1896). On the influence of carbonic acid in the air upon the temperature of the ground. *Philosophical Magazine*, **41**, 237–75.

Bacastow, R. and Maier-Reimer, E. (1990). Ocean circulation model of the carbon cycle. *Climate Dynamics*, **4**, 95–126.

Ball, J. T., Woodrow, I. E. and Berry, J. A. (1986). A model predicting stomatal conductance and its contribution to the control of photosynthesis under different environmental conditions. In *Progress in Photosynthesis Research*, ed. J. Biggins, pp. 221–4. Dordrecht: Martinus-Nijhoff.

Barron, E. J. (1983). A warm equable Cretaceous: the nature of the problem. *Earth Science Reviews*, **19**, 305–38.

Barron, E. J., Fawcett, P. J., Peterson, W. H., Pollard, D. and Thompson, S. L. (1995). A 'simulation' of mid-Cretaceous climate. *Paleoceanography*, **10**, 953–62.

Barron, E. J., Fawcett, P. J., Pollard, D. and Thompson, S. L. (1993a). Model simulations of Cretaceous climates: the role of geography and carbon dioxide. *Philosophical Transactions of the Royal Society of London*, **341**, 307–16.

Barron, E. J. and Peterson, W. H. (1990). Mid-Cretaceous ocean circulation: results from model sensitivity studies. *Paleoceanography*, **5**, 319–37.

Barron, E. J., Peterson, W. H., Pollard, D. and Thompson, S. L. (1993b). Past climate and the role of ocean heat transport: model simulations for the Cretaceous. *Paleoceanography*, **8**, 785–98.

Barron, E. J. and Washington, W. M. (1984). The role of geographic variables in explaining paleoclimates: results from Cretaceous climate model sensitivity studies. *Journal of Geophysical Research*, **89**, 1267–79.

Beerling, D. J. (1994). Modelling paleophotosynthesis: Late Cretaceous to present. *Philosophical Transactions of the Royal Society of London*, **85**, 421–32.

Bergengren, J. C. and Thompson, S. L. (1999). Modeling the effects of global climate change on natural vegetation, Part I. The equilibrium vegetation ecology model. *Climatic Change* (submitted).

Berggren, W. A. (1982). Role of ocean gateways in climatic change. In *Climate in Earth History*, eds. W. H. Berger and J. C. Crowell, pp. 118–25. Washington, DC: National Academy Press.

Berggren, W. A. and Hollister, C. D. (1974). Paleoceanography, paleobiogeography, and the history of circulation of the Atlantic Ocean. In *Studies in Paleoceanography*, ed.

W. W. Hay, pp. 126–86. Society of Economic Paleontologists and Mineralogists Spec. Publ. No. 20. Tulsa: Society for Sedimentary Geology.

Berner, R. A. (1994). GEOCARB II: A revised model of atmospheric CO_2 over Phanerozoic time. *American Journal of Science*, **294**, 56–91.

Bice, K. L. (1997). *An Investigation of Early Eocene Deep Water Warmth Using Uncoupled Atmosphere and Ocean General Circulation Models: Model Sensitivity to Geography, Initial Temperatures, Atmospheric Forcing and Continental Runoff*. PhD thesis, The Pennsylvanian State University, University Park.

Bice, K. L., Barron, E. J. and Peterson, W. H. (1997). Continental runoff and Cenozoic bottom-water sources. *Geology*, **25**, 951–4.

Bice, K. L., Barron, E. J. and Peterson, W. H. (1998). Reconstruction of realistic early Eocene paleobathymetry and ocean GCM sensitivity to specified basin configuration. In *Tectonic Boundary Conditions for Climate Reconstructions*, eds. T. Crowley and K. Burke, pp. 227–47. New York: Oxford University Press.

Blackett, P. M. S. (1961). Comparisons of ancient climate with the ancient latitude deduced from rock magnetic measurements. *Proceedings of the Royal Society of London*, **A263**, 1–30.

Bonan, G. B., Chapin, III, F. S. and Thompson, S. L. (1995). Boreal forest and tundra ecosystems as components of the climate system. *Climatic Change*, **29**, 145–67.

Bonan, G. B., Pollard, D. and Thompson, S. L. (1992). Effects of boreal forest vegetation on global climate. *Nature*, **359**, 716-8.

Boville, B. A. (1991). Sensitivity of simulated climate to model resolution. *Journal of Climate*, **4**, 469–85.

Boville, B. A. and Gent, P. R. (1998). The NCAR Climate System Model, Version One. *Journal of Climate*, **11**, 1115–30.

Box, E. O. (1981). *Macroclimate and Plant Forms: An Introduction to Predictive Modeling in Phytogeography*. The Hague: Dr W. Junk Publishers.

Brady, E. C., DeConto, R. M. and Thompson, S. L. (1998). Deep water formation and Poleward ocean heat transport in the warm climate extreme of the Cretaceous (80 Ma). *Geophysical Research Letters*, **25**, 4205–8.

Brasseur, G. and Granier, C. (1992). Mount Pinatubo aerosols, chlorofluorocarbons, and ozone depletion. *Science*, **257**, 1239–42.

Briegleb, B. P. and Bromwich, D. H. (1998). Polar climate simulation of the NCAR CCM3. *Journal of Climate*, **11**, 1270–86.

Bryan, F. O. (1998). Climate drift in a multi-century integration of the NCAR Climate System Model. *Journal of Climate*, **11**, 1455–71.

Budyko, M. I. (1969). The effect of solar radiation variations on the climate of the Earth. *Tellus*, **21**, 611–9.

Bush, A. B. G. and Philander, G. H. (1997). The Late Cretaceous: simulation with a coupled atmosphere-ocean general circulation model. *Paleoceanography*, **12**, 495–516.

Cande, S. C., LaBrecque, J. L., Larson, R. L., Pitman, III, W. C., Golovchenko, X. and Haxby, W. F. (1989). *Magnetic Lineations of the World's Ocean Basins*. Tulsa: Association of Petroleum Geologists.

Carissimo, B. C., Oort, A. H. and Vonder Harr, T. H. (1985). Estimating the meridional energy transports in the atmosphere and oceans. *Journal of Physical Oceanography*, **15**, 82–91.

Cess, R. D., Potter, G. L., Blanchet, J. P. *et al.* (1989). Interpretation of cloud-climate feedback as produced by 14 atmospheric general circulation models. *Science*, **245**, 513–6.

Charles, C. D., Rind, D., Jouzel, J., Koster, R. D. and Fairbanks, R. G. (1994). Glacial-interglacial changes in moisture sources for Greenland: influences on the ice core record of climate. *Science*, **263**, 508–11.

Collatz, G. J., Ball, J. T., Grivet, C. and Berry, J. A. (1991). Physiological and environmental regulation of stomatal conductance, photosynthesis and transpiration: a model that includes a laminar boundary layer. *Agricultural and Forest Meteorology*, **53**, 107–36.

Collatz, G. J., Ribas-Carbo, M. and Berry, J. A. (1992). Coupled photosynthesis-stomatal conductance model for leaves of C_4 plants. *Australian Journal of Plant Physiology*, **19**, 519–38.

Covey, C. and Barron, E. J. (1988). The role of ocean heat transport in climatic change. *Earth Science Reviews*, **24**, 429–45.

Covey, C. and Thompson, S. L. (1989). Testing the effects of ocean heat transport on climate. *Palaeogeography, Palaeoclimatology, Palaeoecology*, **75**, 331–41.

Cox, B. (1989). An idealized model of the world ocean. Part I: the global water masses. *Journal of Physical Oceanography*, **19**, 1730–52.

Crough, S. T. (1983). The correction for sediment loading on the seafloor. *Journal of Geophysical Research*, **88**, 6449–54.

Crowley, T. J. (1998). Significance of tectonic boundary conditions for paleoclimate simulations. In *Tectonic Boundary Conditions for Climate Reconstructions*, eds. T. J. Crowley and K. Burke, pp. 3–20. New York: Oxford University Press.

Crowley, T. J. and North, G. R. (1991). *Paleoclimatology*. New York: Oxford University Press.

Deque, M. and Piedelievre, J. (1995). High resolution climate simulation over Europe. *Climate Dynamics*, **11**, 321–39.

D'Hondt, S. and Arthur, M. A. (1996). Late Cretaceous oceans and the cool tropical paradox. *Science*, **271**, 1838–41.

Dickinson, R. E., Hendersen-Sellers, A., Kennedy, P. J. and Wilson, M. F. (1986). *Biosphere-Atmosphere Transfer Scheme (BATS) for the NCAR Community Climate Model*. Boulder, Colorado: NCAR Technical Note 275 + STR.

Dutton, J. F. and Barron, E. J. (1996). GENESIS sensitivity to changes in past vegetation. *Paleoclimates*, **1**, 325–54.

England, M. H. (1992). Representing the global-scale water masses in ocean general circulation models. *Physical Oceanography*, **23**, 1523–52.

Farquhar, G. D. and Sharkey, T. D. (1982). Stomatal conductance and photosynthesis. *Annual Review of Plant Physiology*, **33**, 317–45.

Farquhar, G. D., Von Caemmerer, S. and Berry, J. A. (1980). A biogeochemical model of photosynthetic CO_2 assimilation in leaves of C_3 species. *Planta*, **149**, 78–90.

Foley, J. A., Prentice, I. C., Ramankutty, N. *et al.* (1996). An integrated biosphere model of land surface processes, terrestrial carbon balance, and vegetation dynamics. *Global Biogeochemical Cycles*, **10**, 603–28.

Frakes, L. A. and Francis, J. E. (1988). A guide to cold polar climates from high-latitude ice-rafting in the Cretaceous. *Nature*, **339**, 547–9.

Gallimore, R. G. and Kutzbach, J. E. (1996). Role of orbitally induced changes in tundra area in the onset of glaciation. *Nature*, **381**, 503–5.

Gates, W. L., Henderson-Sellers, A., Boer, G. J. *et al.* (1996). Climate models – evaluation. In *Climate Change 1995: The Science of Climate Change*, eds. J. T. Houghton, L. G. Meira Filho, B. A. Callander, N. Harris, A. Kattenberg and K. Maskell, pp. 229–84. Cambridge: Cambridge University Press.

Gill, A. E. and Bryan, K. (1971). Effects of geometry on the circulation in a three-dimensional southern hemisphere ocean model. *Deep-Sea Research*, **18**, 685–721.

Giorgi, F. (1991). Sensitivity of simulated summertime precipitation over the western United States to different physics parameterizations. *Monthly Weather Review*, **119**, 2870–88.

Giorgi, F. and Marinucci, M. R. (1996). An investigation of the sensitivity of simulated precipitation to model resolution and its implications for climate studies. *Monthly Weather Review*, **124**, 148–66.

Giorgi, F., Marinucci, M. R. and Bates, G. T. (1993*a*). Development of a second-generation regional climate model (RegCM2). Part I: Boundary-layer and radiative transfer processes. *Monthly Weather Review*, **121**, 2794–813.

Giorgi, F., Marinucci, M. R., de Canio, G. and Bates, G. T. (1993*b*). Development of a second-generation regional climate model (RegCM2). Part II: Convective processes and assimilation of lateral boundary conditions. *Monthly Weather Review*, **121**, 2814–23.

Giorgi, F. and Mearns., L. O. (1991). Approaches to the simulation of regional climate change: a review. *Reviews of Geophysics*, **29**, 191–216.

Gordon, W. A. (1973). Marine life and ocean surface currents in the Cretaceous. *Journal of Geology*, **81**, 269–84.

Hack, J. J. (1992). Climate system simulation: basic numerical and computational concepts. In *Climate System Modeling*, ed. K. E. Trenberth, pp. 283–318. Cambridge: Cambridge University Press.

Hack, J. J. (1994). Parameterization of moist convection in the NCAR Community Climate Model, CCM2. *Journal of Geophysical Research*, **99**, 5551–68.

Haidvogel, D. B. and Bryan, F. O. (1992). Ocean general circulation modeling. In *Climate System Modeling*, ed. K. E. Trenberth. Cambridge: Cambridge University Press.

Hauglustaine, D. A., Granier, G. P., Brasseur, G. P. and Megie, G. (1994). The importance of atmospheric chemistry in the calculation of radiative forcing on the climate system. *Journal of Geophysical Research*, **99**, 1173–86.

Hay, W. W. (1993). The role of polar deep water formation in global climate change. *Annual Review of Earth and Planetary Sciences*, **21**, 227–54.

Hay, W. W. (1996). Tectonics and Climate. *Geologische Rundschau*, **85**, 409–37.

Hay, W. W., DeConto, R. M. and Wold, C. N. (1997). Climate: Is the past the key to the future? *Geologische Rundschau*, **86**, 471–91.

Henderson-Sellers, A., and McGuffie, K. (1987). *A Climate Modeling Primer*. Chichester: John Wiley & Sons.

Herman, A. B. and Spicer, R. A. (1997). New quantitative palaeoclimate data for the Late Cretaceous Arctic: evidence for a warm polar ocean. *Palaeogeography, Palaeoclimatology, Palaeoeceology*, **128**, 227–51.

Hines, K. M., Bromwich, D. H. and Parish, T. R. (1995). A mesoscale modeling study of the atmospheric circulation of high southern latitudes. *Monthly Weather Review*, **123**, 1146–64.

Holdridge, L. R. (1947). Determination of world plant formations from simple climatic data. *Science*, **105**, 367–8.

Horrell, M. A. (1991). Phytogeography and paleoclimatic interpretation of the Maastrichtian. *Palaeogeography, Palaeoclimatology, Palaeoecology*, **86**, 87–138.

Hostetler, S. F., Giorgi, F., Bates, G. T. and Bartlein, P. J. (1994). Lake-atmosphere feedbacks associated with paleolakes Bonneville and Lahontan. *Science*, **263**, 665–8.

Houghton, J. T. (ed.) (1984). *The Global Climate*, p. 233. Cambridge: Cambridge University Press.

Houghton, J. T., Callander, B. A. and Varney, S. K. (eds.) (1990). *IPCC, Climate Change 1992: The IPCC Scientific Assembly Supplementary Report*. New York: Cambridge University Press.

Huybrechts, P. (1993). Glaciological modelling of the late Cenozoic East Antarctic ice sheet: stability or dynamism? *Geografiska Annaler*, **75**, 221–38.

Huybrechts, P. and Oerlemans, J. (1990). Response of the Antarctic ice sheet to future green-house warming. *Climate Dynamics*, **5**, 93–102.

Johnson, C. C., Barron, E. J., Kauffman, E. G., Arthur, M. A., Fawcett, P. J. and Yasuda, M. K. (1996). Middle Cretaceous reef collapse linked to ocean heat transport. *Geology*, **24**, 376–80.

Joussaume, S., Jouzel, J. and Sadoumy, R. (1984). Water isotope cycles in the atmosphere: First simulation using a general circulation model. *Nature*, **311**, 24–9.

Jouzel, J., Koster, R. D. and Joussaume, S. (1996). Climate reconstruction from water isotopes: what do we learn from isotopic models? In *Climatic Variations and Forcing Mechanisms of the Last 2000 years*, eds. P. D. Jones, R. S. Bradley and J. Jouzel, pp. 213–41. NATO ASI Series I 41. Berlin: Springer-Verlag.

Jouzel, J., Russel, G. L., Suozzo, R. J., Koster, R. D., White, J. W. and Broecker, W. S. (1987). Simulations of the HDO and $H^{18}O$ atmospheric cycles using the NASA GISS general circulation model: The seasonal cycle for present day conditions. *Journal of Geophysical Research*, **92**, 14739–60.

Kattenberg, A., Giorgi, F., Grassl, H. *et al.* (1996). Climate models – predictions of future climate. In *Climate Change 1995: The Science of Climate Change*, eds. J. T. Houghton, L. G. Meira Filho, B. A. Callander, N. Harris, A. Kattenberg and K. Maskell, pp. 285–357. Cambridge: Cambridge University Press.

Kennett, J. P. (1977). Cenozoic evolution of Antarctic glaciation, the circum-Antarctic oceans and their impact on global paleoceanography. *Journal of Geophysical Research*, **82**, 3843–59.

Knoll, M. A. and James, W. C. (1987). Effect of the advent and diversification of vascular land plants on mineral weathering through geologic time. *Geology*, **15**, 1099–102.

Kutzbach, J. E. and Liu, Z. (1997). Response of the African monsoon to orbital forcing and ocean feedbacks in the middle Holocene. *Science*, **278**, 440–2.

Lefèvre, F., Brasseur, G. P., Folkins, I., Smith, A. K. and Simon, P. (1994). Chemistry of the 1991–1992 statospheric winter: three-dimensional model simulations. *Journal of Geophysical Research*, **99**, 8183–95.

Lelieveld, J. and Crutzen, P. J. (1992). Indirect chemical effects of methane on climate warming. *Nature*, **355**, 339–42.

Leslie, L. M. and Fraedrich, K. (1997). A new general circulation model: formulation and preliminary results in a single and multi-processor environment. *Climate Dynamics*, **13**, 35–43.

Lynch, A., Chapman, W. L., Walsh, J. E. and Weller, G. (1995). Development of a regional climate model of the western Arctic. *Journal of Climate*, **8**, 1555–70.

Maier-Reimer, E., Mikolajewicz, U. and Hasselmann, K. (1993). Mean circulation of the Hamburg large-scale geostrophic ocean general circulation model and its sensitivity to the thermohaline surface forcing. *Journal of Physical Oceanography*, **23**, 731–57.

Mathews, E. (1983). Global vegetation and land use: new high resolution data bases for climate studies. *Journal of Climate and Applied Meteorology*, **22**, 478–88.

McGuffie, K. and Henderson-Sellers, A. (1997). *A Climate Modelling Primer*. 2nd ed. Chichester: John Wiley & Sons.

Meehl, G. A. (1992). Global coupled models: atmosphere, ocean, sea ice. In *Climate System Modeling*, ed. K. E. Trenberth, pp. 555–82. Cambridge: Cambridge University Press.

Mikolajewicz, U. and Crowley, T. J. (1997). Response of a coupled ocean/energy balance model to restricted flow through the central American isthmus. *Paleoceanography*, **12**, 429–41.

Mikolajewicz, U., Maier-Reimer, E., Crowley, T. J. and Kim, K.-Y. (1993). Effect of Drake and Panamanian gateways on the circulation of an ocean model. *Paleoceanography*, **8**, 409–26.

Miller, J. R., Russell, G. L. and Caliri, G. (1994). Continental-scale river flow in climate models. *Journal of Climate*, **7**, 914–28.

Moxim, W. J., Levy, II, H. and Kasibhatla, P. S. (1996). Simulated global tropospheric PAN: its tranport and impact on NO_x. *Journal of Geophysical Research*, **101**, 12621–38.

Najjar, R. (1992). Marine biogeochemistry. In *Climate System Modeling*, ed. K. E. Trenberth, pp. 241–80. Cambridge: Cambridge University Press.

Najjar, R. G., Sarmiento, J. L. and Toggweiler, J. K. (1992). Downward transport and fate of organic matter in the ocean: simulations with a general circulation model. *Global Biogeochemical Cycles*, **6**, 45–76.

North, G. R. (1975). Theory of energy balance climate models. *Journal of Atmospheric Science*, **32**, 2033–43.

North, G. R., Cahalan, R. F. and Coakley, J. A. (1981). Energy balance climate models. *Reviews of Geophysics and Space Physics*, **19**, 91–121.

O'Brian, J. J. (1986). *Advanced Physical Oceanographic Numerical Modeling*. Dordrecht: Reidel.

Oerlemans, J. (1982). A model of the Antarctic ice sheet. *Nature*, **297**, 550–3.

Oeschger, H. U., Siegenthaler, U., Schotterer, U. and Gugelmann, A. (1975). A box diffusion model to study the carbon dioxide exchange in nature. *Tellus*, **27**, 168–92.

Oglesby, R. J. and Park, J. (1989). The effect of precessional insolation changes on Cretaceous climate and cyclic sedimentation. *Journal of Geophysical Research*, **94**, 14793–816.

Otto-Bliesner, B. L. (1993). Tropical mountains and coal formation: a climate model study of the Westphalian (306 Ma). *Geophysical Research Letters*, **29**, 1947–50.

Otto-Bliesner, B. L. and Upchurch, G. R. (1997). Vegetation induced warming of high-latitude regions during the Late Cretaceous period. *Nature*, **385**, 804–7.

Parrish, J. T., Ziegler, A. M. and Scotese, C. R. (1982). Rainfall patterns and the distribution of coals and evaporites in the Mesozoic and Cenozoic. *Palaeogeography, Palaeoclimatology, Palaeoecology*, **40**, 67–101.

Peixoto, J. P. and Oort, A. H. (1992). *Physics of Climate*. New York: American Institute of Physics.

Peltier, W. R. and Marshall, S. (1995). Coupled energy balance/ice-sheet model simulations of the glacial cycle: a possible connection between terminations and terrigenous dust. *Journal of Geophysical Research*, **100**, 14269–89.

Pollard, D. and Schulz, M. (1994). A model for the potential locations of Triassic evaporite basins driven by paleoclimatic GCM simulations. *Global and Planetary Change*, **9**, 233–49.

Pollard, D. and Thompson, S. L. (1994). Sea-ice dynamics and CO_2 sensitivity in a global climate model. *Atmosphere-Oceans*, **32**, 449–67.

Pollard, D. and Thompson, S. L. (1995). Use of a Land-Surface-Transfer Scheme (LSX) in a global climate model: the response to doubled stomatal resistance. *Global and Planetary Change*, **10**, 129–61.

Pollard, D. and Thompson, S. L. (1997). Driving a high-resolution dynamic ice-sheet model with GCM climate: ice sheet initiation at 116,000 BP. *Annals of Glaciology*, **25**, 296–304.

Prentice, I. C., Cramer, W., Harrison, S. P., Leemans, R., Monserud, R. A. and Solomon, A. M. (1992). A global biome model based on plant physiology and dominance, soil properties and climate. *Journal of Biogeography*, **19**, 117–34.

Ramanathan, V., Cess, R. D., Harrison, E. F. *et al.* (1989). Cloud radiative forcing and climate: results from the Earth radiation budget experiment. *Science*, **243**, 57–63.

Ramanathan, V., Cicerone, H. B., Singh, H. B. and Kiehl, J. T. (1985). Trace gas trends and their potential role in climate change. *Journal of Geophysical Research*, **90**, 5547–66.

Ramanathan, V. and Collins, W. (1991). Thermodynamic regulation of ocean warming by cirrus clouds deduced from observations of the 1987 El Nino. *Nature*, **351**, 27–32.

Rasch, P. J., Boville, B. A. and Brasseur, G. P. (1995). A three-dimensional general circulation model with coupled chemistry for the middle atmosphere. *Journal of Geophysical Research*, **100**, 9041–71.

Rea, D. K., Leinen, K. M. and Janecek, T. R. (1985). Geologic approach to the long-term history of atmospheric circulation. *Science*, **227**, 721–5.

Retallack, G. J. (1990). *Soils of the Past*. London: Unwin-Hyman.

Revelle, R. and Suess, H. E. (1957). Carbon dioxide exchange between atmosphere and ocean and the question of an increase of atmospheric CO_2 during the past decades. *Tellus*, **9**, 18–27.

Rind, D. and Chandler, M. (1991). Increased ocean heat transports and warmer climate. *Journal of Geophysical Research*, **96**, 7437–61.

Rind, D., Healy, R., Parkinson, C. and Martinson, D. (1995). The role of sea ice in 2x CO_2 climate model sensitivity. Part I: the total influence of sea-ice thickness and extent. *Journal of Climate*, **8**, 379–89.

Robinson, A. R. (ed.) (1983). *Eddies in Marine Science*. Berlin: Springer-Verlag.

Roelofs, G.-J. and Lelieveld, J. (1995). Distribution and budget of O_3 in the troposphere calculated with a chemistry general circulation model. *Journal of Geophysical Research*, **100**, 20983–98.

Sarmiento, J. L. and Bender, M. (1994). Carbon biogeochemistry and climate change. *Photosynthesis Research*, **39**, 209–34.

Sarmiento, J. L., Hughes, T. M. C., Stouffer, R. J. and Manabe, S. (1998). Simulated response of the ocean carbon cycle to anthropogenic climate warming. *Nature*, **393**, 245–9.

Sarmiento, J. L. and LeQuere, C. (1996). Oceanic CO_2 uptake in a model of century-scale global warming. *Science*, **274**, 1346–50.

Sarmiento, J. L. and Toggweiler, J. R. (1984). A new model for the role of the oceans in determining atmospheric pCO_2. *Nature*, **308**, 621–4.

Sausen, R., Barthels, R. K. and Hasselmann, K. (1988). Coupled ocean-atmosphere models with flux correction. *Climate Dynamics*, **2**, 154–63.

Sausen, R. and Voss, R. (1996). Techniques for asynchronous and periodically synchronous coupling of atmosphere and ocean models. Part I. General strategy and application to the cyclo-stationary case. *Climate Dynamics*, **12**, 313–23.

Savin, S. M. (1977). The history of the Earth's surface temperature during the last 100 million years. *Annual Review of Earth and Planetary Sciences*, **5**, 319–56.

Schlesinger, M. E. (ed.) (1988). *Physically Based Modelling and Simulation of Climate and Climatic Change – Part I*. Dordrecht: Klewer.

Schmidt, G. A. and Mysak, L. A. (1996). Can increased poleward oceanic heat flux explain the warm Cretaceous climate? *Paleoceanography*, **11**, 579–93.

Schneider, S. H. and Dickinson, R. E. (1974). Climate modeling. *Reviews of Geophysics and Space Physics*, **12**, 447–93.

Schneider, S. H. and Gal-Chen, T. (1973). Numerical experiments in climate stability. *Journal of Geophysical Research*, **78**, 6182–94.

Schneider, S. H., Thompson, S. L. and Barron, E. J. (1985). Are high CO_2 concentrations needed to simulate above freezing winter conditions? In *The Carbon Cycle and Atmospheric CO_2: Natural Variations Archean to Present*, eds. E. T. Sundquist and W. S. Broecker, pp. 554–60. Washington, DC: American Geophysical Union.

Sellers, P. J., Mintz, Y., Sud, Y. C. and Dalcher, A. (1986). A simple biosphere model (SiB) for use within general circulation models. *Journal of Atmospheric Science*, **43**, 505–31.

Sellwood, B. W. and Price, G. D. (1994). Sedimentary facies as indicators of Mesozoic paleoclimate. In *Paleoclimates and their Modelling: With Special Reference to the Mesozoic*

Era, eds. J. R. L. Allen, B. J. Hoskins, B. W. Sellwood, R. A. Spicer and P. J. Valdes, pp. 17–24. London: Chapman & Hall.

Sellwood, B. W., Price, G. D. and Valdes, P. J. (1994). Cooler estimates of Cretaceous temperatures. *Nature*, **370**, 453–5.

Semazzi, F., Lin, N.-H. and Giorgi, F. (1993). A nested model study of Sahelian climate responses to sea-surface temperature anomalies. *Geophysical Research Letters*, **20**, 2897–900.

Semtner, A. J., Jr. and Chervin, R. M. (1992). Ocean general circulation from a global eddy-resolving simulation. *Journal of Geophysical Research*, **97**, 5493–550.

Simmons, A. J. (1990). Studies of increased horizontal and vertical resolution. Paper read at *Ten Years of Medium Range Weather Forecasting*, at Reading, UK.

Slingo, A. and Slingo, J. M. (1991). Response of the NCAR Community Climate Model to improvements in the representation of clouds. *Journal of Geophysical Research*, **96**, 10942–60.

Sloan, L. C., Crowley, T. J. and Pollard, D. (1996). Modeling of middle Pliocene climate with the NCAR GENESIS general circulation model. *Marine Micropaleontology*, **27**, 51–61.

Sloan, L. C., James, C. G. and Moore, Jr., T. C. (1995). Possible role of ocean heat transport in Early Eocene climate. *Paleoceanography*, **10**, 347–56.

Sloan, L. C. and Rea, D. K. (1995). Atmospheric carbon dioxide and early Eocene climate: A general circulation modeling sensitivity study. *Palaeogeography, Palaeoclimatology, Palaeoecology*, **119**, 275–92.

Sloan, L. C., Walker, J. C. G., Moore, Jr., T. C. and Rea, D. K. (1992). Possible methane-induced polar warming in the early Eocene. *Nature*, **357**, 320–2.

Smith, R. N. B. (1990). A scheme for predicting layer clouds and their water content in a general circulation model. *Quarterly Journal of the Royal Meteorological Society*, **116**, 435–60.

Spicer, R. A. (1989). Physiological characteristics of land plants in relation to environment through time. *Transactions of the Royal Society of Edinburgh: Earth Sciences*, **89**, 571–86.

Spicer, R. A. and Corfield, R. M. (1992). A review of terrestrial and marine climates in the Cretaceous with implications for modelling the 'Greenhouse Earth'. *Geological Magazine*, **129**, 169–80.

Stein, C. and Stein, S. (1992). A model for the global variation in oceanic depth and heat flow with lithospheric age. *Nature*, **359**, 123–9.

Stoll, H. M. and Schrag, D. P. (1996). Evidence for glacial control of rapid sea level changes in the Early Cretaceous. *Science*, **272**, 1772–4.

Taylor, K. E. and Penner, J. E. (1994). Response of the climate system to atmospheric aerosols and greenhouse gases. *Nature*, **369**, 734–7.

Thompson, S. L. and Pollard, D. (1995). A global climate model (GENESIS) with a land-surface-transfer scheme (LSX). Part I: Present-day climate. *Journal of Climate*, **8**, 732–61.

Thompson, S. L. and Pollard, D. (1997). Greenland and Antarctic mass balances for present and doubled atmospheric CO_2 from the GENESIS Version-2 Global Climate Model. *Journal of Climate*, **10**, 158–87.

Tie, X. X., Lin, X. and Brasseur, G. (1994). Two-dimensional coupled dynamical/chemical/microphysical simulation of the global distribution of El Chichon volcanic aerosol. *Journal of Geophysical Research*, **99**, 16779–92.

Toggweiler, J. R. and Samuels, B. (1995). Effect of Drake Passage on the global thermohaline circulation. *Deep-Sea Research*, **42**, 477–500.

Trenberth, K. E. (ed.) (1992). *Climate System Modeling*. Cambridge: Cambridge University Press.

Trenberth, K. E. and Solomon, A. (1994). The global heat balance: heat transports in the atmosphere and ocean. *Climate Dynamics*, **10**, 107-34.

Tzeng, R. Y. and Bromwich, D. H. (1994). NCAR CCM simulation of present-day Arctic climate. Paper read at *Sixth Conference of Climate Variations*, at Nashville, Tennessee.

Upchurch, G. R and Wolfe, J. A. (1987). Mid-Cretaceous to Early Tertiary vegetation and Climate: evidence from fossil leaves and woods. In *The Origin of Angiosperms and Their Biological Consequences*, eds. E. M. Friis, W. G. Chaloner and P. R. Crane, pp. 75–104. Cambridge: Cambridge University Press.

Valdes, P. J., Sellwood, B. W. and Price, G. D. (1996). Evaluating concepts of Cretaceous equability. *Paleoclimates*, **2**, 139–58.

Verbitsky, M. Ya. and Oglesby, R. J. (1995). The CO_2-induced thickening/thinning of the Greenland and Antarctic ice sheets as simulated by a GCM (CCM1) and an ice-sheet model. *Climate Dynamics*, **11**, 247–53.

Verbitsky, M. Ya. and Saltzman, B. (1995). Behavior of the East Antarctic ice sheet as deduced from a coupled GCM/ice-sheet model. *Geophysical Research Letters*, **22**, 2913–16.

Veronis, G. (1973). Large scale ocean circulation. In *Advances in Applied Mechanics*, ed. C.-S. Yih, pp. 1–92. New York: Academic Press.

Voss, R., Sausen, R. and Cubash, U. (1998). Periodically synchronously coupled integrations with the atmosphere-ocean general circulation model ECHAM3/LSG. *Climate Dynamics*, **14**, 249–66.

Walsh, K. and McGregor, J. L. (1995). January and July climate simulations over the Australian region using a limited-area model. *Journal of Climate*, **8**, 2387–403.

Walsh, K. and McGregor, J. L. (1996). Simulation of Antarctic climate using a limited-area model. *Journal of Geophysical Research*, **101**, 19093–108.

Wang, C., Prinn, R. and Sokolov, A. (1998). A global interactive chemistry and climate model: formulation and testing. *Journal of Geophysical Research*, **103**, 3399–417.

Washington, W. M., Meehl, G. A., VerPlank, L. and Bettge, T. W. (1994). A world ocean model for greenhouse sensitivity studies: resolution intercomparison and the role of diagnostic forcing. *Climate Dynamics*, **9**, 321–44.

Washington, W. M. and Parkinson, C. L. (1986). *An Introduction to Three Dimensional Climate Modeling*. Mill Valley: University Science Books and Cambridge University Press.

Weatherly, J. W., Briegleb, B. P., Large, W. G. and Maslanik, J. A. (1998). Sea ice and polar climate in the NCAR CSM. *Journal of Climate*, **11**, 1472–86.

Wolfe, J. A. (1971). Tertiary climatic fluctuations and methods of analysis of Tertiary floras. *Palaeogeography, Palaeoclimatology, Palaeoecology*, **9**, 27–57.

3

Comparison of zonal temperature profiles for past warm time periods

THOMAS J. CROWLEY AND JAMES C. ZACHOS

ABSTRACT

A standard explanation for past warm time periods involves higher atmospheric CO_2 levels. However, model simulations indicate that the zonal temperature response to a CO_2 perturbation is significantly different from observations in the geologic record: model-predicted temperatures are higher in low latitudes and lower in high latitudes than revealed by observations. Although changes in ocean heat transport have been invoked to account for such discrepancies, it is also necessary to subject zonal temperature profiles, particularly tropical sea surface temperature (SST) estimates, to error analysis in order to test for robustness of conclusions. Herein we conduct such an analysis and demonstrate that it is difficult to generalize about the tropical SST pattern in low latitudes during warm periods. Three time periods (Pliocene, Eocene, Cenomanian) indicate tropical SSTs not significantly different from the present, while two time periods (Miocene and Maastrichtian) suggest cooler tropical SSTs. Diagenesis may be responsible for some of the cooler tropical temperature estimates. Analysis of uncertainties in $\delta^{18}O$-based paleotemperature estimates suggests that random non-dissolution-related uncertainties are in the order of 2–3 °C for individual specimens. However, averaging of Holocene core top samples yields zonal mean temperature estimates in the tropics that agree with observations to within 0.5–1.0 °C. Although there is a potential for pre-Pleistocene uncertainties of 2–4 °C in tropical SSTs (due to diagenesis and salinity/alkalinity changes from CO_2 variations), careful screening of samples should produce a smaller non-random uncertainty of +1.5 to −3.0 °C for the zonal average from multiple sites. That is, predicted SSTs could on average be biased toward values slightly cooler than the actual SST. These biases may not apply to short intervals of climate change. This hypothesis is tested with data from the Paleocene–Eocene boundary event; significant tropical warming occurs in this data set. Although more observations are needed to better constrain pre-Pleistocene tropical SSTs, the present assessment continues to support the hypothesis that tropical SSTs did not increase greatly as a response to estimated higher CO_2 levels in the pre-Pleistocene. However, small SST increases (1–2 °C) may not be detectable at the present time.

INTRODUCTION

Geologists have long been fascinated by the concept of an 'ice-free earth.' Although subsequent studies have sometimes raised questions as to whether we can definitively state there was no ice during past time periods, there is little doubt that intervals such as the Cretaceous and Eocene were significantly warmer than the present and had less global ice volume. One attempt to estimate global temperature changes for these intervals determined values of about 5–6 °C and 8–9 °C warmer than present for the early Eocene (55 Ma) and mid- to Late Cretaceous (90–100 Ma), respectively (Crowley and Kim, 1995). Climate modelers have become increasingly interested in warm climates, especially since the seminal work of Eric Barron and colleagues (Barron *et al.*, 1981, 1995; Barron and Washington, 1984, 1985). There are now a number of research groups working on this problem, and this special volume is one more example of the increasing maturity of the field.

Quite early in the investigations of the warm earth phenomenon several important conclusions were reached:

1. Although changing continental positions could have a significant effect on temperatures, changes in geography alone in general cannot explain an ice-free earth. In addition to obtaining small changes in global mean temperatures, modeled winter temperatures still seemed to be significantly colder than estimated by proxy geological data (Barron *et al.*, 1981; Barron and Washington, 1984).

2. Increased carbon dioxide (CO_2) levels reduced the model–data disagreements but did not completely remove them (Barron and Washington, 1985; Sloan and Rea, 1996). The higher CO_2 levels are consistent with geochemical models and a variety of proxy data (e.g., Berner, 1992, 1994, 1997) (Fig. 3.1). The CO_2 proxies are usually based on $\delta^{13}C$ analyses of some carbon species, with the $\delta^{13}C$ varying as a function of pCO_2 in the atmosphere (see references in caption to Fig. 3.1).

3. There appears to be relatively little evidence supporting climate model predictions (Fig. 3.2) for tropical sea surface temperatures (SSTs) warmer than the present during past warm time periods (Crowley, 1991), when carbon dioxide levels were presumed to be higher than the present (e.g., Manabe and Bryan, 1985; Manabe and Stouffer, 1994; Cubasch *et al.*, 1995; Bush and Philander, 1997). The higher model-predicted tropical SSTs are a response to increased downward long-wave radiative forcing by enhanced absorption in the atmosphere of outgoing radiation. If this conclusion can be substantiated it would provide an important constraint on climate models. For example, a substantiated conclusion might imply some significant level of negative cloud feedback in the tropics (Ramanathan and Collins, 1991).

4. By contrast, temperatures in high latitudes seemed to be significantly warmer than predicted by the same climate models. The discrepancy between low and high latitudes (conclusions 3 and 4) has led to solutions such as (a) increased poleward heat transport (Rind and Chandler, 1991; Barron *et al.*,

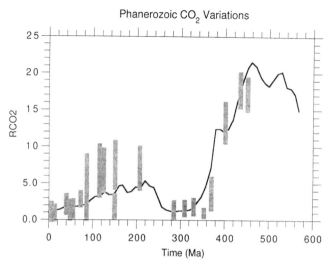

Figure 3.1. Plot of atmospheric CO_2 vs. time for the Phanerozoic (past 600 Ma). The parameter RCO_2 is defined as the ratio of the mass of CO_2 in the atmosphere at some time in the past to that at present (using a pre-industrial value of 300 ppm). The solid line represents the best estimates from the GEOCARB II model (Berner, 1994). Vertical bars represent independent estimates of CO_2 level based on the study of paleosols and other proxy data. These data are from Cerling (1991), Mora *et al.* (1991, 1996), Yapp and Poths (1992, 1996), Sinha and Stott (1994), and Andrews *et al.* (1995). (Modified from Berner, 1997.)

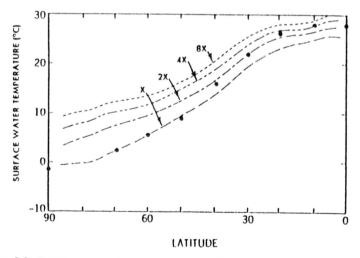

Figure 3.2. Zonally averaged temperatures as calculated by a coupled ocean–atmosphere model with idealized geography for different levels of CO_2 forcing. Dots refer to observations from the US Navy Hydrographic Office (1964). Note higher tropical SSTs with higher CO_2 levels. (Figure after Manabe and Bryan, 1985.)

1995), perhaps due to an increase in the thermohaline circulation; or (b) vegetation feedbacks at high latitudes (Dutton and Barron, 1996; Otto-Bliesner and Upchurch, 1997) which subsequently warm ocean waters. There are problems with both of these suggestions. For example, although changes in the present thermohaline circulation play a key role in warming the North Atlantic, a strong North Atlantic thermohaline circulation causes cooling in the southern hemisphere (Crowley, 1992). So it is difficult to understand how changes in vertical overturn can cause high-latitude warming in all ocean basins. Furthermore, if ocean heat transport is along rather than across isopycnals, acceleration of the major conduit of heat transport (western boundary currents) might actually result in a latitudinal constriction of the gyre due to angular momentum considerations. This latter hypothesis receives some support from recent modeling work of Baum and Crowley (1999).

Although enhanced vegetation cover increases summer warming in high latitudes (e.g., Crowley and Baum, 1994; Foley *et al.*, 1994; Otto-Bliesner and Upchurch, 1997; DeConto *et al.*, Chapter 9, this volume), the effect is very small in winter, when insolation levels are low. Thus the vegetation feedback may not necessarily solve the problem of 'equable' climates (Barron and Washington, 1982; Crowley *et al.*, 1989; Sloan and Barron, 1990), namely that winter cooling at high latitudes is apparently greater than indicated by observations (Taylor *et al.*, 1992; Yemane, 1993; Taylor *et al.*, Chapter 11, this volume). However, Sloan *et al.* (1996) demonstrated that inclusion of SSTs from observations can sometimes result in greater winter freezing line changes than those obtained from mixed layer ocean runs, in which the SSTs are calculated.

Thus understanding the related problems of the zonal temperature gradient and the magnitude of tropical SST changes represents a first-order challenge, whose resolution would be of considerable interest to the climate modeling community. An additional reason to study such times is given by analyses by Hoffert and Covey (1992) and Khesgi and Lapenis (1996) of the Russian paleotemperature data set (Borzenkova, 1992; Borzenkova *et al.*, 1992). These authors cited evidence from such data for a 'universal curve' of climate change in which temperature changes by latitude could be uniformly scaled for different time periods. Unfortunately the raw data used to reconstruct such universal curves are not widely available and it is difficult to evaluate the legitimacy of some of the zonal temperature reconstructions. It is therefore important to test these conclusions against a separate data set in which the raw data are published in readily accessible literature.

In this paper we examine one element of the 'warm climates' problem: the zonal temperature profile for different time periods. The purpose of the exercise is threefold: (1) to develop a composite record of SST fluctuations for a number of different time periods in the past to determine whether any generalizations can be made; (2) to analyze uncertainties (both mean and standard deviation) in the SST estimates in order to estimate their precision; and (3) to examine model predictions of warmer tropical SSTs with higher CO_2 levels in light of the revised estimates of uncertainties.

DATA SOURCES AND ANALYSIS

Below we present results from five time periods when temperatures were significantly warmer than the Holocene: the mid-Pliocene (3 Ma; Dowsett et al., 1996), the early–mid-Miocene (16 Ma; Savin et al., 1985), the early Eocene (55 Ma; Zachos et al., 1994), the Maastrichtian (66 Ma; d'Hondt and Arthur, 1996), and the Cenomanian (90 Ma; Sellwood et al., 1994). All of the data are from published sources and are based on $\delta^{18}O$ measurements of assumed mixed layer dwelling planktonics (see discussion below), except for the Pliocene data, which are based on transfer function estimates. The justification for transferring these $\delta^{18}O$ measurements to SST is clearly open to discussion but we consider it essential to do so in order to more carefully test model predictions.

Conversion of $\delta^{18}O$ to SST incorporated a reduced isotopic composition of seawater (ice volume effect) and a $-0.27‰$ correction for the difference between the $\delta^{18}O$ composition of the carbonate standard and seawater (sw). The assumed ice volume-corrected $\delta^{18}O_{sw}$ for different time periods is $-0.5‰$ (Miocene) and $-1.0‰$ (other time periods). Small uncertainties in these numbers ($< 0.5\%$) do not significantly affect our main conclusions. Underestimating ice volume tends to bias our temperature estimates toward colder values: for example, if $\delta^{18}O_{sw}$ in the Cretaceous was similar to that today, $\delta^{18}O$-based paleotemperature estimates would be too cold by $4\,°C$.

A calibration equation based on Erez and Luz (1983) was used with latitudinally varying $\delta^{18}O$ accounting for regional variations in surface seawater (ssw) due to variation in precipitation minus evaporation (P − E). The latitudinal $\delta^{18}O_{ssw}$ relationship used here was designed by Zachos et al. (1994) primarily as a means of compensating for the effects of the meridional $\delta^{18}O_{ssw}$ gradient on SST estimates between the tropics and high-latitude oceans of the southern hemisphere. They were not as concerned with the northern hemisphere because (1) nearly all of the high-latitude sites in their study were in the southern hemisphere, and (2) they recognized that the error associated with the adjustment is much larger for the more restricted seas of the Arctic and North Atlantic where $\delta^{18}O_{ssw}$ variability (during the Cenozoic) is substantial. This adjustment results in an $\sim3\text{–}4\,°C$ increase in SSTs in tropical regions relative to SSTs computed assuming an isotopically homogeneous ocean (Zachos et al., 1994).

Although the assumption of a fixed latitudinal $\delta^{18}O_{ssw}$ gradient has been criticized (Price et al., 1996), there are two lines of evidence that support the assumption, at least with respect to first-order latitudinal variations in open-ocean salinity. The first line of evidence involves the Phanerozoic distribution of evaporites, which is approximately the same as occurs at present (Gordon, 1975). The second line of evidence involves climate model calculations (Fig. 3.3) indicating that the latitudinal patterns of P − E, which determine the oceanic salinity patterns, are relatively similar for different CO_2 scenarios (the second-order differences will be discussed further below). Even if these arguments cannot conclusively prove that the Zachos et al.

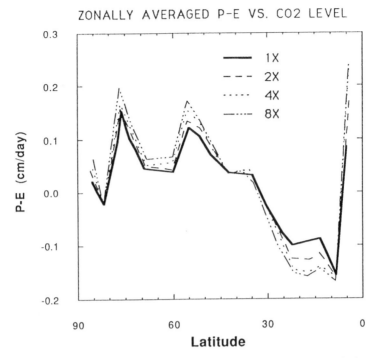

Figure 3.3. Zonally averaged net precipitation-minus-evaporation (P–E) changes for different CO_2 levels from the same model illustrated in Fig. 3.2. (Figure after Manabe and Bryan, 1985.)

(1994) δ_{ssw} correction is correct, they are certainly a justifiable adjustment. Since one of the purposes of the present paper is to test for robustness of conclusions about differences between past and present tropical SSTs, any reasonable adjustment that narrows or eliminates differences cannot be ignored.

The P – E pattern in Fig. 3.3 can be directly attributed to changing water vapor content of the atmosphere with CO_2 and temperature level. As the water vapor content increases, so does net precipitation in areas of rising motion (intertropical convergence zone and mid-latitudes). Increased outflow of air in the upper atmosphere of the inner tropics results in enhanced subsidence on the flanks, and therefore greater net evaporation. This pattern seems to be a very stable feature of climate model simulations (e.g., Manabe and Bryan, 1985; Rind, 1986). Although the latitudinal salinity pattern breaks down somewhat in the subpolar North Atlantic, we would not expect this anomaly to occur for time intervals prior to the late Miocene, when both observations and ocean model results suggest that North Atlantic Deep Water production rates may have been significantly less than those at present (Woodruff and Savin, 1989; Delaney, 1990; Maier-Reimer *et al.*, 1990).

COMPARISON OF WARM PERIOD ZONAL PROFILES

A compilation of zonal SST estimates for different warm time periods is shown in Fig. 3.4. Some comments are necessary for this figure. All data were fit with

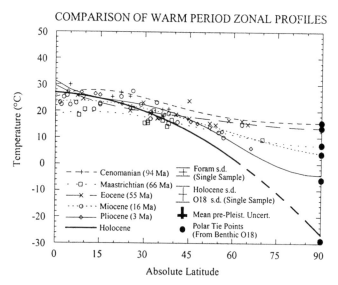

COMPARISON OF WARM PERIOD ZONAL PROFILES

Figure 3.4. Estimated sea surface temperature profiles for five time periods when global temperatures were warmer than the present: the Pliocene (3 Ma), Miocene (16 Ma), Eocene (55 Ma), Maastrichtian (66 Ma), and Cenomanian (94 Ma). The records are compared with the estimates of surface temperature from sediment samples for the present Holocene interglacial, with the latter record extended to the poles for surface air temperature in order to get a complete zonal profile. The polar tie point is based on benthic $\delta^{18}O$ data and the assumption that deep-water temperatures represent the coldest temperature on the planet. Standard deviation (s.d.) units represent one estimate of uncertainties for temperature estimates for single samples. Multiple sample/multiple site uncertainties for the estimates of the Holocene mean value are about 0.5-1.0 °C. The additional pre-Pleistocene $\delta^{18}O$ uncertainty includes bias estimates (discussed in the text) and should be read as indicating that actual SSTs could be 3 °C warmer or 1.5 °C cooler than calculated SSTs. The Pliocene estimate is based on a faunal transfer function regressed against SST. See text for further discussion and data sources.

a third-order polynomial in sine of latitude, with the Holocene fit extended to the poles (dashed line) with surface air temperature for the sake of comparison with warmer time periods. Although this fit is crude, it is about the best that can be done with a small data set. Polar temperatures for different warm time periods were estimated based on the benthic foraminiferal $\delta^{18}O$ record of Douglas and Woodruff (1981), which is similar to other records (e.g., Shackleton, 1986) except that it includes intermittent (but vital) Cretaceous data. The benthic values were used to estimate surface polar temperatures on the assumption that deep waters record the coldest SSTs on the planet. This assumption is supported by the convergence of benthic and planktonic $\delta^{18}O$ values at high latitudes in the intervals for which high-latitude data are available (Shackleton and Kennett, 1975; Zachos *et al.*, 1994). Also shown in Fig. 3.4 is the uncertainty in *individual* SST estimates due to either transfer functions for the Pliocene (1.5 °C; Dowsett *et al.*, 1996) or the standard deviation (2.9 °C) of the larger Holocene core top data base (Zachos *et al.*, 1994; see further discussion below).

Although zonal profiles inevitably conceal important regional differences caused by western boundary currents and upwelling, such profiles capture some essential features of the earth's climate and are used in the study of present climate (e.g., Crowley and North, 1991). Nevertheless, the limitations of zonal profiles should be kept in mind when examining the causes for model–data discrepancies for different time periods. The compilation of profiles (Fig. 3.4) for warm time periods demonstrates the long-established large warming in high latitudes that is consistent with geological data on land and virtual absence of any evidence for permanent ice at high latitudes (e.g., Crowley and North, 1991). Also evident is the virtual lack of any solid evidence for tropical SSTs significantly warmer (>28–$29\,°C$) than the present. A similar conclusion was reached from Coniacian–Santonian (87 Ma) measurements (Huber et al., 1995), although here the total database was smaller ($n = 4$). Norris and Wilson (1998) analyzed exceptionally well-preserved foraminifera from the subtropical North Atlantic and concluded that SSTs might have been about $30\,°C$, but there is sufficient uncertainty in the $\delta^{18}O_{ssw}$ to reframe their conclusions as suggesting SSTs of $28\pm2.5\,°C$. Our conclusion is different from those of Frakes et al. (1994) and Sellwood et al. (1994), who either included nannofossil data or made no latitudinal adjustment for salinity changes (see above).

There is some evidence for cooler temperatures in the mid-Miocene (16 Ma) and late Maastrichtian (66 Ma). These records are as much as 4–8 °C below present tropical SSTs. However, Wilson and Opdyke (1996) reported SSTs comparable to the other warm intervals for earlier in the Maastrichtian (69 Ma). These measurements are from platform guyot samples of rudist bivalves and magnesian calcite cement. For two reasons this latter result is not necessarily in direct variance with the planktonic foraminiferal results. Firstly, there is good evidence for a significant cooling event between the early and late Maastrichtian (e.g., Parrish and Spicer, 1988; Stott and Kennett, 1990; Lécuyer et al., 1993; Barrera, 1994), so the two $\delta^{18}O$ records are not comparable. Secondly, the shallow realm of the guyot could conceivably have had warmer temperatures than open-ocean environments. Possible freshwater dilution effects could also shift $\delta^{18}O$ values to lighter values than the open ocean (cf. Cole et al., 1993). Nevertheless, the Wilson and Opdyke (1996) data are important and still may be of large-scale relevance (see further discussion below).

There is further evidence for cooler tropical SSTs in some intervals we have not mapped: parts of the Pliocene and early Pleistocene (Prentice et al., 1993; Isern et al., 1996), the Oligocene (Savin, 1977; Zachos et al., 1994), and the late Eocene (Zachos et al., 1994). Some of these isotopic results are supported by other data. For example, Feary et al. (1991) document a 5° southward shift of the Great Barrier Reef tropical biota since the onset of Pleistocene glaciations, with present conditions not developing until the late Pleistocene (Brunhes; cf. Isern et al., 1996). Similarly, coral reefs have overall lower diversity and abundance during early Cenozoic warm time periods (Grigg, 1988; Flügel and Flügel-Kahler, 1992). Tropical rainforests are also scarce during warm time periods (Ziegler et al., 1987; Horrell, 1991). However, other tropical biota (mangroves, shallow-water invertebrates) indicate continuous tropical conditions during some of these same intervals (Adams et al., 1990;

Graham, 1992, 1994). The reality of cooler tropical SSTs has yet to be definitively established (see also below).

In conclusion, available data indicate that there is relatively little robust evidence supporting warmer SSTs during times of elevated CO_2 levels but that there is some evidence for cooler tropical SSTs during other time periods that were warmer than the present. Before discussing these results further we now address possible sources of error in our conclusion.

ANALYSIS OF UNCERTAINTIES IN SST ESTIMATES

In order to evaluate the level of agreement/disagreement between theory and observations of tropical SSTs it is necessary to address the uncertainties in the $\delta^{18}O$-based SST estimates. We will break the discussion into two sections, one focusing on random error, using the Holocene core top database as an example, and a second discussing possible systematic errors caused by possible changes in CO_2 levels and the effects of sediment burial on $\delta^{18}O$. We will first address in some detail the random errors associated with individual SST estimates and demonstrate that these numbers can be quite large. However, we will also demonstrate that composite estimates for zonal mean annual SSTs are significantly less than the standard deviation for individual analyses.

Random errors: a case study from Holocene core tops

In this section we will use the Holocene core top database as a constraint on estimating random sources of error in $\delta^{18}O$-based paleotemperature estimates. We will focus mainly on a database compiled by Zachos *et al.* (1994) from various sources, but will also cite an example from another revealing study (Billups and Spero, 1996).

Comparison of Holocene core tops with $\delta^{18}O$-based temperature estimates

A comparison of temperature estimates with Holocene core top data (Fig. 3.5) indicates a 2.9 °C standard deviation in the estimates. This spread is not unusual. As a further illustration for the potential range of such responses, Fig. 3.6 illustrates a range of measurements from only one site in the western equatorial Atlantic (Billups and Spero, 1996). Below we attempt to dissect the sources of this uncertainty.

Analytical error

Although small, the analytical uncertainty of $\delta^{18}O$ measurements (about ±0.4 °C) cannot be entirely ignored; in the Holocene case it accounts for 14% of the total uncertainty.

Oxygen isotope transfer function

Another source of random error is the $\delta^{18}O$ transfer function equation. The error reported by Erez and Luz (1983) for their experimentally derived equation is about ±1.4 °C.

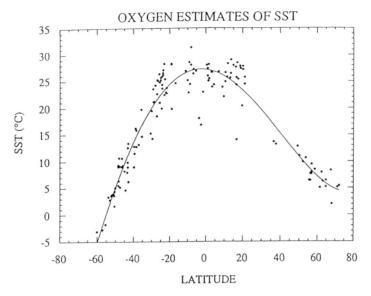

Figure 3.5. Scatter of δ¹⁸O-based SST estimates around the zonal mean value for Holocene core tops (based on compilation of Zachos *et al.*, 1994). The average standard deviation of the zonal average is 2.9 °C.

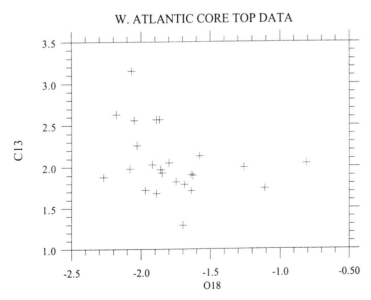

Figure 3.6. Example of scatter of observations from a single well-preserved core top site in the western equatorial Atlantic. (Data from Billups and Spero, 1996.)

Seasonal–centennial scale climate variability

This factor is not always addressed in analyses of paleotemperature estimates, but since a typical sample of deep-sea sediment might be a 1500-year average
of surface water conditions, and foraminifera live for about a month, there are 6000
potential individual climate 'recordings' that could be delivered to the sea floor (this
assumes a season autocorrelation of three months for climate time series). Thus a
small number of single-specimen analyses may not yield an accurate estimate of the
1500-year average even if there were no errors of the type discussed below. The
problem is compounded by bioturbation, which can introduce older samples with
different mean values into a record.

As an illustration of the potential uncertainty in estimates of the long-term
mean, Billups and Spero (1996) note that there is a total range of 2.0 °C in the
observed surface temperature fluctuations at the western equatorial Atlantic site
illustrated in Fig. 3.6. Paleoclimate records (e.g., Bradley and Jones, 1993) indicate
significant centennial–millennial scale variability that almost certainly increased this
range over that of the observed range in the 20th century. Although it is not possible
to estimate accurately the increased spread of potential values, samples from coral
records (Dunbar *et al.*, 1994; Quinn *et al.*, 1998) provide some evidence that the
spread could be 50% greater than in the 20th century, and perhaps even 100%
(especially when upwelling sites are considered). If we take the standard deviation
of the $\delta^{18}O$ values illustrated in Fig. 3.6 as an indication of the uncertainty in the
temperature estimate, we arrive at a value of $\pm 1.5\,°C$. This number is used in our
assessment of the cumulative error in Table 3.1.

Metabolic effects

In principle, to compute temperature from shell $\delta^{18}O$, calcite should be
precipitated in oxygen isotopic equilibrium with seawater. However, as numerous
studies have demonstrated, this is seldom the case. The shells of many marine organisms, including planktonic foraminifers and calcareous nannofossils, are isotopically
offset from predicted equilibrium values. These isotopic displacements appear to
arise primarily from the so-called 'vital' effects which probably include both metabolic and kinetic effects, such as the incorporation of respired or metabolic CO_2 or
changes in the rate of CO_2 hydroxylation and calcification (Spero and DeNiro, 1987;
McConnaughey, 1989). These factors, which may be strongly influenced by growth
rates or the activity of photosymbionts, affect shell isotope values by altering either
the isotopic composition or carbonate ion content of the ambient fluid. While these
effects appear to be relatively constant in space and time, they must be considered in
computing any uncertainty in $\delta^{18}O$-based SST estimates.

To derive values for the uncertainties related to metabolic and kinetic
effects, we rely on data obtained from experimental studies in which mixed layer
foraminifera (photosymbiont and non-symbiont-bearing species) such as
Globigerinoides sacculifer were grown in culture under controlled temperature conditions (Erez and Luz, 1983; Spero and DeNiro, 1987; Spero and Lea, 1996). These
data suggest that metabolic effects on $\delta^{18}O$ fractionation are relatively small, on the

order of a few tenths of a per mil. Nevertheless, the potential effect on temperature estimates is not insignificant (1.0–1.5 °C).

Depth and season of calcification

An additional source of error in using fossil planktonic foraminifera to reconstruct SST arises from uncertainties in depth and season of calcification. Foraminifera collected from plankton tow and sediment traps often show a positive departure from expected equilibrium $\delta^{18}O$ for surface waters. This is true even for supposed 'mixed layer' species (Curry and Matthews, 1981; Kahn and Williams, 1981). Several studies have demonstrated, however, that the higher $\delta^{18}O$ values might reflect calcification in deeper or seasonally cooler waters and not disequilibrium vital effects (Bouvier-Soumagnac and Duplessy, 1985; Deuser, 1987; Sautter and Thunell, 1991). The populations of some species increase substantially during periods of seasonal upwelling, thereby biasing sea surface temperature estimates toward colder values (Sautter and Thunell, 1991). Also, some mixed layer foraminifera may undergo additional shell thickening after they have settled deeper into the water column. If shell thickening occurs within or below the thermocline, the mean oxygen isotopic composition of tropical planktonic foraminifera can shift by more than +0.5‰ (Erez and Honjo, 1981; Duplessy et al., 1987).

These artifacts can be circumvented to some extent with knowledge of the seasonal and depth habitat preferences of planktonic taxa, by avoiding shells with thick walls, and by analyzing a large number of individual specimens. For intervals of time where extant species are available, the uncertainty is probably less than 1 °C. In intervals for which only extinct species are available, the uncertainties may be larger since season and depth of calcification must be inferred, often from isotopic data. A common solution is to assume that the most negative $\delta^{18}O$ values of all species within a given sample are representative of maximum mixed layer temperatures (Shackleton and Boersma, 1981; Poore and Matthews, 1984; Zachos et al., 1994). Although this involves circular reasoning, the inference of a mixed layer habitat for some species can usually be verified with an independent test, for example documentation of photosymbiosis using carbon isotopes (Shackleton et al., 1985; D'Hondt et al., 1994). This approach was used in assembling the Holocene reconstruction shown in Fig. 3.5. This plot is based on data from various species of mixed layer planktonic foraminifera from core top samples (Savin et al., 1985; Zachos et al., 1994). Despite the anomalously low values in some tropical sites, it is clear that the most negative values closely approximate maximum SST across the tropics. We estimate the error associated with depth of calcification uncertainties in extinct foraminifera to be similar, that is less than 2.0 °C.

Summary

Table 3.1 summarizes the results from this section. Most of these numbers are semiquantitative in reliability, and significantly more effort would be required to narrow the uncertainties. Yet it seems from the analysis that some level of understanding can be obtained as to the magnitude of the core top range, once sources of

Table 3.1. *Summary of sources of estimated errors from Holocene core tops (see text for details)*

Analytical error	0.4 °C
$\delta^{18}O$ transfer function	1.5 °C
Climate variability	1.5 °C
Metabolic effects	1.0-1.5 °C
Depth habitat	1.0-1.5 °C
RMS error	2.6-3.0 °C
Observed (Fig. 3.5)	2.9 °C

individual error are identified. Within the uncertainties of the effect of different processes the root mean square (RMS) error agrees with the observed standard deviation.

The Holocene time slice as reconstructed here probably overstates the degree of uncertainty (error) associated with estimating SST from isotopes for the modern oceans. The data were randomly collected from the literature with no consideration of the core top depth, condition of foraminifera, evidence for reworking, etc. Also, some species whose values were included (e.g., *Globigerina bulloides*) are known to show inconsistent relationships with predicted isotopic equilibrium values. It is very likely that the observed estimate of 2.9 °C could be reduced by careful screening for depth habitat and increasing the number of analyses to obtain better estimates of mean SST. If so, such efforts might reduce the RMS error to 2.0 °C, but it seems difficult to reduce it much below that level.

Despite the magnitude of error estimates for individual specimens, comparison of estimated zonal mean temperatures for core tops with observed zonal mean temperatures from Alexander and Mobley (1976) agree to within 0.5–1.0 °C. Since this is the number that is most important with respect to testing the main hypotheses outlined in the Introduction, the good agreement with observations suggests some level of optimism in predicting pre-Pleistocene temperatures, but only if some of the problems addressed below can be better constrained and if enough samples are used from a given time slice to narrow the single-sample uncertainty and obtain a reliable estimate of the maximum zonal SST. Further sampling studies would have to be undertaken on Holocene core tops to determine how many sites are needed to obtain a reliable estimate of this quantity.

Additional uncertainties in pre-Pleistocene records

In addition to random sources of error a number of problems arise when analyzing pre-Pleistocene samples. Some of these errors are also random; however, two may introduce significant biases. Below we discuss the major sources of error in pre-Pleistocene $\delta^{18}O$ analyses.

Kinetic effects

Kinetic isotope effects associated with changes in carbonate ion (CO_3^{2-}) concentration or pH appear to be quite large and responsible for much of the disequilibrium fractionation observed in $\delta^{18}O$. Planktonic foraminifera grown in seawater with CO_3^{2-} content in the range $100–700\,\mu mol\,kg^{-1}$ show a change in fractionation in the order of $0.16‰\ 100\,\mu mol^{-1}\,kg^{-1}$ (Spero *et al.*, 1997). Could this be a significant source of variability in foraminifer shell $\delta^{18}O$ over time? The ocean is an extremely well-buffered system; the CO_3^{2-} content of the modern surface ocean varies by less than $200\,\mu mol\,kg^{-1}$, from 100 to $300\,\mu mol\,kg^{-1}$. Nevertheless, it is still possible that CO_3^{2-} content was higher in the past, particularly during times when pCO_2 was elevated. If we assume an upper limit of $500–600\,\mu mol\,kg^{-1}$ for the last 100 million years, the uncertainty associated with CO_3^{2-}/pH kinetic effects would be about 0.4‰ at most, or roughly 1.5–2.0 °C. This effect should not be considered random; if CO_2 levels were higher in the past the effect would systematically bias SST estimates toward lower values.

Salinity changes

Another possible uncertainty involves changes in zonal P − E due to CO_2 variations. As shown in Fig. 3.3 and discussed above, the modeled zonal pattern of P − E systematically changes with altered CO_2 level. Such changes could have altered zonal $\delta^{18}O$ gradients and therefore affected the 'zonal SST adjustment' developed by Zachos *et al.* (1994) and employed in the present paper.

A systematic relationship between P − E, salinity, and zonal $\delta^{18}O$ can be used to estimate the potential effect of altered P − E patterns on SST estimates for the pre-Pleistocene. For example, enhanced evaporation in the tropics equivalent to $\sim 0.1\ cm\ day^{-1}$ (Fig. 3.7) yields zonally averaged salinity increases of about 1.0‰. If the planetary thermal gradient were flatter, the Cretaceous $\delta^{18}O$/salinity relationship would also have been smaller (closer to 0.3‰). Enhanced subtropical evaporation of $\sim 0.05\ cm\ day^{-1}$ (Fig. 3.3) should therefore yield a zonal salinity increase of $\sim 0.5‰$. Broecker (1989) has shown that on average a 1.0‰ salinity change corresponds to an $\sim 0.5‰$ change in $\delta^{18}O_{sw}(\Delta\delta^{18}O)$. Opposite changes might occur in the equatorial low salinity zone. P − E during warmer time periods could have modified low-latitude $\delta^{18}O$ values by about 0.25‰ (~ 1.0 °C). One important physical constraint that limits surface salinity increases in the open ocean, and hence $\delta^{18}O$ variability, is density (D'Hondt and Arthur, 1986; Zachos *et al.*, 1994). As salinity rises and density increases, the water column becomes unstable, promoting sinking of surface waters. The salinity level at which instability occurs will vary as a function of the temperature and salinity of the underlying water. In general, however, assuming mean ocean salinities similar to present, the salinity at which instability is reached should be lower for times of warmer deep waters. Thus we would not expect open-ocean tropical sea surface salinities noticeably higher than present.

Diagenesis

Dissolution and diagenesis create additional uncertainty for oxygen isotope-based temperature estimates. As plankton shells sink to the sea floor and are

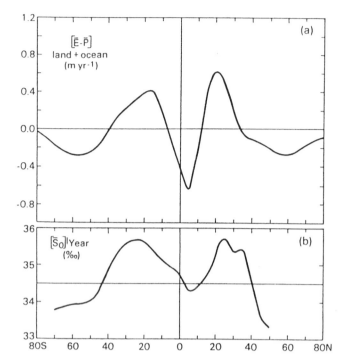

Figure 3.7. Zonal mean profiles of (a) evaporation minus precipitation (E−P) and (b) surface salinity in the world ocean (S_0). (After Peixoto and Oort, 1984.)

gradually buried by sediment, they are subject to three processes: (1) dissolution, (2) secondary calcification, and (3) recrystallization. Each of these can potentially alter shell isotopic composition, although to differing degrees. Of these processes, dissolution probably has the smallest impact on estimates of sea surface temperatures. For an individual mixed layer foraminifer, the effect of partial dissolution on shell $\delta^{18}O$ is negligible. For an assemblage of shells, however, dissolution can create slight biases toward positive $\delta^{18}O$ values since individual shells precipitated in warmer water tend to be more susceptible to dissolution than shells precipitated in cooler water (Erez and Honjo 1981; Wu and Berger, 1989). Studies indicate that for sediments deposited below the lysocline dissolution can increase the values of mixed layer foraminifera by as much as 0.6‰ (2–3 °C) (Wu et al., 1990). No discernible effects have been reported for microfossils deposited above the lysocline.

Recrystallization and secondary calcification are potentially more significant sources of error for $\delta^{18}O$-based estimates of tropical temperatures. This is primarily because oxygen isotopic composition of secondary calcite precipitating in the sediment column can be significantly different from that of primary calcite (Killingley, 1983; Schrag et al., 1995). In general, the $\delta^{18}O$ composition of diagenetic calcite precipitated during early burial in pelagic sediments is significantly enriched in ^{18}O as a result of the cold bottom-water temperatures, while calcite precipitated later in more deeply buried sediment is isotopically depleted because of lower pore water $\delta^{18}O$ compositions and higher *in situ* temperatures.

For most samples overprinted by secondary calcite the effects of diagenesis are easily recognized by either light or scanning electron microscopy. These samples can simply be avoided. In some situations, however, the effects may be more subtle and difficult to detect by either visual or geochemical approaches (Huber and Hodell, 1996; Price *et al.*, 1996). Regardless, it is not yet possible to quantify the effects of diagenetic overprinting on fossil foraminifer $\delta^{18}O$. The primary obstacle has been the inability to specifically quantify either the amount and timing of secondary calcite added or the extent of recrystallization. A numerical model has been developed, however, that estimates the effect of burial diagenesis on bulk carbonate $\delta^{18}O$ (Schrag *et al.*, 1995). The model simulates the effects of oxygen isotope exchange between pore water and sedimentary carbonate over time using fixed pore water $\delta^{18}O$ and thermal gradients. Critical parameters for each site include the age of the sea floor, bottom-water temperatures, and sedimentation and carbonate recrystallization rates. The rate constants for recrystallization rates are computed using a diffusion/advection model of pore water strontium concentration gradients (Richter and Liang, 1993). For tropical sites, the model shows that during early phases of diagenesis overprinting could increase bulk carbonate $\delta^{18}O$ by as much as 1.0‰ over the first 10–20 Ma (Fig. 3.8). By contrast, during later stages of deep burial diagenesis (>50 Ma) values begin to decrease, by as much as 3.0‰ in the more extreme cases. In essence, the model shows that with an input of constant SST the primary long-term changes observed in bulk carbonate $\delta^{18}O$ records of the last 70 Ma can be explained almost entirely by diagenetic overprinting (Schrag *et al.*, 1995).

One important exception to the above conclusion is the warm interval of the early Eocene, in which the assumption of constant SSTs does not agree with the Schrag *et al.* (1995) results; the model suggests that the early Eocene could have been as much as 2–6 °C warmer than present. This is potentially a very important result, for if it could be substantiated it would demonstrate that the geologic record essentially agrees with climate model predictions. The results would also suggest that the spot conclusions of Wilson and Opdyke (1996) are not an artifact of local SST/salinity variations and in fact are representative of the mean tropical SSTs. Resolution of this problem would be of great interest to the climate modeling community.

Can this model be applied to individual microfossils to correct for the effect of diagenesis or estimate diagenesis-related uncertainties? The answer is probably not. Isotopic and textural data clearly indicate that individual components such as foraminifer tests do not recrystallize to the extent suggested for bulk carbonate by the model (Arthur *et al.*, 1989; Norris and Wilson, 1998). For example, such extensive recrystallization of shells would eliminate the large systematic intra- and interspecies carbon isotopic gradients commonly preserved in fossil mixed layer foraminifera from tropical sites (Shackleton *et al.*, 1985; D'Hondt *et al.*, 1994; D'Hondt and Arthur, 1996). Nevertheless, the model still has important implications for the potential errors in estimating tropical SST from foraminifera $\delta^{18}O$. Early

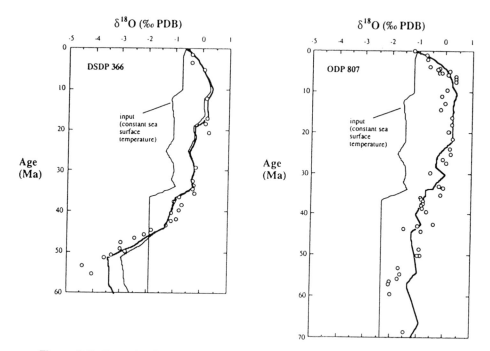

Figure 3.8. Example of the potential effect of diagenesis on carbonate and sea surface values. Figures show results from two cores plotted vs. age in million years (Ma), in which measured whole-rock $\delta^{18}O$ values were compared with a diagenetic model (see text) that assumes sea surface temperatures were constant through time. Two calculations are shown for Site 366 using two different recrystallization rates. For both sites the model does an excellent job of fitting the data, except in the range of 50-60 Ma for which the measured $\delta^{18}O$ values are lower than model predictions. To fit the data over this interval equatorial SSTs would have to be higher by 2-6 °C. See text for further discussion and caveats. (From Schrag *et al.*, 1995.)

diagenetic overprinting would have the greatest impact on the $\delta^{18}O$ isotopic composition of tropical foraminifera since values of the secondary calcite are relatively enriched due to the cold water in the upper portion of the sediment column. The total effect depends primarily on the extent of recrystallization and diagenetic overprinting. If it is as rapid and extensive as indicated by numerical models for bulk carbonate (>50%), the degree of uncertainty will increase as a function of time but in predictable directions. For the first 10–20 Ma, diagenesis will tend to drive microfossil $\delta^{18}O$ compositions toward higher values by as much as 1.2‰. After 30–50 Ma, in most settings with high sedimentation rates diagenesis drives $\delta^{18}O$ toward lower values by as much as −3‰ in extreme cases, assuming nearly complete recrystallization.

While we cannot quantify the uncertainties created by diagenesis in $\delta^{18}O$-based estimates of tropical SST, we can evaluate the potential effects of overprinting on an individual shell at different stages of burial (or diagenesis) with model-consistent end member isotopic values for secondary calcite (sc). As a simple example, we compute the effects of recrystallization on the genetic (size)-related isotopic gradient of individual tests of a mixed layer species in the size range 125–450 μm. Here we will assume only the addition of secondary calcite (with no loss of primary calcite) during early diagenesis when the $\delta^{18}O_{sc}$ is relatively high (2.0‰). The $\delta^{13}C_{sc}$ of the secondary calcite should be similar to that of the bulk carbonate (~3.0‰).

This simple model shows an important result: the impact of diagenetic overprinting under these conditions will be much larger for oxygen than for carbon isotopes. Oxygen isotope values increase by more than 1.0‰ while carbon isotopes shift only slightly. The $\delta^{18}O$ of individual tests of the hypothetical species whose initial isotopic composition ranged from −2.0 to −2.5‰ will increase to −1.0 to −1.5‰ with a 25% increase in mass in the form of secondary calcite. By contrast, because $\delta^{13}C$ of the secondary calcite is similar to the foraminifer values, the $\delta^{13}C$ range of that species changes only slightly. At much higher rates of recrystallization, however, those gradients would be significantly attenuated. This has important implications for the extent of recrystallization of foraminifera. Interspecies carbon isotope differences as large as 3–4‰ are commonly recorded in tropical foraminifera from late Paleocene pelagic sediments (e.g., Shackleton *et al.*, 1985; D'Hondt *et al.*, 1994). These gradients, which are larger than those observed in the modern foraminifera, suggest that the extent of recrystallization/secondary calcification was minimal in these specific cases. As such, it is likely that recorded intra- and interspecies gradients in $\delta^{18}O$ have not changed significantly as well.

This last observation is important since it suggests that like sedimentary components from the same level or over short segments of homogeneous sediment will be equally overprinted by diagenesis, thereby preserving temperature-related differences in $\delta^{18}O$. In such cases one could reconstruct relative temperature changes much more accurately than absolute temperatures. A good example is provided by the Late Paleocene Thermal Maximum (LPTM) event some 55 Ma. This unusual climatic event was characterized by an abrupt 4–6 °C warming of the deep sea and high-latitude oceans, and by a 3‰ negative excursion in mean ocean $\delta^{13}C$ (e.g., Kennett and Stott, 1991). Kelly *et al.* (1996), in an effort to reconstruct equatorial SST change associated with the LPTM, measured the isotopic composition of individual specimens of several species of mixed layer planktonic foraminifera from samples located within the LPTM interval at ODP Site 865. They found within several species a distinct bimodal distribution of $\delta^{13}C$ values into two groups with means separated by roughly 3.0‰, as shown in Fig. 3.9. The mean $\delta^{13}C$ of one group is essentially identical to pre-excursion values, suggesting that these specimens were reworked from below. The mean of the second group is similar to the value expected for excursion taxa (~0.0‰), suggesting that these specimens lived during the LPTM. The range in oxygen isotope values of the two groups varies from 1.0 to 1.5‰, although the minima tend to be similar in each sample (2.2–2.4‰). Because preservation of the inter- and intraspecies isotope gradients implies minimal diagenetic overprinting, the stability of $\delta^{18}O$ minima across the LPTM interval can be viewed as evidence of constant SST. This last observation has potentially important ramifications for climate theory in light of the theoretical and empirical evidence

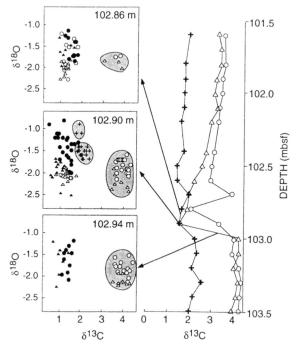

Figure 3.9. Stable isotope values of multiple and single specimens of mixed layer and thermocline-dwelling planktonic foraminifera from the P/E boundary interval at ODP Site 865 (Kelly *et al.*, 1996). The carbon isotope time series in the right-hand panel was constructed from analyses of multiple specimens of three species, while the boxes on the left show the values of individual specimens from three discrete samples within the excursion interval. The individual specimen values are plotted in carbon and oxygen-isotope space; different symbols refer to different species analyzed. Shading denotes reworked pre-excursion specimens. Note that there is little evidence for significant $\delta^{18}O$ changes between the different time intervals. These data therefore indicate no significant tropical warming. (From Kelly *et al.*, 1996.)

for significant increases in greenhouse gas concentrations during this event (Zachos *et al.*, 1993; Dickens *et al.*, 1997).

Thus although 4 °C (due to diagenesis) is a theoretical uncertainty, which if proven would be extremely significant for tropical SSTs, we consider it unlikely that the actual error is this large. This is because careful screening of planktonic foraminifera using visual and geochemical criteria can eliminate many of the extreme cases of diagenesis that might occur in bulk carbonates. Again, it is difficult to estimate the magnitude of the reduced uncertainty. Until further work is done on the problem we can only postulate that the number is closer to 1–3 °C. This is still a large number, and its potential importance will be discussed below.

Summary

Table 3.2 summarizes the error for pre-Pleistocene analyses. Again, these estimates should not be equated to laboratory-derived estimates; they are presented

Table 3.2. *Summary of additional sources of estimated errors for pre-Pleistocene records (see text for details). A negative sign means that calculated SSTs from the $\delta^{18}O$-based transfer functions are colder than actual SSTs*

Kinetic effect	-1.0 to $-2.0\,°C$
Changing salinity	$\pm 1.0\,°C$
Diagenesis	-1.0 to $-2.0\,°C$

here as semiquantitative estimates that will hopefully prompt other investigators to narrow some of the uncertainties further. Results of the assessment of pre-Pleistocene uncertainties indicate that there is a greater uncertainty with respect to estimates of high-end tropical SSTs than to those of low-end tropical SSTs, because the potential effect of alkalinity and diagenesis involves systematic errors (see discussion above). We emphasize that this error applies only to SST estimates based on tropical foraminifers which tend to be more susceptible to preservation effects, and that the estimate pertains to averaging of multiple analyses from different sites; as discussed above, single-specimen uncertainties could be substantially larger.

SUMMARY AND CONCLUSIONS

It is now possible to summarize most of the principal sources of errors in $\delta^{18}O$-based SST estimates by combining the errors assessed in Table 3.1 with the additional errors that could be associated with pre-Pleistocene samples (Table 3.2). The random uncertainties indicate errors in the order of $\pm 2.8\,°C$ for the pre-Pleistocene temperature estimates. The additional bias of 2–$4\,°C$ should be added to this number. As stated in the discussion of Holocene core tops, the random errors can give a misleading estimate of our ability to predict zonal mean temperature changes, as the estimates are actually accurate to within 0.5–$1.0\,°C$ as a result of averaging measurements from different specimens and different sites.

For the pre-Pleistocene there are still uncertainties in the estimate of the mean annual temperature because of uncertainties in salinity and the potential biasing effect of alkalinity/diagenesis changes. Factoring in these additional uncertainties yields an estimate of the uncertainty in the zonal mean annual values of $+1.5\,°C$ to $-3\,°C$ for the zonal average from multiple sites. That is, the calculated temperature estimates may range from being $1.5\,°C$ too warm to $3\,°C$ too cold (compared with the actual SSTs). These uncertainties are large and could probably be reduced to 1–$2\,°C$ if a better handle on diagenesis could be obtained. But it is unlikely that it could be reduced much below that number unless additional paleotemperature estimates are developed or applied.

Given the present level of uncertainty, what can be stated about our ability to constrain climate model predictions of increased tropical SST with higher greenhouse gas concentrations? Firstly, it is difficult to make a robust case that tropical SSTs were significantly different from the present for several past warm time periods.

As stated earlier, this conclusion is dependent on inclusion of the Zachos *et al.* (1994) δ_{ssw} adjustment. However, as also stated earlier, if a 'reasonable' adjustment eliminates a conclusion about SSTs colder than the past, then the latter conclusion cannot be considered robust. Secondly, if CO_2 changes are relatively small (~ 1.5–$2 \times$ increase in CO_2) it may not be possible to test model-predicted tropical SST increases unless there is a large sensitivity in the climate system response to tropical SST changes. However, larger CO_2 changes as predicted for the Cretaceous and early Cenozoic (Berner, 1997; Dickens *et al.*, 1997) should cause changes larger than observed in the geologic record. For example, the 5 °C equatorial increase predicted by Bush and Philander (1997) seems to significantly exceed our present uncertainty in tropical SST estimates.

As discussed above, uncertainties are smaller for abrupt changes in radiative forcing such as the Paleocene Thermal Maximum (Dickens *et al.*, 1997). Lack of evidence for tropical SST changes over this interval (Kelly *et al.*, 1997) further supports the tropical thermostat conclusion. The above analysis reinforces the conclusion that there is little robust geological evidence indicating that tropical SSTs increased as CO_2 increased. If verified by further work this conclusion places substantial constraints on climate models and suggests that some feedbacks, be it clouds or some other process, are not being accurately incorporated into the present generation of climate models.

There is also some evidence for tropical SST cooling for the Miocene and late Maastrichtian, but a diagenetic overprint cannot be eliminated as a cause for such changes. If the effects of diagenesis for these two time intervals could be proven small, then we would have the perplexing dilemma of fundamentally different zonal SST profiles for different warm time periods. If so, it would be hard to argue that the zonal response to a CO_2 increase could be generalized. Such differences would open the door to multiple explanations for zonal temperature profiles for past time periods. This would be a dissatisfying dilemma from a theoretical perspective, and should be invoked as a last resort in order to avoid a proliferation of *ad hoc* explanations for different time periods. Further analysis of these time intervals, and improved estimates of the effect of diagenesis, should help to clarify the magnitude of these problems.

ACKNOWLEDGEMENTS

This research was supported by DOE-National Institute for Global Environmental Changes (NIGEC) Cooperative agreement number DE-FC03-90ER61010 and Tulane University TUL-033-95/96 to T.J.C. and NSF grant OCE-9458367 to J.C.Z. We thank K. Billups for assistance, and K. Bice and G. Price for comments.

REFERENCES

Adams, C. G., Lee, D. E. and Rosen, B. F. (1990). Conflicting isotopic and biotic evidence for tropical sea-surface temperatures during the Tertiary. *Palaeogeography, Paleoclimatology, Palaeoecology*, **77**, 289–313.

Alexander, R. D. and Mobley, R. L. (1976). Monthly average sea-surface temperatures and ice-pack limits on a 1° global grid. *Monthly Weather Review*, **104**, 143–8.

Andrews, J. E., Tandon, S. K. and Dennis, P. F. (1995). Concentration of carbon dioxide in the Late Cretaceous atmosphere. *Journal of the Geological Society (London)*, **152**, 1–3.

Arthur, M. A., Dean, W. E., Zachos, J. C., Kaminiski, M., Hagerty-Reig, S. and Elmstrom, K. (1989). Geochemical expression of early diagenesis in middle Eocene–lower Oligocene pelagic sediments in the southern Labrador Sea, Site 647, ODP Leg 105. In *Proceedings of the ODP, Scientific Results*, vol. 105, eds. S. P. Srivastava, M. A. Arthur, B. Clement *et al.*, pp. 111–36. College Station, TX: Ocean Drilling Program.

Barrera, E. (1994). Global environmental changes preceding the Cretaceous–Tertiary boundary: Early–late Maastrichtian transition. *Geology*, **22**, 877–80.

Barron, E. J., Fawcett, P. J., Peterson, W. H., Pollard, D. and Thompson, S. L. (1995). A "simulation" of mid-Cretaceous climate. *Paleoceanography*, **10**, 953–62.

Barron, E. J., Thompson, S. L. and Schneider, S. H. (1981). An ice-free Cretaceous? Results from climate model simulation. *Science*, **212**, 501–8.

Barron, E. J. and Washington, W. M. (1982). Cretaceous climate: A comparison of atmospheric simulations with the geologic record. *Palaeogeography, Palaeoclimatology, Palaeoecology*, **40**, 103–33.

Barron, E. J. and Washington, W. M. (1984). The role of geographic variables in explaining paleoclimates: Results from Cretaceous climate model sensitivity studies. *Journal of Geophysical Research*, **89**, 1267–79.

Barron, E. J. and Washington, W. M. (1985). Warm Cretaceous climates: High atmospheric CO_2 as a plausible mechanism. In *The Carbon Cycle and Atmospheric CO_2: Natural Variations Archean to Present*, eds. E. T. Sundquist and W. S. Broecker, pp. 546–53. Washington, DC: American Geophysical Union.

Baum, S. K. and Crowley, T. J. (1999). Effect of weaker winds on subtropical gyre circulations (in preparation).

Berner, R. A. (1992). Palaeo-CO_2 and climate. *Nature*, **358**, 114.

Berner, R. A. (1994). GEOCARB II: A revised model of atmospheric CO_2 over Phanerozoic time. *American Journal of Science*, **294**, 56–91.

Berner, R. A. (1997). The rise of plants and their effect on weathering and atmospheric CO_2. *Science*, **276**, 544–6.

Billups, K. and Spero, H. J. (1996). Reconstructing the stable isotope geochemistry and paleotemperatures of the equatorial Atlantic during the last 150,000 years: Results from individual foraminifera. *Paleoceanography*, **11**, 217–38.

Borzenkova, I. (1992). The changing climate during the Cenozoic. St Petersburg: Hydrometeoizdat (in Russian). English translation in press (Dordrecht: Kluwer).

Borzenkova, I., Zubakov, V. A. and Lapenis, A. G. (1992). Global climate changes during the warm epochs of the past. *Meteorologiya i Gidrologiya*, **8**, 25–37 (in Russian).

Bouvier-Soumagnac, Y. and Duplessy, J. (1985). Carbon and oxygen isotopic composition of planktonic foraminifera from laboratory culture, plankton tows and recent sediment: implications for the reconstruction of paleoclimatic conditions and of the global carbon cycle. *Journal of Foraminiferal Research*, **15**, 302–20.

Bradley, R. S. and Jones, P. D. (1993). Little Ice Age summer temperature variations: Their nature and relevance to recent global warming trends. *The Holocene*, **3**, 367–76.

Broecker, W. S. (1989). The salinity contrast between the Atlantic and Pacific Oceans during glacial time. *Paleoceanography*, **4**, 207–12.

Bush, A. B. G. and Philander, S. G. H. (1997). The late Cretaceous: Simulation with a coupled atmosphere–ocean general circulation model. *Paleoceanography*, **12**, 495–516.

Cerling, T. E. (1991). Carbon dioxide in the atmosphere: Evidence from Cenozoic and Mesozoic paleosols. *American Journal of Science*, **291**, 377–400.

Cole, J. E., Rind, D. and Fairbanks, R. G. (1993). Isotopic responses to interannual climate variability simulated by an atmospheric general circulation model. *Quaternary Science Review*, **12**, 387–406.

Crowley, T. J. (1991). Past CO_2 changes and tropical sea surface temperatures. *Paleoceanography*, **6**, 387–94.

Crowley, T. J. (1992). North Atlantic Deep Water cools the southern hemisphere. *Paleoceanography*, **7**, 489–97.

Crowley, T. J. and Baum, S. K. (1994). General circulation model study of late Carboniferous interglacial climates. *Palaeoclimates: Data and Modelling*, **1**, 3–21.

Crowley, T. J., Hyde, W. T. and Short, D. A. (1989). Seasonal cycle variations on the supercontinent of Pangaea. *Geology*, **17**, 457–60.

Crowley, T. J. and Kim, K.-Y. (1995). Comparison of longterm greenhouse projections with the geologic record. *Geophysical Research Letters*, **22**, 933–6.

Crowley, T. J. and North, G. R. (1991). *Paleoclimatology*. New York: Oxford University Press.

Cubasch, U., Hegerl, G. C., Hellbach, A. *et al.* (1995). A climate change simulation starting from 1935. *Climate Dynamics*, **11**, 71–84.

Curry, W. B. and Matthews, R. K. (1981). Equilibrium ^{18}O fractionation in small size fraction planktic foraminifera: Evidence from recent Indian Ocean sediments. *Marine Micropaleontology*, **6**, 327–37.

Delaney, M. L. (1990). Miocene benthic foraminiferal Cd/Ca records: South Atlantic and western equatorial Pacific. *Paleoceanography*, **5**, 743–60.

Deuser, W. G. (1987). Seasonal variations in isotopic composition and deep-water fluxes of the tests of perennially abundant planktonic foraminifera of the Sargasso Sea: Results from sediment-trap collections and their paleoceanographic significance. *Journal of Foraminiferal Research*, **17**, 14–27.

D'Hondt, S. D. and Arthur, M. A. (1996). Late Cretaceous oceans and the cool tropic paradox. *Science*, **271**, 1838–41.

D'Hondt, S., Zachos, J. C. and Schultz, G. (1994). Stable isotope signals and photosymbiosis in late Paleocene planktic foraminifera. *Paleobiology*, **20**, 391–406.

Dickens, G. R., Castillo, M. M. and Walker, J. C. G. (1997). A blast of gas from the latest Paleocene: Simulating first-order effects of massive dissociation of oceanic methane hydrate. *Geology*, **25**, 259–62.

Douglas, R. G. and Woodruff, F. (1981). Deep sea benthic foraminifera. In *The Sea*, vol. 7, ed. C. Emiliani, pp. 1233–327. New York: Wiley-Interscience.

Dowsett, H., Barron, J. and Poore, R. (1996). Middle Pliocene sea surface temperatures: a global reconstruction. *Marine Micropaleontology*, **27**, 13–25.

Dunbar, R. B., Wellington, G. W., Colgan, M. W. and Glynn, P. W. (1994). Eastern Pacific sea surface temperature since 1600 A.D.: The $\delta^{18}O$ record of climate variability in Galápagos corals. *Paleoceanography*, **9**, 291–315.

Duplessy, J.-C., Shackleton, N. J., Fairbanks, R. G., Labeyrie, L., Oppo, D. and Kallel, N. (1987). Deepwater source variations during the last climatic cycle and their impact on the global deepwater circulation. *Paleoceanography*, **3**, 343–60.

Dutton, J. F. and Barron, E. J. (1996). GENESIS sensitivity to changes in past vegetation. *Palaeoclimates: Data and Modelling*, **1**, 325–54.

Erez, J. and Honjo, S. (1981). Comparison of isotopic composition of planktonic foraminifera in plankton tows, sediment traps and sediments. *Palaeogeography, Palaeoclimatology, Palaeoecology*, **33**, 129–56.

Erez, J. and Luz, B. (1983). Experimental paleotemperature equation for planktonic foraminifera. *Geochimica et Cosmochimica Acta*, **47**, 1025–31.

Feary, D. A., Davies, P. J., Pigram, C. J. and Symonds, P. A. (1991). Climatic evolution and control in carbonate deposition in northeast Australia. *Palaeogeography, Palaeoclimatology, Palaeoecology*, **89**, 341–61.

Flügel, E. and Flügel-Kahler, E. (1992). Phanerozoic reef evolution: Basic questions and data base. *Facies*, **26**, 167–278.

Foley, A. J., Kutzbach, J. E., Coe, M. T. and Levis, S. (1994). Feedbacks between climate and boreal forests during the Holocene epoch. *Nature*, **371**, 52–4.

Frakes, L. A., Probst, J.-L. and Ludwig, W. (1994). Latitudinal distribution of paleotemperature on land and sea from early Cretaceous to middle Miocene. *Comptes Rendus de l'Academie des Sciences Paris*, **318**, 1209–18.

Gordon, W. A. (1975). Distribution by latitude of Phanerozoic evaporite deposits. *Journal of Geology*, **83**, 671–84.

Graham, A. (1992). Utilization of the isthmian land bridge during the Cenozoic–paleobotanical evidence for timing, and the selective influence of altitudes and climate. *Reviews of Palaeobotany and Palynology*, **72**, 119–28.

Graham, A. (1994). Neotropical Eocene coastal floras and $^{18}O/^{16}O$-estimated warmer vs. cooler equatorial waters. *American Journal of Botany*, **81**, 301–6.

Grigg, R. W. (1988). Paleoceanography of coral reefs in the Hawaiian-Emperor Seamount Chain. *Science*, **240**, 1737–43.

Hoffert, M. I. and Covey, C. (1992). Deriving global climate sensitivity from paleoclimate reconstructions. *Nature*, **360**, 573–6.

Horrell, M. A. (1991). Phytogeography and paleoclimatic interpretation of the Maastrichtian. *Palaeogeography, Palaeoclimatology, Palaeoecology*, **86**, 87–138.

Huber, B. T. and Hodell, D. A. (1996). Middle–Late Cretaceous climate of the southern high latitudes: Stable isotopic evidence for minimal equator-to-pole thermal gradients: Discussion and reply. *Geological Society of American Bulletin*, **108**, 1192–6.

Huber, B. T., Hodell, D. A. and Hamilton, C. P. (1995). Middle–Late Cretaceous climate of the southern high latitudes: Stable isotopic evidence for minimal equator-to-pole thermal gradients. *Geological Society of America Bulletin*, **107**, 1164–91.

Isern, A. R., McKenzie, J. A. and Feary, D. A. (1996). The role of sea-surface temperature as a control on carbonate platform development in the western Coral Sea. *Palaeogeography, Palaeoclimatology, Palaeoecology*, **124**, 247–72.

Kahn, M. I. and Williams, D. F. (1981). Oxygen and carbon isotopic composition of living planktonic foraminifera from the northeast Pacific Ocean. *Palaeogeography, Palaeoclimatology, Palaeoecology*, **33**, 47–69.

Kelly, D. C., Bralower, T. J., Zachos, J. C., Silva, I. P. and Thomas, E. (1996). Rapid diversification of planktonic foraminifera in the tropical Pacific (ODP Site 865) during the late Paleocene thermal maximum. *Geology*, **24**, 423–6.

Kennett, J. P. and Stott, L. D. (1991). Abrupt deep sea warming, paleoceanographic changes and benthic extinctions at the end of the Palaeocene. *Nature*, **353**, 319–22.

Kheshgi, H. S. and Lapenis, A. G. (1996). Estimating the accuracy of Russian paleotemperature reconstructions. *Palaeogeography, Palaeoclimatology, Palaeoecology*, **121**, 221–37.

Killingley, J. S. (1983). Effects of diagenetic recrystallization on $^{18}O/^{16}O$ values of deep-sea sediments. *Nature*, **301**, 594–7.

Lécuyer, C., Grandjean, P., O'Neil, J. R., Cappetta, H. and Martineau, F. (1993). Thermal excursions in the ocean at the Cretaceous–Tertiary boundary (northern Morocco): $\delta^{18}O$ record of phosphatic fish debris. *Palaeogeography, Palaeoclimatology, Palaeoecology*, **105**, 235–43.

Maier-Reimer, E., Mikolajewicz, U. and Crowley, T. J. (1990). Ocean general circulation model sensitivity experiment with an open Central American isthmus. *Paleoceanography*, **5**, 349–66.

Manabe, S. and Bryan, K. (1985). CO_2-induced change in a coupled ocean–atmosphere model and its paleoclimatic implications. *Journal of Geophysical Research*, **90**, 11689–708.

Manabe, S. and Stouffer, R. J. (1994). Multiple-century response of a coupled ocean-atmosphere model to an increase of atmospheric carbon dioxide. *Journal of Climate*, **7**, 5–23.

McConnaughey, T. (1989). ^{13}C and ^{18}O isotopic disequilibrium in biological carbonates. II. In vitro simulation of kinetic isotope effects. *Geochimica et Cosmochimica Acta*, **53**, 163–71.

Mora, C. I., Driese, S. G. and Colarusso, L. A. (1996). Middle to Late Paleozoic atmospheric CO_2 levels from soil carbonate and organic matter. *Science*, **271**, 1105–7.

Mora, C. I., Driese, S. G. and Seager, P. G. (1991). Carbon dioxide in the Paleozoic atmosphere: Evidence from carbon-isotope compositions of pedogenic carbonate. *Geology*, **19**, 1017–20.

Norris, R. D. and Wilson, P. A. (1998). Low-latitude sea-surface temperatures for the mid-Cretaceous and the evolution of planktic foraminifera. *Geology*, **26**, 823–6.

Otto-Bliesner, B. L. and Upchurch, G. R., Jr. (1997). Vegetation-induced warming of high-latitude regions during the Late Cretaceous period. *Nature*, **385**, 804–7.

Parrish, J. T. and Spicer, R. A. (1988). Late Cretaceous terrestrial vegetation: A near-polar temperature curve. *Geology*, **16**, 22–5.

Peixoto, J. P. and Oort, A. H. (1984). Physics of climate. *Review of Modern Physics*, **56**, 365–429.

Poore, R. Z. and Matthews, R. K. (1984). Oxygen isotope ranking of late Eocene and Oligocene planktonic foraminifers: Implications for Oligocene sea-surface temperatures and global ice volume. *Marine Micropaleontology*, **9**, 111–34.

Prentice, M. L., Friez, J. K., Simonds, G. G. and Matthews, R. K. (1993). Neogene trends in planktonic foraminifera $\delta^{18}O$ from Site 807: Implications for global ice volume and western equatorial Pacific sea surface temperatures. In *Proceedings of the Ocean Drilling Program*, vol. 130, eds. W. H. Berger, L. W. Kroenke and L. A. Mayer, pp. 281–305. College Station, TX: Ocean Drilling Program.

Price, G. D., Sellwood, B. W. and Pirrie, D. (1996). Middle–Late Cretaceous climate of the southern high latitudes: Stable isotopic evidence for minimal equator-to-pole thermal gradients: Discussion and reply. *Geological Society of America Bulletin*, **108**, 1192–6.

Quinn, T. M., Crowley, T. J., Taylor, F. W., Henin, C., Joannot, P. and Join, Y. (1998). A multicentury stable isotope record from a New Caledonia coral: Interannual and decadal sea surface temperature variability in the southwest Pacific since 1657 A.D. *Paleoceanography*, **13**, 412–26.

Ramanathan, V. and Collins, W. (1991). Thermodynamic regulation of ocean warming by cirrus clouds deduced from observations on the 1987 El Nino. *Nature*, **351**, 27–32.

Richter, F. M. and Liang, Y. (1993). The rate and consequences of Sr diagenesis in deep-sea carbonates. *Earth Planetary Science Letters*, **117**, 553–65.

Rind, D. (1986). The dynamics of warm and cold climates. *Journal of Atmospheric Science*, **43**, 3–24.

Rind, D. and Chandler, M. (1991). Increased ocean heat transports and warmer climate. *Journal of Geophysical Research*, **96**, 7437–61.

Sautter, L. R. and Thunell, R. C. (1991). Seasonal variability in the $\delta^{18}O$ and $\delta^{13}C$ of planktonic foraminifera from an up-welling environment: Sediment trap results from the San Pedro Basin, Southern California Bight. *Paleoceanography*, **6**, 307–44.

Savin, S. M. (1977). The history of the earth's surface temperature during the past 100 million years. *Annual Reviews of Earth and Planetary Sciences*, **5**, 319–55.

Savin, S. M., Abel, L., Barrera, E. *et al.* (1985). The evolution of Miocene surface and near-surface marine temperatures: Oxygen isotopic evidence. In *The Miocene Ocean: Paleoceanography and Biogeography. Memoir of Geological Society of America*, **163**, 49–82.

Schrag, D. P., DePaolo, D. J. and Richter, F. M. (1995). Reconstructing past sea surface temperatures: Correcting for diagenesis of bulk marine carbonate. *Geochimica et Cosmochimica Acta*, **59**, 2265–78.

Sellwood, B. W., Price, G. D. and Valdes, P. J. (1994). Cooler estimates of Cretaceous temperatures. *Nature*, **370**, 453–5.

Shackleton, N. J. (1986). Paleogene stable isotope events. *Palaeogeography, Palaeoclimatology, Palaeoecology*, **57**, 91–102.

Shackleton, N. J. and Boersma, A. (1981). The climate of the Eocene ocean. *Journal of Geological Society (London)*, **138**, 153–7.

Shackleton, N. J., Corfield, R. M. and Hall, M. A. (1985). Stable isotope data and the ontogeny of Paleocene planktonic foraminifera. *Journal of Foraminiferal Research*, **15**, 321–36.

Shackleton, N. J. and Kennett, J. P. (1975). Paleotemperature history of the Cenozoic and the initiation of Antarctic glaciation: Oxygen and carbon isotope analysis in DSDP sites 277, 279, and 281. In *Initial Reports Deep-Sea Drilling Project*, vol. 29, eds. J. P. Kennett *et al.*, pp. 743–55. Washington, DC: US Government Printing Office.

Sinha, A. and Stott, L. (1994). New atmospheric pCO_2 estimates from paleosols during the late Paleocene/early Eocene global warming interval. *Global Planetary Change*, **9**, 297–307.

Sloan, L. C. and Barron, E. J. (1990). Equable climates during earth history. *Geology*, **18**, 489–92.

Sloan, L. C., Crowley, T. J. and Pollard, D. (1996). Modeling the Middle Pliocene climate with the NCAR GENESIS general circulation model. *Marine Micropaleontology*, **27**, 51–61.

Sloan, L. C. and Rea, D. K. (1996). Atmospheric carbon dioxide and early Eocene climate: A general circulation modeling sensitivity study. *Palaeogeography, Palaeoclimatology, Palaeoecology*, **119**, 275–92.

Spero, H. J., Bijma, J., Lea, J. W. and Bemis, B. E. (1997). Effect of seawater carbonate chemistry on planktonic foraminiferal carbon and oxygen isotope values. *Nature*, **390**, 497–500.

Spero, H. and DeNiro, M. J. (1987). The influence of symbiont photosynthesis on $\delta^{18}O$ and $\delta^{13}C$ values of planktonic foraminiferal shell calcite. *Symbiosis*, **4**, 213–28.

Spero, H. J. and Lea, D. W. (1996). Experimental determination of stable isotope variability in *Globigerina bulloides*: Implications for paleoceanographic reconstructions. *Marine Micropaleontology*, **28**, 231–46.

Stott, L. D. and Kennett, J. P. (1990). The evolution of Antarctic surface waters during the Paleogene: Inferences from the stable isotopic composition of planktonic foraminifers, ODP Leg 113. *Proceedings of Ocean Drilling Program: Scientific Results*, **113**, 849–63.

Taylor, E. L., Taylor, T. N. and Cúneo, R. (1992). The present is not the key to the past: A polar forest from the Permian of Antarctica. *Science*, **257**, 1675–7.

US Navy Hydrographic Office (1964). *World Atlas of Sea Surface Temperature*. Publication No. 225. Washington, DC.

Wilson, P. A. and Opdyke, B. N. (1996). Equatorial sea surface temperatures for the Maastrichtian revealed through remarkable preservation of metastable carbonate. *Geology*, **24**, 555–8.

Woodruff, F. and Savin, S. M. (1989). Miocene deepwater oceanography. *Paleoceanography*, **4**, 87–140.

Wu, G. and Berger, W. H. (1989). Planktonic foraminifera: Differential dissolution and the Quaternary stable isotope record in the west equatorial Pacific. *Paleoceanography*, **4**, 181–98.

Wu, G., Herguera, J. C. and BERGER, W. H. (1990). Differential dissolution: Modification of Late Pleistocene oxygen isotope records in the western equatorial Pacific. *Paleoceanography*, **5**, 581–94.

Yapp, C. J. and Poths, H. (1992). Ancient atmospheric CO_2 pressures inferred from natural goethites. *Nature*, **355**, 342–4.

Yapp, C. J. and Poths, H. (1996). Carbon isotopes in continental weathering environments and variations in ancient atmospheric CO_2 pressure. *Earth Planetary Science Letters*, **137**, 71–82.

Yemane, K. (1993). Contribution of Late Permian palaeogeography in maintaining a temperate climate in Gondwana. *Nature*, **361**, 51–4.

Zachos, J. C., Lohmann, K. C., Walker, J. C. G. and Wise, S. W. (1993). Abrupt climate change and transient climates during the Paleogene: A marine perspective. *Journal of Geology*, **101**, 191–213.

Zachos, J. C., Stott, L. D. and Lohmann, K. C. (1994). Evolution of early Cenozoic marine temperatures. *Paleoceanography*, **9**, 353–87.

Ziegler, A. M., Raymond, A. L., Gierlowski, T. C., Horrell, M. A., Rowley, D. B. and Lottes, A. L. (1987). Coal, climate and terrestrial productivity: the present and early Cretaceous compared. In *Coal and Coal-bearing Strata: Recent Advances*, ed. A. C. Scott. *Geological Society Special Publication*, **32**, 25–49.

II

Case studies: latest Paleocene–early Eocene

4

Comparison of early Eocene isotopic paleotemperatures and the three-dimensional OGCM temperature field: the potential for use of model-derived surface water $\delta^{18}O$

KAREN L. BICE, LISA C. SLOAN, AND ERIC J. BARRON

ABSTRACT

A high-resolution, global simulation of early Eocene ($\sim$55 Ma) ocean circulation is described. The Parallel Ocean Climate Model is forced by surface temperature, wind stress, and moisture flux predicted by a GENESIS atmospheric general circulation model simulation with CO_2 and oceanic heat transport elevated relative to present-day control values. The ocean model-predicted three-dimensional temperature field is evaluated by comparison against proxy temperature data from benthic and depth-stratified planktonic foraminifera from 28 Deep Sea Drilling Project and Ocean Drilling Program sites. Because of uncertainty in the actual oxygen isotopic composition of early Eocene seawater (δ_w), two methods of reconstructing surface δ_w are used: a paleolatitude-dependent relationship based on modern southern hemisphere data and a salinity-dependent relationship based on correlations of salinity and δ_w across various surface salinity gradients.

Given δ_w values calculated as a function of paleolatitude and corrected for ice-free global conditions, the model temperatures in the uppermost layer predict the isotopic paleotemperatures from *Morozovella* within the estimated error ($\pm2\,^\circ$C) in the isotopic paleotemperature equation. The largest mismatch is observed for data from North Atlantic Sites 548 and 550. At all sites except these, the *Morozovella* paleotemperature falls within the annual temperature cycle simulated by the model. The fit between model and isotopic temperatures is improved for the surface ocean in general and for the North Atlantic in particular when δ_w is calculated from the model-simulated salinity. However, careful consideration must be given to the choice of an empirical salinity–δ_w relationship.

On average, the model under-predicts early Eocene subthermocline temperatures by $\sim$5 $^\circ$C and bottom-water temperatures by $\sim$7 $^\circ$C. The model captures approximately half of the 10–15 $^\circ$C bottom-water warming relative to equivalent depths and latitudes in the present-day ocean. One-third of that warming is due to the lack of continental ice-sheets in the Eocene paleogeography. Additional warming at depth is the result of forcing with increased atmospheric CO_2 and oceanic heat transport specified in the GENESIS simulation. In the simulation described here, no bottom water is formed in the Tethys Seaway. Intermediate and deep waters warm because Southern Ocean upper ocean waters, the source of bottom water, are warmer than in the modern ocean.

INTRODUCTION

Paleoclimatologists infer ancient climatic conditions from geochemical and paleontological data and use general circulation models (GCMs) to understand how the earth system, especially the atmosphere and ocean components, operated to produce the inferred climate. There are uncertainties and limitations to the techniques used both in deducing the paleoclimatic state and in simulating a system with boundary conditions that may have been significantly different from the present day. Discerning how well (or how poorly) proxy data and model output agree requires the comparison of the model prediction with, ideally, widely distributed data of as many types as possible. However, quantitative proxy data are limited for many intervals older than the Quaternary. One of the most abundant paleoclimate data types available for the past 70 million years of earth history is the oxygen stable isotope ($\delta^{18}O$) record of carbonate from foraminifera that inhabited a variety of ocean watermasses. Given certain assumptions concerning isotopic fractionation and the isotopic composition of ancient seawater, $\delta^{18}O_{carbonate}$ is used to infer seawater paleotemperature; however, these assumptions can be a matter of debate. One product of the ocean GCM is a globally consistent, three-dimensional array of water temperatures and salinities that can be compared with geochemical paleotemperature data. A numerical experiment designed to simulate the global ocean circulation of the early Eocene ($\sim$55 Ma) is described here. The simulated ocean temperature field is evaluated by comparison of temperature values at specific model grid points (latitude, longitude, depth dimensions) with isotopic paleotemperatures derived from $\delta^{18}O_{carbonate}$ of early Eocene surface, subthermocline, and bottom-dwelling foraminifera.

The climate of the early Eocene is the subject of a considerable volume of literature. Thomas *et al.* (Chapter 5, this volume) and Thomas and Shackleton (1996) provide good summaries of early Eocene marine conditions. The early Eocene has received much attention in part due to significantly warmer high-latitude and deep-water temperatures relative to modern conditions (Fig. 4.1), moderate winter temperatures in the North American continental interior (Wing and Greenwood, 1993; Markwick, 1994; Sloan, 1994; Greenwood and Wing, 1995), and a dramatic, abrupt change in the global carbon system and temperature maximum during the Paleocene–Eocene transition (Kennett and Barker, 1990; Stott *et al.*, 1990; Thomas, 1990; Kennett and Stott, 1991; Zachos *et al.*, 1993).

The reduced meridional surface temperature gradient and deep-water warmth of the early Eocene have been interpreted to indicate an ocean circulation pattern different from the modern one. Barron (1987) proposed that during the early Eocene the oceans played a greater role in poleward heat transport, cooling tropical surface waters and warming high-latitude waters. The atmosphere would have transported less heat, consistent with an inferred decrease in the intensity of atmospheric circulation (Rea *et al.*, 1985). While numerous climate model studies indicate significant sensitivity of the climate system to specified sea surface temperatures or

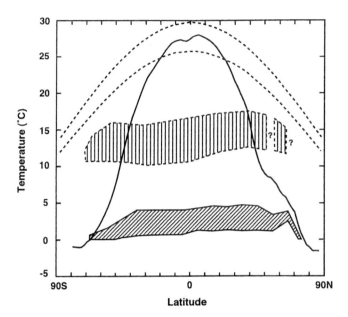

Figure 4.1. Modern (solid line) and estimated early Eocene (dashed lines) zonal sea surface temperatures. Modern (diagonal hatch) and estimated early Eocene (vertical hatch) water temperatures at bottom depths between 1000 m and 5000 m. Modern data are from the *World Ocean Atlas data set* (Levitus and Boyer, 1994). The cooler Eocene SST profile is based on Zachos *et al.* (1994); the warmer SST profile is based on Crowley and Zachos (Chapter 3, this volume).

oceanic heat transport (e.g., Covey and Thompson, 1989; Barron *et al.*, 1993), estimates of the magnitude of heat transport increase required to reproduce the early Eocene warm poles vary widely (Rind and Chandler, 1991; Sloan *et al.*, 1995) and the mechanism for a substantial increase is problematic and remains unidentified (Barron, 1986; Sloan *et al.*, 1995).

Numerous papers discuss the possibility that one site of deep-water formation during the early Eocene was in the low latitudes and that warm, saline bottom water was therefore at times a dominant component of the deep–intermediate circulation (e.g., Shackleton and Boersma, 1981; Brass *et al.*, 1982a,b; Kennett and Stott, 1990; Barron and Peterson, 1991; Oberhänsli, 1992; Pak and Miller, 1992; Zachos *et al.*, 1992). Ocean GCM experiments by Barron and Peterson (1991) indicated that Eocene paleogeography alone could have led to a change in the site of deep-water formation from high latitude to the subtropics. Subsequent experiments (Bice, 1997) have shown that early Eocene geography alone does result in some deep-water warming but that this warming is produced by specifying no continental ice-sheets in either the present-day or the Eocene model geography. In addition, the change to early Eocene paleogeography does not lead to deep convection in the low latitudes if continental runoff is included in the ocean model surface salinity forcing (Bice *et al.*, 1997), an approach that was not used by Barron and Peterson (1991). When low-latitude convection is simulated, surface water is mixed only to intermediate depths, is not a dominant component of the meridional circulation, and results in no net increase in poleward heat transport (Bice, 1997; Bice *et al.*, 1997).

Model simulations performed to date indicate that early Eocene paleogeography alone cannot explain the inferred high-latitude and deep-water warmth. Atmospheric composition may in part account for differences between early Eocene and modern climates (Sloan *et al.*, 1992). Early Cenozoic carbon dioxide and methane concentrations are very poorly constrained (Arthur *et al.*, 1991; Cerling, 1991; Freeman and Hayes, 1992; Berner, 1994; Sinha and Stott, 1994). Based on geochemical cycle models and carbon isotope studies, early Eocene estimates of CO_2 concentration range between 300 and 1200 ppm. The relatively recent recognition of probable destabilization of clathrate hydrates of methane during the late Paleocene–early Eocene boundary interval has directed attention toward quantitative estimates of atmospheric methane (Dickens *et al.*, 1995, 1997; Matsumoto, 1995) for this event and for the 'background' climate.

Sloan and Rea (1995) describe several atmospheric GCM experiments that use 55-Ma geography and the NCAR GENESIS v. 1.02A earth system model (Thompson and Pollard, 1995), which includes a mixed layer ocean component. Sloan and Rea (1995) observed that although CO_2 levels higher than 300 ppm produced a better match with inferred Eocene conditions, increased CO_2 alone did not explain the meridional surface temperature gradient inferred from paleotemperature proxy data. They concluded that CO_2 at least twice the pre-industrial concentration and increased oceanic heat transport are required to produce model results in agreement with the inferred paleoclimate. Atmospheric forcing to the ocean model simulation described below is taken from an experiment performed by Sloan and Rea that includes such increases in $p$$CO_2$ (to 560 ppm) and ocean heat transport.

EARLY EOCENE MODEL-SIMULATED CLIMATE

The models and model boundary conditions

The ocean GCM used in this study is the Parallel Ocean Climate Model (POCM) v. 2.0 (Semtner and Chervin, 1992). POCM is a three-dimensional global primitive equation ocean model derived from the Bryan and Cox model (Bryan, 1969; Cox, 1970; Bryan and Cox, 1972) with modifications by Cox (1975, 1984) and Semtner (1974). The standard equations of motion are integrated over a rigid-lid ocean domain with a realistic basin configuration, allowing for complex shoreline and bottom geometry. POCM allows the specification of variable vertical and horizontal resolution. In the experiment described below, 20 ocean layers and horizontal resolution of 2° by 2° are specified. Early Eocene (~55 Ma) bathymetry was reconstructed using digitized sea-floor magnetic lineations and the age–depth relationship of Parsons and Sclater (1977). Tethyan region paleobathymetry was interpreted from the maps of Dercourt *et al.* (1986). The reconstructed bathymetry and sensitivity of POCM to a realistic vs. a simple bathymetry are described by Bice *et al.* (1998). Present-day POCM simulations using atmospheric forcing derived from the GENESIS model (the technique applied in this study and described below) are described by Bice (1997). Following the approach of Semtner and Chervin (1992), no Arctic Ocean is included in the POCM simulations. The maximum poleward extent of the model Eocene oceans is 72° N and 78° S. The model produces a reasonable match with observed circulation, convection, and water temperatures. Sea surface salinities are also well simulated, with the exception of the North Pacific, where the model salinity tends to be 1‰ greater than observed. This difference is the result of a simulated moisture flux to the region that is lower than observed, due in part to the crude treatment of continental runoff (described below) that is required in a present-day simulation designed to mimic the specification of runoff in a paleo-ocean experiment (Bice, 1997).

Surface forcing to the ocean model is the full seasonal cycle of wind stress, moisture flux, and 2-m air temperature simulated by the GENESIS v. 1.02A atmospheric GCM (Thompson and Pollard, 1995). GENESIS v. 1.02A is a three-dimensional atmospheric GCM consisting of a 12-layer atmosphere model coupled with soil, snow, sea ice, and 50-m mixed layer slab ocean models. The model includes a land surface transfer scheme that simulates the physical effects of vegetation. In the simulation employed in this study, horizontal resolution of 4.5° latitude by 7.5° longitude was used. Model output was then interpolated to POCM resolution.

The GENESIS slab ocean model serves as a source and sink for heat and moisture and allows for a crude parameterization of the meridional transport of heat according to a specified heat flux adjustment (in W m^{-2}) at each slab ocean grid cell. A parameterization that approximates zonally symmetric poleward ocean heat transport values of 1.5 times those described by Sloan and Rea (1995) was specified for the GENESIS mixed layer ocean model. The atmospheric CO_2 concentration is specified as 560 ppm. This represents a doubling of the pre-industrial CO_2 level of 280 ppm. Globally uniform soil color and texture were specified. Vegetation is a reconstruction of the early Eocene distribution (Sloan and Rea, 1995) using the vegetation classification scheme of Dorman and Sellers (1989). In the reconstruction of Sloan and Rea, within 50° of the equator, vegetation type is designated as either broadleaf shrubs or broadleaf trees with perennial groundcover. In the northern hemisphere high-latitudes, vegetation changes poleward from needleleaf-evergreen trees to needleleaf-deciduous trees. In the southern hemisphere high latitude regions, vegetation is specified as either perennial groundcover or tundra vegetation. Paleogeography was interpreted from the paleogeographic reconstructions of Scotese and Golonka (1992). No continental ice-sheets are specified in the model paleogeography. A detailed description of the early Eocene atmospheric simulation is given by Bice (1997).

The ocean surface moisture flux derived from GENESIS output includes simulated continental freshwater runoff. Idealized continental drainage basins were specified based on the GENESIS topography. Net positive continental precipitation was summed for each drainage basin and was added to specified adjacent coastal ocean model cells. The sensitivity of the model to the treatment of continental runoff is described by Bice et al. (1997).

A 1000-year integration of POCM with the GENESIS early Eocene atmospheric forcing was performed. The model was forced with mean annual atmospheric conditions for model years 1 to 300 and with full seasonal cycle conditions for model

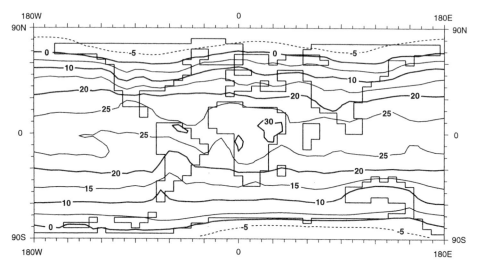

Figure 4.2. Mean annual 2-m air temperature. Contour interval = 5 °C.

years 301 through 1000. After 1000 model years, globally averaged temperatures were at or very near steady state in all model levels. The simulation was run for one additional year with data archiving every three days. These data were used to produce mean annual model statistics and the seasonal cycle of ocean temperatures. Below, the model-simulated atmosphere and ocean for the early Eocene are described in comparison with simulations of the modern system performed using the same uncoupled modeling approach (Bice, 1997).

The GENESIS-simulated atmosphere

The globally averaged, mean annual 2-m air temperature (Fig. 4.2) simulated for the early Eocene is 17.7 °C, 4.4 °C warmer than the present-day simulation. The zonally averaged, mean annual 2-m air temperature is warmer than the present day at all latitudes, with maximum warming at the poles. Mean annual temperatures over land above 50° N latitude are 1 °C (10 °C warmer than the present day). Over Antarctica, the mean annual 2-m air temperature is −3 °C in the Eocene simulation (compared with −32 °C in the modern simulation). No continental snowfall survives summer melting in either hemisphere, although some sea ice survives summer melting in the eastern Arctic Ocean. In the tropical and equatorial regions (23° S–23° N), the model-simulated mean annual air temperature (25.7 °C) is only slightly warmer than that simulated for the present day (25.1 °C). This change is consistent with warming (due to increased CO_2) that is compensated for by increased export of heat out of the low latitudes by higher ocean heat transport specified in the GENESIS ocean model.

The atmosphere–ocean moisture flux, including continent runoff, is shown in Fig. 4.3. The global, mean annual precipitation rate simulated for the early Eocene is approximately 10% greater than that simulated by the model given modern

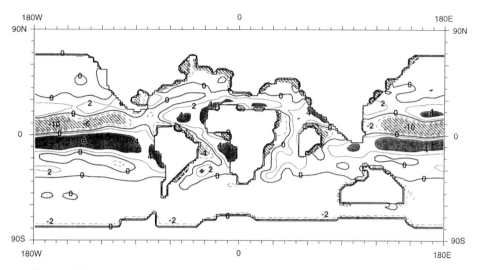

Figure 4.3. Mean annual atmosphere–ocean moisture flux. Contour interval = 2 mm day^{-1}. Moisture flux is calculated as evaporation minus precipitation minus continental runoff. Negative values (dashed contours) indicate regions of net precipitation or runoff; positive values (solid contours) indicate regions of net evaporation. Regions of greater than 4 mm day^{-1} and less than −4 mm day^{-1} are shaded.

boundary conditions. This more active hydrologic cycle is consistent with that observed by Barron *et al.* (1989) in response to changing tropical ocean area. The Eocene land–sea distribution includes approximately 25% more ocean area in the region 10–40° N and 5% more ocean area in equivalent southern hemisphere latitudes, relative to the present-day land–sea configuration. This increase in water vapor, which occurs in response to CO_2-induced warming, also acts as a greenhouse gas, accounting for some of the warming in the early Eocene simulation relative to the present day. Additional warming results from the positive feedback (through surface albedo) between surface temperature and snow cover or sea ice. Relative to the present day, there is a northward shift of at least 5° latitude in the positions of the Intertropical Convergence Zone and the regions of maximum tropical evaporation. The presence of the Tethys Ocean between 10° and 50° N causes increased evaporation in the northern hemisphere tropics, in agreement with sensitivity noted by Barron *et al.* (1989). The local change in precipitation rate, however, is not as large as that observed at other latitudes. The maximum zonal average net precipitation beneath the Intertropical Convergence Zone more than doubles relative to the present-day simulation. Changes in the position of precipitation and evaporation maxima due to changes in the distribution of land masses are consistent with the modeling results of others (Barron *et al.*, 1989; Fawcett, 1994).

Zonally averaged, mean annual surface eastward wind stresses are generally reduced in the early Eocene simulation relative to those simulated for the present day. This reduction in zonal winds is consistent with the reduced meridional temperature gradient resulting from increased poleward ocean heat transport and the absence of continental ice-sheets in the Eocene experiment. However, in the region 30–45° N, model-predicted early Eocene eastward wind stress is greater than that simulated for the present day. The largest geographic changes at this latitude between the early Eocene and today are the closing of the Tethys Seaway and the uplift of the Tibetan Plateau.

The POCM-simulated ocean

The model-simulated surface circulation is illustrated in Fig. 4.4. In the Atlantic Ocean, the well-defined gyres observed in the modern ocean are less well developed in the narrower Eocene ocean. Currents analogous to the Gulf Stream (GS), Labrador Current, and Canary Current (CC) are evident in the North Atlantic. The North Atlantic Drift (NAD), however, is highly zonal in the early Eocene simulation in response to a smaller ocean basin above 40° N latitude. In the South Atlantic Ocean, the Brazil Current (BC) is simulated for the early Eocene. There is northward flow along the Argentina coast in the region of the present-day Falkland Current (FC). However, in the region of the present northward-flowing Benguela Current, model-simulated flow is southward along the coast in the early Eocene. This change is due to the position of the African continent at least 5° farther

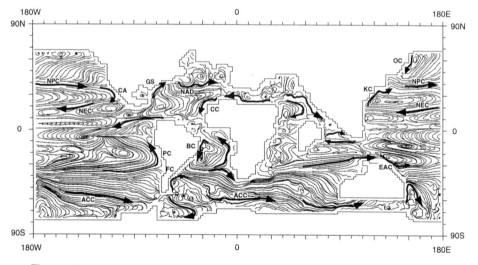

Figure 4.4. Mean annual surface circulation streamlines simulated for the uppermost 75 m. The major currents are highlighted. Although surface current systems simulated by the model for the early Eocene may not be strictly analogous to modern currents, where the position and direction of flow of a surface current is similar to one observed in the present ocean the same name has been applied to the early Eocene current. Currents are abbreviated as follows: ACC, proto-Antarctic Circumpolar Current; BC, Brazil Current; CA, California current; CC, Canary Current; EAC, Eastern Australia Current; FC, Falkland Current; GS, Gulf Stream; KC, Kuroshio Current; NAD, North Atlantic Drift; NEC, Northern Equatorial Current; NPC, North Pacfic Current; OC, Oyashio Current; PC, Peru Current.

south. This shift moves the western coast of South Africa south of the coastal divergence of the South Atlantic gyres at 55 Ma.

Early Eocene Pacific Ocean gyres strongly resemble those simulated for the modern ocean. The North Pacific (NPC), California (CA), Northern Equatorial (NEC), Kuroshio (KC), and Oyashio (OC) currents are all evident in the early Eocene simulation. The Alaska Gyre is less well defined in the early Eocene simulation with flow in the northeastern Pacific Ocean predominantly northeastward. In the southern Pacific Ocean, the model predicts the existence of a Peru Current (PC). A westward-flowing current comparable to the Southern Equatorial Current is also evident but is positioned further south than that simulated for the present day. Because Australia occupied a higher southern-latitude position in the early Eocene, the poleward flow that presently defines the Eastern Australia Current (EAC) was located along the northeasternmost coast of the Australian continent. A well-defined current transports mixed layer water westward from the western Pacific into the eastern proto-Indian Ocean. Less than 10° southward, a returning, eastward-flowing current indicates transport across the proto-Indian Ocean and into the western Pacific north of Australia.

One well-developed gyre is evident in the Tethys, centered at about 18° N. Model-predicted flow through the opening between India and Asia is southeastward when averaged over model layers 2 and 3 (depth interval 26–75 m). However, in the surface layer (uppermost 25 m), flow through this passageway is northwestward. At the western opening of the Tethys Ocean, flow is westward out of the Tethys.

Because of restriction of the Southern Ocean passageways, no true circum-Antarctic current (ACC) is predicted for the Eocene. The eastward-flowing current system driven by the southern hemisphere westerly winds is shallow and not strongly zonal. No circum-Antarctic throughflow exists below 400 m depth because of barriers to flow in the shallow, narrow Southern Ocean gateways. The global meridional overturning (Fig. 4.5) is dominated by a counterclockwise cell that transports water out of the Southern Ocean and into the northern hemisphere as far as 40° N. This cell represents cool, saline deep water formed along the coast of Antarctica, roughly analogous to Antarctic Bottom Water (AABW) in the modern ocean. Unlike the modern ocean, the lack of a strong meridional circumpolar gyre allows strong outflow of cold, saline water. However, the maximum transport predicted in the gyre (~55 Sv) is less than that believed to exist for modern AABW. The decrease is due to weaker zonal winds and warmer high-latitude sea surface temperatures in the Eocene case. A much weaker (but still deep) northern hemisphere clockwise cell is also simulated in the global overturning. However, convection of high-latitude cool saline water in the northern hemisphere occurs in the northwestern Pacific Ocean, not in the North Atlantic, as today (Bice *et al.*, 1997).

The simulated surface circulation is in general agreement with the Haq (1984) circulation in the northern Atlantic, equatorial Atlantic, and North Pacific with the exception of the Labrador and Oyashio Currents, which Haq did not reconstruct. In the southern hemisphere, Haq did not infer the distinct gyres simulated for the early Eocene. Instead, he proposed a single large anticyclonic circulation pattern in each ocean basin extending from the equator to the Southern Ocean

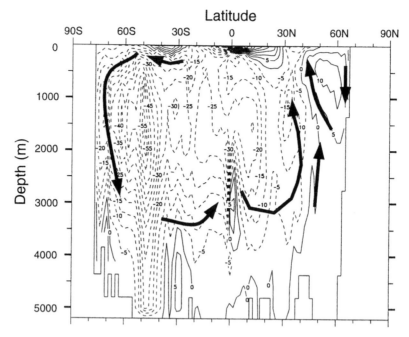

Figure 4.5. Global meridional overturning (Sv; 1 Sv = 10^6 m^3 s^{-1}) simulated for the early Eocene.

boundary. In the Pacific and proto-Indian Oceans, this inferred system would have carried equatorial waters from the western Pacific across the southern proto-Indian ocean, southward along the eastern coast of South America, returning high-altitude Southern Ocean water to the western Pacific by a route north of Australia. This looping system of surface circulation is not predicted by POCM. The primary difference in paleogeography in this part of the globe between the late Paleocene (Haq's reconstruction) and the early Eocene (that used in the present study) is the extent of rifting between Australia and Antarctica. Previous modeling studies employing 60-Ma and 40-Ma paleogeographies (Barron and Peterson, 1991; Fawcett, 1994) do not support a change in the general structure of the atmospheric circulation that would be required to give rise to the difference between Haq's southern hemisphere circulation and that shown in Fig. 4.4.

The largest discrepancy between the surface circulation simulated in the early Eocene experiment and that inferred by Haq (1981, 1984) for the late Paleocene occurs in the Tethys Seaway. Haq proposed a 'Tethys Current' that carried water from the low-latitude southern hemisphere region south of Indonesia northwestward through the Tethys and into the North Atlantic. Details of surface flow through narrow gateways are sensitive to atmospheric forcing and specified bathymetry (Bice *et al.*, 1998), but Haq's 'Tethys Current' is not reproduced by Barron and Peterson (1991), nor in numerous POCM sensitivity tests (Bice, 1997).

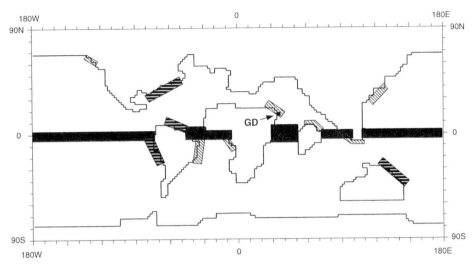

Figure 4.6. Regions of upwelling predicted by the model. The solid gray stripe indicates the width of the region of strong equatorial upwelling. Strong year-round Ekman-driven upwelling (bold horizontal shading) and seasonal upwelling regions (fine diagonal shading) are indicated. The approximate paleoposition of the Gebel Duwi section (GD) is shown.

Surface winds specified to the ocean model result in strong Ekman-driven upwelling on a mean annual basis (Fig. 4.6) along the eastern North American margin, the Peru margin, the northeastern coast of South America, and the northeastern coast of Australia. Other regions exhibiting less intense, seasonal upwelling are shown in Fig. 4.6, as is the width of the band of strong equatorial upwelling. Seasonally strong (June–November) upwelling is simulated for the Gebel Duwi section of eastern Egypt. Speijer *et al.* (1996) present evidence for late Paleocene–early Eocene variability in the strength of upwelling of nutrient-rich waters in the Gebel Duwi section. Based on only one model experiment, it is not possible to confirm or deny a change in wind-driven upwelling in any region. However, the strength of upwelling in a region subject to seasonally varying winds, such as at the paleoposition of Gebel Duwi, could be expected to exhibit sensitivity to changes in the strength of tradewinds, which may have occurred in response to changes in continental-to-marine atmospheric contrasts (Schneider *et al.*, 1985).

The simulated early Eocene mean annual global sea surface temperature (uppermost 25 m) is 19.1 °C, 3.3 °C warmer than that simulated for the modern ocean. The mean annual temperature in the uppermost 25 m of the early Eocene ocean is shown in Fig. 4.7. Temperatures range from 28.8 °C in the northern hemisphere equatorial waters of the western Tethys to 5 °C at high latitudes in all oceans. The zonal average temperature is warmer at all latitudes than the present day, with the greatest warming at high latitudes. Maximum warming of +7.8 °C occurs at 73° in the southern hemisphere. Between 10° S and 5° N, the zonally averaged mean annual temperature is only 0.2–0.5 °C warmer than the present day. The early

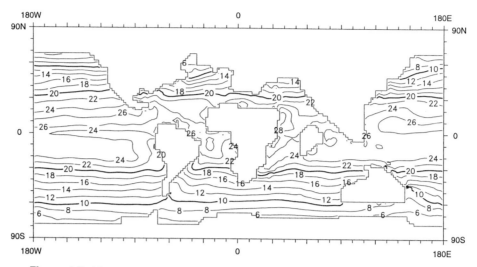

Figure 4.7. Mean annual temperature in the uppermost 25 m. Contour interval = 2 °C.

Eocene model ocean is warmer than the modern ocean at all model depths, with maximum warming (>4.2 °C) below 2000 m depth. Below 2000 m, Eocene temperatures average approximately 6.3 °C.

The vertical temperature (Fig. 4.8) and salinity (Fig. 4.9) structures of the early Eocene ocean reflect the dominant meridional transport of Southern Ocean water northward below 3000 m depth. The primary site of bottom-water formation is the Southern Ocean. This is confirmed by numerous temperature–salinity–depth plots (not shown) that were constructed from globally distributed sites in the model ocean. The bottom water at all sites has a salinity and temperature very near that of Southern Ocean shallow water. The warming observed at depth in the model therefore does not result from a change in the site of deep-water formation from polar to low-latitude regions: deep water is formed in the Southern Ocean and surface waters there are warmer than in the present ocean.

The distinct 'plume' of saline water in the upper 1000 m in the southern hemisphere (Fig. 4.9) is the result of strong net evaporative conditions centered on ~10° S (Fig. 4.3) enhanced by the well-developed, strongly zonal southern hemisphere Pacific subequatorial gyre (Fig. 4.4). Similarly, a tongue of warm, saline water in this region (Fig. 4.8) is associated with surface convergence and salinity-enhanced downwelling within the region of equatorial easterly and mid-latitude westerly surface winds. However, deep convection that would lead to the existence of warm, saline intermediate or deep water is not predicted by the model.

Unlike earlier Eocene ocean model simulations (Barron and Peterson, 1991), the simulation described here includes an explicit treatment of riverine delivery of continental runoff to the ocean. Continental drainage basins are reconstructed from paleotopography and net precipitation (precipitation minus evaporation) in each

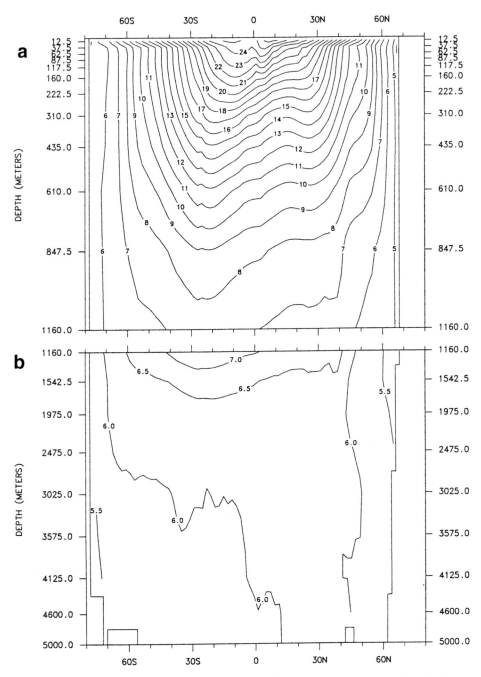

Figure 4.8. Meridional cross-section of the global mean annual temperature structure. (a) Upper 1160 m, contour interval = 1 °C. (b) Depths below 1160 m, contour interval = 0.5 °C.

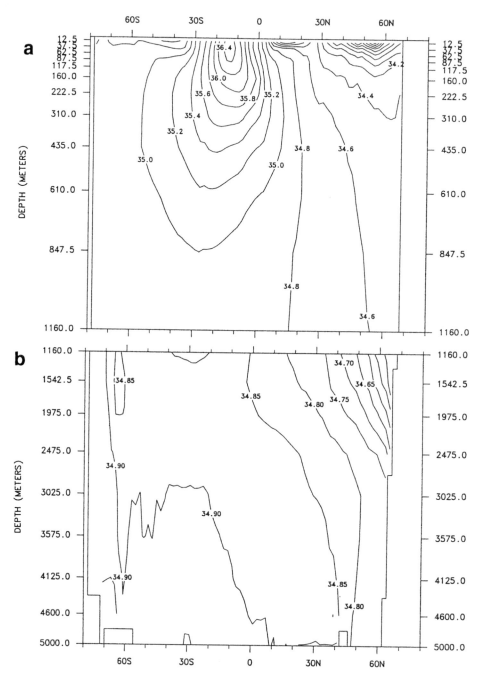

Figure 4.9. Meridional cross-section of the global mean annual salinity structure. (a) Upper 1160 m, contour interval = 0.2‰. (b) Depths below 1160 m, contour interval = 0.05‰.

drainage basin is routed to specified adjacent coastal ocean model cells. In order to maintain a global balance in moisture flux, it was necessary to assume no interior drainage and no drainage to the Arctic Ocean. This approach maximizes coastal runoff. In two other sensitivity experiments (Bice *et al.*, 1997), the treatment of continental runoff was designed to replicate other salinity forcing approaches that have been used in uncoupled model experiments: runoff is ignored in the moisture flux calculation, and runoff is distributed evenly to all surface ocean model cells. In these tests, deep convection occurs in the northern Tethys Seaway, as well as in high-latitude regions. The experiments indicate that the positions and relative importance of sites of early Eocene deep-water formation are highly sensitive to the manner in which continental runoff is treated in the ocean model salinity forcing. In the sensitivity experiments in which convection occurs in the northern Tethys, convection extends only to intermediate depths and the flux of this warm, saline source is minor. The sensitivity of the model to changes in the treatment of continental runoff do suggest, however, that changes in the atmosphere–ocean moisture flux are a feasible mechanism for varying the relative importance of high-latitude and low-latitude intermediate water sources (Bice *et al.*, 1997).

PALEOTEMPERATURES FROM OXYGEN STABLE ISOTOPE MEASUREMENTS

The model-simulated three-dimensional temperature field was evaluated by comparison against paleotemperatures calculated using oxygen stable isotope analyses of early Eocene foraminifera. Individual published isotope analyses from Deep Sea Drilling Project (DSDP) and Ocean Drilling Program (ODP) cores were compiled. The remainder of this paper is devoted to calculation of marine paleotemperatures, comparison of the model-predicted and isotopic temperatures, and consideration of the potential use of the model-predicted salinity field to calculate the isotopic composition of surface seawater.

The oxygen isotope 'paleothermometer'

Following the suggestion of Urey (1947) that the oxygen isotopic composition (δ^{18}O) of carbonate materials might be an indicator of paleotemperature, Epstein *et al.* (1951, 1953) showed that the difference between the oxygen isotopic composition of mollusc shell carbonate and that of the ambient water is a function of water temperature. The correlation observed in culture experiments yielded a paleotemperature equation of the form

$$T(^{\circ}C) = a - b(\delta_c - \delta_w) + c(\delta_c - \delta_w)^2, \tag{4.1}$$

where a, b, and c are empirical coefficients, δ_c is the oxygen isotopic composition of the calcite sample (expressed relative to the Pee Dee Belemnite standard, or PDB), and δ_w is the oxygen isotopic composition of the ambient water (expressed relative to Standard Mean Ocean Water, or SMOW). The symbol δ^{18}O denotes the ratio

$$\delta^{18}O(\%) = \left(\frac{(^{18}O/^{16}O)_{\text{sample}}}{(^{18}O/^{16}O)_{\text{standard}}} - 1 \right) \times 1000. \tag{4.2}$$

Different values for the coefficients in Equation (4.1), based on inorganic and organic carbonate (either calcite or aragonite), have been proposed (McCrea, 1950; Craig and Gordon, 1965; O'Neil *et al.*, 1969; Horibe and Oba, 1972; Shackleton, 1974; Anderson and Arthur, 1983; Erez and Luz, 1983; Grossman and Ku, 1986).

Emiliani (1954, 1955, 1966) applied the equations of Epstein *et al.* (1953) and Craig and Gordon (1965) to planktonic foraminifera from deep-sea cores and concluded that planktonic foraminifera secrete calcium carbonate in equilibrium with seawater and that marine paleotemperatures can therefore be calculated if the isotopic composition of ancient seawater can be estimated. This technique has been widely used during the past 40 years, although the issue of equilibrium fractionation has been reconsidered (see discussion below). In the present study, published stable isotope analyses of early Eocene planktonic and benthic foraminifera from 28 deep DSDP and ODP sites (Table 4.1) were compiled in order to create a data set from which paleotemperature calculations could be made. Data were included if the sediments analyzed were from core assigned to nannoplankton zone CP9, or foraminifera zone P.6, P.7, or P.8, or were assigned an absolute age between 54.6 and 57.0 Ma (timescale of Berggren *et al.*, 1985).

Analyses from an admittedly broad age range have been included in this compilation in an attempt to obtain a data set with as much global and water paleodepth coverage as is feasible, within the early Eocene. Figure 4.10 shows the positions of drilling program sites that have recovered core identified as early Eocene. The circled sites are those for which isotopic paleotemperatures are calculated in this study. Even given the relatively broad range of samples included using the criteria given above, data coverage is generally poor except in the South Atlantic. Within the South Atlantic, more than twice as many sites have data from planktonic foraminifera as from benthic species. Along a west–east transect within any ocean basin, only the sites along paleolatitude 30° S in the Atlantic provide reasonable data coverage, and then only for planktonic species. However, there is no doubt that the sampling approach required to generate reasonably good global coverage introduces the problem of 'oversampling' the data, resulting in isotopic paleotemperatures that represent averages over a time interval on the order of at least 1 million years.

Correction factors applied to benthic foraminifera

It has been shown that the stable isotopic composition of several species of modern benthic foraminifera is not in equilibrium with ambient waters (Shackleton *et al.*, 1984; Keigwin and Corliss, 1986; Zachos *et al.*, 1992). In many species, observed $\delta^{18}O$ values are depleted relative to expected equilibrium values (Boersma *et al.*, 1979). This disequilibrium fractionation effect is generally accounted for by application of a 'correction factor' to measured $\delta^{18}O$ values. For the two benthic taxa used in this study, *Nuttalides* and *Cibicidoides*, published correction factors vary from +0.35‰ to +0.6‰ (Shackleton *et al.*, 1984; Keigwin and Corliss, 1986; Kennett and Stott, 1990; Barrera and Huber, 1991; Zachos *et al.*, 1992, 1994). Pak and Miller (1992) assert that disequilibrium fractionation is not

Table 4.1. *Drilling program sites for which foraminiferal stable isotope data were compiled*

Site	Location at 55 Ma		Approximate depth at 55 Ma (source)	Oxygen isotope data sources
	Latitude (°N)	Longitude (°E)		
20	−29.0	−13.2	2900 m (2)	4, 12
21	−29.1	−17.0	2100 m (2)	3, 4
94	27.1	−74.9	2400 m (2)	4
152	14.2	−69.0	4000 m (20)	4
213	−29.6	79.0	4000 m (19)	19
215	−35.6	65.3	4000 m (19)	19
277	−70.4	−167.0	4200 m (20)	4, 15
356	−28.9	−27.5	3100 m (3)	4
357	−30.6	−22.0	1800 m (2)	4
363	−26.0	−0.6	1500 m (12)	12
364	−18.5	3.7	›1500 m (3)	4
384	37.3	−35.6	2900 m (10)	3
401	43.0	−9.3	1800 m (13); 3000 m (8)	13
525	−34.4	−8.7	1600 m (5); 3390 m (11)	4, 16
527	−33.2	−9.8	3400 m (18)	4, 16
528	−33.7	−9.3	3400 m (18)	4, 16
529	−34.2	−8.9	3400 m (18)	4, 16
548	44.5	−12.4	1100 m (9)	4, 14
550	44.1	−13.5	4000 m (9)	4
577	14.6	−163.5	1900 m (18)	13
690	−62.5	−2.2	2100 m (18)	7, 17
698	−52.0	−19.2	950 m (6)	6
699	−52.1	−16.8	2850 m (6)	6
700	−52.0	−16.4	2400 m (6)	6
702	−51.4	−12.4	1850 m (6)	6, 19
738	−62.0	74.1	1800 m (18)	1
757	−44.8	61.8	›250 m (19)	19
865	−0.5	−147.3	1300 m (5)	5

Source references: 1, Barrera and Huber, 1991; 2, Boersma and Premoli Silva, 1983; 3, Boersma *et al.*, 1979; 4, Boersma *et al.*, 1987; 5, Bralower *et al.*, 1995; 6, Katz and Miller, 1991; 7, Kennett and Stott, 1990; 8, Miller and Curry, 1982; 9, Miller *et al.*, 1985; 10, Miller *et al.*, 1987*a*; 11, Miller *et al.*, 1987*b*; 12, Oberhänsli *et al.*, 1991; 13, Pak and Miller, 1992; 14, Poag *et al.*, 1985; 15, Shackleton and Kennett, 1975*b*; 16, Shackleton *et al.*, 1984; 17, Stott *et al.* 1990; 18, Zachos *et al.*, 1993; 19, Zachos *et al.*, 1994; 20, paleodepth estimate based on basement age at 55 Ma and the equation of Parsons and Sclater (1977).

understood well enough to warrant use of any correction factors. In order to facilitate comparison with most published estimates, the correction values of Zachos *et al.* (1994) were applied to benthic foraminifera oxygen isotope data compiled in this study. The correction factors are $+0.4‰$ for measurements of *Nuttalides* and $+0.6‰$ for *Cibicidoides*.

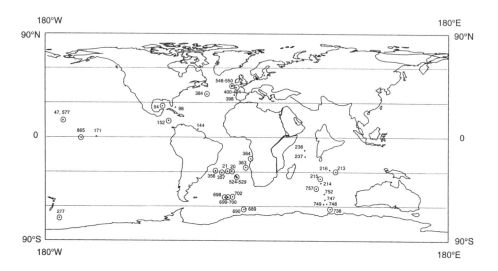

Figure 4.10. Map showing the paleopositions (at 55 Ma) of Deep Sea Drilling Project (DSDP) and Ocean Drilling Program (ODP) sites from which core assigned an age of early Eocene has been recovered. The sites for which foraminiferal oxygen stable isotope measurements have been compiled are circled. The present-day shorelines (rotated to their positions at 55 Ma) are shown for reference. Plate and site rotations were made using the plate rotation stage poles of Barron (1987).

Intertaxa differences and depth stratification of planktonic foraminifera

The importance of the use of monospecific stable isotope analyses, where possible, is illustrated by the isotope records of the subthermocline-dwelling planktonic foraminifera *Subbotina triangularis* and *Subbotina eocaenica* from Site 690 (Stott *et al.*, 1990) shown in Fig. 4.11. Through the interval 55–53 Ma, the offset between the two calculated isotopic paleotemperature curves averages approximately 1.5 °C. The peak warmth of the Late Paleocene Thermal Maximum (LPTM) indicated by the two species of *Subbotina* differs by at least 2 °C. The variation in $\delta^{18}O$, and therefore the paleotemperature calculated from these data, may be the result of differential isotopic fractionation, differences in water depth preference or differences in seasonality of reproduction (Deuser *et al.*, 1981; Deuser, 1987; Zachos *et al.*, 1994). Ideally, a global study would avoid error introduced by intertaxa differences through use of $\delta^{18}O$ measurements of only one species and for a restricted test size range (Corfield, 1994). However, various species (and genera) exhibit environmental preferences that restrict their distribution. In addition, the availability of isotope data for any particular time interval is limited by the number of sites that have been successfully sampled or for which size fraction analyses were performed. In order to obtain a global database against which to compare model-simulated temperatures, it is not possible to use a monospecific approach. Instead, the approach taken here is a monogeneric one in which, as much as possible, comparisons are made between measurements of species within one genus. This approach has been applied when the aim was to reconstruct a global picture of seawater temperatures

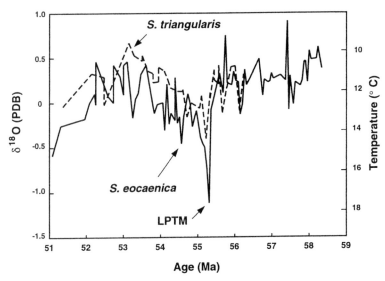

Figure 4.11. ODP Site 690 oxygen stable isotope measurements of two species of the planktonic foraminifer *Subbotina* (data from Stott *et al.*, 1990). The timescale used by Stott *et al.* (1990) has been revised to that of Cande and Kent (1992) as noted by Kennett and Stott (1995). LPTM, Late Paleocene Thermal Maximum.

for a specific time interval (e.g., Shackleton and Kennett, 1975*a*; Shackleton and Boersma, 1981; Keigwin and Corliss, 1986; Boersma *et al.*, 1987; Zachos *et al.*, 1994).

    In the same way that the regional distribution of foraminifera may be controlled in part by environmental tolerances and optimal conditions, the water depth inhabited by planktonic foraminifera may be controlled by ontogeny, temperature, salinity, light level, nutrient supply, and other climatological factors. Depth habitat segregation is the cause of much of the intertaxa variation in $\delta^{18}O$ (Deuser *et al.*, 1981; Deuser, 1987). D'Hondt and Arthur (1995) showed that, for Maastrichtian planktonic foraminifera, the depth habitat of an individual species or genus may have varied as a function of latitude. As yet, no detailed global study has been performed for the early Eocene that indicates how foraminiferal depth habitat may have varied as a function of latitude, and so a generalized depth ranking will be used here.

    Boersma *et al.* (1979) and Zachos *et al.* (1994) have proposed depth rankings for extinct Paleogene planktonic foraminifera. These rankings are based on two assumptions: (1) organisms living in the uppermost mixed layer experience the warmest water temperatures and should therefore have the most depleted oxygen isotope values, and (2) seawater bicarbonate $\delta^{13}C$ is highest near the surface and decreases rapidly with depth. Upper ocean bicarbonate is enriched in ^{13}C relative to ^{12}C as a result of the export of isotopically depleted organic carbon from the surface to the deep ocean. This vertical gradient is greatest in regions of high net organic matter production. In addition, near-surface-dwelling planktonic foraminifera are more likely to have algal photosymbionts that, through the preferential extraction of ^{12}C, further increase the $\delta^{13}C$ of bicarbonate in the surrounding water (Spero *et al.*, 1991; D'Hondt *et al.*, 1994; Zachos *et al.*, 1994). Calcareous foraminifera extract seawater bicarbonate to form calcium carbonate. Near-surface species should exhibit more enriched carbon isotope values than planktonic species that inhabit deeper depths. Therefore, foraminifera living in the uppermost mixed layer should exhibit the highest $\delta^{13}C$ and the lowest $\delta^{18}O$ values.

Boersma *et al.* (1979) note that the relative positions of various planktonic genera changed within the Paleocene, perhaps as the result of phylogenetic changes in response to changes in water masses and nutrient availability. It appears then that, for any particular time interval of interest (in this case the early Eocene), the depth stratification of foraminifera must be redefined. Among planktonic foraminifera with stratigraphic ranges that include the early Eocene, Zachos *et al.* (1994) delineate a shallow (mixed layer) ranking and an intermediate to deep (subthermocline) ranking. In order of increasing depth, their mixed layer species include: *Morozovella subbotina, M. velascoensis, M. aragonensis, Acarinina nitida, A. primitiva, A. mckannai, Chiloguembelina* spp., and *Muricoglobigerina* spp. Zachos *et al.* (1994) rank subthermocline species as follows: *Subbotina triangularis, S. velascoensis*, and *S. patagonica*. In order to obtain a better global distribution of calculated paleotemperatures, we have followed the depth stratification ranking of Zachos *et al.* (1994) but have assumed that all species of the genus *Morozovella* are uppermost mixed layer dwellers, even those not specifically cited as such by Zachos *et al.* (1994). The same assumption was made for the planktonic taxa *Acarinina, Chiloguembelina, Muricoglobigerina*, and *Subbotina*. Table 4.2 shows how planktonic foraminifer species, as identified at sites listed in Table 4.1, were grouped in order to calculate mixed layer and subthermocline paleotemperatures. The issue of a depth stratification ranking is confused by the fact that the depth at which planktonic foraminifera live may vary seasonally (Deuser, 1987). But in the absence of some independent proxy record for the seasonal variability of local and regional conditions, it is assumed that planktonic foraminifera record some average of the range of conditions in the water column through which they migrate.

Compiled oxygen isotope data were averaged by depth ranking in order to obtain an average $\delta^{18}O$ value for various depth ranges at each site. The mean and standard deviation values for these data are given in Table 4.3. (Standard deviations are not reported for sites where only one published $\delta^{18}O$ value is available.) The mean $\delta^{18}O$ and $\delta^{13}C$ values for all sites (Fig. 4.12) show that, based on the relative $\delta^{18}O$ and $\delta^{13}C$ values, the data compiled for this study generally follow the depth ranking of Zachos *et al.* (1994). The slope of the surface-to-bottom carbon isotope gradient observed in the compiled data agrees well with that identified within planktonic foraminiferal zone P.8 (early Eocene; Berggren *et al.*, 1985) at Site 356 (Boersma *et al.*, 1987). Subthermocline and benthic data for North Atlantic Sites 548 and 550 exhibit $\delta^{18}O$ values lower than other sites in the compilation. Possible reasons for this are discussed below.

Paleotemperatures were calculated using the mean $\delta^{18}O$ values shown in Table 4.3 and the calcite–water temperature relationship of Erez and Luz (1983):

Table 4.2. *Depth stratification of planktonic foraminifera*

Mixed layer '1' taxa
Morozovella acuta
Morozovella aragonensis
Morozovella formosa
Morozovella gracilis
Morozovella lehneri
Morozovella marginodentata
Morozovella subbotinae
Morozovella velascoenis
Morozovella spp.

Mixed layer '2' taxa
Acarinina bullbrooki
Acarinina coalingensis
Acarinina mckannai
Acarinina nitida
Acarinina primitiva
Acarinina pseudotop.
Acarinina soldadoensis
Acarinina wilcoxensis
Acarinina/Truncorotaloides
Acarinina nitida
Acarinina (spiral)
Acarinina spp.

Mixed layer '3' taxa
Chiloguembelina midwayensis
Chiloguembelina wilcoxensis
Chiloguembelina spp.
Muricoglobigerina mckannai
Muricoglobigerina spp.

Subthermocline taxa
Subbotina eocaenica
Subbotina patagonica
Subbotina triangularis
Subbotina trilocularis
Subbotina spp.

$$T(^{\circ}\mathrm{C}) = 16.998 - 4.52(\delta_c - \delta_w) + 0.028(\delta_c - \delta_w)^2, \tag{4.3}$$

where δ_c is the oxygen isotopic composition (δ^{18}O) of the calcite foraminifer sample (relative to PDB) and δ_w is the oxygen isotopic composition of the water (relative to SMOW). Equation (4.3) is derived from planktonic foraminifera grown in cultures with a temperature range of 14–30 °C. Zahn and Mix (1991) showed that the equation of Erez and Luz (1983) provided the best fit to northeastern Atlantic Ocean core top benthic foraminifera δ^{18}O at water depths between 2500 and 5100 m where bottom-water *in situ* temperatures range from 1.8 to 0.5 °C. The equation of Erez

Table 4.3. *Mean $\delta^{18}O$ and standard deviation for stable isotope values*

Site	Mixed layer '1' $\delta^{18}O$	σ	Mixed layer '2' $\delta^{18}O$	σ	Mixed layer '3' $\delta^{18}O$	σ	Subthermocline $\delta^{18}O$	σ	Nuttalides $\delta^{18}O$	σ	Cibicidoides $\delta^{18}O$	σ
20	−1.28						−1.07	0.32				
21	−0.94	0.16	−0.84	0.05	−0.08		−0.28	0.05	0.44			
94	−1.55		−1.56									
152	−2.19	0.03										
213	−1.17	0.07	−1.13	0.17	−1.26	0.11	−0.82	0.33	−0.44	0.18		
215	−1.13				−0.92	0.06	−0.49	0.10				
277					−0.70							
356	−1.84		−1.83	0.03								
357	−1.34		−1.82	0.23								
363			−0.46	0.54			−0.60	0.21				
364	−1.42		−1.83									
384	−1.16	0.02					−0.14					
401											−0.24	0.19
525	−0.86	0.24			−0.44	0.15	−0.53	0.19	−0.18	0.15		
527	−0.77	0.14			−0.59		−0.27	0.24	−0.38			
528	−0.82	0.19	−0.78				−0.45		−0.14			
529	−1.06	0.01										
548	−2.12						−1.59				−0.90	0.04
550	−2.07											
577												
690			−1.16	0.97	−0.95	0.16	0.24	0.22	−0.21	0.23	0.30	0.15
698									0.14	0.14	−0.06	0.14
699									0.18	0.11		
700									−0.81	0.31		
702			−0.77	0.42			−0.37	0.26	0.33		−0.16	0.11
738			−1.12	0.27	−0.39	0.16	−0.46	0.27	−0.26	0.16	0.19	0.11
757	−0.78	0.01	−1.05	0.12			−0.71	0.13			0.09	0.16
865	−1.96		−2.10				−1.05		0.14		0.34	

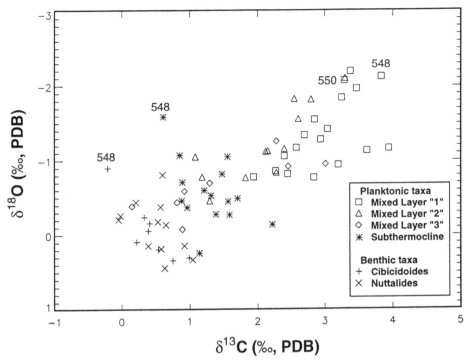

Figure 4.12. Cross-plot of DSDP and ODP site δ^{18}O and δ^{13}C averages for planktonic and benthic taxa. See text for discussion of Sites 548 and 550.

and Luz (1983) is used in the present study because it has been shown to apply to a wide range of water temperatures.

Seawater oxygen isotopic composition

Use of Equation (4.3) requires knowledge of the oxygen isotopic composition of the water (δ_w) in which the organism grew. This value is a function of several factors that cannot be determined independently for paleo-oceanographic conditions. In order to calculate paleotemperatures from isotope data it is therefore necessary to make certain assumptions about seawater δ^{18}O and to keep in mind possible error associated with these assumptions.

On the timescale of greater than 10 million years, the mean δ^{18}O value of seawater is controlled by the balance between processes that increase δ^{18}O (hydrothermal alteration and dehydration of oceanic basalt) and those that decrease the δ^{18}O of seawater (low temperature submarine weathering of basalts and continental weathering) (Holland, 1984). There is evidence that, during the last 400 million years, some change in the relative importance of these processes may have caused a shift of ~ 2‰ (SMOW) or more in seawater δ^{18}O (Popp et al., 1986; Veizer et al., 1986; Lohmann and Walker, 1989). However, the cause of such a shift is as yet very poorly understood. The rate of hydrothermal exchange of oxygen may have been higher at 55 Ma because of increased hydrothermal alteration associated with late Paleocene–early Eocene volcanism and seafloor spreading (Oberhänsli and Hsü, 1986; Olivarez and Owen, 1989; Larson, 1991; Ritchie and Hitchen, 1996). Assuming no ice volume effect (discussed below), L. Kump and M. Arthur (unpublished data) estimate that seawater $\delta^{18}O$ could have been 0.10–0.15‰ higher than the modern ocean at 55 Ma. In the isotopic paleotemperature equation of Erez and Luz (1983), this change would result in temperatures 0.5–0.7 °C higher. However, the results of Kump and Arthur have only recently become available and have not been considered in the present study. A larger difference between modern and early Eocene seawater oxygen isotopic composition is attributed to the ice-free conditions of the warm early Eocene.

On the timescale of less than 10 million years, the global mean composition of seawater is dependent primarily on the global volume of continental ice-sheets. In general, evaporation preferentially removes the lighter isotope of oxygen (^{16}O) from oceans. Precipitation preferentially removes the heavier isotope (^{18}O) from the atmospheric water vapor reservoir. Along its atmospheric path, which tends to be from low latitudes to high, water vapor becomes increasingly depleted with respect to ^{18}O (Craig and Gordon, 1965). Precipitation at high latitudes (where significant ice-sheets develop) is today −24 to −35‰ (SMOW) over northern and central Greenland and −18 to −55‰ on Antarctica (Dansgaard, 1964; Rozanski *et al.*, 1993). Continental ice-sheets can be viewed as reservoirs of ^{16}O. During time periods of large continental ice-sheets, the global ocean is enriched with respect to ^{18}O. To determine the global mean isotopic composition of seawater during periods of small or no continental ice-sheets, the present-day mean value must be decreased by some value that is a function of ice-sheet volume and isotopic composition. The value widely used is that of Shackleton and Kennett (1975b) who estimated that, for time periods of ice-free global conditions, a mean seawater $\delta^{18}O$ value of approximately −1.0‰ (SMOW) was appropriate for paleotemperature calculations. To a first order, over most of the earth, the isotopic composition of seawater below the mixed layer can be assumed to equal the global mean value. In paleotemperature calculations using subthermocline and benthic $\delta^{18}O$ data, δ_w in the paleotemperature equation (Eqn 4.3) is therefore assumed to be equal to −1‰ (SMOW).

The local or regional composition of surface seawater is a function of the global mean composition as well as continental runoff (in coastal regions), precipitation and evaporation, and ocean circulation patterns. Because the same processes that control salinity control the $\delta^{18}O$ of surface water, there is a strong correlation observed in the present-day open ocean between surface salinity and $\delta^{18}O$ (Broecker, 1989; Fairbanks *et al.*, 1992). In order to account crudely for the effects of evaporation and atmospheric transport of moisture on regional $\delta^{18}O$ seawater variation, Zachos *et al.* (1994) proposed the use of an equation relating ^{18}O and paleolatitude. They derived an expression for the present-day latitudinal variation of seawater $\delta^{18}O$ by fitting a curve to southern hemisphere open ocean data collected as part of the Geochemical Ocean Sections study (GEOSECS):

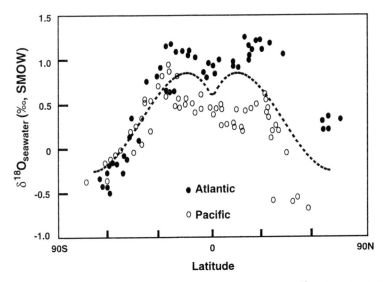

Figure 4.13. Present-day latitudinal variation of seawater δ^{18}O for the Atlantic (closed circles) and Pacific (open circles) oceans (GEOSECS data from Broecker, 1989). The curve fit by Zachos et al. (1994, dashed line) to southern hemisphere data is transposed for the northern hemisphere.

$$\delta_w = 0.576 + 0.041y - 0.0017y^2 + 1.35 \times 10^{-5}y^3, \tag{4.4}$$

where δ_w is the oxygen isotopic composition of the seawater (δ^{18}O, SMOW) and y is absolute latitude in the range of 0–70°. The meridional surface water curve calculated using Equation (4.4) is shown in Fig. 4.13 in comparison to the GEOSECS data (Broecker, 1989) for both hemispheres. Zachos et al. (1994) propose use of the present-day δ_w–latitude relationship (with correction for continental ice volume) to paleotemperature calculations. Application of Equation (4.4) to early Eocene isotopic paleotemperature calculations assumes that the meridional gradient in open ocean surface water δ^{18}O was not significantly different from the present day. There may be error in this assumption, in light of the differences in model-simulated surface salinity gradients for the present-day and Eocene simulations (Bice, 1997). However, as a first approximation, the use of the paleolatitude-dependent δ_w is tried here.

For all sites for which planktonic foraminifera isotope analyses were compiled, δ_w values were calculated as a function of paleolatitude and then decreased by 1.0‰ to approximate early Eocene ice-free conditions (Table 4.4). The resulting paleotemperatures calculated from mixed layer foraminifera are also given in Table 4.4. Table 4.5 summarizes the subthermocline and bottom-water isotopic paleotemperatures for each site. At sites where bottom-water temperature calculations can be made from both *Nuttalides* and *Cibicidoides* data (Sites 690, 698, 702, 738, and 865), the isotopic paleotemperatures are within 1.6 °C of each other. The

Table 4.4. *Seawater composition as a function of paleolatitude and resulting mixed layer isotopic paleotemperatures*

Site	$\delta^{18}O_{seawater}$ (SMOW)	Mean isotopic paleotemperatures (°C)		
		Mixed layer '1'	Mixed layer '2'	Mixed layer '3'
20	−0.34	21.3		
21	−0.34	19.7	19.3	15.8
94	−0.29	22.7	22.8	
152	−0.15	26.3		
213	−0.35	20.7	20.6	21.2
215	−0.51	19.8		18.8
277	−1.25		19.6	14.5
356	−0.33	23.9	23.8	
357	−0.37	21.4		
363	−0.27		17.9	
364	−0.16	22.7	24.6	
384	−0.56	19.7		
525	−0.48	18.7		16.8
527	−0.44	18.5		17.7
528	−0.46	18.6	18.5	
529	−0.47	19.7		
548	−0.78	23.1		
550	−0.76	22.9		
690	−1.21		16.8	15.9
702	−0.98		16.1	
738	−1.20		16.6	13.4
757	−0.78	17.0	18.2	
865	−0.41	24.1	24.7	

difference is not systematic: at three of the five sites, *Nuttalides* yields a warmer temperature estimate than *Cibicidoides*.

Estimated error in the isotopic paleotemperature equation

Erez and Luz (1983) cite four sources of error in the experimental data used to derive the paleotemperature equation used here (Eqn 4.3): (1) ±0.15 °C in the culture temperature, (2) ±0.5 °C from the measurement of the isotopic composition of the foram test, (3) ±1 °C resulting from application of a disequilibrium fractionation correction factor to the planktonic foraminifer *Globigerinoides sacculifer*, and (4) 0.5 °C from the determination of the oxygen isotopic composition of the culture water. Assuming that the general forms of the first and second laws of the propagation of error apply to their data and assuming no error in the coefficients of the temperature equation, the error associated with the use of the Erez and Luz equation is a function of the calculated value of $\delta_c - \delta_w$ in Equation (4.3) and, for the data compiled here, varies around ±1.8 °C.

In the absence of a proxy indicator for early Eocene surface water $\delta^{18}O$, an approximation based on the present-day seawater $\delta^{18}O$–latitude relationship is a

Table 4.5. *Subthermocline and bottom-water temperatures*

	Mean isotopic paleotemperatures (°C)		
		Bottom water	
Site	Subthermocline water	*Nuttalides*	*Cibicidoides*
20	17.3		
21	13.8	10.6	
213	16.2	14.5	
215	14.7		
363	15.2		
384	13.1		
401			13.6
525	14.9	13.3	
527	13.7	14.2	
528	14.5	13.1	
548	19.7		16.6
577		13.5	
690	11.4	11.9	11.2
698		11.7	12.8
699		16.2	
700			13.2
702	14.1	11.0	11.7
738	14.6	13.7	12.1
757	15.7		
865	17.2	11.9	11.0

reasonable first attempt. However, given the spread of the data (Fig. 4.13) fit by Zachos *et al.* (1994), there is clearly error in this assumption. For example, between about 40° S and 40° N, the scatter about the curve defined by Equation (4.4) is ±0.5‰, as a minimum. This is equal to a temperature error of 2.3 °C in Equation (4.3). If, in earth history, atmospheric water vapor transport and ocean circulation gave rise to salinity contrasts similar to that observed between the present Atlantic and Pacific oceans (Broecker, 1989), the assumption of a paleolatitude-dependent δ_w value could introduce considerable uncertainty in isotopic paleotemperatures calculated for the two basins. As mentioned above, there may also be error in the assumption that the meridional gradient in open ocean surface water δ¹⁸O was not significantly different from the present day. In order to determine at least a minimum error, we assume the error in the seawater δ¹⁸O to be only 0.1‰, which equals an error of ±0.5 °C in Equation (4.3). This does not include potential error introduced by assuming a global mean ocean δ¹⁸O of −1.0‰ (SMOW) during times of no significant continental ice-sheets. The estimated error in δ_w assumed here is clearly a conservative value. As will be shown for Sites 548 and 550, there is good evidence that in the North Atlantic, early Eocene surface seawater δ¹⁸O values may be underpredicted by as much as 1.3‰ by using the equation of Zachos *et al.* (1994).

Reported values of the precision of replicate analyses of foraminiferal carbonate δ¹⁸O measurements vary between ±0.05‰ (Shackleton and Kennett,

Table 4.6. *Model layers against which isotopic paleotemperatures are compared*

Foraminiferal depth ranking	Model layer	Model layer mid-point depth (m)
Mixed layer '1' taxa	1	12.5
Mixed layer '2' taxa	2	37.5
Mixed layer '3' taxa	3	62.5
Subthermocline taxa	9	435.0
Benthic taxa	Varies by site	Depth closest to site paleodepth given in Table 4.1

1975*a,b*; Pak and Miller, 1992) and ±0.11‰ (Boersma and Shackleton, 1977), with 0.1‰ (PDB) the most commonly reported precision (Douglas and Savin, 1975; Miller *et al.*, 1985; Poag *et al.*, 1985; Oberhänsli *et al.*, 1991; Zachos *et al.*, 1994). An analytical uncertainty of ±0.1‰ in δ_c (Eqn 4.3) equals an uncertainty in the paleotemperature of ±0.5 °C. When the three sources of error derived above (that for the empirical equation, δ_w, and δ_c) are considered simultaneously, the error in the isotopic paleotemperatures compiled here is at least ±2.0 °C.

In their discussion of estimated meridional temperature profiles for warm time periods, Crowley and Zachos (Chapter 3, this volume) provide a more detailed quantitative analysis of error in the isotopic paleotemperatures. The ±2 °C error estimate derived above is compatible with the error in zonal averaged temperatures estimated by Crowley and Zachos (1 °C too warm to 3 °C too cold).

COMPARISON OF ISOTOPIC PALEOTEMPERATURES AND MODEL-PREDICTED TEMPERATURES

In order to evaluate the quality of the model-simulated three-dimensional temperature field, isotopic paleotemperatures are compared against model-simulated temperatures. The model layers chosen for comparison with the depth-stratified paleotemperatures are indicated in Table 4.6. Because the actual depth at which the subthermocline taxon *Subbotina* lived at any particular site is unknown, it would be difficult to defend a decision to compare isotopic paleotemperatures against model-predicted temperatures for different levels at different sites. The decision to compare subthermocline isotopic paleotemperatures to the model-predicted temperatures at 435 m is based on the approximate mid-point depth of the region of maximum temperature decrease at the sites for which data from *Subbotina* are available. With the exception of Sites 690, 702, and 737 where there is virtually no vertical gradient in temperature, model temperatures in the layer above 435 m (at 310 m) are higher by 1.2–2.8 °C; model temperatures in the layer below (at 610 m) are lower by 1.5–3.3 °C.

MIXED LAYER TEMPERATURES

The model-predicted layer 1 temperature (12.5 m mid-point depth) falls within 2.5 °C of the mixed layer '1' isotopic paleotemperatures at all sites except

Site 356, a coastal site in the South Atlantic, and Sites 548 and 550 in the northern North Atlantic (Fig. 4.14a). At Site 356, the model under-predicts the isotopic paleotemperature by approximately 3 °C. At Sites 548 and 550, the model under-predicts the isotopic paleotemperature by about 8 °C. With the exception of these sites, which are discussed below, and given error of at least ±2.0 °C in the isotopic paleotemperature calculation, there is good agreement between the model temperatures and early Eocene paleotemperatures in the uppermost mixed layer. At Site 356, continental runoff may supply water depleted in ^{18}O. The local $\delta^{18}O$ of seawater would then be lower than open ocean values resulting in a higher apparent temperature at Site 356. Taking this effect into account would move the isotopic paleotemperature at Site 356 toward the model-predicted temperature rather than away from it. The possible sources of error in the Site 548 and 550 isotopic paleotemperatures are considered in a later section.

The fit between isotopic paleotemperatures and the model-predicted site temperatures is worse in general for the mixed layer '2' and '3' taxa (Figs. 4.14b and c), especially for the cooler ocean regions. The model-predicted temperature is within 2 °C of the isotopic paleotemperature at fewer than half the sites. Given the good correlation of mixed layer '1' values, and assuming that the model layers chosen for comparison to each group of foraminifera are valid (for example, that the depth habitat of *Acarinina* species was approximately 26–50 m), this result could indicate that the ocean model simulation of the heat flux between layer 1 and layer 2 is inadequate. The error likely in the *ad hoc* assumption that, for example, paleotemperatures based on *Acarinina* species should be comparable to model layer 2 temperatures could not negate the observation that, below the uppermost layer, the model under-predicts mixed layer temperatures as a whole.

To within 2 °C, the model-predicted temperatures agree with approximately 70% of the isotopic paleotemperatures from mixed layer '1' taxa and only 38% of the isotopic paleotemperatures from deeper mixed layer foraminifera. In general, the worst match between model-predicted temperatures in the uppermost three model layers and the isotopic paleotemperatures calculated from taxa designated as 'mixed layer' occurs poleward of 40° latitude (Fig. 4.15). This suggests that the atmospheric forcing temperatures (GENESIS) are too cool at high latitudes (also noted by Sloan and Rea, 1995), the model-simulated poleward heat transport is too weak, or some combination of these factors. Alternatively, the error introduced by use of the latitude-dependent δ_w value may be greater at high latitudes. In fact, as is shown in a later section, δ_w values calculated from model-predicted surface salinities yield high-latitude isotopic paleotemperatures that are in better agreement with those predicted by this simulation.

The mixed layer seasonal temperature cycle

Should the mixed layer isotopic paleotemperatures be compared with the model-simulated mean annual temperatures? Foraminifera are short-lived and reproduction and growth are seasonal. Although interannual variability is observed, the timing of phytoplankton and zooplankton blooms varies primarily as a function of season and taxa (Deuser *et al.*, 1981). Because the early Eocene foraminifera are

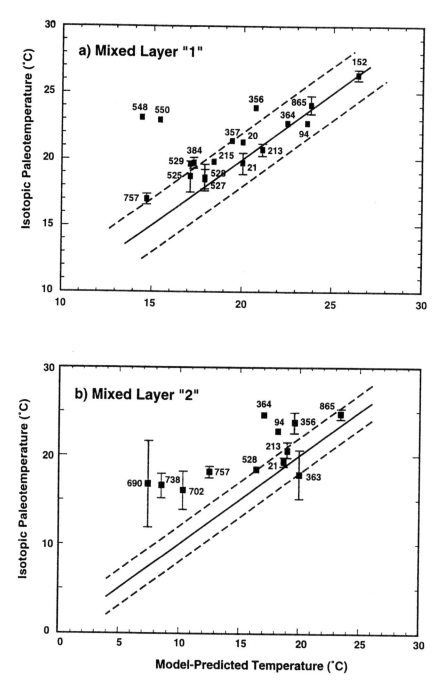

Figure 4.14. Comparison of isotopic paleotemperatures with ocean model grid-point values of water temperature. (a) Taxa identified as mixed layer '1'; (b) taxa identified as mixed layer '2.' (c) Taxa identified as mixed layer '3'; (d) taxa identified as sub-thermocline and benthic. The solid line marks a 1:1 correspondence between the isotopic temperature and the model temperature. In (a), (b) and (c) the dashed lines

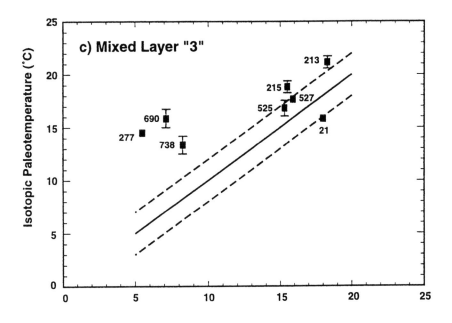

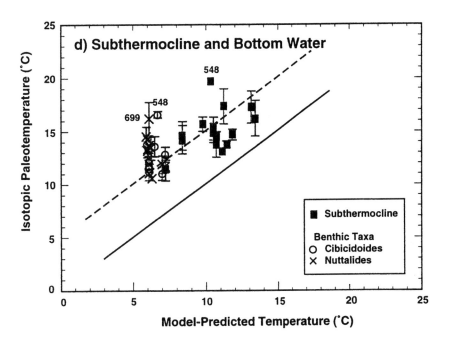

Caption for Fig. 4.14 (cont.)
demarcate a ±2°C mismatch between the two sets of values. In (d), the dashed line
denotes an under-prediction by 5°C of the isotopic paleotemperature. (Error bars on
the isotopic paleotemperatures in Figs. 4.14–4.18 represent the standard deviations
at sites where more than one data value is available.)

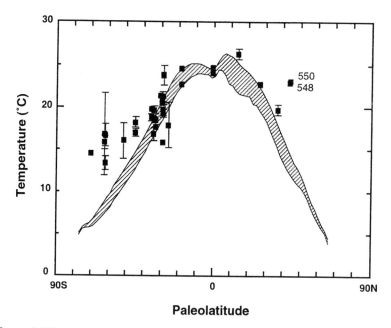

Paleolatitude

Figure 4.15. Isotopic paleotemperatures (boxes) calculated from foraminifera designated as mixed layer '1', '2', and '3' taxa plotted against paleolatitude. The shaded area indicates the range of zonal mean annual temperatures simulated for model layers 1 through 3 (mid-point depths 12.5-62.5 m). The isotopic paleotemperature calculation assumes the latitude-dependent δ_w values given in Table 4.4.

extinct taxa, any assertion as to the absolute or relative timing of reproduction and maturation of the mixed layer taxa discussed here would be unconstrained. It is reasonable to expect that the isotopic paleotemperature should fall within the model-simulated seasonal cycle. In Fig. 4.16, the model-simulated seasonal temperature cycle is shown for each site from which *Morozovella* δ^{18}O data were compiled. The isotopic paleotemperature is indicated by an arrow on the left side of each graph. At all sites except 548 and 550, the isotopic paleotemperatures from *Morozovella* fall within the annual temperature range simulated for the uppermost model layer.

The nature of the mismatch between shallow-water model and isotopic temperatures at Sites 548 and 550 in the North Atlantic deserves further examination because of its bearing on the probable inadequacy of a latitude-dependent δ_w estimation, especially in the northern hemisphere.

North Atlantic shallow water temperatures

In the comparison of model-predicted and isotopic paleotemperatures, the largest discrepancy occurs in the data for Sites 548 and 550. The uppermost model layer temperature under-predicts the isotopic paleotemperatures based on δ^{18}O measurements of *Morozovella* by 8.7 and 7.5 °C, respectively (Fig. 4.14a). Mixed layer '1' isotopic paleotemperatures at these sites (23 °C at ~44 ° N paleolatitude) are nearly equivalent to that inferred for Site 865 (24 °C at ~0.5° S paleolatitude). They are 6 °C higher than temperatures from *Morozovella* δ^{18}O at the same latitude in southern hemisphere Site 757 (17 °C at 44° S paleolatitude). The large mismatch at Sites 548 and 550 suggests (1) that the model provides a very poor simulation of the North Atlantic for the Eocene, (2) that there is some systematic error in foraminiferal δ^{18}O measurements for these sites (perhaps diagenetic alteration), (3) that the isotope analyses from Poag *et al.* (1985) and Boersma *et al.* (1987) are not from the same time interval as other data used, or (4) that the δ_w values assumed here in the paleotemperature equation are significantly in error.

There is no good evidence that the model does not provide as good a simulation of conditions in the North Atlantic as elsewhere. At Site 384 in the mid-North Atlantic, Eocene model temperatures are in good agreement with paleotemperatures from *Morozovella* and the subthermocline genus *Subbotina* (Fig. 4.14a). Unfortunately, however, no other mixed layer isotopic values that could be used to further evaluate the Eocene model are available for the western North Atlantic. Considering the deeper environment, benthic foraminiferal δ^{18}O values are available at Site 401, as well as Site 548. Although the Eocene simulation also under-predicts the bottom-water temperature inferred from *Cibicidoides* measurements at Site 401, the magnitude of the difference (7.0 °C) is close to the average difference for all bottom-water temperatures (6.5 °C) and is 2.8 °C less than the Site 548 isotopic temperature-model temperature difference (9.8 °C). The model does not under-predict present-day surface or bottom-water temperatures in the North Atlantic by more than 2 °C (Bice, 1997).

The history of Site 548 suggests that diagenetic alteration is a possible source of error. At Site 548 the early Eocene specimens directly underlie an unconformity that represents a hiatus of at least 2 Ma (Poag *et al.*, 1985). Aubry (1995) and Aubry *et al.* (1996) describe several unconformities near the Paleocene/Eocene boundary at Sites 549 and 550. Poag *et al.* (1985) believe that Site 548, located on the present-day Goban Spur, experienced substantial shallow marine or subaerial erosion during the global eustatic sea level drop during the mid- to late–early Eocene (Haq *et al.*, 1987). Poag *et al.* (1985) also describe evidence of bioturbation. Post-depositional recrystallization under warmer, shallower water conditions during the sea level lowstand could have decreased the δ^{18}O values of early Eocene calcareous foraminifera at this site. The cross-plot of foraminiferal δ^{18}O/δ^{13}C (Fig. 4.12) shows that Site 548 data for *Subbotina* and *Cibicidoides* plot outside the overall data field and that the δ^{18}O values are more depleted than other data with low (0–1‰, PDB) carbon isotope values, suggesting possible alteration. However, diagenesis might be expected to affect most or all tests within a narrow core section, but carbon and oxygen measurements for *Morozovella* at Sites 548 and 550 fall well within the general field of the early Eocene data (Fig. 4.12). Additionally, neither Poag *et al.* (1985) nor Boersma *et al.* (1987) note evidence of recrystallization of these samples.

Another possible diagenetic effect is the alteration of volcanic ash at these sites followed by minor secondary calcite cementation. Rhyolitic ash and bentonites

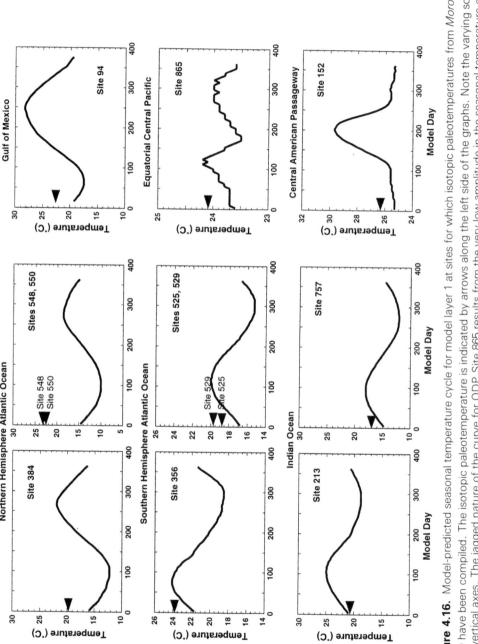

Figure 4.16. Model-predicted seasonal temperature cycle for model layer 1 at sites for which isotopic paleotemperatures from *Morozovella* $\delta^{18}O$ have been compiled. The isotopic paleotemperature is indicated by arrows along the left side of the graphs. Note the varying scales of the vertical axes. The jagged nature of the curve for ODP Site 865 results from the very low amplitude in the seasonal temperature cycle at this equatorial site.

are common in the upper Paleocene–lower Eocene section at Sites 549 and 550 (Aubry et al., 1996). The alteration of volcanic ash to clays preferentially removes ^{18}O, leaving pore waters more depleted than the original fluids. A small amount of secondary calcite precipitated from these depleted fluids in the foram test would yield a carbonate $\delta^{18}O$ value suggestive of artificially high temperatures. Unfortunately, pore water geochemical data needed to examine this problem were not collected. While possible diagenesis cannot be ruled out, it is not clearly the cause of the depleted foraminiferal carbonate at these sites.

The possibility of error in the assignment of ages for any particular core in the early Eocene marine stratigraphic section is good. Hiatuses occur throughout the late Paleocene–early Eocene interval (Aubry, 1995; Aubry et al., 1996; Berggren and Aubry, 1996). In fact, the development of a global magnetobiochronology for the late Paleocene–early Eocene interval has required considerable attention over the last decade and remains problematic today (Berggren et al., 1985, 1995; Berggren and Miller, 1988; Cande and Kent, 1992, 1995; Berggren and Aubry, 1996; Aubry et al., 1998). The occurrence of abrupt, short-lived climate change and multiple hiatuses within the same interval make it more likely that error in age assignment might result in apparent temperatures too warm or too cool compared with coeval sediments. Although there is likely to be error in the assignment of ages to North Atlantic cores, there is no evidence that error within the upper Paleocene–lower Eocene section could result in inferred shallow ocean paleotemperatures at 44° N paleolatitude that are within 1 °C of equatorial surface temperatures (e.g., Site 865).

Given the above considerations and the scatter in the δ_w-latitude plot for the modern surface ocean, it is possible that $\delta^{18}O$ of surface water in the eastern North Atlantic may have been substantially more depleted than the latitude-dependent δ_w equation of Zachos et al. (1994) allows. Nearly all of the sites for which *Morozovella* data were compiled exhibit isotopic paleotemperatures within 2 °C of the model-predicted temperatures (Fig. 4.14a). Assuming that the model provides as good a simulation of North Atlantic conditions as elsewhere and assuming no diagenetic alteration of Site 548 and 550 data, it is possible to estimate a seawater isotopic composition that would result in isotopic paleotemperatures within 2° of those predicted by the model. The ocean model predicts mean annual shallow ocean temperatures of about 15 °C at Sites 548 and 550 (Fig. 4.14a). To yield an isotopic paleotemperature of 17 °C (i.e., a match within 2° of the model), δ_w (relative to SMOW) in Equation (4.3) must approximately equal the $\delta^{18}O$ (relative to PDB) of the foraminiferal carbonate, or approximately −2.1‰ at Sites 548 and 550. This seawater value is ∼1.3‰ more depleted than that derived using the latitude-dependent δ_w equation with the correction for ice-free conditions (∼−0.77‰; Table 4.4). Mean annual salinity in the eastern North Atlantic at 44° latitude today is 35.7‰ (Levitus and Boyer, 1994). The estimated mean annual δ_w value for the area is 0.66–0.8‰ (SMOW), assuming three slightly different δ_w–salinity relationships published for the Atlantic Ocean surface waters (Broecker, 1989; Duplessy et al., 1991; Fairbanks et al., 1992). The early Eocene estimated value of −2.1‰ is therefore about 2.8‰ more depleted than modern surface seawater in the area. Only about 1.0‰ of this difference could be attributed to the ice volume effect (Shackleton and Kennett, 1975*a*).

Some additional decrease in local surface water $\delta^{18}O$ could result from an increased flux of isotopically depleted precipitation and/or continental runoff. The GENESIS model results show some support for this case. Over the eastern North Atlantic region at 44° latitude and adjacent continent, the largest difference between the warm Eocene and the present day is an increase in continental precipitation (Bice, 1997). Such a change implies a greater flux of freshwater to the ocean through continental runoff.

Alternatively, it is possible that local freshwater fluxes could have been unchanged but that the local precipitation and runoff were more depleted than the present day. However, a decrease in the $\delta^{18}O$ of local freshwater is difficult to explain given simple assumptions of equilibrium ^{18}O fractionation and a reduced meridional temperature gradient. Let us assume that equatorial waters (the source area for vapor) were 2 °C cooler and that surface air temperatures at 44° latitude were no more than 2° higher in the early Eocene. In the low-latitude temperature range 23–28 °C, a 2° decrease in water surface temperature results in a ∼0.16 decrease in the $\delta^{18}O$ of the resulting vapor (Faure, 1986). Considering this effect alone, slightly more depleted water vapor might be reasonable for the early Eocene. However, a larger change (in the opposite direction) results from a 2 °C increase in the temperature of air from which condensation occurs (i.e., that at 44° latitude). Assuming a range of observed $\delta^{18}O$ precipitation–temperature relationships (Rozanski *et al.*, 1993), a 2 °C increase in surface air temperature would result in at least a 1.16‰ *increase* in the $\delta^{18}O$ of local precipitation, overwhelming any decrease in the source vapor $\delta^{18}O$. Additionally, if relatively depleted freshwater must be invoked to account for North Atlantic surface water compositions, the same argument could be applied globally, but such a global 'correction' is not warranted by the match between the model and the isotopic data in other regions.

The reason for the large mismatch between model and isotopic temperatures at Sites 548 and 550 cannot be determined unequivocally, but there is no evidence that the ocean model is largely in error in the North Atlantic near 45° latitude. The best explanation for the apparent under-prediction of surface temperatures there is that use of the latitude-dependent relationship for seawater $\delta^{18}O$ yields δ_w values that are too high. In the isotopic paleotemperature equation of Erez and Luz (1983), the resulting surface ocean paleotemperatures for the early Eocene at Sites 548 and 550 are too warm by approximately 6–9 °C, relative both to the model temperatures and to the more abundant mid-latitude southern hemisphere isotopic data. The most likely mechanism for more depleted seawater is an increased flux of continental runoff to the eastern North Atlantic. Below, it will be shown that the model-predicted salinities in the region support this argument.

Subthermocline and bottom-water temperatures

In subthermocline and bottom-water temperatures (Fig. 4.14*d*), the model under-predicts the isotopic paleotemperatures (which assume a uniform δ_w value of −1.0‰) by from 2 °C to nearly 10 °C. For both subthermocline and bottom-water

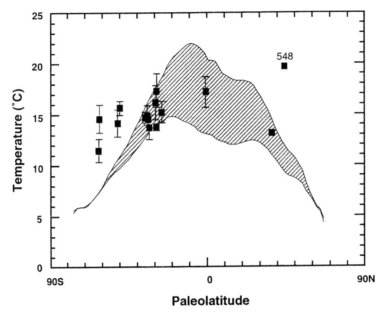

Figure 4.17. Isotopic paleotemperatures (boxes) calculated from foraminifera desig-nated as subthermocline taxa plotted against paleolatitude. The shaded area indicates the range of zonal mean annual temperatures simulated for model layers 7 through 10. The isotopic paleotemperature calculation assumes a δ_w value of −1.0‰ (SMOW).

temperatures, the mismatch is greatest at Site 548. The fit for most subthermocline calculations is better than bottom-water calculations. The mean difference (isotopic temperature minus model temperature) is 4.5 °C for subthermocline calculations and 6.5 °C for bottom-water calculations. On a zonal mean basis, the fit between sub-thermocline temperatures is slightly better when compared with model-predicted temperatures in the layer mid-point depth range 222.5–610 m (Fig. 4.17). This depth range encompasses most of the model-simulated mean annual thermocline at the DSDP and ODP sites.

Figure 4.18 shows the comparison of bottom-water isotopic paleotempera-tures from individual sites and the model-predicted zonal mean temperature range between 1000 m and 4500 m, the approximate depth range represented by the sites for which *Nuttalides* and *Cibicidoides* δ¹⁸O measurements have been compiled. In comparison with Fig. 4.1, it is apparent that the model captures only approximately 6 °C of the 10–15 °C warming relative to equivalent depths and latitudes in the present-day ocean. Approximately one-third of that warming is due to the absence of high-latitude continental ice-sheets (Bice, 1997). Additional warming at depth is the result of forcing with increased atmospheric CO_2 and oceanic heat transport specified in the atmospheric GCM simulation. There is, however, no change in the site of deep-water formation. Bottom waters warm because Southern Ocean upper ocean waters, the source of bottom water, are warmer than the present day.

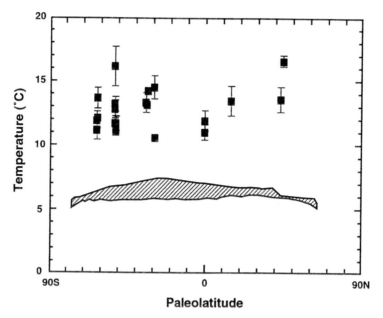

Figure 4.18. Isotopic paleotemperatures (boxes) calculated from benthic foraminifera plotted against paleolatitude. The shaded area indicates the range of zonal mean
annual temperatures simulated for model layer depths 1000-4500 m, the approximate
depth range represented by the DSDP and ODP sites for which benthic foraminiferal
$\delta^{18}O$ data were compiled. The isotopic paleotemperature calculation assumes a δ_w
value of $-1.0‰$ (SMOW).

SEAWATER ISOTOPIC COMPOSITION AS A FUNCTION OF SALINITY

The isotopic composition of seawater (δ_w) must be known in order to calculate paleotemperatures using equations of the form shown in Equation (4.2).
However, there is presently no proxy for paleo-ocean δ_w. In the open ocean surface
water $\delta^{18}O$ and salinity are both controlled by evaporation, precipitation, and atmospheric vapor transport. The oxygen isotopic composition of open ocean surface
water therefore correlates well with salinity. Using the GEOSECS Atlantic and
Pacific ocean data, Broecker (1989) derived a linear relationship between $\delta^{18}O$ and
salinity:

$$\delta_w = 0.5S - 17.12, \qquad (4.5)$$

where δ_w is the seawater oxygen isotopic composition relative to SMOW and S is
salinity in parts per thousand (‰). Fairbanks *et al.* (1992), using the same data set,
computed a set of seven δ_w–salinity equations encompassing all major surface salinity
gradients in the Atlantic and eastern tropical Pacific oceans. Two of the equations of
Fairbanks *et al.* (1992) will be considered here, the eastern equatorial Atlantic:

$$\delta_w = 0.08S - 1.86, \qquad (4.6)$$

and the western equatorial Atlantic:

$$\delta_w = 0.19S - 5.97. \tag{4.7}$$

Because δ_w and salinity correlate strongly in the modern ocean, it is reasonable to assume that δ_w could be better estimated if paleosalinity were known. However, there is presently no known quantitative proxy indicator for salinity. Because the atmospheric and ocean general circulation models include the effects of evaporation, precipitation, atmospheric water vapor transport, and surface ocean circulation, the potential exists to use the ocean model-simulated salinity field to estimate paleo-ocean δ_w values. Use of the model-predicted salinity to calculate seawater $\delta^{18}O$ has the added advantage of allowing longitudinal variations in $\delta^{18}O$ due to surface ocean circulation and zonal atmospheric transport. In order to show that the model-predicted salinity might be a reliable tool, a more rigorous evaluation of the ability of the model to simulate surface salinities (and their seasonal variation) must be made. That exercise is planned as future work. For the present study, only a rudimentary analysis has been made. Given the quality of the crude comparison made to date, however, the potential for estimating reasonable δ_w values appears promising.

Using the δ_w–salinity relationship of Broecker (1989) (Eqn 4.5), the zonally averaged POCM salinity simulated for the modern ocean predicts the Zachos *et al.* (1994) latitude-dependent δ_w values to within 0.5‰ (Fig. 4.19). In the northern

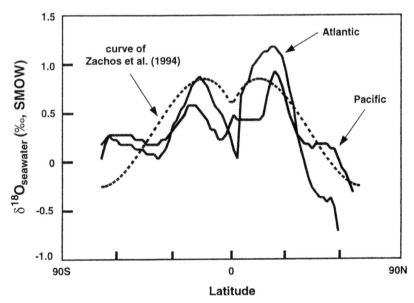

Figure 4.19. Comparison of seawater $\delta^{18}O$ values calculated from the model-simulated present-day zonal Atlantic and Pacific ocean salinities (Bice, 1997) using the equation of Broecker (1989) (solid lines) and the curve of Zachos *et al.* (1994) (dashed line).

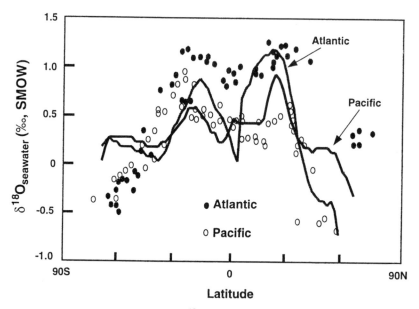

Figure 4.20. Comparison of seawater $\delta^{18}O$ values calculated from the model-simulated present-day zonal Atlantic and Pacific ocean salinities (Bice, 1997) using the equation of Broecker (1989) (solid lines) and the GEOSECS data (circles).

hemisphere, the zonal model-simulated salinity predicts the modern latitude-dependent δ_w values to within 0.5‰ everywhere except poleward of 55° N. Between 70° S and 55° N, the mean difference between the two estimates of surface seawater $\delta^{18}O$ (model salinity-dependent minus latitude-dependent) is −0.18‰. Between 70° S and 0° latitude, the mean difference is −0.11‰; between 0° and 55° N the mean difference is −0.18%.

Between 35° S and 50° N in the Pacific Ocean, the model salinity-dependent δ_w produces a much better match with the GEOSECS data than does the latitude-dependent δ_w curve (compare Figs. 4.13 and 4.20). Between 30° S and 40° N, the model captures the general shape of the Atlantic Ocean curve but tends to underpredict the observed values, except in the region 10–30° N. Poleward of 30° N, neither technique provides a good match with GEOSECS Atlantic Ocean values.

Assuming some linear error between the present-day simulation and the Eocene simulation, the model-predicted salinity can be used to estimate paleo-ocean surface water isotopic composition to within at least 0.5‰, in other words at least as well as the curve of Zachos *et al.* but with the added advantage that longitudinal variability can be predicted. Therefore, for each site for which *Morozovella* $\delta^{18}O$ values have been compiled, the oxygen isotopic composition of surface seawater was calculated from the model salinity in the uppermost model layer at the site using Equation (4.5) (Table 4.7). The resulting value was decreased by 1.0‰ to correct for ice-free global conditions (Shackleton and Kennett, 1975*b*) and was used in Equation (4.3) to generate new isotopic paleotemperatures. These

Table 4.7. *Seawater composition as a function of model-predicted salinity and resulting isotopic paleotemperatures*

Site	$\delta^{18}O_{seawater}$ (SMOW)	Mean isotopic paleotemperatures (°C): Mixed layer '1'
20	−0.62	20.0
21	−0.57	18.7
94	−1.77	16.0
152	−0.87	23.0
213	−0.72	19.0
215	−0.87	18.2
356	−0.67	22.3
357	−0.57	20.5
364	−0.12	22.9
384	−0.87	18.3
525	−0.97	16.5
527	−0.87	16.6
528	−0.87	16.8
529	−0.97	17.4
548	−2.17	16.8
550	−2.02	17.2
757	−0.97	16.2
865	−0.17	25.2

results are shown compared with the model-simulated site temperatures in Fig. 4.21. Using the model-predicted salinity and the δ_w–salinity equation of Broecker (1989), the model-predicted temperature is within 2 °C of the isotopic paleotemperature at all sites except Site 94 in the Gulf of Mexico (paleolatitude 27° N) and Site 152 in the Central American passageway (paleolatitude 14° N). Because Sites 94 and 152 are low-latitude sites, δ_w values at these sites and all sites located equatorward of 28° paleolatitude were recalculated using the Atlantic Ocean eastern equatorial (Eqn 4.6) and western equatorial (Eqn 4.7) δ_w–salinity equations of Fairbanks *et al.* (1992) (Figs. 4.22 and 4.23).

The fit between isotope and model temperatures is improved for low-latitude Sites 94 and 152 using either of the equatorial Atlantic Ocean δ_w–salinity equations of Fairbanks *et al.* (1992). At the same time, the isotopic paleotemperature at sites equatorward of 28° (Sites 364 and 865) remains within or very nearly within 2 °C of the model-predicted temperature. In both cases, the overall fit between isotope and model temperatures is improved by use of the δ_w–salinity relationship as compared with the calculation using a latitude-dependent δ_w value. These results suggest that the model-simulated salinity may be a viable tool for estimating the isotopic composition of surface seawater for past time periods, given the best possible reconstruction of boundary conditions. However, careful consideration will have to be given to the choice of a δ_w–salinity relationship. The approach taken here in the choice of an equation relating salinity and δ_w is highly subjective. A more

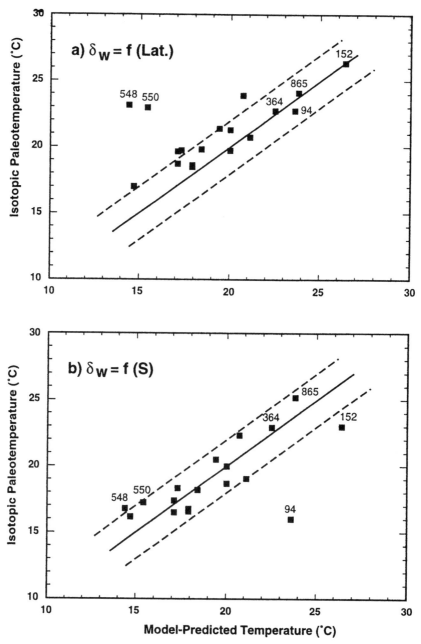

Figure 4.21. Comparison of the match between model-predicted temperature and isotopic paleotemperatures using two different methods of estimating δ_w. (a) δ_w calculated as a function of paleolatitude using the equation of Zachos *et al.* (1994) (repeated from Fig. 4.14a for ease in comparing the two data sets); (b) δ_w calculated as a function of the model-predicted salinity using the relationship of Broecker (1989). The solid line marks a 1:1 correspondence between the isotopic temperature and the model temperature. Dashed lines above and below demarcate a $\pm 2\,°C$ mismatch between the two sets of values.

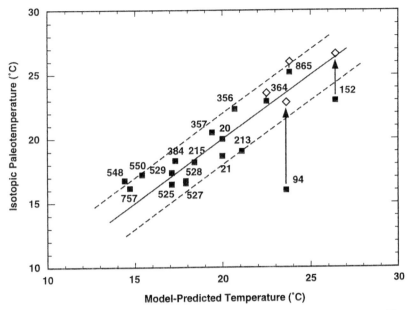

Figure 4.22. First recalculation of the isotopic paleotemperatures shown in Fig. 4.21b. Values of δ_w at sites below 28° latitude were determined using the eastern equatorial Atlantic Ocean δ_w–salinity equation of Fairbanks et al. (1992). The recalculated paleotemperatures are indicated by the open diamonds with the shift in temperature shown by the arrows. The solid line marks a 1:1 correspondence between the isotopic temperature and the model temperature. Dashed lines above and below demarcate a ±2°C mismatch between the two sets of values.

objective approach would be modification of the models to include an oxygen stable isotope tracer. The feasibility of this modification is under consideration for POCM.

Calculation of seawater δ¹⁸O from model-predicted salinity

If we accept that the model-predicted surface temperatures compare well with isotopic paleotemperatures calculated using a seawater $\delta^{18}O$ value derived from the model-predicted surface salinity, then it is also reasonable to assume that the model-predicted surface salinity field can be used to produce a map of surface seawater isotopic composition for the early Eocene. Figure 4.24 (see color plate) shows the surface seawater $\delta^{18}O$ distribution calculated using the mean annual salinity predicted by the ocean GCM for the early Eocene. Regions above 28° paleolatitude use the δ_w–salinity equation of Broecker (1989); regions below 28° use the western equatorial Atlantic equation of Fairbanks et al. (1992). All calculated seawater values were adjusted by $-1.0‰$ to correct for ice-free conditions.

That surface seawater $\delta^{18}O$ may have been $-3‰$ or lower in the early Eocene northern Tethys or northern Atlantic would be difficult to validate in the absence of a δ_w or salinity proxy. Values as low as $-2‰$ are observed today in the East Greenland Sea (Fairbanks et al., 1992). The discussion of mixed layer

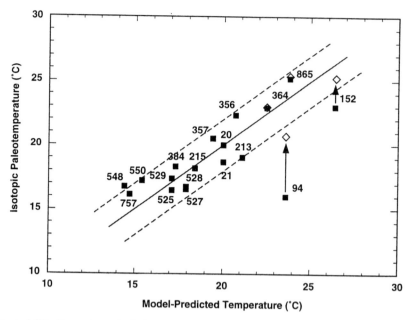

Figure 4.23. Second recalculation of the isotopic paleotemperatures shown in Fig. 4.21b. Values of δ_w at sites below 28° latitude were determined using the western equatorial Atlantic Ocean δ_w–salinity equation of Fairbanks *et al.* (1992). The recalculated paleotemperatures are indicated by the open diamonds with the shift in temperature shown by the arrows. The solid line marks a 1:1 correspondence between the isotopic temperature and the model temperature. Dashed lines above and below demarcate a ±2°C mismatch between the two sets of values.

temperatures at Sites 548 and 550 supports the depleted δ_w values predicted for the Eocene North Atlantic; however, the low salinities in that region are the result in part of continental runoff from a large portion of northeastern Europe and Greenland. In a POCM sensitivity experiment (Bice *et al.*, 1997) that includes the Arctic Ocean, runoff to the Arctic was permitted, decreasing surface salinities at northern high latitudes. In that experiment, the average salinity in the northeastern Atlantic was increased by ~0.3‰. The resulting sensitivity in δ_w is relatively small, changing from an average of −2.1 to −1.9‰, including the assumption of a −1‰ adjustment for ice-free conditions.

SUMMARY

In subthermocline and bottom-water temperatures, the model under-predicts the isotopic paleotemperatures (which assume a seawater isotopic composition of −1.0‰) by from 2 to 10 °C. The fit for most subthermocline calculations is better than bottom-water calculations. On average, the model under-predicts subthermocline temperatures by 4.5 °C and bottom-water temperatures by 6.5 °C. The model captures approximately 6 °C of the 10–15 °C bottom-water warming relative to equivalent depths and latitudes in the present-day ocean. One-third of that warming is due to lack of high-latitude continental ice-sheets (Bice, 1997). Additional warming at depth is the result of forcing with increased atmospheric CO_2 and oceanic heat transport specified in the GENESIS simulation. When continental runoff is routed to coastal ocean model cells using a crude reconstruction of Eocene drainage basins, there is no deep convection in the Tethys Seaway. Bottom waters warm because Southern Ocean upper ocean waters, the source of bottom water, are warmer than in the modern ocean.

Early Eocene mixed layer ocean paleotemperatures were calculated using two different assumptions about the ambient seawater $\delta^{18}O$ value (δ_w). The δ_w value used in the paleotemperature equation was first calculated as a function of paleolatitude and was corrected for ice-free global conditions. At 15 of 18 sites, the model temperatures in the uppermost layer predict the isotopic paleotemperatures from *Morozovella* within the error ($\pm 2\,^\circ C$) in the isotopic paleotemperature equation. (Error in the paleotemperature calculation is estimated to be at least $2\,^\circ C$.) The largest mismatch is observed for data from North Atlantic Sites 548 and 550. This is perhaps not surprising given the fact that the latitude-dependent δ_w equation of Zachos *et al.* (1994) is a curve fit to present-day southern hemisphere data alone. (The large amount of scatter in modern northern hemisphere seawater $\delta^{18}O$ values prevents fitting a single curve to the data.) At all sites except Sites 548 and 550, the *Morozovella* paleotemperature falls within the annual temperature cycle simulated by the model. Similar comparisons of model and isotope temperatures in mixed layer depths below the uppermost 25 m do not yield as good a fit. The model-predicted temperature falls within $2\,^\circ C$ of the isotopic temperature at only 38% of these sites. In general, below the uppermost ocean layer, the model tends to under-predict the mixed layer isotopic temperature, with the worst match at high latitudes.

The fit between model-predicted and isotopic paleotemperatures is improved for the surface ocean if the model-predicted surface salinities are used to calculate the seawater oxygen isotopic composition. However, in this approach, the choice of one (or more) empirical seawater salinity relationship is subjective. Using the δ_w–salinity equation of Broecker (1989), isotopic paleotemperatures for North Atlantic Sites 548 and 550 fall within approximately $2\,^\circ C$ of the model-predicted temperatures at those sites. However, Sites 94 (Gulf of Mexico, paleolatitude 27° N) and 152 (Central American passageway, paleolatitude 14° N) no longer fall within $2\,^\circ C$. Both are low-latitude Atlantic Ocean sites, suggesting that one Atlantic and Pacific Ocean average δ_w–salinity relationship is inadequate for calculating paleoocean seawater $\delta^{18}O$ values.

When the western equatorial Atlantic Ocean δ_w–salinity equation of Fairbanks *et al.* (1992) is applied for all sites equatorward of 28° (the latitude of Site 94), the fit between isotope and model temperatures is improved for Sites 94 and 152, while paleotemperatures at Sites 364 and 865 remain within or very nearly within $2\,^\circ C$ of the model-predicted temperatures. The overall fit between isotope and model temperatures is improved by use of the δ_w–salinity relationship as compared with the correlation made using a latitude-dependent δ_w value. A similar improved fit results from the use of the eastern equatorial Atlantic Ocean δ_w–salinity equation of Fairbanks *et al.* (1992). These results suggest that the model-simulated salinity is a viable tool for estimating the isotopic composition of surface seawater for past time periods, given the best possible reconstruction of boundary conditions. However, careful consideration will have to be given to the choice of a δ_w–salinity relationship. The approach taken here in the choice of an equation relating salinity and δ_w is highly subjective. A more objective approach would be modification of the models to include an oxygen stable isotope tracer. This approach awaits ocean model improvements.

ACKNOWLEDGEMENTS

The Parallel Ocean Climate Model was developed by Albert Semtner of the Naval Postgraduate School and Robert Chervin of the National Center for Atmospheric Research. The GENESIS Earth systems model was developed by Starley Thompson and Dave Pollard of the Interdisciplinary Climate Systems Section at the National Center for Atmospheric Research. We thank Bill Peterson for invaluable assistance in both ocean and atmosphere simulations. The manuscript was greatly improved by the comments of Mike Arthur, Lee Kump, Ellen Thomas and Paul Valdes. Ferret v. 4.4, developed by the Thermal Modeling and Analysis Project at NOAA/PMEL, was used in data analysis and visualization. This research was made possible by support from NSF grant ATM-9113944 to Eric J. Barron.

REFERENCES

Anderson, T. F. and Arthur, M. A. (1983). Stable isotopes of oxygen and carbon and their application to sedimentologic and environmental problems. In *Stable Isotopes in Sedimentary Geology*, SEPM Short Course Notes No. 10, eds. M. A. Arthur, T. F. Anderson, I. R. Kaplan, J. Veizer and L. S. Land, pp. 1–151. Tulsa, OK: Society of Economic Paleontologists and Mineralogists, p. 1-151.

Arthur, M. A., Hinga, K. R., Pilson, M. E., Whitaker, E. and Allard, D. (1991) Estimates of $p\mathrm{CO}_2$ for the last 120 Ma based on $\delta^{13}\mathrm{C}$ of marine phytoplanktic organic matter (Abstract). *Eos*, **72**, 166.

Aubry, M.-P. (1995). From chronology to stratigraphy: Interpreting the lower and middle Eocene stratigraphic record in the Atlantic Ocean. In *Geochronology, Time Scales and Global Stratigraphic Correlation*, SEPM Special Publication No. 54, eds. W. A. Berggren, D. V. Kent, M.-P. Aubrey and J. Hardenbol, pp. 213–274. Tulsa, OK: Society of Economic Paleontologists and Mineralogists.

Aubry, M.-P., Berggren, W. A. and Lucas, S. (1998). *The Paleocene/Eocene Boundary (IGC Project 308)*. Palisades, New York: Eldigio Press.

Aubry, M.-P., Berggren, W. A., Stott, L. and Sinha, A. (1996). The upper Paleocene–lower Eocene stratigraphic record and the Paleocene–Eocene boundary carbon isotope excursion: Implications for geochronology. In *Correlation of the Early Paleogene in Northwest Europe*, Geological Society Special Publication No. 101, eds. R. W. Know, R. Corfield and R. E. Dunay, pp. 353–80. London: Geological Society.

Barrera, E. and Huber, B. T. (1991). Paleogene and Early Neogene oceanography of the southern Indian Ocean: Leg 119 foraminifer stable isotope results. In *Proceedings of the Ocean Drilling Program: Scientific Results*, vol. 119, eds. J. Barron *et al.*, pp. 693–717. College Station, TX: Ocean Drilling Program.

Barron, E. J. (1986). Physical paleoceanography: A status report. In *Mesozoic and Cenozoic Oceans*, ed. K. J. Hsü, pp. 1–9. Washington, DC: American Geophysical Union.

Barron, E. J. (1987). Eocene equator-to-pole surface ocean temperatures: A significant climate problem? *Paleoceanography*, **2**, 729–39.

Barron, E. J., Hay, W. W. and Thompson, S. (1989). The hydrologic cycle: A major variable during Earth history. *Palaeogeography, Palaeoclimatology, Palaeoecology*, **75**, 157–74.

Barron, E. J. and Peterson, W. H. (1991). The Cenozoic ocean circulation based on ocean General Circulation Model results. *Palaeogeography, Palaeoclimatology, Palaeoecology*, **83**, 1–28.

Barron, E. J., Peterson, W. H., Thompson, S. and Pollard, D. (1993). Past climate and the role of ocean heat transport: Model simulations for the Cretaceous. *Paleoceanography*, **8**, 785–98.

Berggren, W. A. and Aubry, M.-P. (1996). A late Paleocene–early Eocene NW European and North Sea magnetobiochronological correlation network. In *Correlation of the Early Paleogene in Northwest Europe*, Geological Society Special Publication No. 101, eds. R. W. Knox, R. Corfield and R. E. Dunay, pp. 309–52. London: Geological Society.

Berggren, W. A., Kent, D. V., Flynn, J. J. and Van Couvering, J. A. (1985). Cenozoic geochronology. *Geological Society of America Bulletin*, **96**, 1407–18.

Berggren, W. A., Kent, D. V., Swisher, C. C., III and Aubry, M.-P. (1995). Geochronology, time scales and global stratigraphic correlation. In *Geochronology, Time Scales and Global Stratigraphic Correlation*, SEPM Special Publication No. 54, eds. W. A. Berggren, D. V. Kent, M.-P. Aubry and J. Hardenbol, pp. 129–212. Tulsa, OK: Society of Economic Paleontologists and Mineralogists.

Berggren, W. A. and Miller, K. G. (1988). Paleogene tropical planktonic foraminiferal biostratigraphy and magnetobiochronology. *Micropaleontology*, **34**, 362–80.

Berner, R. A. (1994). GEOCARB II: A revised model of atmospheric CO_2 over Phanerozoic time: *American Journal of Science*, **294**, 56–91.

Bice, K. L. (1997). *An Investigation of Early Eocene Deep Water Warmth using Uncoupled Atmosphere and Ocean General Circulation Models: Model Sensitivity to Geography, Initial Temperatures, Atmospheric Forcing and Continental Runoff*. Technical Report No. 97-002 Pennsylvania State University, University Park: Earth System Science Center.

Bice, K. L., Barron, E. J. and Peterson, W. H. (1997). Continental runoff and early Eocene bottom-water sources. *Geology*, **25**, 951–4.

Bice, K. L., Barron, E. J. and Peterson, W. H. (1998). Reconstruction of realistic early Eocene paleobathymetry and ocean GCM sensitivity to specified basin configuration. In *Tectonic Boundary Conditions for Climate Reconstructions*, eds. T. Crowley and K. Burke, pp. 227–47. New York: Oxford University Press.

Boersma, A. and Premoli Silva, I. (1983). Paleocene planktonic foraminiferal biogeography and the paleoceanography of the Atlantic Ocean. *Micropaleontology*, **29**, 355–81.

Boersma, A., Premoli Silva, I. and Shackleton, N. J. (1987). Atlantic Eocene planktonic foraminiferal paleohydrographic indicators and stable isotope paleoceanography. *Paleoceanography*, **2**, 287–331.

Boersma, A. and Shackleton, N. (1977). Tertiary oxygen and carbon isotope stratigraphy, Site 357 (mid Latitude South Atlantic). In *Initial Reports of the Deep Sea Drilling Project*, vol. 39, eds. P. R. Supko *et al.*, pp. 911–24. Washington, DC: US Government Printing Office.

Boersma, A., Shackleton, N., Hall, M. and Given, Q. (1979). Carbon and oxygen isotope records at DSDP Site 384 (North Atlantic) and some Paleocene paleotemperatures and carbon isotope variations in the Atlantic Ocean. In *Initial Reports of the Deep Sea Drilling Project*, vol. 43, eds. B. E. Tucholke, P. R. Vogt *et al.*, pp. 695–717. Washington, DC: US Government Printing Office.

Bralower, T. J., Zachos, J. C., Thomas, E. *et al.* (1995). Late Paleocene to Eocene paleocean-ography of the equatorial Pacific Ocean: Stable Isotopes recorded at Ocean Drilling Program Site 865, Allison Guyot. *Paleoceanography*, **10**, 841–65.

Brass, G. W., Saltzmann, E., Sloan, J. L., II *et al.* (1982*a*). Ocean circulation, plate tectonics, and climate. In *Climate in Earth History*, pp. 83–9. Washington, DC: National Academy Press.

Brass, G. W., Southam, J. R. and Peterson, W. H. (1982*b*). Warm saline bottom water in the ancient ocean. *Nature*, **296**, 620–3.

Broecker, W. S. (1989). The salinity contrast between the Atlantic and Pacific Oceans during glacial time. *Paleoceanography*, **4**, 207–12.

Bryan, K. (1969). A numerical model for the study of the world ocean. *Journal of Computational Physics*, **4**, 347–76.

Bryan, K. and Cox, M. D. (1972). The circulation of the world ocean: A numerical study. Part I, A homogeneous model. *Journal of Physical Oceanography*, **2**, 319–35.

Cande, S. C. and Kent, D. V. (1992). A new geomagnetic polarity time scale for the Late Cretaceous and Cenozoic. *Journal of Geophysical Research*, **97**, 13917–51.

Cande, S. C. and Kent, D. V. (1995). Revised calibration of the geomagnetic polarity time scale for the Late Cretaceous and Cenozoic. *Journal of Geophysical Research*, **100**, 6093–5.

Cerling, T. (1991). Carbon dioxide in the atmosphere: evidence from Cenozoic and Mesozoic paleosols. *American Journal of Science*, **291**, 377–400.

Corfield, R. M. (1994). Palaeocene oceans and climate: An isotopic perspective. *Earth-Science Reviews*, **37**, 225–52.

Covey, C. and Thompson, S. T. (1989). Testing the effects of ocean heat transport on climate: *Palaeogeography, Palaeoclimatology, Palaeoecology*, **75**, 331–41.

Cox, M. D. (1970). A mathematical model of the Indian Ocean. *Deep-Sea Research*, **17**, 47–75.

Cox, M. D. (1975). A baroclinic numerical model of the world ocean: Preliminary results. In *Numerical Models of Ocean Circulation*, pp. 107–20. Washington, DC: National Academy of Sciences.

Cox, M. D. (1984). *A Primitive Equation Three-dimensional Model of the Ocean.* GFDL Ocean Group Technical Report No. 1. Princeton: GFDL/NOAA, Princeton University.

Craig, H. and Gordon, L. I. (1965). Isotopic oceanography: Deuterium and oxygen 18 varia-tions in the ocean and marine atmosphere. In *Marine Geochemistry*, Occasional Publication No. 3, pp. 277–374. RI: University of Rhode Island.

Dansgaard, W. (1964). Stable isotopes in precipitation. *Tellus*, **16**, 436–68.

Dercourt, J. *et al.* (1986). Geologic evolution of the Tethys belt from the Atlantic to the Pamirs since the Lias. In Evolution of the Tethys, eds. J. Aubouin, X. Le Pinchon and A. S. Monin. *Tectonophysics*, **123**, 241–315.

Deuser, W. G. (1987). Seasonal variations in isotopic composition and deep-water fluxes of the tests of perennially abundant planktonic foraminifera of the Sargasso Sea: Results from sediment trap collections and their paleoceanographic significance. *Journal of Foraminiferal Research*, **17**, 14–27.

Deuser, W. G., Ross, E. H., Hemleben, C. and Spindler, M. (1981). Seasonal changes in species composition, numbers, mass, size, and isotopic composition of planktonic foraminifera settling into the deep Sargasso Sea. *Palaeogeography, Palaeo-climatology, Palaeoecology*, **33**, 103–27.

D'Hondt, S. and Arthur, M. A. (1995). Interspecies variation in stable isotopic signals of Maastrichtian planktonic foraminifera. *Paleoceanography*, **10**, 123–35.

D'Hondt, S., Zachos, J. C. and Schultz, G. (1994). Stable isotopic signals and photosymbiosis in late Paleocene planktic foraminifera. *Paleobiology*, **20**, 391–406.

Dickens, G. R., Castillo, M. M. and Walker, J. C. G. (1997). A blast of gas in the latest Paleocene; simulating first-order effects of massive dissociation of oceanic methane hydrate. *Geology*, **25**, 259–62.

Dickens, G. R., O'Neil, J. R., Rea, D. K. and Owen, R. M. (1995). Dissociation of oceanic methane hydrate as a cause of the carbon isotope excursion at the end of the Paleocene. *Paleoceanography*, **10**, 965–71.

Dorman, J. L. and Sellers, P. J. (1989). A global climatology of albedo, roughness length and stomatal resistance for atmospheric general circulation models as represented by the Simple Biosphere Model (SIB). *Journal of Applied Meteorology*, **28**, 833–55.

Douglas, R. G., and Savin, S. M. (1975). Oxygen and carbon isotope analyses of Tertiary and Cretaceous microfossils from Shatsky Rise and other sites in the north Pacific Ocean. In *Initial Reports of the Deep Sea Drilling Project*, vol. 32, eds. R. L. Larson, R. Moberly *et al.*, pp. 509–20. US Government Printing Office: Washington, DC.

Duplessy, J.-C., Labeyrie, L., Juillet-LeClerc, A., Maitre, F., Duprat, J. and Sarnthein, M. (1991). Surface salinity reconstruction of the North Atlantic Ocean during the last glacial maximum. *Oceanologica Acta*, **14**, 311–24.

Emiliani, C. (1954). Depth habitats of some pelagic foraminifera as indicated by oxygen isotope ratios. *American Journal of Science*, **252**, 149–58.

Emiliani, C. (1955). Pleistocene temperatures. *Journal of Geology*, **63**, 538–78.

Emiliani, C. (1966). Paleotemperature analysis of Caribbean cores P6304-8 and P6304-9 and a generalized temperature curve for the past 425,000 years. *Journal of Geology*, **74**, 102–22.

Epstein, S. R., Buchsbaum, R., Lowenstam, H. A. and Urey, H. C. (1951). Carbonate-water isotopic temperature scale. *Geological Society of America Bulletin*, **63**, 417–26.

Epstein, S. R., Buchsbaum, R., Lowenstam, H. A. and Urey, H. C. (1953). Revised carbonate-water isotopic temperature scale. *Geological Society of America Bulletin*, **64**, 1315–26.

Erez, B. and Luz, J. (1983). Experimental paleotemperature equation for planktonic foraminifera. *Geochimica et Cosmochimica Acta*, **47**, 1025–31.

Fairbanks, R. G., Charles, C. D. and Wright, J. D. (1992). Origin of global meltwater pulses. In *Radiocarbon After Four Decades: An Interdisciplinary Perspective*, eds. R. E. Taylor, A. Long, and R. S. Kra, pp. 473–500. New York: Springer-Verlag.

Faure, G. (1986). *Principles of Isotope Geology*, 2nd edn. New York: John Wiley.

Fawcett, P. J. (1994). *Simulation of Climate-sedimentary Evolution: A Comparison of Climate Model Results of the Geologic Record for India and Australia*. Technical Report No. 94-001. Pennsylvania State University, University Park: Earth System Science Center.

Freeman, K. H. and Hayes, J. M. (1992). Fractionation of carbon isotopes by phytoplankton and estimates of ancient CO_2 levels. *Global Biogeochemical Cycles*, **6**, 185–98.

Greenwood, D. R. and Wing. S. L. (1995). Eocene continental climates and latitudinal temperature gradients. *Geology*, **23**, 1044–48.

Grossman, E. L. and Ku, T.-L. (1986). Oxygen and carbon isotope fractionation in biogenic aragonite: Temperature effects. *Chemical Geology*, **59**, 59–74.

Haq, B. U. (1981). Paleogene paleoceanography: Early Cenozoic oceans revisited. In Geology of Oceans, eds. X. Le Pichon, J. Debyser and F. Vine. *Oceanologica Acta*, **4**, 71–82.

Haq, B. U. (1984). Paleoceanography: A synoptic overview of 200 million years of ocean history. In *Marine Geology and Oceanography of Arabian Sea and Coastal Pakistan*, eds. B. U. Haq and J. D. Milliman, pp. 201–31. New York: Van Nostrand Reinhold.

Haq, B. U., Hardenbol, J. and Vail, P. R. (1987). Chronology of fluctuating sea levels since the Triassic. *Science*, **235**, 1156–67.

Holland, H. D. (1984). *The Chemical Evolution of the Atmosphere and Oceans.* Princeton: Princeton University Press.

Horibe, Y. and Oba, T. (1972). Temperature scales of aragonite-water and calcite-water systems. *Fossiles,* **23/24,** 69–74.

Katz, M. E. and Miller, K. G. (1991). Early Paleocene benthic foraminiferal assemblages and stable isotopes in the Southern Ocean. In *Proceedings of the Ocean Drilling Program: Scientific Results,* vol. 114, eds. P. F. Ciesielski, Y. Kristoffersen *et al.,* pp. 481–99. College Station, TX: Ocean Drilling Program.

Keigwin, L. D. and Corliss, B. H. (1986). Stable isotopes in late middle Eocene to Oligocene foraminifera. *Geological Society of America Bulletin,* **97,** 335–45.

Kennett, J. P. and Barker, P. F. (1990). Latest Cretaceous to Cenozoic climate and oceanographic developments in the Weddell Sea, Antarctica: An ocean-drilling perspective. In *Proceedings of the Ocean Drilling Program: Scientific Results,* vol. 113, eds. P. F. Barker, J. P. Kennett *et al.,* pp. 937–60. College Station, TX: Ocean Drilling Program.

Kennett, J. P. and Stott, L. D. (1990). Proteus and Proto-Oceanus: Ancestral Paleogene oceans as revealed from Antarctic stable isotopic results; ODP Leg 113. In *Proceedings of the Ocean Drilling Program: Scientific Results,* vol. 113, eds. P. F. Barker, J. P. Kennett *et al.,* pp. 865–80. College Station, TX: Ocean Drilling Program.

Kennett, J. P. and Stott, L. D. (1991). Abrupt deep-sea warming, paleoceanographic changes and benthic extinctions at the end of the Palaeocene. *Nature,* **353,** 225–9.

Kennett, J. P. and Stott, L. D. (1995). Terminal Paleocene mass extinction in the deep sea: Association with global warming. In *Effects of Past Global Change on Life,* pp. 94–107. Washington, DC: National Academy Press.

Larson, R. L. (1991). Geological consequences of superplumes. *Geology,* **19,** 963–90.

Levitus, S. and Boyer, T. P. (1994). *World Ocean Atlas 1994,* vol. 4, *Temperature, NOAA Atlas NESDIS 4.* Washington, DC: US Government Printing Office.

Lohmann, K. C. and Walker, J. C. G. (1989). The $\delta^{18}O$ record of Phanerozoic abiotic marine calcite cements. *Geophysical Research Letters,* **16,** 319–22.

Markwick, P. J. (1994). "Equability," continentality, and Tertiary "climate": The crocodilian perspective. *Geology,* **22,** 613–6.

Matsumoto, R. (1995). Causes of the $\delta^{13}C$ anomalies of carbonates and a new paradigm 'Gas Hydrate Hypothesis' (Japanese with English abstract and figure captions). *Journal of the Geological Society of Japan,* **11,** 902–24.

McCrea, J. M. (1950). On the isotopic chemistry of carbonates and a paleotemperature scale. *Journal of Chemical Physics,* **18,** 849–57.

Miller, K. G. and Curry, W. B. (1982). Eocene to Oligocene benthic foraminiferal isotopic record in the Bay of Biscay. *Nature,* **296,** 347–50.

Miller, K. G., Curry, W. B. and Ostermann, D. R. (1985). Late Paleogene (Eocene to Oligocene) benthic foraminiferal oceanography of the Goban Spur region, Deep Sea Drilling Project Leg 80. In *Initial Reports of the Deep Sea Drilling Project,* vol. 80, eds. P. C. Graciansky, C. W. Poag *et al.,* pp. 505–31. Washington, DC: US Government Printing Office.

Miller, K. G., Fairbanks, R. G. and Mountain, G. S. (1987a). Tertiary oxygen isotope synthesis, sea level history, and continental margin erosion. *Paleoceanography,* **2,** 1-19.

Miller, K. G., Janecek, T. R., Katz, M. E. and Keil, D. J. (1987b). Abyssal circulation and benthic foraminiferal changes near the Paleocene/Eocene boundary. *Paleoceanography,* **2,** 741–61.

Oberhänsli, H. (1992). The influence of the Tethys on the bottom waters of the early Tertiary ocean. In *The Antarctic Paleoenvironment: A Perspective on Global Change: Antarctic*

Research Series, vol. 56, eds. J. P. Kennett and D. A. Warnke, pp. 167–84. Washington, DC: American Geophysical Union.

Oberhänsli, H. and Hsü, K. J. (1986). Paleocene–Eocene paleoceanography. In *Mesozoic and Cenozoic Oceans*, ed. K. J. Hsü, pp. 85–100. Washington, DC: American Geophysical Union.

Oberhänsli, H., Müller-Merz, E. and Oberhänsli, R. (1991). Eocene paleoceanographic evolution at 20°–30°S in the Atlantic Ocean. *Palaeogeography, Palaeoclimatology, Palaeoecology*, **83**, 173–215.

Olivarez, A. M. and Owen, R. M. (1989). Plate tectonic reorganizations: Implications regarding the origin of hydrothermal ore deposits. *Marine Mining*, **8**, 123–38.

O'Neil, J. R., Clayton, R. N. and Mayeda, T. K. (1969). Oxygen isotope fractionation in divalent metal carbonates. *Journal of Chemical Physics*, **51**, 5547–58.

Pak, D. K. and Miller, K. G. (1992). Paleocene to Eocene benthic foraminiferal isotopes and assemblages: Implications for deepwater circulation. *Paleoceanography*, **7**, 405–22.

Parsons, B. and Sclater, J. G. (1977). An analysis of the variation of ocean floor bathymetry and heat flow with age. *Journal of Geophysical Research*, **82**, 803–27.

Poag, C. W., Reynolds, L. A., Mazzullo, J. M. and Keigwin, L. D. (1985). Foraminiferal, lithic, and isotopic changes across four major unconformities at Deep Sea Drilling Project Site 548, Goban Spur. In *Initial Reports of the Deep Sea Drilling Project*, vol. 80, eds. P. C. Graciansky, C. W. Poag *et al.*, pp. 539–55. Washington, DC: US Government Printing Office.

Popp, B. N., Anderson, T. F. and Sandberg, P. A. (1986). Brachiopods as indicators of original isotopic compositions in some Paleozoic limestones. *Geological Society of America Bulletin*, **97**, 1262–9.

Rea, D. K., Leinen, M. and Janecek, T. R. (1985). Geological approach to the long-term history of atmospheric circulation. *Science*, **227**, 721–5.

Rind, D. and Chandler, M. (1991). Increased ocean heat transports and warmer climate. *Journal of Geophysical Research*, **96**, 7437–61.

Ritchie, J. D. and Hitchen, K. (1996). Early Paleogene offshore igneous activity to the northwest of the UK and its relationship to the North Atlantic igneous province. In *Correlation of the Early Paleogene in Northwest Europe*, Geological Society Special Publication No. 101, eds. R. W. Knox, R. Corfield and R. E. Dunay, pp. 63–78. London: Geological Society.

Rozanski, K., Araguás-Araguás, L. and Gonfiantini, R. (1993). Isotopic patterns in modern global precipitation. In *Climate Change in Continental Isotopic Records*, Geophysical Monograph No. 78, eds. P. K. Swart, K. C. Lohmann *et al.*, pp. 1–36. Washington, DC: American Geophysical Union.

Schneider, S. H., Thompson, S. L. and Barron, E. J. (1985). Mid-Cretaceous continental surface temperatures: Are high CO_2 concentrations needed to simulate above-freezing winter conditions? In *The Carbon Cycle and Atmospheric CO_2: Natural Variations Archean to Present*, American Geophysical Union Monograph No. 32, eds. E. T. Sundquist and W. S. Broecker, pp. 554–9. Washington, DC: American Geophysical Union.

Scotese, C. R. and Golonka, J. (1992). *PALEOMAP Paleogeographic Atlas, PALEOMAP Progress Report No. 20*. Arlington, TX: Department of Geology, University of Texas.

Semtner, A. J., Jr. (1974). *An Oceanic General Circulation Model with Bottom Topography*. Technical Report No. 9. Los Angeles, CA: Department of Meteorology, University of California.

Semtner, A. J., Jr. and Chervin, R. M. (1992). General circulation from a global eddy-resolving model. *Journal of Geophysical Research*, **97**, 5493–550.

Shackleton, N. J. (1974). Attainment of isotopic equilibrium between ocean water and the benthonic foraminifera genus *Uvigerina*: Isotopic changes in the ocean during the last glacial. *Colloques Internationaux du CNRS*, **219**, 203–9.

Shackleton, N. J. and Boersma, A. (1981). The climate of the Eocene ocean. *Geological Society of London Journal*, **138**, 153–7.

Shackleton, N. J., Hall, M. A. and Boersma, A. (1984). Oxygen and carbon isotope data from Leg 74 foraminifers. In *Initial Reports of the Deep Sea Drilling Project*, vol. 74, eds. T. C. Moore, Jr., P. D. Rabinowitz *et al.*, pp. 599–612. Washington, DC: US Government Printing Office.

Shackleton, N. J. and Kennett, J. P. (1975*a*). Paleotemperature history of the Cenozoic and the initiation of Antarctic glaciation: Oxygen and carbon isotope analyses in DSDP Sites 277, 279, and 281. In *Initial Reports of the Deep Sea Drilling Project*, vol. 29, eds. J. P. Kennett, R. E. Houtz *et al.*, pp. 743–55. Washington, DC: US Government Printing Office.

Shackleton, N. J. and Kennett, J. P. (1975*b*). Late Cenozoic oxygen and carbon isotopic changes at DSDP Site 284: Implications for glacial history of the northern hemisphere and Antarctica. In *Initial Reports of the Deep Sea Drilling Project*, vol. 29, eds. J. P. Kennett, R. E. Hout *et al.*, pp. 801–7. Washington, DC: US Government Printing Office.

Sinha, A. and Stott, L. D. (1994). New atmospheric pCO_2 estimates from paleosols during the late Paleocene/early Eocene global warming interval. *Global and Planetary Change*, **9**, 297–307.

Sloan, L. C. (1994). Equable climates during the early Eocene: Significance of regional paleogeography for North American climate. *Geology*, **22**, 881–4.

Sloan, L. C. and Rea, D. K. (1995). Atmospheric carbon dioxide and early Eocene climate: A general circulation modeling sensitivity study. *Palaeogeography, Palaeoclimatology, Palaeoecology*, **119**, 275–92.

Sloan, L. C., Walker, J. C. G. and Moore, T. C. (1995). Possible role of ocean heat transport in early Eocene climate. *Paleoceanography*, **10**, 347–56.

Sloan, L. C., Walker, J. C. G., Moore, T. C., Rea, D. K. and Zachos, J. C. (1992). Possible methane-induced polar warming in the early Eocene. *Nature*, **357**, 320–2.

Speijer, R. P., van der Zwaan, G. J. and Schmitz, B. (1996). The impact of Paleocene/Eocene boundary events on middle neritic benthic foraminiferal assemblages from Egypt: *Marine Micropaleontology*, **28**, 99–132.

Spero, H. J., Lerche, I. and Williams, D. F. (1991). Opening the carbon isotope "vital effect" black box. 2. Quantitative model for interpreting foraminiferal carbon isotope data. *Paleoceanography*, **6**, 639–55.

Stott, L. D., Kennett, J. P., Shackleton, N. J. and Corfield, R. M. (1990). The evolution of Antarctic surface waters during the Paleogene: Inferences from the stable isotopic composition of planktonic foraminifers, ODP Leg 113. In *Proceedings of the Ocean Drilling Program: Scientific Results*, vol. 113, eds. P. F. Barker, J. P. Kennett *et al.*, pp. 849–63. College Station, TX: Ocean Drilling Program.

Thomas, E. (1990). Late Cretaceous through Neogene benthic foraminifers (Maud Rise, Weddell Sea, Antarctica). In *Proceedings of the Ocean Drilling Program: Scientific Results*, vol. 113, eds. P. F. Barker, J. P. Kennett *et al.*, pp. 571–94. College Station, TX: Ocean Drilling Program.

Thomas, E. and Shackleton, N. J. (1996). The Paleocene–Eocene benthic foraminiferal extinction and stable isotope anomalies. In *Correlation of the Early Paleogene in Northwest Europe*, Geological Society Special Publication No. 101, eds. R. W. Knox, R. Corfield and R. E. Dunay, pp. 401–41. London: Geological Society.

Thompson, S. L. and Pollard, D. (1995). A global climate model (GENESIS) with a land-surface transfer scheme (LSX), Part I: Present climate simulation. *Journal of Climate*, **8**, 732–61.

Urey, H. C. (1947). The thermodynamic properties of isotopic substances. *Journal of the Chemical Society*, **1947**, 562–81.

Veizer, J., Fritz, P. and Jones, B. (1986). Geochemistry of brachiopods: oxygen and carbon isotopic records of Paleozoic oceans. *Geochimica et Cosmochimica Acta*, **50**, 1679–96.

Wing, S. L. and Greenwood, D. R. (1993). Fossils and fossil climate: The case for equable continental interiors in the Eocene. *Philosophical Transactions of the Royal Society of London*, **B341**, 243–52.

Zachos, J. C., Lohmann, K. C., Walker, J. C. G. and Wise, S. W., Jr. (1993). Abrupt climate change and transient climates during the Paleogene: A marine perspective. *Journal of Geology*, **101**, 191–213.

Zachos, J. C., Rea, D. K., Seto, K., Nomura, R. and Niitsuma, N. (1992). Paleogene and Early Neogene deep water paleoceanography of the Indian Ocean as determined from benthic foraminifer stable carbon and oxygen isotope records. In *Synthesis of Results of Scientific Drilling in the Indian Ocean*, Geophysical Monograph No. 70, eds. R. A. Duncan, D. K. Rea *et al.*, pp. 351–85. Washington, DC: American Geophysical Union.

Zachos, J. C., Stott, L. D. and Lohmann, K. C. (1994). Evolution of early Cenozoic marine temperatures. *Paleoceanography*, **9**, 353–87.

Zahn, R. and Mix, A. C. (1991). Benthic foraminiferal $\delta^{18}O$ in the ocean's temperature–salinity–density field: Constraints on ice-age thermohaline circulation. *Paleoceanography*, **6**, 1–20.

5

Deep-sea environments on a warm earth: latest Paleocene–early Eocene

ELLEN THOMAS, JAMES C. ZACHOS, AND TIMOTHY J. BRALOWER

ABSTRACT

Latest Paleocene–early Eocene high-latitude surface and global deep-ocean waters were warmer than those of today by up to 15 °C; planktonic foraminiferal and nannofossil assemblages suggest that primary oceanic productivity was low. Low oceanic productivity is also indicated by geochemical evidence that the supply of nutrients to the oceans may have been low. Climate modeling suggests that oceanic and atmospheric circulation may have been sluggish at low temperature gradients, leading to low rates of upwelling of nutrients. Benthic foraminiferal data, by contrast, suggest that the food supply to the deep sea floor in open-ocean settings was larger than that in Recent oceans, in agreement with the speculation that a larger fraction of organic carbon was buried. The benthic foraminiferal evidence might be explained by more efficient food transfer to the bottom in poorly oxygenated, warm deep waters. Possibly the pelagic microbial loop was more active at the higher temperatures, leading to enhanced zooplankton productivity and thus enhanced food supply. Or possibly the benthic faunas do not indicate a high average food supply, but a more continuous and less seasonally pulsed supply than that today. Environmental interpretation of early Eocene benthic foraminiferal faunas is difficult not only because they differ substantially from Recent ones but also because the faunas had been decimated by a massive extinction during an episode of rapid warming, the Late Paleocene Thermal Maximum (LPTM), with a duration of between 25 and 200 000 ka. During the LPTM carbon isotope values of the atmospheric and oceanic carbon reservoir decreased by 2–3‰, a sign of major upset in the global carbon cycle. The carbon isotope excursion could be explained by dissociation of methane hydrates as a consequence of warming of deep water masses, which occurred when dominant formation of deep–intermediate waters shifted from high to low latitudes. Methane dissociation in combination with changes in ocean circulation offers a possible mechanism for climatic instability in the absence of polar ice caps. We lack the high-resolution, stratigraphically complete biostratigraphical and isotope data sets necessary to evaluate whether the early Eocene climate was unstable, but high average temperatures could reflect a warm background climate with superimposed 'hyperthermals': intervals of extremely high temperatures and very low latitudinal sea surface temperature gradients, during which the deep–intermediate oceans were dominated by waters derived from subtropical latitudes.

INTRODUCTION

A time interval of about 5 million years in the latest Paleocene through early Eocene was the warmest part of the Cenozoic, and deep-ocean waters reached temperatures of up to 15 °C (see reviews by Zachos et al., 1993; Seto, 1995; Oberhänsli, 1997; Fig. 5.1). Latitudinal temperature gradients in surface waters were low (e.g., Zachos et al., 1994), warm-water pelagic marine organisms extended their geographic range into polar latitudes, and their associations indicate low productivity (see reviews by Aubry, 1992, 1998b). Thermophilic vertebrates occurred in the Arctic (e.g., Markwick, 1994). Vegetation and soil-types suggest warm high latitudes in both hemispheres (e.g., Askin, 1992; Basinger et al., 1994; Wolfe, 1994), and clay mineral associations in oceanic sediments indicate high humidity and intense chemical weathering in the Antarctic (Robert and Kennett, 1992), on the New Jersey margin (Gibson et al., 1993), in the North Sea region (Knox, 1996), and in New

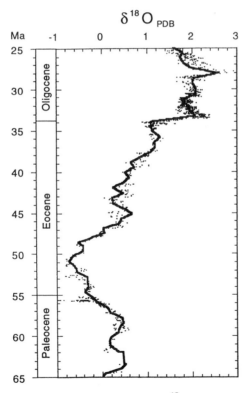

Figure 5.1. Smoothed compilation of deep-sea $\delta^{18}O$ data, modified after Zachos et al. (1993); numerical ages after Berggren et al. (1995). Note small dots indicating single data points.

Zealand (Kaiho *et al*., 1996). The land regions around eastern Tethys were very warm and arid (Oberhänsli, 1992). It has been generally assumed that polar ice sheets were either small or absent (e.g., Zachos *et al*., 1993), although there is some evidence that they existed for at least some time intervals (e.g., Leckie *et al*., 1995).

The unusual warmth of this period has been commonly explained by high atmospheric CO_2 levels (e.g., Freeman and Hayes, 1992). Possible causes of these higher levels of atmospheric CO_2 include massive volcanism in the North Atlantic Igneous Province during the initial opening of the North Atlantic, decarbonation of limestone or oxidation of organic-rich sediments resulting from the India–Asia continental collision, and high hydrothermal activity along mid-oceanic ridges (see review by Thomas and Shackleton, 1996). Climate modeling indicates, however, that at high CO_2 levels tropical temperatures would be much higher than deduced from paleoceanographic data. Mechanisms for highly increased heat transport from low to high latitudes at the low latitudinal temperature gradients remain unexplained (Barron and Moore, 1994; Sloan and Rea, 1995; Sloan *et al*., 1995).

The earth's climate during warm periods has been said to be more stable than that during cold periods such as the Plio-Pleistocene. Climate fluctuations on Milankovitch and sub-Milankovitch timescales during cold periods could have been amplified through feedback loops involving climatically sensitive factors (e.g., sea ice) which are absent in a 'greenhouse' world (e.g., Rind and Chandler, 1991). Rapid climate change, however, did occur when polar ice caps were small or absent during the Late Paleocene Thermal Maximum (LPTM; Zachos *et al*., 1993), when intermediate to deep oceans globally warmed by 4–6 °C over less than a few thousand years (Kennett and Stott, 1991; Pak and Miller, 1992; Thomas and Shackleton, 1996), while mid-latitude to tropical surface temperatures did not change much (Stott, 1992; Bralower *et al*., 1995a,b; Lu and Keller, 1995b; Pardo *et al*., 1996; Bralower *et al*., 1997). Estimates of the duration of the LPTM range from 25 ka (Cramer *et al*., 1997; Norris, 1997b) to 200 000 ka (Kennett and Stott, 1991; Lu *et al*., 1996), but the transition into the LPTM is generally agreed to have occurred over less than 10 ka (e.g., Kennett and Stott, 1991; Lu *et al*., 1996; Thomas and Shackleton, 1996).

During the LPTM carbon isotope values in the oceans and the atmosphere decreased globally by about 2–3‰ (Kennett and Stott, 1991; Koch *et al*., 1992, 1995; Pak and Miller, 1992; Lu and Keller, 1993, 1995a,b; Canudo *et al*., 1995; Aubry *et al*., 1996; Kaiho *et al*., 1996; Schmitz *et al*., 1996; Stott *et al*., 1996; Thomas and Shackleton, 1996; Bralower *et al*., 1997). This rapid decrease is superimposed on a long-term decrease that started in the middle Paleocene (see review by Corfield, 1995). Mass-balance equations show that this carbon isotope excursion was so large that it could not have been caused by transfer of terrestrial biomass into the ocean–atmosphere system or by eruption of volcanogenic CO_2; it was so rapid that it probably could not have been caused by a change in deposition or erosion rates of carbon in carbonate as compared with carbon in organic matter (Thomas and Shackleton, 1996). Such large, rapid excursions in the isotopic composition of the global carbon reservoirs require causes that are not included in the commonly used models of the carbon cycle at various timescales (e.g., Walker and Kasting, 1992; Berner, 1995): the input rates of isotopically light carbon during the transition into the LPTM excursion are similar to rates of anthropogenic fossil fuel burning. Such an excursion could have been caused by massive dissociation of isotopically light methane hydrates in oceanic sediments as a result of the deep-ocean warming (Dickens *et al.*, 1995, 1997; Matsumoto, 1995; Kaiho *et al.*, 1996).

This paper combines a review of the rapidly growing information on the LPTM and early Eocene warm period with new data. We will review latest Paleocene–early Eocene deep-ocean habitats as compared with those in the present deep oceans, discuss the nature and duration of the environmental changes during the LPTM, and consider the possibility that global climate during warm periods was highly unstable.

METHODS

Faunal data are presented from Ocean Drilling Program (ODP) Site 865 and isotope data from ODP Site 690 and Deep Sea Drilling Project (DSDP) Site 215 (Table 5.1); see Fig. 5.2 for location of sites. Samples for benthic foraminiferal faunal analysis were dried overnight at 50 °C and weighed, then soaked overnight in distilled water. Most samples disaggregated readily and could be washed over a 63-μm screen. Benthic foraminifera for faunal analysis were picked from the >63-μm size fraction, following Thomas (1990*a,b*). All specimens were picked and mounted in cardboard slides. All samples contained sufficient specimens (>250) for analysis. Taxonomy is as in Thomas (1990*a*) and Thomas and Shackleton (1996), and largely follows Van Morkhoven *et al.* (1986). In order to obtain a measure of diversity independent of the number of specimens counted, we calculated the number of species that would be present if only 100 specimens had been counted. We used the rarefaction method developed for Recent metazoan deep-sea faunas, which typically have high species diversity and many rare species, like deep-sea benthic foraminiferal faunas (Sanders, 1968).

Isotope measurements of benthic foraminifera were performed at the University of California at Santa Cruz. Samples were reacted in a common phosphoric acid bath at 90 °C. Average precision as determined from replicate analyses of the laboratory standards NBS-19 and Carrera marble was better than ±0.10‰ for both $\delta^{18}O$ and $\delta^{13}C$.

WARM DEEP-OCEAN ENVIRONMENTS

Discussions of links between climate and oceanic ecology and biota have characterized 'greenhouse' periods as having low average oceanic productivity coupled with high species diversity (Fischer and Arthur, 1977; Brasier, 1995*a,b*; Norris, 1997*a*). Low nutrient supply and slow oceanic circulation have been most widely discussed as causal factors for low productivity, but no explanation has been widely accepted. Nutrient supply from land to the oceans may have been low as a result of the low topographic relief, combined with high sea levels and thus relatively small continental area. Together these factors could have caused low weathering and erosion rates as deduced from the strontium isotopic record of marine carbonates

Table 5.1. *Isotope data for benthic foraminiferal taxa at Sites 690, 689, and 215*

Sample	Taxon	$\delta^{13}C$	$\delta^{18}O$
690B 19-3, 51-53	*B. ovula*	−1.55	−1.17
690B 19-3, 51-53	*N. truempyi*	1.29	−0.02
690B 19-3, 60-62	*B. ovula*	−1.10	−0.89
690B 19-3, 66-68	*B. ovula*	−1.05	−0.83
690B 19-3, 72-74	*N. truempyi*	1.57	−0.05
690B 19-3, 72-74	*B. ovula*	−1.15	−0.98
689B 23-1, 80-82	*B. thanetensis*	0.80	0.32
689B 23-1, 87-89	*Lenticulina* sp.	−1.10	−0.42
689B 23-1, 94-96	*B. thanetensis*	0.96	0.43
689B 23-1, 94-96	*N. truempyi*	1.30	0.17
689B 23-1, 94-96	*N. truempyi*	1.36	0.22
689B 23-1, 94-96	*Lenticulina* sp.	−0.84	−0.21
689B 23-1, 104-106	*B. thanetensis*	0.52	0.40
689B 23-1, 106-108	*Lenticulina* sp.	−0.84	−0.18
689B 23-1, 104-106	*N. truempyi*	1.24	0.27
215 11-6, 32-34	*N. truempyi*	0.98	−0.31
215 11-6, 32-34	*N. truempyi*	0.92	−0.27
215 11-6, 129-131	*N. truempyi*	0.74	−0.45
215 11-6, 129-131	*N. truempyi*	0.63	−0.53
215 11-7, 79-81	*N. truempyi*	0.97	−0.19
215 11-CC, 0-2	*N. truempyi*	0.77	−0.42
215 12-1, 10-12	*N. truempyi*	1.15	0.06
215 12-1, 12-14	*N. truempyi*	1.21	0.02
215 12-1, 38-40	*N. truempyi*	1.24	−0.09
215 12-1, 70-72	*N. truempyi*	1.32	−0.06
215 12-1, 94-96	*N. truempyi*	1.41	−0.01
215 12-1, 125-127	*N. truempyi*	1.27	−0.05
215 12-2, 6-8	*N. truempyi*	1.16	−0.07

(e.g., François and Walker, 1992) and the osmium isotope record from deep-sea clays (Turekian and Pegram, 1997). By contrast, higher temperatures, possibly higher atmospheric $p\mathrm{CO_2}$, higher humidity at high latitudes, and lack of ice cover on the Antarctic continent may have worked in an opposite direction, causing higher rates of weathering during 'greenhouse' periods (see review in Berner, 1995). In addition, the question of ocean nutrient supply involves more than just continental weathering rates. For instance, less oxygen dissolves in the warmer ocean waters, possibly leading to more efficient recycling of phosphorus in the oceans, and hence to higher productivity on long (10^6 years) timescales (Van Cappellen and Ingall, 1994). We therefore do not know whether the Paleogene supply of nutrients to the oceans was higher or lower than that today, or whether various differences with the present world canceled each other (see papers in Ruddiman, 1997).

Slow oceanic turnover rates caused by the low latitudinal thermal gradients have also been cited as a cause of the low average oceanic productivity during

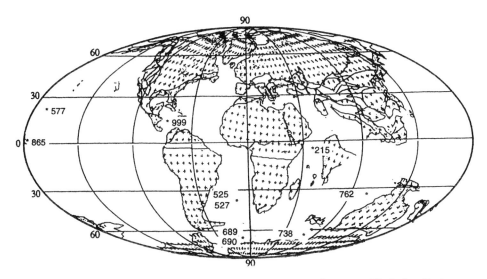

Figure 5.2. Location of drill sites mentioned in the text; figure modified after Zachos *et al.* (1994).

'greenhouse' periods, because such low rates of turnover would cause slow recycling of nutrients into the surface waters by upwelling (e.g., Fischer and Arthur, 1977). A possible corollary of sluggish ocean circulation might have been that large amounts of nutrients were sequestered in the deep oceans, resulting in low average productivity, but high productivity locally where these nutrient-rich waters welled up (Hallock, 1987; Hallock *et al.*, 1991).

It is not clear, however, whether oceanic circulation was indeed sluggish during greenhouse periods. During warm periods in the Cretaceous, sediments suggesting low oxygen conditions in the oceans ('black shales') were commonly deposited, and this deposition has been interpreted as resulting from sluggish circulation. But it has also been suggested that the low oxygenation resulted from high productivity during vigorous ocean overturn (see reviews by Hay, 1995; Parrish, 1995).

INTERPRETATION OF BENTHIC FORAMINIFERAL FAUNAS

Recent faunas

Benthic foraminiferal faunas reflect combined effects of deep-water physicochemical parameters (dominantly oxygen content) and the flux of organic matter to the bottom. There is considerable controversy over which of these factors dominantly determines faunal composition (see reviews by Gooday, 1994; Schnitker, 1994; Murray, 1995; and modeling by Jorissen *et al.*, 1995). One of the reasons for this uncertainty is the fact that in the present oceans high productivity in the surface waters is usually the cause of low oxygenation of bottom waters so that it is difficult to deconvolve the influence of these two factors.

Presently regions of higher productivity are mostly along the continental margins (e.g., Berger, 1989). Species using the abundant food supply below oxygen minimum zones must be able to survive in relatively low oxygen conditions. They commonly migrate through the sediment vertically, following seasonally fluctuating oxygen gradients which result from seasonal fluctuations in productivity, and 'prefer' to stay away from extreme dysoxia to anoxia (Jorissen *et al.*, 1995; Kitazato and Ohga, 1995; Bernhard and Alve, 1996), even though some recent species have been documented to be facultative anaerobes (Bernhard and Reimers, 1991; Bernhard, 1993; Alve and Bernhard, 1995). Lack of oxygen could become a limiting factor to the faunas, or at least strongly influence faunal composition (e.g., Bernhard, 1992, 1996; Sen Gupta and Machain-Castillo, 1993).

The oligotrophic open-ocean deep-sea environment (away from hydrothermal vent systems) is in stark contrast to high-productivity regions, and populations of metazoan deep-sea organisms are strongly limited by the food supply (Gooday *et al.*, 1992). Under such conditions we expect that benthic foraminiferal faunas would be dominantly controlled by the food supply (e.g., Jorissen *et al.*, 1995), and we see clear evidence of benthic–pelagic coupling at many locations (e.g., Corliss and Chen, 1988; Gooday, 1993; Mackensen *et al.*, 1993; Rathburn and Corliss, 1994; Smart *et al.*, 1994; Thomas *et al.*, 1995). Some authors suggest that both present faunas (Kaiho, 1994*a*) and faunas throughout the last 100 million years of earth history have been strongly influenced by oxygen levels ($> 44 \mu M\ O_2$, or $>1\ ml\ l^{-1}\ O_2$) even in such overall oxygenated oceans (Thomas, 1989, 1990*a*; Kaiho, 1991, 1994*b*; Loubere, 1994), while others doubt this possibility (e.g., Rathburn and Corliss, 1994; Thomas *et al.*, 1995; Thomas, 1998).

Recent open-ocean calcareous benthic faunas living below shelf depth can be grouped, with much simplification, into three assemblages (Gooday, 1993): (1) relatively eutrophic faunas with high percentages of infaunally dwelling taxa (*Bolivina, Bulimina, Pullenia, Cassidulina, Melonis*), many of which belong to the superfamily Bolivinacea, and which are common in the bathyal reaches of continental margins; (2) faunas in open-ocean, oligotrophic regions with a seasonal high food supply (common *Epistominella exigua, Alabaminella weddellensis*; also called 'phytodetritus species'); these species react opportunistically to the pulsed food influx; (3) faunas in open-ocean, oligotrophic, $CaCO_3$-corrosive waters including abundant *Nuttallides umbonifera* (Bremer and Lohmann, 1982; Gooday, 1993; Loubere, 1994). The occurrence of these three assemblages is depth dependent, because less food arrives at greater depths: food supply at the ocean floor is a function of primary productivity, preservation, and depth (Herguera and Berger, 1991). Large agglutinated foraminifera in the abyssal oceans show a similar grouping, with Komokiacea in the most oligotrophic central oceanic regions, and astrorhizaceans and hippocrepinaceans in more eutrophic continental margin regions (Gooday, personal communication, 1998). This simplistic subdivision of calcareous benthic faunas into three large assemblages does not reflect the great variety and variability of living deep-sea benthic foraminiferal faunas. *Cibicidoides wuellerstorfi*, for instance, appears to be capable of suspension feeding, sometimes living on objects sticking out above the sediment–water interface. Common occurrence of this and similar taxa has been linked to active bottom currents, which can supply additional food (Lutze and Thiel, 1989; Linke and Lutze, 1993; Schnitker, 1994).

Late Paleocene–early Eocene faunas

We cannot interpret late Paleocene–early Eocene deep-sea faunas using these observations on Recent faunas, because two of the three groups of deep-sea calcareous benthic foraminifera have been common only since the latest Eocene (Fig. 5.3), when the psychrosphere (cold deep-ocean environment) was established (e.g., Benson, 1975). In addition, common Recent species such as *C. wuellerstorfi* and miliolids first occurred in the deep sea in the middle Miocene (Thomas and Vincent, 1987). The Paleogene counterpart of the three Recent groups listed above would read: (1) faunas with common taxa belonging to the superfamily Bolivinacea (e.g., *Brizalina, Bulimina, Coryphostoma, Tappanina, Bolivinoides, Uvigerina*), resembling the Recent faunas along continental margins; (2) faunas dominated by cylindrical taxa (e.g., *Pleurostomella* spp., *Stilostomella* spp., uniserial lagenids); (3) faunas dominated by *Nuttallides truempyi*. There is no early Paleogene counterpart of the faunas dominated by opportunistic, 'phytodetritus' species (Fig. 5.3).

In analogy with Recent faunas, we speculatively interpret faunas with common Bolivinacea as indicative of a continuous, fairly high food supply. Such faunas occurred in the Paleogene at deep open-ocean locations in the equatorial Pacific and Southern oceans, where they presently do not occur. This observation suggests that in the early Paleogene generally more food reached the ocean floor than after the establishment of the psychrosphere (Fig. 5.3). While we cannot be certain that early Paleogene taxa had the same trophic preferences and requirements as their Recent relatives, the early Paleogene Bolivinacea species were generally large, and had many-chambered, decorated tests, suggesting that they were fairly long-lived and non-opportunistic, requiring a fairly high food supply.

Faunas dominated by cylindrical taxa (e.g., *Pleurostomella* spp., *Stilostomella* spp., uniserial lagenids) have no modern analog (Fig. 5.3). These taxa declined strongly in abundance during the late Eocene/early Oligocene cooling, and further during the Miocene cooling (Thomas, 1987, 1992; Thomas and Vincent, 1987). *Stilostomella* spp. and possibly *Pleurostomella* spp. became extinct during the Pliocene in the Atlantic and Indian oceans, and their demise has been interpreted as resulting from increased ventilation of the oceans (Weinholz and Lutze, 1989; Gupta, 1993). Kaiho (1994b) tentatively classified such species as low-oxygen taxa. We cannot interpret these faunas with confidence, but we suggest that they might indicate a richer food source than what is common today on the open-ocean floor. This suggestion is supported by observations that lagenid taxa are presently much more common at shelf to upper bathyal depths (where the food supply is overall greater) than in deeper waters.

We do not know the environmental preferences of *N. truempyi*, which became extinct in the late–middle Eocene. We are not even certain about its Recent descendant species, *N. umbonifera*, which has been said to be an indicator of Antarctic Bottom Water in the present oceans, but has also been linked to corrosivity of bottom waters (Bremer and Lohmann, 1982), or to extreme oligotrophy

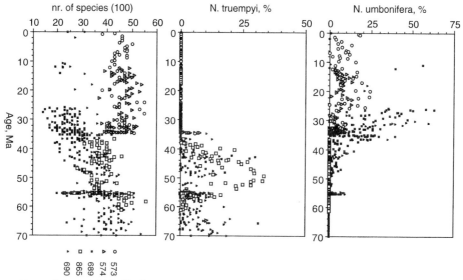

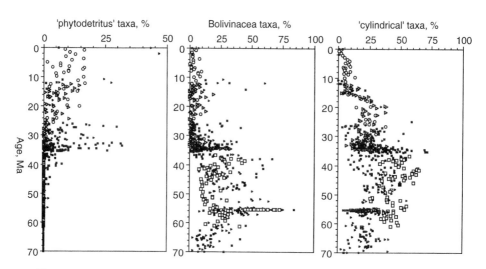

Figure 5.3. Benthic foraminiferal faunal data plotted vs. the timescales of Berggren *et al.* (1995). Sites 573 and 574: eastern equatorial Pacific (Thomas, 1985); Site 865: equatorial Pacific (Thomas, unpublished data); Sites 689 and 690: Maud Rise, Weddell Sea (Thomas, 1990*a*).

(Gooday, 1993; Loubere, 1994). We argue that *N. truempyi* may have been an oligotrophic species, as supported by its common occurrence at the deepest sites (Van Morkhoven *et al.*, 1986; Müller-Merz and Oberhänsli, 1991; Thomas, 1998). This species shows its highest Cenozoic abundance in the middle Eocene at many locations in various oceans (Fig. 5.3; Miller *et al.*, 1992; Oberhänsli, 1997), and became extinct during the period of gradual cooling of deep waters in the middle–late Eocene. We speculate that both *N. truempyi* and *N. umbonifera* indicate relatively oligotrophic conditions, but that *N. truempyi* could not survive in the more corrosive waters that filled the deep oceans from the late Eocene onwards.

We summarize tentatively that *N. truempyi* faunas may have been similar in food requirements to the present *N. umbonifera* faunas, and reflect the most oligotrophic conditions in open ocean (in the depth range where calcareous species occur). The cylindrical taxa might have had somewhat higher food preferences, and the species belonging to the Bolivinacea the highest. Both the Bolivinacea group and the cylindrical-species group were more common at southern high latitudes and in the tropical equatorial Pacific before the establishment of the psychrosphere, which suggests that overall more food reached the sea floor before the establishment of the psychrosphere than afterwards. This is amazing because there is strong evidence that oceanic productivity increased at the establishment of the Antarctic ice sheets (see reviews in Brasier, 1995*a,b*; Diester-Haass and Zahn, 1996). In addition, metabolic rates of foraminifera increase with increasing temperatures; therefore, an assemblage at higher temperatures should be more oligotrophic in character than an assemblage with the same food supply at lower temperatures (Hallock *et al.*, 1991).

We can suggest several reasons why lower surface water productivity in the early Paleogene resulted in increased food supply to the benthic faunas. The most obvious explanation is that lower overall oxygenation of the oceans resulting from the higher temperatures caused better preservation of organic matter. In the present oceans only about 1% of organic productivity reaches the sea floor (Jahnke, 1996), and relatively small changes in preservation could thus have a major effect on the total flux of organic matter to the bottom. There is evidence that preservation is linked to oxygenation through indirect links involving bacterial action (e.g., Kristensen *et al.*, 1995; Stigebrant and Djurfeldt, 1996). Better preservation of organic matter in the early Paleogene agrees with suggestions that the fraction of organic matter preserved in the sediments was larger in the early Paleogene (e.g., Kump and Arthur, 1997). There is, however, no clear correlation between oxygenation and organic matter preservation in the present-day oceans (e.g., Hedges and Keil, 1995), and delivery of food to the deep-sea floor in present environments with low productivity and warm waters (thus low oxygenation – Red Sea) is very low (Thiel *et al.*, 1987).

Bacterial action might have played a role in delivering more food to foraminiferal faunas at higher oceanic temperatures. At elevated temperatures the overall metabolic activity of the deep-sea bacteria would have been higher: even psychrophilic bacteria exhibit growth optima at temperatures between 8 and 16 °C (Jannasch, 1994). Higher rates of activity of the bacteria could have resulted in conversion of more dissolved organic carbon into particulate organic carbon

(Deming and Yager, 1992; Jannasch, 1994). Output from the 'bacterial loop' in the pelagic system would thus have resulted in an enhanced food supply for zooplankton, which in turn could result in more food for benthic foraminifera (e.g., through enhanced delivery of fecal pellets).

Possibly not just the quantity of food but also its quality and fluctuations in supply were important. Foraminifera consume their food in many different ways and from many different sources (e.g., Gooday *et al.*, 1992; Kitazato and Ohga, 1995), and species vary in their food preferences. The present 'phytodetritus species' – for which there was no early Paleogene counterpart – consume fresh, little-altered organic material, whereas infaunal taxa consume slightly more degraded material (Kitazato and Ohga, 1995). The mechanism for rapid delivery of little-degraded phytodetrital material might be dependent on high seasonal variability, which may not have existed in the warmer oceans of the early Paleogene (Thomas and Gooday, 1996). Early Paleogene deep-sea species thus may have depended dominantly on sedimentary organic material or on bacteria, rather than on non-degraded fresh phytodetrital material.

It might also be possible that the apparently high delivery of food to the sea floor during the early Paleogene is a sampling artifact. Hallock *et al.* (1991) proposed that sluggish ocean circulation resulted in higher concentrations of nutrients in the intermediate to deep ocean waters. In this model, oceanic productivity would have been low overall, but enhanced in the few regions with upwelling. Uneven site coverage in terms of geography and depth (Thomas, 1998) could thus cause an overestimate of the food supply to the sea floor. This does not appear to be the case for equatorial Pacific Site 865, which was probably located in an oligotrophic region for most of the Paleocene and early–middle Eocene (Bralower *et al.*, 1995a,b).

THE LPTM EXTINCTION EVENT: FAUNAS AND ISOTOPES

Benthic foraminifera

During the LPTM, calcareous benthic foraminiferal faunas at middle bathyal and greater depths show high rates of extinction (30–50% of species; see Thomas (1998) for a review). At upper bathyal through neritic depths faunas exhibit significant, but temporary, changes in species composition, and extinction was less severe (Thomas, 1998). Abyssal agglutinated foraminiferal faunal change has not been described in much detail, but appears to have been less catastrophic (Kaminski *et al.*, 1996; Kuhnt *et al.*, 1996). After the extinction, low-diversity, high-dominance deep-sea benthic foraminiferal faunas, typical for perturbed communities, were ubiquitous (Thomas, 1998). Thin-walled benthic foraminifera and ostracodes were common (Steineck and Thomas, 1996). Organisms belonging to non-related phyla (foraminifera and ostracodes) were similarly affected, suggesting that the formation of thin-walled shells was probably caused by increasing corrosivity of the deep waters. Comparison of data on productivity and dissolution in open-ocean environments indicates that the high corrosivity was not caused by globally increased productivity (Thomas, 1998).

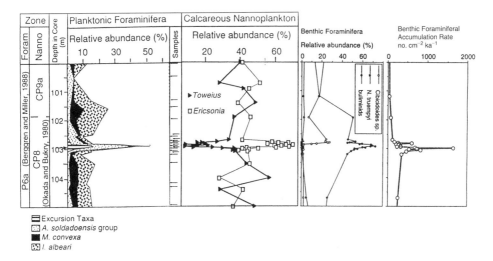

Figure 5.4. Data on the relative abundance of oligotrophic planktonic foraminiferal species and nannofossil species (Kelly *et al.*, 1996), combined with data on relative abundance of benthic foraminifera belonging to the superfamily Bolivinacea (bi/tri-serial taxa) and benthic foraminiferal accumulation rate (indicator of high food supply). Note co-occurrence of indicators of low productivity in surface water and high food-supply/low oxygenation indicators in the bottom waters just after the latest Paleocene benthic foraminiferal extinction. *A. soldadoensis*: *Acarinina soldadoensis*; *M. convexa*: *Morozovella convexa*; *I. albeari*: *Igorina albeari*.

Thomas and Shackleton (1998) argued that carbon isotopes as well as benthic faunal composition suggest increased upwelling and increased productivity during the LPTM in the Weddell Sea. Runoff, weathering, and possibly nutrient supply from the Antarctic continent probably could have increased during the LPTM (Robert and Kennett, 1994). Post-extinction ostracode faunas in this region had abundant opportunistic taxa, indicative of ephemeral, food-rich environments (Steineck and Thomas, 1996). By contrast, productivity during the LPTM was probably low in the Southern Atlantic (Thomas and Shackleton, 1996). At ODP Site 865 (tropical Pacific), planktonic foraminifera indicate decreased primary productivity during the LPTM (Kelly *et al.*, 1996), but benthic foraminiferal accumulation rates and the relative abundance of Bolivinacea taxa increased, suggesting that more food arrived at the sea floor (Fig. 5.4). A similar increase in Bolivinacea taxa was observed at equatorial Pacific Site 577 (Miller *et al.*, 1987). The abundant *Cibicidoides* species occurring just after the increased abundance of Bolivinacea at Site 865 resemble recent *C. wuellerstorfi* in shape (large and flat), and they may likewise indicate increased bottom-current activity, as also suggested by increased winnowing and very low sediment accumulation rates. At Caribbean Sea Site 999 (Bralower *et al.*, 1997), lower bathyal laminated LPTM sediments do not contain calcareous benthic foraminifera, but only tubular agglutinated benthic foraminifera, which are dominated by suspension feeders (Kaminski, personal communication, 1997). These faunas suggest enhanced bottom-water circulation and low productivity at that site. Both low oxygen conditions and high productivity during the LPTM may have occurred along many continental margins, as interpreted from benthic foraminiferal data (e.g., in the eastern Tethys, Speijer *et al.*, 1995, 1996; in Spanish sections, Ortiz, 1995, and Coccioni *et al.*, 1994; in the Bay of Biscay Site 401, Pardo *et al.*, 1996; in New Zealand sections, Kaiho *et al.*, 1996).

The occurrence of both oligotrophic and strongly eutrophic faunas during the LPTM after the extinction suggests that the trophic resource continuum may have expanded (Thomas, 1998). Globally, productivity decreased especially in open-ocean settings, but locally this effect was counteracted by lower oxygen levels in the water column, resulting in delivery of a larger fraction of organic matter to the sea floor (as at Site 865), or by increased productivity at locations where nutrient-enriched deep waters welled up to the surface (Sites 689 and 690).

Biogeographic differences in deep-sea benthic foraminifera persisted from the LPTM through the early Eocene (Fig. 5.3; Thomas, 1998). Throughout this interval *N. truempyi* was generally more common in the equatorial Pacific than at the Weddell Sea sites, whereas the species belonging to the Bolivinacea were less abundant; there were no clear differences in the abundance of the cylindrical species. This pattern can be interpreted as suggesting higher and/or more sustained delivery of food to the ocean floor in the Weddell Sea during the early Eocene, but not during the late Paleocene. This would suggest a different biogeographic pattern during the early Eocene than has been proposed by Widmark (1995) and Widmark and Speijer (1997) for the Late Cretaceous.

Isotope data

Isotope data on benthic foraminifera from several sites (Figs. 5.5–5.7) demonstrate the difficulties of working with records from short-lived events. The carbon isotope event is at most sites present in 10–20 cm of sediment (Fig. 5.5). *Nuttallides truempyi* is commonly used for isotope analysis, but is rare or absent at sites where Bolivinacea-dominated faunas occur after the extinction (Fig. 5.3). At Site 865, *Bulimina semicostata* specimens from the extinction interval show excursion values, whereas the few specimens of *N. truempyi* in the same samples do not, and thus appear to be reworked (Fig. 5.6). At Site 689 *N. truempyi* is rare in the upper Paleocene throughout lower Eocene, and rare specimens in the excursion interval gave non-excursion values; *Lenticulina* sp. specimens from the same samples gave excursion values (Table 5.1). At Site 690 *N. truempyi* is fairly common and specimens give a typical excursion signature about 20 cm above the last appearance of *Gavelinella beccariiformis*, indicator of the extinction (Kennett and Stott, 1991; Thomas and Shackleton, 1996). *Nuttallides truempyi* is rare or absent in the 20-cm-thick interval just above the extinction (690B-19-4, 51–53 cm through 690B-19H-4, 72–74 cm; Fig. 5.7). Rare *N. truempyi* from this interval gave non-excursion isotope values (Table 5.1), whereas *Bulimina ovula* specimens from these samples gave excursion values. Analyses from ODP Site 215 (Indian Ocean) demonstrated that the excursion is not represented in the recovered sediments, either because of an unconformity or because of non-recovery between Cores 215-11 and 215-12 (Fig. 5.6).

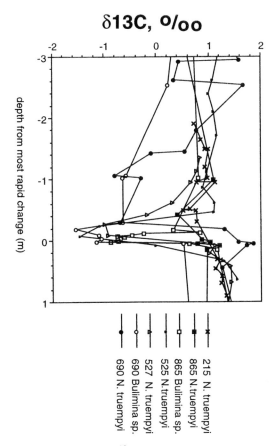

Figure 5.5. Benthic foraminiferal $\delta^{13}C$ record plotted vs. depth in meters above (negative values) or below (positive values) the benthic foraminiferal extinction. Data on *N. truempyi* from Sites 690, 525, and 527 (Thomas and Shackleton, 1996); data on *N. truempyi* and *Bulimina* sp. from Site 865 (Bralower *et al.*, 1995a,b); data on *N. truempyi* from Site 215 (Zachos *et al.*, 1994; see Table 5.1); data on *Bulimina* sp. from Site 690 (see Table 5.1).

It is therefore extremely difficult or impossible to shed light on possible paleoceanographic differences during the LPTM by comparing isotope records from different sites (e.g., temperature, salinity, productivity) (Figs. 5.5 and 5.8), even at two sites in close proximity such as 689 and 690 (Fig. 5.7). Isotope records can be fairly compared only if we can assume that the records have similar time resolution, which requires extremely detailed work (e.g., Kelly *et al.*, 1996). Work at lower resolution may easily lead to aliasing (e.g., compare Pak and Miller (1992) and Pardo *et al.* (1996) for Site 401). But high-resolution work suffers under the possibility that specimens of different sizes and shapes have been mixed to different depths by bioturbation (Thomson *et al.*, 1995), especially at sites with low sedimentation rates (Kelly *et al.*, 1996).

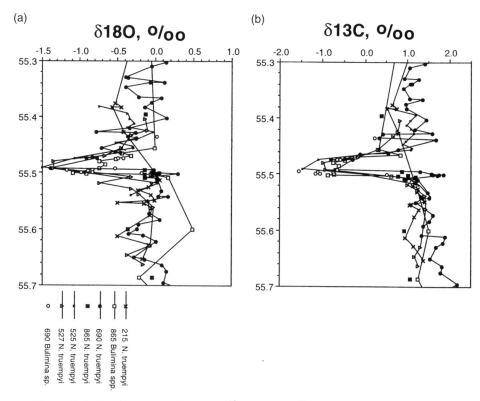

Figure 5.6. Benthic foraminiferal (a) $\delta^{18}O$ and (b) $\delta^{13}C$ data plotted vs. numerical age, using the age model in Thomas and Shackleton (1996). For data sources see Fig. 5.5.

In addition, at many sites different species must be used to capture the isotope excursion across an interval of severe extinction (Fig. 5.8), and for carbon isotopes we cannot simply adjust for the vital effects of different species because the difference in carbon isotope values of epifaunal species (e.g., *N. truempyi*) and infaunal species (e.g., *Bulimina* spp., *Lenticulina* spp.) has been demonstrated to vary according to productivity (e.g., Thomas and Shackleton, 1996).

At Site 690 we have sufficient planktonic and benthic foraminiferal isotope data to evaluate the timing of events (Fig. 5.9). In the interval lacking *N. truempyi* the planktonic faunas contain common *Morozovella aequa*, a shallow-dwelling species that is extremely rare at polar latitudes (Stott and Kennett, 1990; Stott *et al.*, 1990). The benthic foraminifera *Bulimina ovula* in these samples have a typical excursion isotope signature, as do *Acarinina* species (Table 5.1; Fig. 5.9). Specimens of *M. aequa*, however, have excursion $\delta^{18}O$ values but non-excursion $\delta^{13}C$ values, thus differing from specimens of *A. mckannai* in the same samples, which have excursion values for both isotopes (Thomas and Shackleton, 1996). The observation that the *Morozovella* specimens have excursion values for oxygen, but not for carbon, suggests that at some time the surface waters were unusually warm (as also

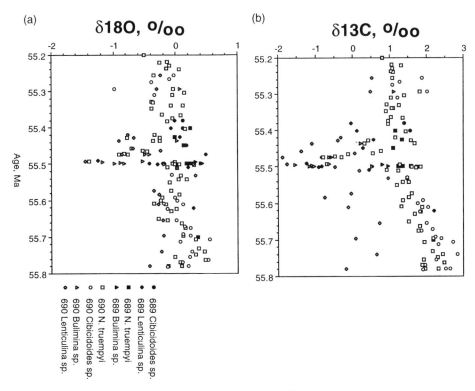

Figure 5.7. Benthic foraminiferal (a) $\delta^{18}O$ and (b) $\delta^{13}C$ data plotted vs. numerical age for Sites 689 (paleodepth 1100 m) and 690 (paleodepth 1900 m; Maud Rise, Weddell Sea). Data after Thomas and Shackleton (1996); see Table 5.1.

demonstrated by the presence of keeled planktonic foraminifera), but did not yet have the anomalous LPTM carbon isotope signature. The warming of surface waters at high latitudes thus started before the major change in carbon isotope values, but we cannot compare these data with those on other specimens (including benthics), because of the problems in mixing specimens by bioturbation (see Kelly *et al.*, 1996).

We provide the following speculative scenario to explain the available data. In the middle Paleocene a long-term period of warming started (e.g., Rea *et al.*, 1990; Corfield, 1995), possibly in response to massive CO_2 fluxes from North Atlantic flood basalts (e.g., Eldholm and Thomas, 1993). High latitudes warmed more than the tropics, resulting in very shallow latitudinal temperature gradients. When a threshold in warming (thus low density) at high latitudes was reached (Zachos *et al.*, 1994), waters could no longer sink to bathyal or abyssal depths, and the oceans at these depths were filled with relatively warm waters derived from subtropical latitudes (e.g., Thomas, 1989; Kennett and Stott, 1991). Alternatively, low density of surface waters at these latitudes may have been caused by increased precipitation and runoff at high latitudes (Bice *et al.*, 1997; Bice *et al.*, Chapter 4, this volume), or the deep-water circulation change might have been triggered by ephemeral low-latitude

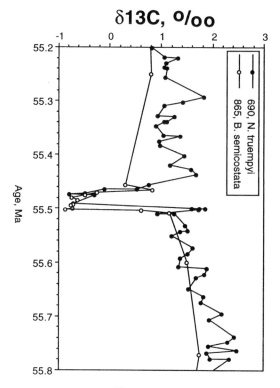

Figure 5.8. Benthic foraminiferal $\delta^{13}C$ data vs. numerical age for equatorial Pacific Site 865 (Bralower *et al.*, 1995a,b) and Weddell Sea Site 690 (Thomas and Shackleton, 1996).

cooling resulting from explosive volcanic eruptions in the Caribbean (Bralower *et al.*, 1997). The change in circulation resulted in rapid warming of the bathyal oceans, which caused dissociation of methane hydrate at depths of about 900–1500 m (Dickens *et al.*, 1995; Matsumoto, 1995; Kaiho *et al.*, 1996). This dissociation released a large amount of isotopically very light carbon in the oceans (Dickens *et al.*, 1997); the methane was oxidized, possibly leading to oceanic oxygen depletion and corrosivity for carbonate (Thomas, 1998).

The episode might have ended when greenhouse gas levels declined, or when heat transport to higher latitudes was no longer efficient at the very shallow temperature gradients, leading to high-latitude cooling and renewed formation of inter-mediate–deep waters at these latitudes. In this scenario, we would thus see an alternation between two modes of deep-sea circulation: in one mode, deep and intermediate waters would form dominantly at high latitudes; in another mode, deep–intermediate waters would dominantly form at subtropical latitudes, but the deepest ocean basins could still be ventilated from high-latitude regions, as proposed by Wilde and Berry (1982). Isotope data compilations suggest that deep–intermedi-ate waters generally formed to at least some extent at high southern latitudes even

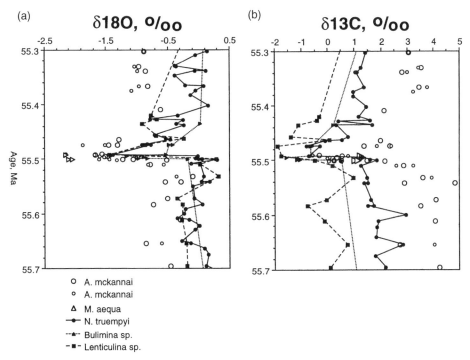

Figure 5.9. Benthic and planktonic foraminiferal (a) $\delta^{18}O$ and (b) $\delta^{13}C$ data for Site 690. Data after Stott *et al.* (1990) and Thomas and Shackleton (1996); see Table 5.1. The small circles indicate specimens of *A. mckannai* in the size range 212–250 mm; larger circles indicate specimens in the range 250–350 mm.

during the warm early Paleogene (e.g., Seto, 1995; Oberhänsli, 1997), and climate modeling indicates that deep convection occurs close to the Antarctic under most conditions (Bice *et al.*, Chapter 4, this volume). But information on the LPTM (e.g., Pak and Miller, 1992, 1995) suggests that relatively large volumes of deep–intermediate waters may have been derived from low latitudes for at least part of the latest Paleocene through early Eocene.

In such a scenario the benthic foraminiferal extinction was caused by multiple, geographically variable, factors. Methane oxidation in the oceans could have increased corrosivity while lowering oxygenation, and the degree of oxygen use could have been geographically variable (Dickens *et al.*, 1997). In some oceanic regions lowered productivity at a time of extreme warming and low windspeeds (Rea, 1994) could have been a factor, but at other locations increased preservation of organic matter counteracted the lowered productivity (e.g., equatorial Pacific Site 865). At yet other locations productivity might have increased because of increased weathering, precipitation, and nutrient runoff into the oceans (e.g., Tethys: Charisi and Schmitz, 1995; Speijer *et al.*, 1995, 1996; Speijer and Schmitz, 1998; New Zealand: Kaiho *et al.*, 1996). And at other locations productivity might have increased as a result of changing patterns of deep-oceanic circulation, causing increased upwelling of deeper waters with more nutrients (e.g., Maud Rise: Thomas and Shackleton, 1996).

At first sight, lower productivity appears unlikely to have played a role in the late Paleocene benthic extinction, because deep-sea benthic foraminifera did not suffer major extinction during the collapse of surface-ocean productivity at the end of the Cretaceous (Thomas, 1990*b*). We suggest that the end-Cretaceous collapse had less effect because it probably occurred on such a short term that detritus feeders were not affected. Comparatively, the decrease in productivity during the latest Paleocene and earliest Eocene was a long-term effect (e.g., Corfield, 1995; Aubry, 1998*b*). In the benthic foraminiferal faunas this long-term decline manifests itself in the gradually decreasing upper depth limit of the (inferred oligotrophic) *N. truempyi* assemblages (Tjalsma and Lohmann, 1983).

HOW MANY WARM EVENTS (HYPERTHERMALS)?

The occurrence of the LPTM demonstrates that short-term, rapid climate change is possible during a period of greenhouse climate in the absence of climate feedbacks that depend on the presence of polar ice. An obvious question is whether such rapid climate change occurred only once, or whether it occurred more often, possibly modulated by Milankovitch forcing. This question is of major importance for the use of paleoceanographic data to set boundary conditions for climate models: if short-term climate fluctuations between very warm periods (which we will call 'hyperthermals') and background warm climate were common, we must be extremely careful in data selection and time correlation to prevent aliasing.

There are some indications in isotope records that there may have been more than one hyperthermal (Kennett and Stott, 1990; Lu and Keller, 1993; Seto, 1995; Fig. 5.10; see also Fig. 5.1). In all these records, however, the possible 'events' are documented by one data point only, in some cases next to unconformities. For several reasons it remains difficult to obtain records with the resolution required to document events of such short duration. Firstly, the thickness of the sediments of the LPTM is in many sites on the order of 10–20 cm (Fig. 5.5). Such a small amount of material may not be recovered in deep-sea drilling sites, even at good recovery rates (e.g., Site 577, Aubry, 1998*a*; Site 215, this paper). Secondly, the lower Eocene record has proven to be less complete and more commonly riddled with short-term unconformities than had been assumed (Aubry, 1995, 1998*a*). We thus need high-resolution biostratigraphy in order to ascertain the completeness of each record. Thirdly, we cannot use benthic foraminiferal events to indicate at which level we will find an isotopic event even if the record is complete and we sample at high resolution, because the benthic faunas after the extinction show major fluctuations, probably resulting from the fluctuating abundances of 'disaster' or 'opportunistic' taxa (Thomas, 1998). In short, we must obtain highly detailed records from sites with excellent recovery and high-quality biostratigraphic data, where we can demonstrate the absence of unconformities. The highly detailed records would have to be obtained on some parameter that is quick, easy, and cheap to measure, because hundreds of observations are required. Possibly bulk isotopic records might be used (Shackleton and Hall, 1990), and downhole logging information on recently

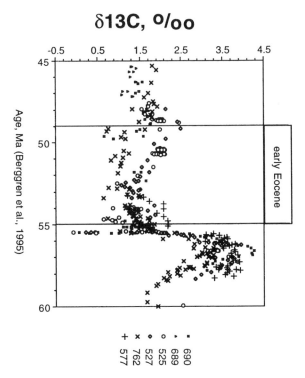

Figure 5.10. Bulk carbon isotope records from several sites, with numerical ages adjusted to Berggren *et al*. (1995). Sites 689 and 690: Shackleton and Hall (1990); Site 762: Thomas *et al*. (1991); Sites 525 and 527: Shackleton and Hall (1984); Site 577: Shackleton *et al*. (1985).

drilled ODP holes has great promise because recovery problems are circumvented (Norris, 1997*b*). Until we have such records, we cannot know whether the LPTM event was unique, or one of a series of events (Fig. 5.11).

CONCLUSIONS

1. Late Paleocene–early Eocene deep-sea environments cannot easily be characterized using deep-sea benthic foraminiferal faunas, because the most common Recent deep-sea faunal assemblages in open ocean originated in the late Eocene. There was no early Paleogene analog for faunas dominated by 'phytodetritus' species, i.e., opportunistic taxa, which in the present oceans are common in regions with a fluctuating primary productivity and supply of little-altered organic matter to the sea floor.

2. Early Paleogene benthic foraminiferal faunas may be indicative of overall higher supply to the ocean floor, even at lower primary productivity. The higher supply might have resulted from increased preservation of organic matter at lower oxygenation.

3. At least one short-term climate upheaval occurred during the late Paleocene–early Eocene 'greenhouse': the Late Paleocene Thermal

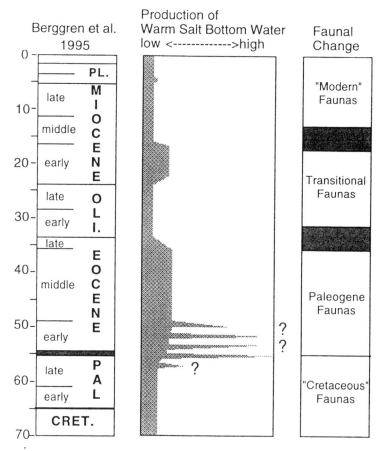

Figure 5.11. Speculation on possible occurrence of hyperthermal events in the early Eocene. During hyperthermals, deep-intermediate oceans were dominated by warm, high-salinity waters derived from subtropical sources, and climate was very warm, with extremely low latitudinal temperature gradients.

Maximum. We speculate that long-term high-latitude warming resulted in reversal of deep-ocean circulation so that waters at bathyal depths were derived from subtropical latitudes, resulting in increased deep-sea temperatures, leading to methane hydrate dissociation.

4.   The benthic foraminiferal extinction was a complex event. Global productivity may have declined, while oxygenation decreased as a result of methane dissociation and warming of deep waters. Productivity increased locally as a result of increased runoff and nutrient supply, or upwelling; locally more food reached the ocean floor even as productivity declined.

5.   We do not know whether other similar warm periods ('hyperthermals') occurred or were common in the early Paleogene; we thus do not know whether a warm climate would be inherently stable or unstable.

ACKNOWLEDGEMENTS

This paper improved much because of reviews by Karen Bice, Andy Gooday, Dottie Pak, and Joen Widmark. We thank the Ocean Drilling Program for samples. Research support for Thomas and Zachos was provided by NSF (EAR-94-06099), and for Bralower by NSF (EAR-94-05784).

REFERENCES

Alve, E. and Bernhard, J. M. (1995). Vertical migratory response of benthic foraminifera to controlled oxygen concentrations in an experimental mesocosm. *Mar. Ecol. Prog. Ser.*, **116**, 137–51.

Askin, R. A. (1992). Late Cretaceous–early Tertiary Antarctic Outcrop evidence for past vegetation and climate. *Antarctic Res. Ser.*, **56**, 61–73.

Aubry, M.-P. (1992). Late Paleogene calcareous nannoplankton evolution; a tale of climatic deterioration. In *Eocene–Oligocene Climatic and Biotic Evolution*, eds. D. R. Prothero and W. A. Berggren, pp. 272–309. Princeton: Princeton University Press.

Aubry, M.-P. (1995). From chronology to stratigraphy: interpreting the lower and middle Eocene stratigraphic record in the Atlantic Ocean. *SEPM Spec. Publ.*, **54**, 213–74.

Aubry, M.-P. (1998*a*). Stratigraphic (dis)continuity and temporal resolution in the upper Paleocene–lower Eocene deep-sea record. In *Late Paleocene–early Eocene Biotic and Climatic Events in the Marine and Terrestrial Records*, eds. M.-P. Aubry, S. Lucas and W. A. Berggren, pp. 37–66. New York: Columbia University Press.

Aubry, M.-P. (1998*b*). Early Paleogene calcareous nannoplankton evolution; a tale of climate amelioration. In *Late Paleocene–early Eocene Biotic and Climatic Events in the Marine and Terrestrial Records*, eds. M.-P. Aubry, S. Lucas and W. A. Berggren, pp. 158–203. New York: Columbia University Press.

Aubry, M.-P., Berggren, W. A., Stott, L. D. and Sinha, A. (1996). The upper Paleocene–lower Eocene stratigraphic record and the Paleocene/Eocene boundary carbon isotope excursion. In *Correlation of the Early Paleogene in Northwest Europe*, Geol. Soc. Spec. Publ. No. 101, eds. R. W. Knox, R. Corfield and R. E. Dunay, pp. 353–70. Princeton, NJ: Geological Society.

Barron, E. J. and Moore, G. T. (1994). Climate model applications in paleoenvironmental analysis. *SEPM Short Course*, **33**, 1–339.

Basinger, J. F., Greenwood, D. R. and Sweda, T. (1994). Early Tertiary vegetation in Arctic Canada and its relevance to Tertiary climate reconstructions. In *Cenozoic Plants and Climate of the Arctic*, NATO ASI Series No. 124, eds. M. C. Boutter and H. C. Fisher, pp 175–98. Heidelberg: Springer-Verlag.

Benson, R. H. (1975). The origin of the psychrosphere as recorded in changes in deep-sea ostracode assemblages. *Lethaia*, **8**, 69–83.

Berger, W. H. (1989). Appendix: Global Maps of Ocean Productivity. In *Productivity of the Oceans: Present and Past*, eds. W. H. Berger, V. S. Smetacek and G. Wefer, pp. 429–55. New York: John Wiley & Sons.

Berggren, W. A., Kent, D. V., Swisher, C. C., III and Aubry, M.-P. (1995). A revised Cenozoic geochronology and chronostratigraphy. *SEPM Spec. Vol.*, **54**, 129–212.

Berggren, W. A. and Miller, K. G. (1988). Paleogene tropical planktonic foraminiferal biostratigraphy and magnetostratigraphy. *Micropaleontol.*, **34**, 326–80.

Berner, R. A. (1995). Chemical weathering and its effect on atmospheric CO_2 and climate. *Rev. Mineral.*, **31**, 565–83.

Bernhard, J. M. (1992). Benthic foraminiferal distribution and biomass related to pore water oxygen content: Central California Continental Slope and Rise. *Deep-Sea Res.*, **39**, 585–605.

Bernhard, J. M. (1993). Experimental and field evidence of Antarctic foraminiferal tolerance to anoxia and hydrogen sulfide. *Micropaleontol.*, **20**, 203–13.

Bernhard, J. M. (1996). Microaerophilic and facultative anaerobic benthic foraminifera: a review of experimental and ultrastructural evidence. *Rev. Paleobiol.*, **15**, 261–75.

Bernhard, J. M. and Alve, E. (1996). Survival, ATP pool, and ultrastructural characterization of benthic foraminifera from Drammensfjord (Norway): response to anoxia. *Mar. Micropaleontol.*, **28**, 5–17.

Bernhard, J. M. and Reimers, C. E. (1991). Benthic foraminiferal population fluctuations related to anoxia: Santa Barbara Basin. *Biogeochemistry*, **15**, 127–45.

Bice, K. L., Barron, E. J. and Peterson, W. H. (1997). Continental run-off and early Cenozoic bottom water sources. *Geology*, **25**, 951–4.

Bralower, T. J., Parrow, M., Thomas, E. and Zachos, J. C. (1995*a*). Data Report: Stable isotope stratigraphy of the Paleogene pelagic cap at Site 865, Allison Guyot. *Proc. ODP, Scientific Results*, **143**, 581–6.

Bralower, T. J., Thomas, D. J., Zachos, J. C. *et al.* (1997). High-resolution records of the Late Paleocene Thermal Maximum and circum-Caribbean volcanism: is there a causal link? *Geology*, **25**, 963–5.

Bralower, T. J., Zachos, J. C., Thomas, E. *et al.* (1995*b*). Late Paleocene to Eocene paleoceanography of the equatorial Pacific Ocean: stable isotopes recorded at ODP Site 865, Allison Guyot. *Paleoceanography*, **10**, 841–65.

Brasier, M. D. (1995*a*). Fossil indicators of nutrient levels. 1. Eutrophication and climate change. *J. Geol. Soc., Spec. Publ.*, **83**, 113–32.

Brasier, M. D. (1995*b*). Fossil indicators of nutrient levels. 2. Evolution and extinction in relation to oligotrophy. *J. Geol. Soc., Spec. Publ.*, **83**, 133–50.

Bremer, M. L. and Lohmann, G. P. (1982). Evidence for primary control of the distribution of certain Atlantic Ocean benthonic foraminifera by degree of carbonate saturation. *Deep-Sea Res.*, **29**, 987–98.

Canudo, J. I., Keller, G., Molina, E. and Ortiz, N. (1995). Planktic foraminiferal turnover and $\delta^{13}C$ isotopes across the Paleocene–Eocene transition at Caravaca and Zumaya, Spain. *Palaeogeogr., Palaeoclim., Palaeoecol.*, **114**, 75–100.

Charisi, S. D. and Schmitz, B. (1995). Stable ($\delta^{13}C$, $\delta^{18}O$) and strontium ($^{87}Sr/^{86}Sr$) isotopes through the Paleocene at Gebel Aweina, eastern Tethys region. *Palaeogeogr., Palaeoclim., Palaeoecol.*, **116**, 103–29.

Coccioni, R., Di Leo, R., Galeotti, S. and Monechi, S. (1994). Integrated biostratigraphy and benthic foraminiferal faunal turnover across the Paleocene–Eocene boundary at Trabuaka pass section, Northern Spain. *Palaeopelagos*, **4**, 87–100.

Corfield, R. M. (1995). Palaeocene oceans and climate: an isotopic perspective. *Earth Science Rev.*, **37**, 225–52.

Corliss, B. H. and Chen, C. (1988). Morphotype patterns of Norwegian deep-sea benthic foraminifera and ecological implications. *Geology*, **16**, 716–19.

Cramer, B. S., Aubry, M.-P., Olsson, R. K., Miller, K. G. and Wright, J. D. (1997). Stratigraphic and climatic implications of a continuous, thick Paleocene/Eocene boundary, Bass River, NJ (ODP Leg 174AX). *EOS, Trans. AGU, Suppl.*, **78**, F363.

Deming, J. W. and Yager, P. L. (1992). Natural bacterial assemblages in deep-sea sediments; towards a global view. In *Deep-Sea Food Chains and the Global Carbon Cycle*, eds. G. T. Rowe and V. Pariente, pp. 11–28. Dordrecht: Kluwer Academic.

Dickens, G. R., O'Neil, J. R., Rea, D. K. and Owen, R. M. (1995). Dissociation of oceanic methane hydrate as a cause of the carbon isotope excursion at the end of the Paleocene. *Paleoceanography*, **10**, 965–71.

Dickens, G. R., Castillo, M. M. and Walker, J. C. G. (1997). A blast of gas in the latest Paleocene: simulating first-order effects of massive dissociation of oceanic methane hydrate. *Geology*, **25**, 259–64.

Diester-Haass, L. and Zahn, R. (1996). Eocene–Oligocene transition in the Southern Ocean: History of water mass circulation and biological productivity. *Geology*, **24**, 163–6.

Eldholm, E. and Thomas, E. (1993). Environmental impact of volcanic margin formation. *Earth Planet. Sci. Lett.*, **117**, 319–29.

Fischer, A. G. and Arthur, M. A. (1977). Secular variations in the pelagic realm. *SEPM Spec. Publ.*, **25**, 19–50.

François, L. M. and Walker, J. C. G. (1992). Modeling the Phanerozoic carbon cycle and climate: constraints from the $^{87}Sr/^{86}Sr$ isotopic ration of seawater. *Am. J. Sci.*, **292**, 81–135.

Freeman, K. H. and Hayes, J. M. (1992). Fractionation of carbon isotopes by phytoplankton and estimates of ancient CO_2 levels. *Global Biogeochemical Cycles*, **6**, 185–98.

Gibson, T. G., Bybell, L. M. and Owens, J. P. (1993). Latest Paleocene lithologic and biotic events in neritic deposits of southwestern New Jersey. *Paleoceanography*, **8**, 495–514.

Gooday, A. J. (1993). Deep-sea benthic foraminiferal species which exploit phytodetritus: Characteristic features and controls on distribution. *Mar. Micropaleontol.*, **22**, 187–206.

Gooday, A. J. (1994). The biology of deep-sea foraminifera: A review of some advances and their applications in paleoceanography. *Palaios*, **9**, 14–31.

Gooday, A. J., Levin, L. A., Linke, P. and Heeger, T. (1992). The role of benthic foraminifera in deep-sea food webs and carbon cycling. In *Deep-Sea Food Chains and the Global Carbon Cycle*, eds. G. T. Rowe and V. Pariente, pp. 63–91. Dordrecht: Kluwer Academic.

Gupta, A. K. (1993). Biostratigraphic vs. paleoceanographic importance of *Stilostomella lepidula* (Schwager) in the Indian Ocean. *Micropaleontol.*, **39**, 47–51.

Hallock, P. (1987). Fluctuations in the trophic resource continuum: a factor in global diversity cycles? *Paleoceanography*, **2**, 457–71.

Hallock, P., Premoli-Silva, I. and Boersma, A. (1991). Similarities between planktonic and larger foraminiferal evolutionary trends through Paleogene paleoceanographic changes. *Palaeogeogr., Palaeoclim., Palaeoecol.*, **83**, 49–64.

Hay, W. H. (1995). Paleoceanography of marine organic-carbon-rich sediments. *AAPG Studies in Geology*, **40**, 21–59.

Hedges, J. I. and Keil, R. G. (1995). Sedimentary organic matter preservation: an assessment and speculative synthesis. *Marine Chemistry*, **49**, 81–115.

Herguera, J. C. and Berger, W. H. (1991). Paleoproductivity from benthic foraminiferal abundance: glacial to postglacial changes in the west equatorial Pacific. *Geology*, **19**, 1173–6.

Jahnke, R. A. (1996). The global ocean flux of particulate organic carbon: areal distribution and magnitude. *Global Biogeochem. Cycles*, **10**, 71–88.

Jannasch, H. (1994). The microbial turnover of carbon in the deep-sea environment. *Glob. Planet. Change*, **9**, 289–95.

Jorissen, F. J., De Stigter, H. C. and Widmark, J. (1995). A conceptual model explaining benthic foraminiferal microhabitats. *Mar. Micropaleontol.*, **26**, 3–15.

Kaiho, K. (1991). Global changes of Paleogene aerobic/anaerobic benthic foraminifera and deep-sea circulation. *Palaeogeogr., Palaeoclim., Palaeoecol.*, **83**, 65–86.

Kaiho, K. (1994a). Benthic foraminiferal dissolved-oxygen index and dissolved oxygen levels in the modern ocean. *Geology*, **22**, 719–22.

Kaiho, K. (1994b). Planktonic and benthic foraminiferal extinction events during the last 100 m.y. *Palaeogeogr., Palaeoclim., Palaeoecol.*, **111**, 45–71.

Kaiho, K., Arinobu, T., Ishiwatari, R. *et al.* (1996). Latest Paleocene benthic foraminiferal extinction and environmental changes at Tawanui, New Zealand. *Paleoceanography*, **11**, 447–65.

Kaminski, M. A., Kuhnt, W. and Radley, J. (1996). Palaeocene–Eocene deep water agglutinated foraminifera from the Numidian Flysch (Rif, Northern Morocco): their significance for the palaeoceanography of the Gibraltar Gateway. *J. Micropalaeontol.*, **15**, 1–19.

Kelly, D. C., Bralower, T. J., Zachos, J. C., Premoli-Silva, I. and Thomas, E. (1996). Rapid diversification of planktonic foraminifera in the tropical Pacific (ODP Site 865) during the late Paleocene thermal maximum. *Geology*, **24**, 423–6.

Kennett, J. P. and Stott, L. D. (1990). Proteus and Proto-Oceanus, Paleogene Oceans as revealed from Antarctic stable isotopic results: ODP Leg 113. *Proc. ODP, Scientific Results*, **113**, 865–80.

Kennett, J. P. and Stott, L. D. (1991). Abrupt deep-sea warming, paleoceanographic changes and benthic extinction at the end of the Paleocene. *Nature*, **353**, 225–9.

Kitazato, H. and Ohga, T. (1995). Seasonal changes in deep-sea benthic foraminiferal populations: results of long-term observations at Sagami Bay, Japan. In *Biogeochemical Processes and Ocean Flux in the Western Pacific*, eds. H. Sakai and Y. Nozaki, pp. 331–42. Tokyo: Terra Scientific Publishing Company.

Knox, R. W. O'B. (1996). Correlation of the early Paleogene in northwest Europe: an overview. *Geol. Soc. Spec. Publ.*, **101**, 1–11.

Koch, P. L., Zachos, J. C. and Dettman, D. L. (1995). Stable isotope stratigraphy and palaeoclimatology of the Paleogene Bighorn Basin (Wyoming, USA). *Palaeogeogr., Palaeoclim., Palaeoecol.*, **115**, 61–89.

Koch, P. L., Zachos, J. C. and Gingerich, P. D. (1992). Coupled isotopic change in marine and continental carbon reservoirs at the Palaeocene/Eocene boundary. *Nature*, **358**, 319–22.

Kristensen, E., Ahmed, S. I. and Devol, A. H. (1995). Aerobic and anaerobic decomposition of organic matter in marine sediment: which is fastest? *Limnol. Oceanogr.*, **40**, 1430–7.

Kuhnt, W., Moullade, M. and Kaminski, M. A. (1996). Ecological structuring and evolution of deep-sea agglutinated foraminifera – a review. *Rév. Micropaleontol.*, **39**, 271–81.

Kump, L. R. and Arthur, M. A. (1997). Global chemical erosion during the Cenozoic: weatherability balances the budgets. In *Tectonic Uplift and Climate Change*, ed. W. F. Ruddiman, pp. 399–426. New York: Plenum Press.

Leckie, D. A., Morgans, H., Wilson, G. J. and Edwards, A. R. (1995). Mid-Paleocene dropstones in the Whangai Formation, New Zealand – evidence of mid-Paleocene cold climate? *Sedim. Geol.*, **97**, 119–29.

Linke, P. and Lutze, G. F. (1993). Microhabitats of benthic foraminifera – a static concept or a dynamic adaptation to optimize food acquisition? *Mar. Micropaleontol.*, **20**, 215–33.

Loubere, P. (1994). Quantitative estimation of surface ocean productivity and bottom water oxygen concentration using benthic foraminifera. *Paleoceanography*, **9**, 723–37.

Lu, G. and Keller, G. (1993). Climatic and oceanographic events across the Paleocene–Eocene transition in the Antarctic Indian Ocean: inference from planktic foraminifera. *Mar. Micropaleontol.*, **21**, 101–42.

Lu, G. and Keller, (1995*a*). Planktic foraminiferal faunal turnovers in the subtropical Pacific during the late Paleocene to early Eocene. *J. Foram. Res.*, **25**, 97–116.

Lu, G. and Keller, (1995*b*). Ecological stasis and saltation: species richness change in planktic foraminifera during the late Paleocene to early Eocene, DSDP Site 577. *Palaeogeogr., Palaeoclim., Palaeoecol.*, **117**, 211–27.

Lu, G., Keller, G., Adatte, T., Ortiz, N. and Molina, E. (1996). Long-term (10^5) or short-term (10^3) $\delta^{13}C$ excursion near the Paleocene–Eocene transition: Evidence from the Tethys. *Terra Nova*, **8**, 347–55.

Lutze, G. F. and Thiel, H. (1989). *Cibicidoides wuellerstorfi* and *Planulina ariminensis*, elevated epibenthic foraminifera. *J. Foram. Res.*, **19**, 153–8.

Mackensen, A., Fütterer, D. K., Grobe, H. and Schmiedl, G. (1993). Benthic foraminiferal assemblages from the eastern Southern Atlantic Polar Front region between 35 and 75°S: distribution, ecology and fossilization potential. *Mar. Micropaleontol.*, **22**, 33–69.

Markwick, P. J. (1994). Equability, continentality, and Tertiary climate: the crocodilian perspective. *Geology*, **22**, 613–6.

Matsumoto, R. (1995). Causes of the $\delta^{13}C$ anomalies of carbonates and a new paradigm 'Gas-Hydrate Hypothesis'. *J. Geol. Soc. Japan*, **111**, 902–24.

Miller, K. G., Janecek, T. R., Katz, M. R. and Keil, D. J. (1987). Abyssal circulation and benthic foraminiferal changes near the Paleocene/Eocene boundary. *Paleoceanography*, **2**, 741–61.

Miller, K. G., Katz, M. E. and Berggren, W. A. (1992). Cenozoic deep-sea benthic foraminifera: a tale of three turnovers. In *BENTHOS '90*, pp. 67–75. Sendai: Tokai University Press.

Müller-Merz, E. and Oberhänsli, H. (1991). Eocene bathyal and abyssal benthic foraminifera from a South Atlantic transect at 20°–30°S. *Palaeogeogr., Palaeoclim., Palaeoecol.*, **83**, 117–72.

Murray, J. W. (1995). Microfossil indicators of ocean water masses, circulation and climate. *Geol. Soc. Spec. Publ.*, **83**, 245–64.

Norris, R. D. (1997a). Symbiosis as an evolutionary innovation in the radiation of Paleocene planktic foraminifera. *Paleobiology*, **22**, 461–80.

Norris, R. D. (1997b). Chronology and climate of the Paleocene–Eocene boundary and the early Eocene warm period: ODP Sites 1051 and 1050, Western North Atlantic. *EOS, Trans. AGU, Suppl.*, **78**, F364.

Oberhänsli, H. (1992). The influence of the Tethys on the bottom waters of the early Tertiary Oceans. *Antarctic Res. Ser.*, **56**, 167–83.

Oberhänsli, H. (1997). Klimatische und ozeanographische Veränderungen im Eozän. *Z. dt. Geol. Ges.*, **147**, 303–413.

Okada, H. and Bukry, D. (1980). Supplementary modification and introduction of code numbers to the low-latitude coccolith biostratigraphic zonation (Bukry, 1973; 1975). *Mar. Micropaleontol.*, **5**, 321–5.

Ortiz, N. (1995). Differential patterns of benthic foraminiferal extinctions near the Paleocene/ Eocene boundary in the North Atlantic and the western Tethys. *Mar. Micropaleontol.*, **26**, 341–60.

Pak, D. K. and Miller, K. G. (1992). Late Palaeocene to early Eocene benthic foraminiferal stable isotopes and assemblages: implications for deep-water circulation. *Paleoceanography*, **7**, 405–22.

Pak, D. K. and Miller, K. G. (1995). Isotopic and faunal record of Paleogene deep-water transitions, Leg 145. *Proc. ODP, Scientific Results*, **145**, 265–81.

Pardo, A. Keller, G., Molina, E. and Canudo, J. I. (1996). Planktic foraminiferal turnover across the Paleocene–Eocene transition at DSDP Site 401, Bay of Biscay, North Atlantic. *Mar. Micropaleontol.*, **29**, 129–58.

Parrish, J. T. (1995). Paleoceanography of C_{org}-rich rocks and the preservation versus production controversy. *AAPG Studies in Geology*, **40**, 1–20.

Rathburn, A. E. and Corliss, B. H. (1994). The ecology of living (stained) benthic foraminifera from the Sulu Sea. *Paleoceanography*, **9**, 87–150.

Rea, D. K. (1994). The paleoclimatic record provided by eolian deposition in the deep-sea: the geological record of wind. *Rev. Geophys.*, **32**, 159–95.

Rea, D. K., Zachos, J. C., Owen, R. M. and Gingerich, D. (1990). Global change at the Paleocene-Eocene boundary: climatic and evolutionary consequences of tectonic events. *Palaeogeogr., Palaeoclim., Palaeoecol.*, **79**, 117–28.

Rind, D. and Chandler, M. A. (1991). Increased ocean heat transports and warmer climate. *J. Geophys. Res.*, **96**, 7437–61.

Robert, C. and Kennett, J. P. (1992). Paleocene and Eocene kaolinite distribution in the South Atlantic and Southern Ocean: Antarctic climatic and paleoceanographic implications. *Mar. Geol.*, **103**, 99–110.

Robert, C. and Kennett, J. P. (1994). Antarctic subtropical humid episode at the Paleocene–Eocene boundary: clay mineral evidence. *Geology*, **22**, 211–14.

Ruddiman, W. F. (ed.) (1997). *Tectonic Uplift and Climate Change*. New York: Plenum Press.

Sanders, H. L. (1968). Marine benthic diversity: a comparative study. *Amer. Nat.*, **102**, 243–82.

Schmitz, B., Speijer, R. and Aubry, M.-P. (1996). Latest Paleocene benthic extinction event on the southern Tethyan shelf (Egypt): foraminiferal stable isotope (δ^{13}C, δ^{18}O) record. *Geology*, **24**, 347–50.

Schnitker, D. (1994). Deep-sea benthic foraminifers: Food and bottom water masses. In *Carbon Cycling in the Glacial Ocean: Constraints on the Ocean's Role in Global Change*, eds. R. Zahn, T. F. Pedersen, M. A. Kaminski and L. Labeyrie, pp. 539–54. New York: Springer.

Sen Gupta, B. K. and Machain-Castillo, M. L. (1993). Benthic foraminifera in oxygen-poor habitats. *Mar. Micropaleontol.*, **20**, 183–201.

Seto, K. (1995). Carbon and oxygen isotopic paleoceanography of the Indian and South Atlantic Oceans – paleoclimate and paleo-ocean circulation. *J. Sci. Hiroshima Univ., Ser. C*, **10**, 393–485.

Shackleton, N. J. and Hall, M. A. (1984). Carbon isotope data from Leg 74 sediments. *Initial Reports DSDP*, **74**, 613–19.

Shackleton, N. J. and Hall, M. A. (1990). Carbon isotope stratigraphy of bulk sediments, ODP Sites 689 and 690. *Proc. ODP, Scientific Results*, **113**, 985–9.

Shackleton, N. J., Hall, M. A. and Bleil, U. (1985). Carbon isotope stratigraphy, Site 577. *Initial Reports DSDP*, **86**, 503–11.

Sloan, L. C. and Rea, D. K. (1995). Atmospheric CO_2 and early Eocene climate: a general circulation study. *Palaeogeogr., Palaeoclim., Palaeoecol.*, **119**, 275–92.

Sloan, L. C., Walker, J. G. and Moore, T. C., Jr (1995). The possible role of oceanic heat transport in early Eocene climate. *Paleoceanography*, **10**, 347–56.

Smart, C. W., King, S. C., Gooday, A., Murray, J. W. and Thomas, E. (1994). A benthic foraminiferal proxy of pulsed organic matter paleofluxes. *Mar. Micropaleontol.*, **23**, 89–99.

Speijer, R. P. and Schmitz, B. (1998). A benthic foraminiferal record of Paleocene sea level and trophic/redox conditions at Gebel Aweina, Egypt. *Palaeogeogr., Palaeoclim., Palaeoecol.*, **137**, 79–101.

Speijer, R. P., Schmitz, B., Aubry, M.-P. and Charisi, S. (1995). The latest Paleocene benthic extinction event: punctuated turnover in outer neritic foraminiferal faunas from Egypt. *Israel J. Earth Sci.*, **44**, 207–22.

Speijer, R. P., Van Der Zwaan, G. J. and Schmitz, B. (1996). The impact of Paleocene/Eocene boundary events on middle neritic benthic foraminiferal assemblages from Egypt. *Mar. Micropaleontol.*, **28**, 99–132.

Steineck, P. L. and Thomas, E. (1996). The latest Paleocene crisis in the deep-sea: ostracode succession at Maud Rise, Southern Ocean. *Geology*, **24**, 583–6.

Stigebrant, A. and Djurfeldt, L. (1996). Control of production of organic matter in the ocean on short and long terms by stratification and remineralization. *Deep-Sea Res. II*, **43**, 23–35.

Stott, L. D. (1992). Higher temperatures and lower oceanic pCO$_2$: a climate enigma at the end of the Paleocene epoch. *Paleoceanography*, **7**, 395–404.

Stott, L. D. and Kennett, J. P. (1990). Antarctic Paleogene planktonic foraminiferal biostratigraphy: ODP Leg 113, Sites 689 and 690. *Proc. ODP, Scientific Results*, **113**, 549–70.

Stott, L. D., Kennett, J. P., Shackleton, N. J. and Corfield, R. M. (1990). The evolution of Antarctic surface waters during the Paleogene: inferences from the stable isotopic composition of planktonic foraminifera, ODP Leg 113. *Proc. ODP, Scientific Results*, **113**, 849–64.

Stott, L. D., Sinha, A., Thiry, M., Aubry, M.-P. and Berggren, W. A. (1996). The transfer of ^{12}C changes from the ocean to the terrestrial biosphere across the Paleocene/Eocene boundary: criteria for terrestrial–marine correlations. *Geol. Soc. Spec. Publ.*, **101**, 381–99.

Thiel, H., Pfannkuche, O., Theeg, R. and Schriever, G. (1987). Benthic metabolism and standing stick in the Central and Northern Red Sea. *Marine Ecol.*, **8**, 1–20.

Thomas, E. (1985). Late Eocene to Recent deep-sea benthic foraminifers from the central equatorial Pacific Ocean. *Initial Reports DSDP*, **85**, 655–79.

Thomas, E. (1987). Late Oligocene to Recent benthic foraminifers from DSDP Sites 608 and 610, northeastern Atlantic Ocean. *Initial Reports DSDP*, **94**, 997–1032.

Thomas, E. (1989). Development of Cenozoic deep-sea benthic foraminiferal faunas in Antarctic waters. *Geol. Soc. Spec. Publ.*, **47**, 283–96.

Thomas, E. (1990a). Late Cretaceous through Neogene deep-sea benthic foraminifers, Maud Rise, Weddell Sea, Antarctica. *Proc. ODP, Scientific Results*, **113**, 571–94.

Thomas, E. (1990b). Late Cretaceous-early Eocene mass extinctions in the deep sea. In *Global Catastrophes, Geol. Soc. Am., Spec. Publ.*, **247**, 481–96.

Thomas, E. (1992). Cenozoic deep-sea circulation: evidence from deep-sea benthic foraminifera. *Antarctic Res. Ser.*, **56**, 141–65.

Thomas, E. (1998). The biogeography of the late Paleocene benthic foraminiferal extinction. In *Late Paleocene–early Eocene Biotic and Climatic Events in the Marine and Terrestrial Records*, eds. M.-P. Aubry, S. Lucas and W. A. Berggren, pp. 214–43. New York: Columbia University Press.

Thomas, E., Booth, L., Maslin, M. and Shackleton, N. J. (1995). Northeastern Atlantic benthic foraminifera during the last 45,000 years: productivity changes as seen from the bottom up. *Paleoceanography*, **10**, 545–62.

Thomas, E. and Gooday, A. J. (1996). Deep-sea benthic foraminifera: tracers for changes in Cenozoic oceanic productivity. *Geology*, **24**, 355–8.

Thomas, E. and Shackleton, N. J. (1996). The late Paleocene benthic foraminiferal extinction and stable isotope anomalies. *Geol. Soc. Spec. Publ.*, **101**, 401–41.

Thomas, E., Shackleton, N. J. and Hall, M. A. (1991). Data Report: Carbon isotope stratigraphy of Paleogene bulk sediments, Hol 762C (Exmouth Plateau, Eastern Indian Ocean). *Proc. ODP, Scientific Results*, **122**, 897–901.

Thomas, E. and Vincent, E. (1987). Major changes in benthic foraminifera in the equatorial Pacific before the middle Miocene polar cooling. *Geology*, **15**, 1035–9.

Thomson, J., Cook, G. T., Anderson, R., MacKenzie, A. B., Harkness, D. D. and McCave, I. N. (1995). Radiocarbon age and off-sets in different-sized carbonate components of deep-sea sediments. *Radiocarbon*, **137**, 90–4.

Tjalsma, R. C. and Lohmann, G. P. (1983). Paleocene–Eocene bathyal and abyssal benthic foraminifera from the Atlantic Ocean. *Micropaleontology Spec. Publ.*, **4**, 1–94.

Turekian, K. K. and Pegram, W. J. (1997). Os isotope record in a Cenozoic deep-sea core: its relation to global tectonics and climate. In *Tectonic Uplift and Climate Change*, ed. W. F. Ruddiman, pp. 383–97. New York: Plenum Press.

Van Cappellen, P. and Ingall, E. (1994). Benthic phosphorus regeneration, net primary production, and ocean anoxia: a model of the coupled marine biogeochemical cycles of carbon and phosphorus. *Paleoceanography*, **9**, 677–92.

Van Morkhoven, F. P. C., Berggren, W. A. and Edwards, A. S. (1996). *Cenozoic Cosmopolitan Deep-water Benthic Foraminifera*. Pau, France: Elf-Aquitaine.

Walker, J. C. G. and Kastin, J. F. (1992). Effects of fuel and forest conservation on future levels of atmospheric carbon dioxide. *Palaeogeogr., Palaeoclim., Palaeoecol.*, **97**, 151–89.

Weinholz, P. and Lutze, G. F. (1989). The *Stilostomella* extinction. *Proc. ODP, Scientific Results*, **108**, 113–17.

Widmark, J. G. V. (1995). Multiple deep water sources and trophic regimes in the latest Cretaceous deep-sea: evidence from benthic foraminifera. *Mar. Micropaleontol.*, **26**, 361–84.

Widmark, J. G. V. and Speijer, R. P. (1997). Benthic foraminiferal ecomarker species of the terminal Cretaceous (late Maastrichtian) deep-sea Tethys. *Mar. Micropaleontol.*, **31**, 135–55.

Wilde, P. and Berry, W. B. N. (1982). Progressive ventilation of the oceans – potential for return to anoxic conditions in the post-Paleozoic. In *Nature and Origin of Cretaceous Carbon-rich Facies*, eds. S. O. Schlanger and M. B. Cita, pp. 209–24. London: Academic Press.

Wolfe, J. A. (1994). Alaskan Palaeogene climates as inferred from the CLAMP database. In *Cenozoic Plants and Climate of the Arctic*, NATO ASI Series No. 124, eds. M. C. Boutter and H. C. Fisher, pp. 223–38. Heidelberg: Springer-Verlag.

Zachos, J. C., Lohmann, K. C., Walker, J. C. G. and Wise, S. W. (1993). Abrupt climate change and transient climates during the Paleogene: a marine perspective. *J. Geol.*, **101**, 191–213.

Zachos, J. C., Stott, L. D. and Lohmann, K. C. (1994). Evolution of early Cenozoic temperatures. *Paleoceanography*, **9**, 353–87.

6

Mountains and Eocene climate

RICHARD D. NORRIS, RICHARD M. CORFIELD, AND KAREN HAYES-BAKER

ABSTRACT

Mountains produce local changes in climate through their control of vegetation and precipitation, but they may also have significant effects on hemispheric climate by setting up long-period waves in the atmosphere and preventing the simple zonal circulation that would sharply limit latitudinal heat transport through the atmosphere. As such, accurate estimates of paleotopography are an important boundary condition in global climate models of past warm periods. Detailed reconstruction of mountain belts, particularly their average elevation and aerial extent, is needed to address issues of high latitude warmth during the last 'hyperthermal' in the early Eocene. A variety of techniques based on sedimentology, structural geology, basalt vesicularity, stable isotopes, and paleotemperature estimates from fossil plant assemblages have been devised to reconstruct the elevations of ancient mountain systems. We present a new paleoaltimeter to estimate the difference in relative elevation between intermontane basins and the high elevations of ranges near the tree line. Application of this paleoaltimeter to the Eocene Green River Formation supports recent evidence that the Laramide mountains of the western United States were as high as or higher than the modern Rocky Mountains and suggests that many recent global climate simulations have prescribed elevations that are substantially too low.

INTRODUCTION

The early Eocene was perhaps the warmest period in the past 100 million years. Floral and faunal data suggest that warm conditions extended to much higher latitudes than today both in the oceans and on the continents. Winter freeze lines were significantly north of their present locations in North America, while no parts of lowland Australia or South America appear to have experienced significant freezing conditions (Greenwood and Wing, 1995). Likewise, high latitude surface waters were about 10–15 °C in the early Eocene, considerably higher than the <0–5 °C temperature range of modern polar waters (Zachos et al., 1994). Elevated temperatures at high latitudes reflect shallowing of the equator-to-pole temperature gradient compared with today, because low latitude surface water temperatures do not appear to have been any warmer than modern temperatures (Zachos et al., 1994; Crowley and Zachos, Chapter 3, this volume).

Global climate models have typically had trouble simulating the low latitudinal temperature gradients indicated by paleontological or geochemical data. Climate models can be made to match proxy data by assuming large increases in atmospheric CO_2 and heat transport, or by considering the regional effects of large lake systems (Hutchison, 1982; Sloan and Barron, 1992a; Sloan, 1994; Greenwood and Wing, 1995). Still, model calculations suggest that heat transport from the low latitudes to the poles would have needed to be ~30% higher than today – a figure difficult to explain by increased oceanic heat transport and one seemingly requiring large changes in high latitude albedo or atmospheric circulation (Sloan *et al.*, 1995).

Part of the mismatch between general circulation models (GCMs) and proxy data may also reside within inaccuracies in the model vegetation and orographic effects on mid-latitude atmospheric circulation. Estimates of surface temperatures are strongly affected by the distribution and type of vegetation, which traps solar radiation by lowering albedo and increasing soil moisture, and through transpiration-related effects on water content of the surface atmosphere.

It is the contribution of mountains to global climate and methods for the estimation of mountain elevation that are the focus of this paper. Mountains produce local changes in climate through their control of vegetation and precipitation, but they may also have significant effects on hemispheric climate by setting up long-period waves in the atmosphere and preventing the simple zonal circulation that would sharply limit latitudinal heat transport through the atmosphere (Bolin, 1950; Manabe and Terpstra, 1974; Hahn and Manabe, 1975). In addition, mountains have significant effects on the concentration of greenhouse gases – the denudation of rapidly rising mountain systems is implicated in the regulation of atmospheric CO_2 while the atmospheric water content is regulated in part by the disposition and height of mountains (Raymo *et al.*, 1988; Raymo, 1991). Many of these insights have come from numerical climate models which have begun to explore the climatological effects of large mountain systems in the late Cenozoic (Hahn and Manabe, 1975; Kutzbach *et al.*, 1989; Ruddiman *et al.*, 1989; Ruddiman and Kutzbach, 1990; Beck *et al.*, 1995). Attention has also been paid to the climatological role of mountain chains under ancient plate configurations as well as hypothetical plate geometries (Hay *et al.*, 1990; Moore *et al.*, 1992; Otto-Bliesner, 1993).

Climate model simulations of the climatological effects of mountain systems are generally based upon relatively low spatial resolution GCMs. Consequently, mountain systems are commonly represented by broad plateaus rather than narrow ranges and often represent average elevations – thus omitting the lofty peaks of many mountain belts (Manabe and Terpstra, 1974; Kutzbach *et al.*, 1993). Despite these model limitations, the broad climatological effects of mountain systems are reasonably well known. Considerably less well known are the heights and elevation histories of most ancient mountain belts. This deficiency is not surprising given the erosional nature of mountain peaks, but represents a severe limitation for paleoclimate models.

Here we (1) discuss the role of large mountain systems in controlling regional climate through effects on atmospheric dynamics, albedo, and atmospheric CO_2 as well as local weather systems and vegetation; (2) discuss the types of proxy data used to estimate mountain elevations in the geologic past; (3) present a new method for determining paleotopography as well as new data on the paleoaltitude of the Eocene Rocky Mountains, and (4) discuss the significance of Eocene paleotopography for the climate of North America.

OROGRAPHY AND ATMOSPHERIC DYNAMICS

Primary climatic effects of mountain chains

Pressure systems

Land exposed at high altitude gives up heat to the cold air, heating the atmosphere and cooling the mountain surface. Heat transferred from the land to the atmosphere at high elevations has a larger effect on atmospheric circulation than heat transferred at low elevations because the relatively small mass of air at high elevation is more easily heated than the much larger mass of air over low-lying regions. Heating or cooling air over mountain tops can therefore produce strong density contrasts in the atmosphere compared with air masses over the adjacent lowlands and result in strong rising or sinking motions in the atmosphere. For example, summer heating over a large mountain range produces a rising, low density air mass, with low atmospheric pressure over the mountain and strong winds blowing toward the mountain front (Fig. 6.1). In the northern hemisphere, the winds blow counterclockwise around the range in response to the Coriolis force. During the winter, cooling of the mountain surface produces dense air masses which sink and spread out away from the mountain front (Fig. 6.1). In this case, northern hemisphere winds blow clockwise around the mountains. The change in wind direction from winter to summer is typical of monsoon systems and is often accompanied by periods of heavy rain.

The strength of the winds around mountains depends partly upon the breadth of the range and the strength of heating and cooling produced by radiation from the land surface. Broad, high-elevation plateaus produce stronger monsoon systems than high, but narrow, mountain ranges (Meehl, 1992). Broad, high plateaus can control surface climate far from the mountain front because the atmospheric pressure systems they form can have strong effects on the upper atmosphere in addition to the deflection of surface winds caused by the physical topographic barrier. Narrow ranges, such as the modern Andes, do not produce monsoon conditions *per se*. Most of the heating of the Altiplano appears to be due to convective circulation around the mountain range rather than sensible heating of the mountain surface (Meehl, 1992).

Stationary eddies and the jet stream

Narrow mountain belts will typically lack strong monsoon systems owing to the relatively small land surface at high elevation, but any type of high barrier can deflect major air masses and produce large downstream changes in climate. Climate simulations run without mountains generally find zonal circulation with air masses staying at nearly constant latitude as they transit the globe (Bolin, 1950; Manabe and

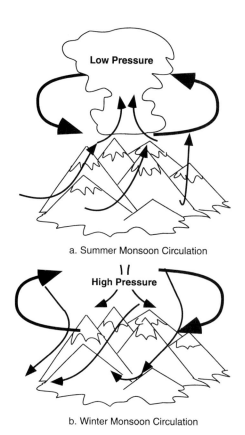

a. Summer Monsoon Circulation

b. Winter Monsoon Circulation

Figure 6.1. Monsoon circulation over a large, mid-latitude, northern hemisphere mountain range (a) in summer, and (b) in winter. Note that rising (a) and sinking (b) air masses (thin arrows) spiral up or down the mountain fronts as a result of the Coriolis force and produce the high-level circulation directions shown by the thick shaded arrows.

Terpstra, 1974; Barron and Washington, 1984; Barron, 1985; Broccoli and Manabe, 1992; Hay and Brock, 1992; Kutzbach *et al.*, 1993; Barron and Moore, 1994). Topographic highs introduce a wave-like behavior to atmospheric circulation, particularly in mid-latitudes where air flow is dominantly west–east in the northern hemisphere (Figs. 6.2b and c). The flow of air has a clockwise component over topographic highs and a counterclockwise component over adjacent lowlands. Consequently, air masses flowing over a topographic barrier will tend to veer toward the equator on the lee side of a large mountain range and then shift toward higher latitude over the lowland (Figs. 6.2b and c). This process results in the southward deflection of the jet stream east of the Rocky Mountains and the northward shift in the jet stream over eastern North America (Bolin, 1950 and Fig. 6.3).

Large mountain systems can anchor atmospheric waves and the associated storm systems. Bolin (1950) discussed the effects of circular mountains in a simplified GCM and found that westerly flow over the mountain front produced a trough in the lee of the range and a strong zonal wind system on the eastern limb of the trough. The stationary waves propagate far downstream of a mountain front. For example, the winter high pressure ridge that extends from the Azores to France and the northerly position of storm tracks over the eastern Atlantic that warm Europe in winter are at least partly controlled by the influence of the Rocky Mountains (Fig. 6.3; Bolin, 1950). Manabe and Terpstra (1974) obtained a similar result for simulations of the northern hemisphere. They found that the intensity of the jet stream increased in the southeastern United States in the lee of the Rocky Mountains as well as on the eastern side of the Tibetan Plateau with effects propagating far downstream.

Topography also modifies the distribution of eddy kinetic energy between stationary and transient systems. Mountain belts increase the poleward transports of heat by stationary systems, which are fairly efficient, and reduce the effects of energy-inefficient, transient eddy systems. The result is a net increase in heat transport compared with simulations run without mountains (Manabe and Terpstra, 1974).

Cyclogenesis and precipitation

Large cyclones are common in the lee of many large mountain systems such as the Tibetan Plateau and the Rocky Mountains. Bolin (1950), Manabe and Terpstra (1974), and Kutzbach et al. (1989) demonstrated that the intensity of the jet stream is greatest in the downstream trough formed as air masses are deflected toward the equator after passing over a topographic high. Manabe and Terpstra (1974) suggest that the increased baroclinicity associated with the intensified jet stream partly controls the distribution of major storm systems. These authors note that simulations with mountains produce much more realistic storm tracks than simulations without high ranges. Cyclones are scattered widely over the mid-latitudes in no-mountain simulations but clearly track the wave-like features of the jet stream in model runs that include mountains.

In turn, the positions of the major storm systems play a major role in the distribution of surface precipitation. Precipitation is heaviest on the eastern side of stationary wave troughs such as over eastern North America and the North Atlantic and the eastern side of China and the adjacent northwestern Pacific – both areas where strong atmospheric disturbances result from an increase in jet stream velocity and vertical uplift (Figs. 6.2b and c, and 6.3; Broccoli and Manabe, 1992). Heavy rains are also associated with the western sides of northern hemisphere mountain fronts owing to uplift and cooling of air masses.

By contrast, large dry areas are associated with the eastern limbs of the stationary waves in mid-northern hemisphere latitudes (Fig. 6.2b). In some instances these dry regions are partly controlled by rain shadow effects where air masses have been forced to give up much of their moisture by vertical compression and cooling of the air as it rises on the western side of the range. Precipitation is inhibited on the lee side of the mountain belt because the air both expands vertically and warms as it descends after passing over the range.

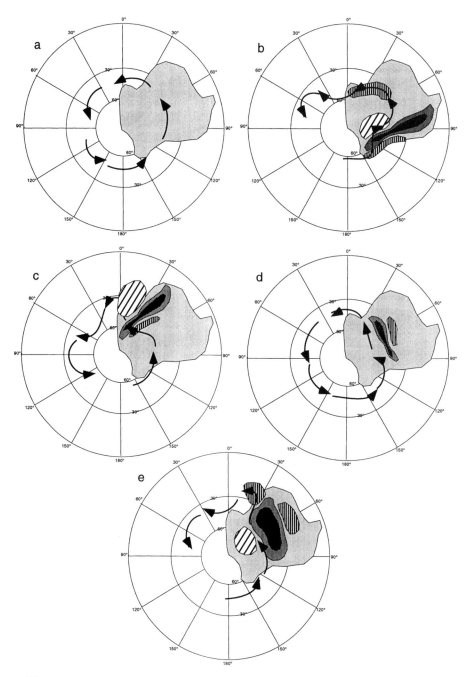

Figure 6.2. Atmospheric and precipitation effects of (a) no mountain, (b) western north-south mountain, (c) eastern north-south mountain, (d) east-west narrow mountain, and (e) east-west wide mountain in mid-latitudes (warm season circulation pattern). Arrows denote major flow paths of the westerlies and dark shaded areas are mountain belts. Diagonally ruled regions are areas of aridity whereas wavy ruled areas are regions of heavy rain fall. Note nearly zonal flow associated with no-mountain

However, rain shadow effects cannot explain the large dry areas of Mongolia and central Asia where there are no large blocking mountains in the path of the westerlies (Fig. 6.2b). Indeed, the westerlies are located well north of the Tibetan Plateau in summer. Aridity in central Asia appears to be partly a function of monsoon circulation around the Tibetan Plateau in which the atmospheric effects of the mountains extend well away from the mountains themselves (Manabe and Broccoli, 1990; Chase *et al.*, 1998). Summer heating of the plateau surface results in the formation of a zone of low pressure and rising air. Monsoon circulation produces cyclonic circulation around the base of the plateau and introduction of subsiding dry air across central Asia. In effect, the strong, low pressure cell over the Tibetan Plateau draws relatively cool, dry, Arctic air masses in from the north and contributes to aridification of central Asia and the stability of the Siberian high pressure system (Hahn and Manabe, 1975; Kutzbach *et al.*, 1989; Manabe and Broccoli, 1990) .

Importance of geography and elevation

Mountain orientation and paleogeography

The orientation and paleogeographic position of mountains play an important role in determining their climatological effects. Simulations run with Carboniferous and Mesozoic paleogeographies as well as hypothetical geographies have demonstrated that mountain belts modify global climate partly by altering the strength and position of atmospheric pressure systems and partly by anchoring eddies in the atmosphere (Moore *et al.*, 1992; Otto-Bliesner, 1993). Direct blocking of atmospheric circulation is also important in cases where mountains are in the path of major weather systems like the mid-latitude westerlies and the tropical easterlies.

Several studies have simulated Mesozoic and Paleozoic paleotopography when the orientation of mountain belts was much different from the modern case. Otto-Bliesner (1993) examined the climatological effects of the late Carboniferous Central Pangea Mountains which formed an east–west-oriented mountain range of Himalayan scale along the equator. The Central Pangea Mountains produced a low pressure zone in the middle of the continent which drew in moisture from adjacent oceans and maintained a belt of heavy precipitation in the tropics throughout the year. The mountain front effectively blocked cross-equatorial flow from the southern

Caption for Fig. 6.2 *(cont.)*
condition whereas north-south ranges produce strong equatorward deflections of the jet stream in the lee of the mountains. The amount of rain associated with the western side of the range in (c) is dependent upon how dry the air is after transiting the continent. Scenario (d) may develop monsoonal rain fall depending upon the latitude of the mountain belt, particularly if the range is near enough to the equator to anchor the position of the ITCZ as discussed in the text. Scenario (e) produces monsoonal precipitation both in the lee of the range and toward the equator owing to both the influence on jet stream path and the monsoonal circulation produced by strong summer heating and winter cooling over the plateau.

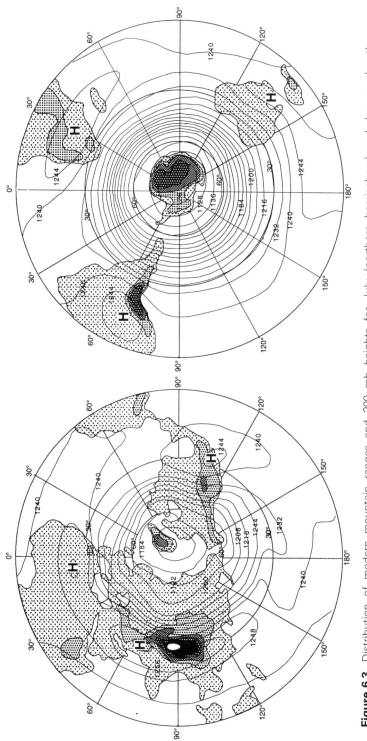

Figure 6.3. Distribution of modern mountain ranges and 200 mb heights for July (northern hemisphere) and January (southern hemisphere). Air circulation roughly follows mb height contours with clockwise flow around highs (labeled H) in the northern hemisphere and counterclockwise circulation around highs in the southern hemisphere. Note that most large mountains in the northern hemisphere block the westerlies (centered between 30 and 60° N) whereas there are no large ranges to block westerlies in the southern hemisphere. Accordingly there are strong high-pressure cells associated with northern hemisphere atmospheric circulation of the westerlies, but nearly zonal circulation in the southern hemisphere. Mountain elevations contoured in 1 km increments. Elevations after Manabe and Terpstra (1974) and 200 mb heights from Meehl (1992).

hemisphere to the northern and anchored precipitation despite the continued migration of the Intertropical Convergence Zone (ITCZ). Hence simulated tropical rain fall was far heavier than it would have been without tropical mountains (Otto-Bliesner, 1993).

Moore et al. (1992) simulated late Jurassic atmospheric dynamics using a variety of mountain elevations. They found enhanced monsoonal precipitation associated with a mountain range located along the mid-latitude eastern coast of Asia. There was also elevated precipitation on the eastern coast of North America apparently related to a deflection of the jet stream in a fashion analogous with that produced by the modern Rocky Mountains (Moore et al., 1992). By contrast, high ranges in Gondwana between 0 and 30° S and near 80° S had relatively few discernible effects on the Jurassic simulation relative to models run without high mountains, perhaps because neither range was situated within regions of strong wind velocities like those associated with the tropical easterlies or the mid-latitude westerlies. Likewise, Manabe and Terpstra (1974) found relatively modest effects of high mountains in the warm subtropics and polar regions. Their simulation of the modern southern hemisphere found that neither the high elevations of the Andes nor the high ice sheet of Antarctica had major effects on the nearly zonal circulation of westerlies in the southern hemisphere (Manabe and Terpstra, 1974). The highest parts of the Andes lie primarily between 0 and 30° S and so do not intersect the main belt of the westerlies between 40 and 60° S. Conversely, Antarctica is located entirely above 60° S latitude, and is also outside the main region of the westerlies. Accordingly, it appears that even very high mountains such as the Andes and the Antarctic ice sheet may have relatively slight effects on global atmospheric dynamics because they are located outside the main flow paths of surface air masses. This is not to say that these ranges are unimportant, however. The Andes clearly deflect atmospheric flow toward the equator in a fashion analogous with the trough formed in the jet stream downwind of the Rocky Mountains (Bolin, 1950; Meehl, 1992).

Hypothetical paleotopographies reveal some of the same features observed in more realistic model runs. Hay et al. (1990) modeled the effects of north–south mountain ranges on the eastern and western sides of continents. Their simulations showed that precipitation and storm tracks penetrate the continental interior when tropical mountains are present on the western side of a continent, much like the modern Amazon Basin. However, tropical mountains on the eastern side of a continent completely block the easterlies flowing off the ocean and resulted in no tropical precipitation anywhere in the continental interior (Hay et al., 1990). Such a range would have relatively little effect on the climate of the continental interior if located in mid-latitudes since it would not block the westerlies passing over the continent until they reached the western side of the range. However, by analogy with the modern Rocky Mountains, we might expect a high pressure cell to develop in the lee of the range over the coast. There is no modern analog for tall eastern mountain ranges, although the Jurassic East Asian range of Moore et al. (1992) may provide an example.

In summary, mid-latitude mountains and those located near the equator have major effects on global climate since they disrupt zonal flow in the major subtropical and tropical circulation systems. Tall, narrow ranges oriented along zonal flow paths have relatively few effects on atmospheric circulation since they do not oppose the flow or set up asymmetries in the atmosphere (Fig. 6.2d). The width of large plateaus is of great importance, since heating of a broad plateau can set up pressure cells in the overlying atmosphere that can produce very widespread disruptions of zonal flow (Fig. 6.2e). However, mountains concentrated in one hemisphere typically do not exert strong effects in the other hemisphere owing to the decoupling nature of the ITCZ between hemispheres. The position of mountains on the eastern or western coast of a continent is also important: mountains on a northern hemisphere western coast produce strong rain shadow effects on the west coast and cause a major deflection of the jet stream and storm tracks downstream of the range, whereas mountains concentrated on a northern hemisphere east coast do not have a similar effect on continental climate because they oppose air masses that have already been dehydrated during passage across the continent (Figs. 6.2b and c).

Mountain elevation

Kutzbach *et al.* (1989) found that there was an approximately linear response in the climatological effects of mountain ranges with an increase in their altitude. Their model runs suggest that the intensity of the Indian monsoon is related fairly directly to the height of the Tibetan Plateau, although other factors such as atmospheric CO_2, glacial and interglacial boundary conditions, and orbital geometry also play important roles in the intensity of precipitation in the Indian monsoon (Kutzbach *et al.*, 1989; Prell and Kutzbach, 1992). Model runs by other workers have found that there is probably some critical height at which paleotopography begins to produce detectable changes in climate on the scale considered by current models. For example, Barron and Washington (1984) found relatively modest effects of topography on global temperature and precipitation in models of Cretaceous and Eocene climate, which may be related to the relatively low (<1 km) elevations specified for most mountain ranges in their models.

Secondary climatic effects of mountain chains

Albedo

Large mountain chains strongly affect surface albedo through regulation of snow cover, vegetation types, and soil moisture. Different types of vegetation can have markedly different effects on albedo and surface humidity (Prentice *et al.*, 1992; Manzi and Planton, 1994; McGuffie *et al.*, 1995). In turn, albedo regulates the amount of heat absorbed by a mountain or plateau and thereby the return heating of the atmosphere with consequent effects on atmospheric circulation. Snow cover, being highly reflective, serves to keep a mountain system cool and reduces the strength of low pressure systems that might form over a mountain as the land surface heats the overlying atmosphere. Vegetation can also modify albedo as different types of plant assemblages have different degrees of reflectivity, particularly when covered with snow (Foley *et al.*, 1994). Grassland and tundra are nearly as reflective as bare rock when snow-covered whereas coniferous forest is much less so. Mountains control the distribution of plant communities by regulating surface temperature and precipitation, and seasonal variability in these factors. Conversely, vegetation types can strongly affect regional climate both within a mountain belt and in the downstream areas affected by the atmospheric modifications produced by a range. Finally, topographic highs produce feedbacks between soil moisture and precipitation. Broccoli and Manabe (1992) showed that the reduction in precipitation associated with the lee sides of many ranges results in a decrease in soil moisture. Heating of these soils in the summer can produce a decrease in evaporation from the soil and a drop in atmospheric humidity that feeds back into a further drop in precipitation. Calculations suggest that the soil moisture feedback may account for more than half the summer variability in precipitation within arid regions associated with mountain belts (Broccoli and Manabe, 1992). Not only does soil moisture affect regional precipitation, but through the link to atmospheric humidity it also affects the ability of the atmosphere to retain and transport heat.

Atmospheric CO_2 budget

The formation of major mountain belts within humid regions produces high rates of chemical weathering. Chemical weathering of silicate rocks involves the absorption of CO_2 from the atmosphere to form carbonates and silicates. Raymo (1991) and Raymo et al. (1988) have speculated that elevated rates of weathering associated with the formation of large topographic highs like the Himalayas and Andes should contribute to a drawdown in the concentration of CO_2 in the atmosphere. By their calculations, weathering could partly account for late Cenozoic cooling as greenhouse gas was taken up during weathering of the Himalayan or Andean detritus.

Climate model resolution of mountain chains

The spatial resolution of global climate simulations plays a critical role in determining the results that they produce. As Kutzbach et al. (1993) note, most global climate models do not resolve paleotopography very realistically. Narrow ranges tend to be portrayed as broad plateaus at the average elevation of a range, thereby reducing the total elevation of a given mountain chain and emphasizing the role of mountains in producing changes in atmospheric pressure systems over their role in blocking air masses. This 'plateau bias' may also produce inaccuracies in regions, like the modern Basin and Range, where tall but narrow ranges alternate with deep basins, since the modeled 'plateau' may have very different effects of atmospheric circulation from that of the actual topography.

Broccoli and Manabe (1992) note that simulation of precipitation and mid-latitude aridity is strongly dependent upon model resolution. Moore et al. (1992) note that proper simulation of paleotopography is of equal importance to accuracy in paleogeography when it comes to reconstructing realistic climates. Likewise, nested climate models which analyze particular regions of the globe tend to do a better job of simulating observed conditions (Giorgi et al., 1994) than do GCMs.

At the same time, global climate simulations have also been severely hampered by the absence of information on local paleotopography. Although we can reconstruct the elevations of individual mountain ranges based on a variety of techniques, it is a vastly greater job to make such reconstructions across the large areas that are needed to average out local effects in a GCM. Hence it seems likely that GCMs will continue to be limited to horizontal resolution of 5° or more if only because of inadequacies in our ability to reconstruct regional paleotopography.

Representation in the fossil record

Large mountain systems are naturally areas of substantial erosion, but they are also associated with high rates of sedimentation within intermontane basins. Accordingly, the distribution of paleofloras and other fossil localities – indeed, much of the terrestrial fossil evidence for ancient climates – tends to be associated with mountainous regions and volcanic centers. For example, many important megaflora sites for the Eocene of North America are located within the cordillera and much the same is true for Europe and Asia (Greenwood and Wing, 1995). Hence it is to be expected that many of our fossil proxies for continental climate may record mountain climates that may not be representative of interior lowlands. Furthermore, we may expect to observe greater variability in paleoclimate estimates based upon intermontane fossil assemblages than from lowland areas given the large variability in exposure, precipitation, soil moisture, and other factors such as soil types that is found in mountainous regions.

METHODS OF DETERMINING PALEOALTITUDE

The climatological importance of mountains is not matched by a wealth of information on paleotopography. To date most, if not all, techniques for estimating ancient topography have major limitations or large errors. Some techniques such as sedimentological evidence of mountain glaciers or clast diameters in conglomerates provide little data on absolute elevation. Others, such as the global climate simulations mentioned above, could provide rough estimates of paleotopography. In principle, mountains of different elevation and orientation should produce distinctive wind directions, precipitation patterns, or upwelling conditions that could be validated with existing paleoclimate proxies. However, present GCMs are probably not sufficiently sophisticated to provide more than sensitivity tests of different topographic scenarios. Here we consider several techniques that provide qualitative and quantitative estimates of paleotopography.

Methods from sedimentology, tectonics, and landform ages

A variety of methods have been employed to derive qualitative estimates of paleotopography, and these are discussed in somewhat more detail by Ziegler *et al.* (1997) and Chase *et al.* (1998). Sedimentological methods include analyses of clast size and composition to determine the size of adjacent highlands and their unroofing history. Coupled with data on the age and thickness of sedimentary sections adjacent to range-bounding faults, it is also possible to determine the volume of rock removed from a highland. This sediment can then be added back to the mountain to estimate the paleoelevations after an isostatic correction is made for the sediment loading. Unfortunately the history of denudation of mountain belts also involves climatic variables, changes in crustal or lithospheric thickness, and changes in uplift rates, all of which may be difficult to quantify with much precision.

Ziegler *et al.* (1985, 1997) have utilized the tectonic setting of ancient mountain belts to provide general estimates of paleotopography. In their scheme, a combination of paleomagnetic reconstructions of continental position and lithologies of rocks in ancient mountain systems and the types of plate boundaries (convergent, spreading, and strike–slip) is used to estimate the tectonic setting in which ancient mountain belts were formed. Analogies with modern mountain systems can then be used to estimate paleoaltitudes in ancient mountain systems. For instance, continent–continent suture belts associated with high temperature, high pressure metamorphic rocks are estimated to have average heights of over 4 km by analogy with the Himalayan–Tibetan Plateau (Ziegler *et al.*, 1985). By contrast, ocean–continent collision belts like the Andes and mountainous American West are associated with andesitic volcanism and average elevations over 2 km.

Basalt vesicularity technique

It has been known for some time that vesicle diameter and number in basaltic rocks are related partly to water pressure. Lava flows in Hawaii are nearly free of vesicles at 3000 m water depth but are highly vesiculated at the surface (Moore, 1965). Vesicle volume decreases approximately exponentially with increasing water depth (Moore, 1977). Sahagian and Maus (1994) adapted this relationship between pressure and vesicularity to deduce the paleoaltitude of basaltic eruptions. Their technique involves measuring vesicle diameter and then calculating the volumes of the modal bubble sizes – a quantity that is related to the atmospheric pressure. The volumes of the modal vesicle sizes are related to atmospheric pressure by the relationship

$$V_t/V_b = (P + pgH)/P,$$

where V_t and V_b are the volumes of the modal vesicle sizes at the top and bottom of a flow, p is the lava density, g is the acceleration due to gravity, H is the flow thickness, and P is the atmospheric pressure. The atmospheric pressure at sea level must be known or assumed to estimate the emplacement elevation of a basalt erupted at unknown elevation, and the method cannot account for changes in flow thickness during emplacement. Under the most favorable circumstances, the method has a reported error of about 0.1 bar which corresponds to a resolution in the elevation estimate of $\sim$1–1.4 km (Sahagian and Maus, 1994).

Mean annual temperature

Several workers have used paleobotanical data sets to quantitatively estimate paleoelevation (Gregory and Chase, 1992; Meyer, 1992; Povey *et al.*, 1994; Wolfe, 1994, 1995; Chase *et al.*, 1998). The technique involves analysis of the physiognomy (size distribution and morphology) of plants in fossil floras from contemporaneous highland sites and sea level sites. Plant physiognomy is used to estimate

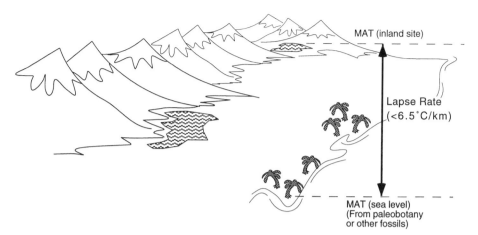

Figure 6.4. Estimating paleoelevation using the MAT technique. Mean annual temperatures are calculated for sites at sea level and interior, intermontane basins. Application of a lapse rate provides an estimate of the elevation of the interior site after correction for differences in paleolatitude of the various sites and differences in continentality.

mean annual temperature (MAT) for the two sites, which can be converted to a paleoelevation estimate for the upland site by application of a terrestrial lapse rate (Fig. 6.4).

Estimates of MAT are based upon a moderately large data set of modern floras and have an error of ~0.8 °C for floras with more than 15 taxa (e.g., Povey *et al.*, 1994). However, much larger errors are associated with the estimate of paleo-elevation since the method requires two corrections which are difficult to test. In the first case, part of the difference in temperature between the upland and sea level sites could be related to any latitudinal difference between them, requiring a correction for paleoaltitude (Gregory and Chase, 1992, 1994; Meyer, 1992; Povey *et al.*, 1994). Secondly, the temperature at the coast is not the same as the temperature at sea level for an inland site. Hence a correction also needs to be made for the difference in sea level MAT between coastal and inland areas that results from differences in humidity, continentality, and elevated base level, among other factors, (e.g. Gregory and Chase, 1992; Meyer, 1992). Both the latitudinal correction and that for sea level MAT between coastal and upland sites are known empirically for modern regions, but these relationships probably do not hold for ancient times when latitudinal thermal gradients may have been smaller than today and when coastal mountain ranges may have been lower. Furthermore, the method is strongly dependent upon the lapse rate that is employed. Wolfe (1992) calculated a mean regional environmental lapse rate of 3 °C km^{-1} while Meyer (1992) estimated the modern-day environmental lapse rate for many areas and showed that it can vary between 3.6 °C km^{-1} and over 8 °C km^{-1}. Clearly, large differences in estimated elevation can result just from the lapse rate used in the calculation (see below).

Enthalpy technique

Enthalpy, or the heat content of a body, is related to atmospheric pressure and hence to altitude. Forest et al. (1995) note that, for a given latitude, enthalpy (H) is fairly simply related to altitude (Z) by the equation

$$Z = (H_{\text{sea level}} - H_{\text{unknown}})/g,$$

where g is the gravitational acceleration and H_{unknown} is the enthalpy of a site at unknown elevation. These authors showed that enthalpy can be derived from plant physiognomy in much the same way that mean annual temperatures can be calculated using a multiple linear regression analysis of foliar characteristics for fairly large fossil floras.

This method is attractive because enthalpy is not dependent upon continentality or extrapolations to mean annual temperature at sea level for inland sites. The method also does not rely on environmental lapse rates and all their attendant uncertainties (see below). However, the method is sensitive to differences in paleo-latitude between sites at sea level and inland sites, particularly for regions where there are large departures from zonal air flow (as is the case in most mountainous regions). The method also assumes that atmospheric circulation patterns and humidity have not varied significantly over geologic time. Furthermore, the original data set used by Forest et al. (1995) did not include high humidity sites from Japan, Puerto Rico, and Panama which is unfortunate since these areas have floristic characteristics similar to those of many Cenozoic paleofloras (Chase et al., 1998). Nonetheless, for sites at similar latitude and approximately zonal air movement in the overlying atmosphere the method has an error of ~700 m (Forest et al., 1995).

The oxygen isotope paleoaltimeter

Drummond et al. (1993) used isotopic data of modern precipitation from eastern California (taken from Smith et al., 1979) and a compilation of $\delta^{18}O$ data from International Atomic Energy Agency–World Meteorological Organization (IAEA–WMO) stations to estimate the variation in $\delta^{18}O$ of precipitation as a function of altitude in a band between 38 and 48° N latitude. Regressions fit through the two separate sources of data both produce slopes of -4.2‰ km^{-1} but with different intercepts: $\sim 7\text{‰}_{(\text{SMOW})}$ for the (Smith et al., 1979) data set, and $\sim -10\text{‰}_{(\text{SMOW})}$ for the IAEA–WMO data set. Drummond et al. (1993) used stable isotopic measurements of lacustrine carbonates to estimate elevation of upper Miocene lake sediments using this modern $\delta^{18}O$/altitude relationship.

This method includes some significant sources of error that limit its applicability to studies of relatively high mountain systems. The modern isotopic data that provide the $\delta^{18}O$/elevation display considerable scatter ($\sim 10‰$ in the low-elevation sites). In addition, there are few modern stations at elevations above 2 km and only two stations above 3 km elevation. Consequently, the relationship between $\delta^{18}O$ and elevation is strongly controlled by a small number of high-elevation data points that may not be representative of large regions. Even if we take the regression fits as an accurate representation of the data, the modern data still yield errors of at least ~ 1 km (Drummond et al., 1993) and this is likely an underestimate of the error considering the small number of high-altitude modern stations in the modern calibration data set.

Furthermore, ^{18}O-enrichment through evaporation in lakes and streams will cause the method to underestimate the actual elevation in geologic data sets since the method assumes that the $\delta^{18}O$ of ancient lake sediments yields an estimate of the $\delta^{18}O$ in precipitation. One must also assume that the sources of moisture have not changed substantially over time since the moisture source can profoundly influence the isotopic composition of rainfall. For instance, clouds that have lost relatively little moisture since their formation over the ocean will produce precipitation with $\delta^{18}O$ only a little more negative than seawater whereas clouds that have lost a lot of water in transit will produce precipitation with strongly depleted $\delta^{18}O$ (see below).

A new isotopic technique: the snowline/MAT paleoaltimeter

Both the MAT and enthalpy estimates of paleoaltitude rely on paleobotanical data. Thus these methods are best at estimating the elevations of intermontane basins and coastal sediments in which many floras are preserved. The basalt vesicularity approach can measure any paleoelevation at which basalts were erupted, but has such large errors that it is only practical to study samples extruded at very high altitudes.

We have developed a method of estimating the difference in relative elevation between intermontane basins and the high elevations of ranges near the tree line. This method requires three types of data: (1) isotopic evidence for snow pack, (2) paleobotanical or other estimates of MAT for some site well below the snowline, and (3) the lapse rate (Fig. 6.5). In short, if there is evidence that snow accumulated as a snow pack and the mean annual temperature of a region below the snowline is

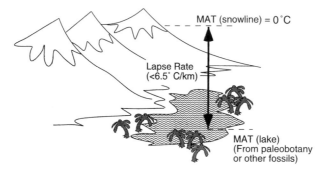

Figure 6.5. Method for determination of paleoelevation using isotopic methods. The method relies on possessing (1) mean annual temperature for an intermontane basin (generally from fossil evidence), and (2) evidence for snow pack (and evidence that snow did not accumulate in the floor of the intermontane basin). We can then calculate the elevation difference between the basin and the lower elevation of the snow pack by using a lapse rate, the MAT estimated for the basin, and the 0 °C isotherm for the mean annual temperature of snow pack.

known, such as an intermontane basin, a lapse rate can be used to estimate the difference in elevation between the basin and the snowline.

The method is useful since we have few methods for estimating the height of mountains that surround intermontane basins. Our method, in combination with enthalpy-based or MAT-based calculations of intermontane basin elevation, can examine the range of absolute paleoaltitudes within a mountain chain.

The most critical quantity for isotopic determination of paleoaltitude is identification of snow pack. Snow has distinctly negative $\delta^{18}O$ ratios that may be preserved in lake sediments or the shell of organisms that lived in lakes and their feeder streams. For low to mid-latitude regions, snow precipitated at high elevations typically has a $\delta^{18}O$ of less than about $-17‰$ to $-20‰$ whereas rain is much less negative with typical ratios of $-10‰$ or more (Fig. 6.6; Dansgaard, 1964; Friedman *et al.*, 1964; Gat, 1996). The $\delta^{18}O$ of both rain and snow decreases with increasing latitude as a result of Rayleigh distillation but the difference in their isotopic chemistries remains pronounced. Still, below $\sim45°$ N, $\delta^{18}O$ of less than $-17‰$ to $-20‰$ in lake water implies a significant percentage of snow melt in the lake (Fig. 6.6; Friedman *et al.*, 1964). In such a case, either the snow accumulated in snow pack or most of the precipitation occurred in the winter by snowfall at low elevations.

The $\delta^{18}O$ of rainfall can reach ratios of $-17‰$ under unusual circumstances, but generally the $\delta^{18}O$ of surface waters is considerably more positive than this. Areas with heavy rainfall (e.g., the Amazon Basin) can experience considerable removal of ^{18}O from the vapor (Grootes, 1993). As a result, the $\delta^{18}O$ of precipitation becomes progressively more negative as rain clouds move inland from the coast, particularly during the heavy rains of the wet season. This 'amount effect' (Dansgaard, 1964) or 'rain-out effect' dramatically reduces the $\delta^{18}O$ of precipitation across Europe toward the Urals (Rozanski *et al.*, 1993). The average depletion of ^{18}O due to 'rain-out' is about $2‰$ to $5‰_{(SMOW)}$ in the Amazon and $4‰$ to $5‰_{(SMOW)}$ across Europe, with extreme ^{18}O depletions of up to $-17‰$ in both regions (Dansgaard, 1964; Rozanski *et al.*, 1993). However, even in the Amazon, where rain-out effects can be intense, average $\delta^{18}O$ of precipitation is typically no less than $-5‰$ to $-6‰$ because extremely ^{18}O-depleted rain is a small proportion of total seasonal rain fall (Rozanski *et al.*, 1993).

We cannot easily distinguish between snow fall at different elevations using isotopic techniques; however, it is probably reasonable to consider snow pack as the main source of snow melt in lakes because only large volumes of snow melt could substantially modify the chemistry of a large body of water. Today about 70% of the runoff in the western United States is derived from melting snow in the mountains (Barry, 1990, p. 168), suggesting that low-altitude snow fall might not be sufficiently voluminous to strongly affect the $\delta^{18}O$ of surface waters. Still, other geological evidence is desirable to verify the ancient occurrence of snow pack. In particular, MAT estimates of above-freezing temperatures, or the occurrence of warm-loving species of plants or animals in the lake sediments, would strongly suggest that snow melt filling a lake did not fall at the lake elevation. For example, palms, turtles, and crocodilians are very sensitive to low temperatures and are limited to areas that do not experience prolonged freezing temperatures (Hutchison, 1982; Chaloner and

a. Rain - Temperate Latitudes

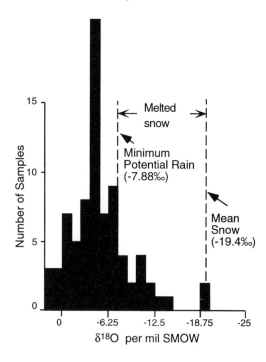

b. Snow - Temperate Latitudes

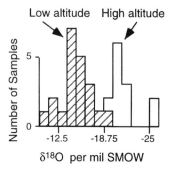

Figure 6.6. Range of $\delta^{18}O$ ratios (SMOW) for temperate-latitude precipitation: (a) rain fall, and (b) snow fall. Ratios in rain fall less than −7.88‰ represent rain formed by melting snow during its transit to the ground. Low-altitude snow fall (~<1 km) has more positive $\delta^{18}O$ than snow accumulating at higher elevations. Data from Friedman *et al.* (1964).

Creber, 1990; Wing and Greenwood, 1993; Greenwood and Wing, 1995). The presence of such organisms in lake or stream sediments would suggest that the snow melt filling these water bodies was derived either from high altitudes or from a distant highland. Furthermore, many intermontane basins are hydrologically closed; if closed-basin hydrology can be demonstrated for an ancient lake system then it is likely that any snow melt reaching the lake is locally produced from snow pack on surrounding mountains (Norris *et al.*, 1996).

Hence the isotopic technique is based on three types of geologic and paleontological data: (1) isotopic evidence that snow melt is present in lake water; (2) evidence that this snow cannot have fallen at the elevation of the lake itself based on paleobotanical or faunal proxies for above-freezing conditions in or near the lake; and (3) evidence that the lake was in a closed basin so that snow melt was unlikely to have been introduced from some distant highland. Once the existence of a snow pack feeding a lake has been confirmed, a lapse rate that defines the temperature change with change in altitude can be used to estimate the difference in elevation between the lake surface and snowline.

Estimates of snowline temperature

A further requirement is some knowledge of the warmest temperatures at which snow pack can accumulate. We have previously used the 0 °C isotherm for the warmest mean annual temperature at which snow pack can exist (Norris *et al.*, 1996). Karl *et al.* (1993) compiled data on snow depth and mean maximum annual temperature for snow stations throughout North America. They found that the mean maximum annual temperature of all snow stations in the contiguous United States was about 0 °C with a range of ±1 °C during the 40-year span between 1950 and 1990 (Karl *et al.*, 1993). The mean maximum annual temperature is obviously greater than mean annual temperature, suggesting that average snow accumulation occurs at mean annual temperatures below 0 °C. Similar results were reported by Karl *et al.* (1993) for compilations of stations in southern Canada and Alaska.

In the Rocky Mountains, 0 °C is the mean annual temperature at the minimum elevation of significant snow pack (~3000 m; M. Williams, personal communication, 1995; Williams *et al.*, 1996) and is the mean annual temperature at which about 50% of precipitation falls as snow for a global compilation of stations near sea level (Lauscher, 1976, p. 152; Barry, 1981, p. 191). Because snow contains about 10% of the water in an equivalent volume of rain (e.g., Karl *et al.* 1993, p. 1335), snow fall does not begin to have a significant impact on runoff until the ratio of snow fall to rain reaches well over 50%. Hence the estimate of 0 °C for the minimum elevation of snow pack is probably conservative.

Seltzer (1990) noted that the elevation of the snowline is influenced by cloud cover and aridity as well as temperature. Snowline studies in the Peruvian–Bolivian Andes show that in areas of high to moderate precipitation the snowline is close to the 0 °C atmospheric isotherm whereas areas under dry air masses may have snowlines as much as 1000 m above the summer 0 °C atmospheric isotherm (Seltzer, 1990). For example, snowlines occur at ~ 4800 m in much of Peru and northern Bolivia because of the persistence of the 0 °C isotherm near that altitude. However, in southern Bolivia ablation and dry air maintain snowlines at higher elevations than the 0 °C isotherm and snow pack can even be absent from peaks over 6000 m high owing to extreme aridity (Seltzer, 1990). Snowlines and the 0 °C isotherm can dip below 4800 m in areas where cloud cover blocks incoming solar radiation (Seltzer, 1990).

What lapse rate to use?

The free air lapse rate of 6.5 °C km^{-1} reflects the decrease in air temperature with increasing altitude in the atmosphere (Battan, 1979). Lapse rate is strongly dependent upon where it is measured. Close to the surface of a mountain, the 'terrestrial' lapse rate is typically lower than the free air lapse rate and has values of between 3 °C km^{-1} to more than 6 °C km^{-1} (Meyer, 1992). For example, stations in the Front Range of Colorado display environmental lapse rates of between 2.3 °C km^{-1} to 4.8 °C km^{-1} depending upon exposure to wind, humidity, and cloud cover (Barry, 1990). A regression analysis of 50 climate stations from Andean sites in Ecuador and Peru yielded an environmental lapse rate of 6.5 °C km^{-1} between 2500 and 4500 m (Johnson, 1976). Similar studies in Bolivia produced summer lapse rates of 6.5 °C km^{-1} between 2500 and 3600 m and $\sim$7.5 °C km^{-1} 1000 m above this (Graf, 1981). Meyer (1992) calculated terrestrial lapse rates for a large number of localities around the globe and determined a global mean lapse rate of 5.89 °C km^{-1} but also found wide variance around this mean. By contrast, Wolfe (1992) calculated a mean regional lapse rate of 3 °C km^{-1} for modern stations that were used to calibrate foliar physiognomy and mean annual temperatures.

The wide range of environmental lapse rates translates into large differences in paleoaltitude. For an 18 °C difference in temperature between a highland station and a lowland station, an environmental lapse rate of 3 °C km^{-1} suggests an elevation difference between the stations of 6 km whereas a lapse rate of 5.86 °C km^{-1} implies only a 3 km difference. Clearly, the choice of lapse rate is critical to the estimate of paleoaltitude. The most conservative approach is to calculate paleoaltitude with a range of lapse rates. The free air lapse rate will generally provide the measure of minimum paleoaltitude, while the altitude estimate will increase as lapse rate declines.

Error analysis of the snowline/MAT technique

The major sources of error include determining the appropriate lapse rate, the error of MAT estimates from paleofloras, and variations in snowline elevation around the 0 °C isotherm.

As noted above, lapse rate varies tremendously making it prudent to use maximal lapse rates (the free air rate of 6.5 °C km^{-1}) in order to make conservative estimates of elevation while recognizing that lower lapse rates like those estimated by Meyer (1992) may be more appropriate for vegetated slopes at relatively low elevations. Use of the free air lapse rate may then provide a minimum estimate of paleo-topography, but have errors of +2 km or more on the maximum elevation of the 0 °C isotherm.

By contrast, the error associated with MAT estimates is considered to be no better than $\sim$0.8 °C for paleofloras with more than 15 taxa (Povey *et al.*, 1994), resulting in a minimum error of ±120 to ±140 m for lapse rates of 6.5 °C km^{-1} and 5.89 °C km^{-1}, respectively. Wolfe (1994, 1995) estimated errors for MAT at 0.7–1 °C for a more modern version of his modern leaf character database, which implies errors similar to those estimated by Povey *et al.* (1994).

Finally, the relationship between snowline and the 0 °C isotherm is reasonably good for mountain belts bathed in relatively moist air masses but the snowline can move as much as several kilometers above the 0 °C isotherm in extremely dry conditions, as noted above. For practical purposes the 0 °C isotherm is probably no more than a few hundred meters below the snowline in cases where the snowline/MAT technique might be applied, because the method relies on having large enough runoff from snow pack to affect the chemistry of lake carbonates and such a supply would be problematic in very arid conditions. Finally, it is obvious that the method does not calculate total mountain elevation but only the height of the 0 °C isotherm. Accordingly, the snowline/MAT technique is of little value for regions where the basin is close to the snowline.

Errors in determination of the presence of snow melt are not considered in detail here since it is necessary to make a convincing case for the presence of substantial snow melt in the first place, by whatever technique, before a data set can be used for paleoelevation analysis. Still, analytical precision for $\delta^{18}O$ analysis is typically better than 0.1‰, which implies that the largest errors in snow melt determination have to do with sediment diagenesis, alteration of the original snow $\delta^{18}O$ by evaporation, or dilution by rainfall or ground water, and 'vital effects' caused by non-equilibrium fractionation during precipitation of biogenic carbonates.

Case study: isotopic estimates of the paleoelevation in Eocene Laramide mountains

During the late Paleocene to middle Eocene, large lakes occupied several of the intermontane basins in Colorado, Wyoming, and Utah (Fig. 6.7; Bradley, 1929; Bradley, 1964; Surdam and Wolfbauer, 1975; Sullivan, 1985; Roehler, 1993; Matthew and Permutter, 1994). These lakes contain rich vertebrate and floral assemblages that can provide estimates of paleotemperature at lake level. In addition, the lake deposits in the Green River Formation are largely calcareous sediments that are suitable for isotopic study of the chemistry of the lake water. Therefore, we have used our isotopic technique described above to estimate the elevation of the mountain ranges surrounding the Green River lakes.

Previous estimates of Laramide mountain elevation

Climate models of Eocene North America have generally estimated the average elevation of the Eocene cordillera at 500 m to perhaps 1500 m above sea level (Bradley, 1963; Bradley, 1964; Barron and Washington, 1984; Sloan and Barron, 1992b; Sloan 1994). However, the altitude of the highlands and the plains they rise from is relatively speculative. Most authors have followed Bradley (1964) in proposing that the Laramide basins were 150–300 m above sea level while the adjacent mountains may have reached elevations of ∼1500 m. Bradley (1929, 1963) suggested the mountains reached altitudes above the basins similar to the modern Uintas, which suggests an elevation of ∼2100 m. The best of these estimates are based on combining temperatures derived from paleofloras with lapse rate calculations (e.g., Bradley, 1963). However, there are no data to indicate the minimum

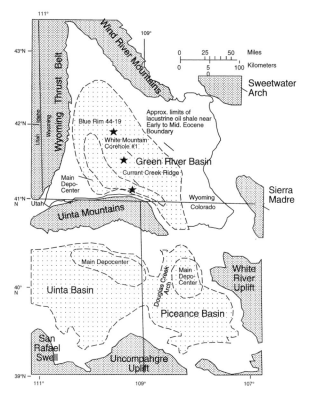

Figure 6.7. Distribution of Green River Formation in Colorado, Wyoming, and Utah.

temperatures that were present on the mountain peaks, which leaves considerable uncertainty in the maximum estimated altitude.

More recently, several workers have used the MAT and enthalpy techniques to suggest that at least some of the Laramide basins formed at elevations of over 2000 m. Although it is not a Laramide basin (having been deposited in a river valley on a Laramide highland), the Florissant lake beds have also been used in paleoelevation studies. Gregory and Chase (1992), Meyer (1992), Gregory (1994) and Forest *et al.* (1995) have proposed that the late Eocene Florissant Beds in the Front Range of Colorado were deposited at an elevation of 2.4–2.9 km – an altitude apparently similar to that of the Denver Basin based on the distribution of ash flow deposits (Chase *et al.*, 1998). Their work suggests that the Laramide intermontane basins had similar elevation to today and implies a much earlier uplift of the southern Rocky Mountains than was proposed by previous studies (Gregory and Chase, 1992, 1994; Gregory, 1994).

Climate of the Green River lakes

The Green River lakes fluctuated greatly in size and salinity and ranged from deep, freshwater bodies, to saline lakes that may have dried up completely now and then (Bradley, 1929, 1964; Surdam and Wolfbauer, 1975; Sullivan, 1985;

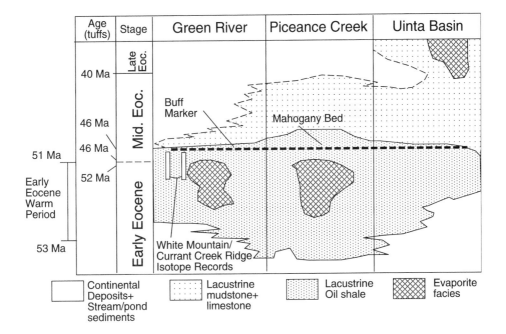

Figure 6.8. Generalized stratigraphic sequences in the various Green River depocenters. All the basins record an evaporite sequence in the middle of the formation sandwiched between intervals of extensive oil shale deposition.

Roehler, 1993). Thick trona and nahcolite beds in parts of the lake sequence suggest that the lakes became highly saline, while oil shales throughout much of the Green River Formation are consistent with an interpretation that the lakes were often highly stratified with anoxic bottom waters (Fig. 6.8; summarized in Roehler, 1993). The climate of the region surrounding the lakes apparently varied from temperate, arid conditions to warm, subtropical conditions judging from paleobotanical and fossil vertebrate finds (MacGinitie, 1969; Leopold and MacGinitie, 1972; Hutchison, 1982; Wing and Greenwood, 1993; Markwick, 1994). Faunal and floral data suggest that temperatures rarely dipped below freezing at the elevation of the lakes (summarized in Greenwood and Wing, 1995). Analysis of growth rings in fossil wood also suggests that the climate near lake level was seasonal, but not extreme (Berry, 1925).

Isotopic chemistry of Green River lake deposits
     We analyzed the stable isotopic composition of bulk carbonate samples in several cores from the early to middle Eocene in southwestern Wyoming near the center of Eocene Lake Gosiute (Fig. 6.7). Our data come from the Energy Research and Development Administration (ERDA) White Mountain No. 1 core, the United States Department of Energy (DOE) Currant Creek Ridge No. 1 core and the Union Pacific Railroad Company Blue Rim No. 44–19 core. The correlation between these

cores is based on numerous distinctive tuffs, and oil shale cycles (Roehler, 1991). All our data come from the upper third of the Wilkins Peak Member and lowermost part of the Laney Member of the Green River Formation (Fig. 6.8).

Both the DOE Currant Creek Ridge No. 1 core and the ERDA White Mountain No. 1 core display approximately correlative intervals of highly depleted $\delta^{18}O$ in the interval between bed 67 and the main tuff (Fig. 6.9), and ERDA White Mountain No. 1 also has a series of highly negative single point excursions in the vicinity of bed 72. Typical $\delta^{18}O$ is between -8 and $-10‰$ within the unusually depleted intervals and reaches extreme ratios of nearly $-16‰$ in the ERDA White Mountain No. 1 core near 540 ft (160 m) depth. Re-sampling the highly ^{18}O-depleted intervals in the White Mountain No. 1 core produced a similar range of $\delta^{18}O$ to that originally identified in this core (Norris *et al.*, 1996). The average background $\delta^{18}O$ of the calcareous mudstones and calcareous oil shales measured here is about $-3.5‰$.

The intervals of very negative $\delta^{18}O$ are unlikely to record very late stage diagenesis or contamination by modern ground water. The excursions occur in a number of different lithologies such as oil shale, gray micritic limestone, green calcareous mudstone, and shortite-bearing calcareous mudstone, which suggests that diagenesis is not the source of the negative $\delta^{18}O$ spikes (Fig. 6.10). Analysis of shortite crystals and calcareous mudstone from the same sample levels in the Blue Rim core shows that the shortite tends to have more negative $\delta^{18}O$ than the mudstone by 2–3‰ (Fig. 6.11). Shortite crystallizes from brines in the burial environment following compaction of the sediments and has been shown experimentally to grow authigenically from calcite at temperatures of $>90\,°C$ (Bradley and Eugster, 1969). However, we have avoided analysis of shortite other than our test in the Blue Rim core. The shortite crystals occur as coarse crystals (>2–3 mm to 1 cm) within green calcareous mudstone and are easy to separate from the matrix. Furthermore, shortite is uncommon within other lithologies like oil shale and limestone that have also produced sharp negative $\delta^{18}O$ excursions.

Some of the variability in our isotopic records could reflect the isotopic difference between calcite and dolomite. Weber (1964) reported differences between the two carbonate phases of as much as $-4.6‰$ $\delta^{18}O$ for a small set of paired samples from Lake Flagstaff, a lake system in central Utah contemporaneous with the Green River lakes. However, a large suite of samples from the Green River Formation in Colorado and Wyoming display fairly small differences between calcite and dolomite in paired samples (delta $\delta^{18}O = -0.98‰$; $n = 73$: J. Pitman, United States Geological Survey, personal communication, 1995). Both Weber (1964) and Pitman (personal communication, 1995) show that calcite has more negative $\delta^{18}O$ than dolomite. Hence one cannot explain the large negative excursions in our data by isotopic depletion during dolomitization.

Isotopic evidence for snow melt?

We believe that the sharp negative $\delta^{18}O$ excursions in the Wyoming Green River cores reflect the presence of snow melt in the lake system. Carbonates with $\delta^{18}O$ of $-15.8‰_{(PDB)}$ (our most ^{18}O-depleted ratio) would be in equilibrium with

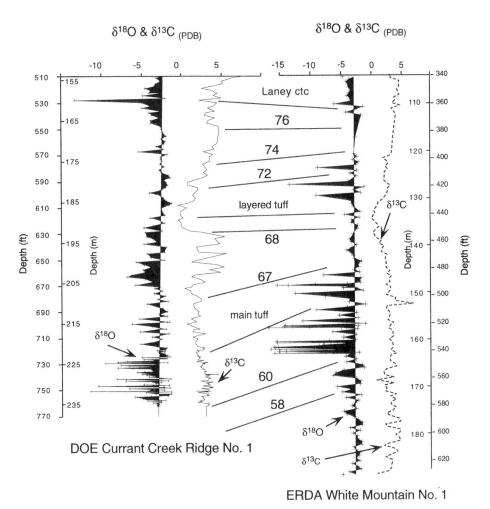

Figure 6.9. $\delta^{18}O$ and $\delta^{13}C$ of calcareous mudstones and oil shale in the DOE Currant Creek Ridge No. 1 core and the ERDA White Mountain No. 1 core. Correlation lines and bed numbers are those of Roehler (1991) and are based on distinctive tuff beds and oil shale cycles. Note intervals of extremely depleted $\delta^{18}O$ in correlative parts of both cores which are interpreted here as intervals when snow melt became a major source of lake water. Depth scale presented in feet as well as meters since all the original well logs and core descriptions are in feet. Laney ctc is contact between Laney member and Wilkins Peak Member of the Green River Formation.

water of $-12.9‰$ to $-19.8‰_{(SMOW)}$ assuming precipitation temperatures between 0 and 30 °C, respectively, and using the paleotemperature equation for the carbonate–water system (Friedman and O'Neil, 1977). The estimated $-12.9‰$ to $-19.8‰$ range of lake water $\delta^{18}O$ is at, or above, the upper range of ratios expected for rain fall and is consistent with snow melt that has been evaporatively enriched in ^{18}O within the lake.

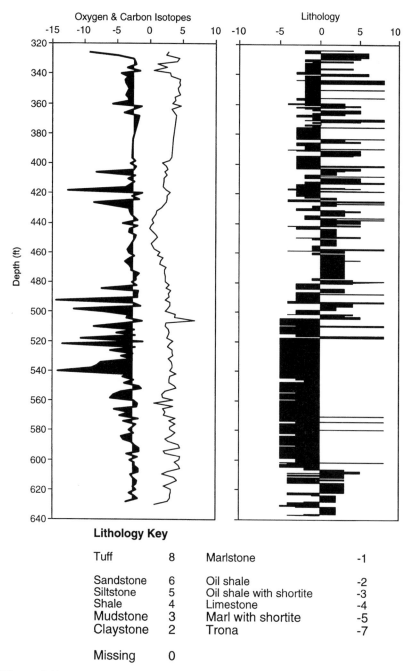

Figure 6.10. Comparison of the ERDA White Mountain No. 1 isotope record with lithology of the core. Extremely negative δ^{18}O excursions are associated with most of the major carbonate lithologies in the core.

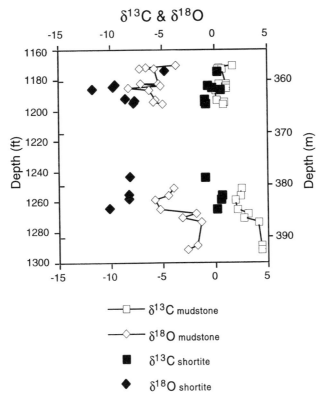

Figure 6.11. Comparison of $\delta^{18}O$ data for green calcareous mudstone and shortite crystals in the same samples over a short interval of the Union Pacific Railroad Blue Rim 44-19 core. Note that shortite constantly yields more negative $\delta^{18}O$ than the micritic matrix. Accordingly, we have avoided sampling shortite whenever possible in the other cores described here.

The $\delta^{18}O$ of rain or snow fall is strongly controlled by its condensation temperature so that lower temperatures produce precipitation depleted in ^{18}O (Friedman *et al.*, 1964). A result is that snow has much lower $\delta^{18}O$ than rain fall from the same clouds. In turn, the isotopic composition of surface water is strongly affected by the ratio of rain fall to snow melt. Surface waters from the Mississippi River drainage have $\delta^{18}O$ of about −5‰ to −6‰(SMOW) due to mixing <50% snow melt with isotopically more positive rain fall (Friedman *et al.*, 1964). By contrast, surface waters in the mountain states have $\delta^{18}O$ typically between −10‰ and −17‰(SMOW) which reflects the substantial contribution of high-altitude snow fall to surface waters in this region (Dansgaard, 1964; Friedman *et al.*, 1964).

In turn, the isotopic composition of lake sediments is controlled by the composition of lake source waters and by the residence time of water in the lake (Friedman *et al.*, 1964; Talbot, 1990; Talbot and Kelts, 1990; Gat, 1996). Source water composition reflects a combination of rain fall and snow melt as well as water from springs and seeps. Isotope ratios of lake water can be altered further by eva-poration which removes the lighter, more reactive, isotopes of oxygen. A key point is that lakes produce an integrated $\delta^{18}O$ signal that reflects not only the composition of precipitation but also the effects of processes occurring in the lake.

Post-precipitation evaporation, whether in the drainage basin or in the lake, results in a shift in $\delta^{18}O$ toward more positive ratios (Talbot and Kelts, 1990; Gat, 1996). Rain and snow will nearly always have more negative $\delta^{18}O$ than lake waters, particularly if waters remain in the lake for a long time and the lake has no outlet. For example, runoff entering the Great Salt Lake has a $\delta^{18}O$ of $-20‰$ that reflects the high content of snow melt; however, the lake water itself has a $\delta^{18}O$ of about $-12.4‰$ as a result of high evaporation in the lake basin (Friedman *et al.*, 1964). The Great Salt Lake is not unusual: the difference in $\delta^{18}O$ between modern lakes in the western USA and their feeder rivers ranges between 4.7 and 11.9‰ (average delta $^{18}O_{lake-runoff} = 8.4‰$, $n = 7$; Friedman *et al.*, 1964). Hence carbonates formed in equilibrium with lake waters will tend to have more positive $\delta^{18}O$ than the original precipitation.

Our measurements of Green River lake water composition must underesti-mate the true ^{18}O depletion of the original precipitation by between 4.7 and 11.9‰ if we take the isotopic enrichment in modern lakes as representative of possible Eocene lakes. The Green River lakes were undoubtedly saline bodies, particularly during deposition of the Wilkins Peak Member, as suggested by the extensive accumulations of trona and authigenic salts and paleoclimate data suggesting a semi-arid environ-ment (e.g., Bradley and Eugster, 1969).

Accordingly, if we apply a corection of $>5‰$ $\delta^{18}O$ to our measured values of Green River lake water, we should arrive at an estimate of the composition of the runoff that originally fed the lake. When we apply a $>5‰$ correction to the 'back-ground' $\delta^{18}O$ we have measured in our cores ($\sim -2.5‰_{(PDB)}$ or 0.49 to $-6.5‰_{(SMOW)}$ over a 0–30 °C temperature range), average lake source water $\delta^{18}O$ should lie between -4.5 and $-11.5‰$. A similar correction to the most negative $\delta^{18}O$ measured in our cores ($-15.8‰_{(PDB)}$ or -12.9 to $-19.8‰_{(SMOW)}$ over a 0–30 °C temperature range) should have originated from runoff with $\delta^{18}O < -17.9‰_{(SMOW)}$ to $-24.8‰_{(SMOW)}$. Note that the $-17.9‰$ estimate for lake water $\delta^{18}O$ is very conser-vative since it assumes carbonate precipitation at 0 °C and minimal evaporative enrichment of water in a lake (use of the mean salinity correction of 8.4‰ would suggest that runoff $\delta^{18}O$ was $\sim -21.3‰_{(SMOW)}$.

Our estimates of the average $\delta^{18}O$ of Eocene lake water (-4.5 to $-11.5‰$) are similar to estimates for late Paleocene to early Eocene meteoric waters from the Bighorn Basin (-5 to $-14‰_{(SMOW)}$) made on isotopic analyses of bivalves and mammal teeth (Koch *et al.*, 1995). Likewise, the negative excursions in the Green River Formation (<-17.9 to $-24.8‰$) are comparable to ratios for meteoric water estimated from individual bivalve shells from the late Paleocene of the Bighorn Basin ($\delta^{18}O$ of -16.8 to $-20‰_{(SMOW)}$; Dettman and Lohmann, 1993). Dettman and Lohmann (1993) attributed their very negative $\delta^{18}O$ to the growth of bivalves in water enriched in snow melt, noting that only snow fall can account for isotopic depletions as extreme as those preserved in their bivalve carbonate.

We would not expect the isotopic signature of extremely ^{18}O-depleted rain to be preserved in large lakes like those in the Green River basins because the lakes should record the average chemistry of all the runoff during the year. Hence the more negative estimates of Eocene meteoric water $\delta^{18}O$ require either that wet season precipitation completely dominated the isotopic composition of annual rain fall or that rain fall was mixed with snow melt. Since intervals with very negative $\delta^{18}O$ are relatively rare in the Green River samples, it may be that 'snow-melt' events are relatively short-lived in the lake record and that most of the sequence reflects either a rain-out process or strong evaporative aging of snow melt in lake water.

Paleotopography of the Eocene Rocky Mountains

Evidence for substantial amounts of snow melt reaching the Green River lakes suggests that the snow originally accumulated in snow pack. There is abundant paleobotanical and faunal evidence that temperatures rarely reached freezing temperatures near lake level. As just one example, Chaloner and Creber (1990) and Greenwood and Wing (1995) note that palms are essentially restricted to areas with cold season mean temperatures of $>5\,°C$, yet palms are present in paleofloras as far as 650 km north of the Green River Basin. Hence snow cannot have fallen in any abundance at lake level and must have originated either at high elevation or on a distant highland. Geological evidence that the Green River Formation in southwest Wyoming (Lake Gosiute) was largely deposited in a closed basin (summarized in Roehler, 1993) suggests that snow melt reaching the lake was probably derived from the mountains surrounding the basin rather than a distant source. Therefore we can estimate the minimum elevation of these mountains by combining the isotopic evidence for significant snow melt, paleontological estimates of the mean annual temperature at lake level, and the lapse rate for temperature decrease with increasing altitude.

As noted above, snow pack probably does not accumulate at mean annual temperatures above about $0\,°C$. When mean annual temperatures climb above $0\,°C$, snow melt becomes increasingly diluted by rain fall (Barry, 1981), rendering it unlikely that a strong isotopic signal of snow melt will be preserved in surface waters.

Using a lapse rate of $6.5\,°C\,km^{-1}$ (Battan, 1979) and a mean annual temperature of $18\,°C$ at Lake Gosiute during deposition (Wing and Greenwood, 1993), the mean annual temperature was $0\,°C$ at 2800 m above the lake. The use of smaller environmental lapse rates ($5.89\,°C\,km^{-1}$ to $3\,°C\,km^{-1}$) suggested by Gregory and Chase (1992), Meyer (1992), and Wolfe (1990, 1992) results in much larger elevation estimates that range from 3070 m to 6000 m.

Although the paleoaltitude of the lakes is unknown, estimates of 150–300 m above sea level have been reported (Bradley, 1929; Bradley and Eugster, 1969; Sloan and Barron, 1992b; Wing and Greenwood, 1993) which imply that the snow fields may have accumulated at elevations of >3000 m. MAT-based paleoelevation estimates from six early Eocene paleofloras in North Dakota, Wyoming, and Montana range from ~1 km to 2.2 km (using a $5.5\,°C\,km^{-1}$ lapse rate) and are broadly similar to the modern elevations of these sites (summarized by Chase et al., 1998). Their work suggests that the Rocky Mountain region reached substantial elevation well before the late Neogene uplift suggested by earlier workers, and opens the possibility that the Green River basins may have had comparable elevations during the early Eocene. Indeed, a number of lower Eocene paleofloras in Wyoming (Green River flora – Green River and Washakie basins; Wind River flora – Wind River Basin) probably formed at an elevation of 2.3±1.3 km to 3.1±1.5 km depending upon the lapse rate used (Chase *et al.*, 1998). The combination of the minimum elevation estimate proposed by Chase *et al.* (1998) of ~2 km to 3 km and our minimum elevation for the surrounding mountains above the Green River lakes suggests a total elevation above sea level of at least 4–5 km. All of these calculations are considerably higher than previous estimates for mountain elevations of 500–1500 m above sea level.

Estimated errors in our calculation of mountain height

The errors attendant to our calculation include ~100–200 m for errors in the MAT estimate of lake-level temperature, uncertainties of several hundred meters in the elevation of the snowline above the 0 °C isotherm, and an error of +1 km or more if the lapse rate is actually lower than 6.5 °C km^{-1}. The last two errors show only that the snowline in the Eocene mountains could have been a kilometer or more higher than our conservative estimate. A liberal error in the MAT method represents less than 7% of the calculated 3 km elevation above lake level. Furthermore, the presence of snow melt in the Green River lake system requires that the mountains rose sufficiently above the snowline to accumulate a large snow pack, so the 3 km mountain height above the lake is conservative, indeed. By far the largest two-way error in the elevation calculation is the ~ ±1.3 km error in the MAT-based calculation of the elevation of the Green River lake system, which suggests, that at a minimum, the 0 °C isotherm could have been at as little as 4 km elevation above sea level.

Implications for Eocene continental climate

Large mountains during the Eocene would have substantial effects on the regional climate of North America. Large mountains (>3000 m) resting upon a moderately high plateau (>2000 m) should anchor the jet stream and stationary eddy over the mid-continent. A result should be a broad semi-arid belt to the east of the range and a sharp temperature contrast between the western and eastern side of the cordillera. Storm tracks should be concentrated over southeastern North America by the southward deflection of the jet stream as they are today (Manabe and Terpstra, 1974; Manabe and Broccoli, 1990). Heat transport to high latitudes should be pronounced compared with a scenario with low Eocene mountains because atmospheric circulation should be considerably less zonal in the presence of a large mountain chain than without. There might also be a modest effect on albedo both from increased cloudiness over the mountains themselves and from the relatively high albedo of open vegetation present across the mid-continent.

More local effects would also be pronounced. High mountains should pro-duce a strong rain shadow in the immediate lee of the range and could account for moderately arid conditions implied by the evaporite facies in the Green River Formation. Differences in exposure between the various intermontane basins might partly explain the variation in the species composition of paleofloras in the region (S. Wing, personal communication, 1995).

A major uncertainty is the latitudinal extent of large mountains in the Eocene, particularly to the north of Wyoming where tall mountains could disrupt the jet stream. Many of the present mountain ranges in the Rocky Mountains are part of the same tectonic province and represent uplifts that formed along fault systems active during the Paleogene (Chase *et al.*, 1998). Hence our results may have more than local significance. Indeed, Meyer (1992) reports MAT estimates of paleoelevation for Eocene and Oligocene floras in New Mexico that range from 3000 m to 3200 m and are only a little higher than the 2500–2900 m elevation proposed for the Eocene Florissant beds of Colorado (Gregory and Chase, 1992; Forest *et al.*, 1995). Although the geographic distribution of large mountain ranges in the Eocene is poorly constrained, there is good evidence that many model simulations may proscribe elevations that are substantially too low.

None of the present paleotopography data for the Eocene of western North America support the presence of a large plateau with average elevations much over 2 km. In this respect, present GCMs like the $4.5° \times 7.5°$ atmospheric module in the GENESIS GCM and proscribed elevations of ~ 1.5 km have probably not missed major effects of surface heating produced by the Laramide surface. Still, the blocking effects of tall north–south mountains within the Laramide chain have yet to be considered.

ACKNOWLEDGEMENTS

We thank Katherine Gregory-Wodznicki, Karen Bice, and an anonymous reviewer for insightful and encouraging reviews that significantly improved this paper. Our research was supported by the Department of Earth Sciences at the University of Oxford and grants from the Geology and Paleontology program of the National Science Foundation.

REFERENCES

Barron, E. J. (1985). Explanations of the Tertiary global cooling trend. *Palaeogeography, Palaeoclimatology, Palaeoecology*, **50**, 45–61.

Barron, E. J. and Moore, G. T. (1994). Climate model application in paleoenvironmental analyses. *SEPM Short Course*, **33**, 1–339.

Barron, E. J. and Washington, W. M. (1984). The role of geographic variables in explaining paleoclimates: results from Cretaceous climate model sensitivity studies. *Journal of Geophysical Research*, **89**, 1267–79.

Barry, R. G. (1981). *Mountain Weather and Climate*. London: Methuen.

Barry, R. G. (1990). Changes in mountain climate and glacio-hydrological responses. *Mountain Research and Development*, **10**, 161–70.

Battan, L. P. (1979). *Fundamentals of Meteorology*. Englewood Cliffs, NJ: Prentice Hall.

Beck, R. A., Burbank, D. W., Sercombee, W. J., Olson, T. L. and Kahn, A. M. (1995). Organic carbon exhumation and global warming during the early Himalayan collision. *Geology*, **23**, 387–90.

Berry, R. W. (1925). Flora and ecology of so-called Bridger beds of Wind River Basin, Wyoming. *Pan-American Geologist*, **44**, 357–68.

Bolin, B. (1950) On the influence of the Earth's orography on the general character of the westerlies. *Tellus*, **20**, 184–95.

Bradley, W. H. (1929). The varves and climate of the Green River epoch. *United States Geological Survey Professional Paper*, **158**, 87–110.

Bradley, W. H. (1963). Paleolimnology. In *Limnology in North America*, ed. D. G. Frey, pp. 621–52. Madison, WI: Wisconsin University Press.

Bradley, W. H. (1964). Geology of Green River Formation and associated Eocene rocks in southwestern Wyoming and adjacent parts of Colorado and Utah. *United States Geological Survey Professional Paper*, **496-A**, 1–86.

Bradley, W. H. and Eugster, H. P. (1969). Geochemistry and paleolimnology of the trona deposits and associated authigenic minerals of the Green River Formation, Wyoming. *United States Geological Survey Professional Paper*, **496-B**, B1–64.

Broccoli, A. J. and Manabe, S. (1992). The effects of orography on midlatitude northern hemisphere dry climates. *Journal of Climate*, **5**, 1181–201.

Chaloner, W. G. and Creber, G. T. (1990). Do fossil plants give a climatic signal? *Journal of the Geological Society, London*, **147**, 343–50.

Chase, C. G., Gregory-Wodzicki, K. M., Parrish-Jones, J. T. and DeCelles, P. G. (1998). Topographic history of the western Cordillera of North America and controls on climate. In *Tectonic Boundary Conditions for Climate Model Simulations*, eds. T. J. Crowley and K. Burke, pp. 73–99. Oxford Monographs on Geology and Geophysics. Oxford: Oxford University Press.

Dansgaard, W. (1964). Stable isotopes in precipitation. *Tellus*, **16**, 436–67.

Dettman, D. L. and Lohmann, K. C. (1993). Seasonal change in Paleogene surface water $\delta^{18}O$: Fresh-water bivalves of western North America. In *Climate Change in Continental Isotopic Records*, Geophysical Monograph, vol. 78, eds. P. K. Swart, K. C. Lohmann, J. McKenzie and S. Savin, pp. 153–63. Washington, DC: American Geophysical Union.

Drummond, C. N., Wilkinson, B. H., Lohmann, K. C. and Smith, G. R. (1993). Effect of regional topography and hydrology on the lacustrine isotopic record of Miocene paleoclimate in the Rocky Mountains. *Palaeogeography, Palaeoclimatology, Palaeoecology*, **101**, 67–79.

Foley, J. A., Kutzbach, J. E., Coe, M. T. and Levis, S. (1994). Feedbacks between climate and boreal forests during the Holocene epoch. *Nature*, **317**, 52–4.

Forest, C. E., Molnar, P. and Emanuel, K. A. (1995). Palaeoaltimetry from energy conservation principles. *Nature*, **374**, 347–50.

Friedman, I. and O'Neil, J. R. (1977) Compilation of stable isotope fractionation factors of geochemical interest. *United States Geological Survey Professional Paper*, **440-KK**, 1–12.

Friedman, I., Redfield, A. C., Schoen, B. and Harris, J. (1964). The variation of the deuterium content of natural waters in the hydrologic cycle. *Reviews of Geophysics*, **2**, 177–224.

Gat, J. R. (1996). Oxygen and hydrogen isotopes in the hydrologic cycle. *Annual Reviews of Earth and Planetary Sciences*, **24**, 225–62.

Giorgi, F., Shields Brodeur, C. and Bates, G.T. (1994). Regional climate change scenarios over the United States produced with a nested regional climate model. *Journal of Climate*, **7**, 375–99.

Graf, K. (1981). Zum Höhenverlauf der Subnivalstufe in den Tropischen Anden, insbesondere in Bolivien und Ecuador. *Zeitschrift für Geomorphologie*, **37**, 1–24.

Greenwood, D. R. and Wing, S. L. (1995) Eocene continental climates and latitudinal temperature gradients. *Geology*, **23**, 1044–8.

Gregory, K. M. (1994). Palaeoclimate and palaeoelevation of the 35 Ma Florissant flora, Front Range, Colorado. *Palaeoclimate*, **1**, 23–57.

Gregory, K. M. and Chase, C. G. (1992). Tectonic significance of paleobotanically estimated climate and altitude of the late Eocene erosion surface, Colorado. *Geology*, **20**, 581–5.

Gregory, K. M. and Chase, C. G. (1994). Tectonic and climatic significance of a late Eocene low-relief, high-level geomorphic surface, Colorado. *Journal of Geophysical Research*, **99**, 20141–60.

Grootes, P. M. (1993). Interpreting continental oxygen isotope records. In *Climate Change in Continental Isotopic Records*, Geophysical Monography, vol. 78, eds. P. K. Swart, K. C. Lohmann, J. McKenzie and S. Savin, pp. 37–46. Washington, DC: American Geophysical Union.

Hahn, D. G. and Manabe, S. (1975). The role of mountains in the South Asian Monsoon circulation. *Journal of Atmospheric Sciences*, **32**, 1515–41.

Hay, W. W., Barron, E. J. and Thompson, S. L. (1990). Results of global atmospheric circulation experiments on an Earth with a meridional pole-to-pole continent. *Journal of the Geological Society, London*, **147**, 385–92.

Hay, W. W. and Brock, J. C. (1992). Temporal variation in intensity of upwelling off southwest Africa. In *Upwelling systems: Evolution since the Early Miocene*, Geological Society of London Special Publication, vol. 63, eds. C. P. Summerhayes, W. L. Prell and K. C. Emis, pp. 463–97. London: Geological Society of London.

Hutchison, J. H. (1982). Turtle, crocodilian, and champosaur diversity changes in the Cenozoic of the north-central region of western United States. *Palaeogeography, Palaeoclimatology, Palaeoecology*, **37**, 149–64.

Johnson, A. M. (1976) The climate of Peru. In *World Survey of Climatology*, vol. 12, ed. W. Schwerdtfeger, pp. 147–218. New York: Elsevier.

Karl, T. R., Groisman, P. Y., Knight, R. W. and Heim, R. R., Jr. (1993). Recent variations of snow cover and snowfall in North America and their relation to precipitation and temperature variations. *Journal of Climate*, **6**, 1327–44.

Koch, P. L., Zachos, J. C. and Dettman, D. L. (1995). Stable isotope stratigraphy and paleoclimatology of the Paleogene Bighorn Basin (Wyoming, USA). *Palaeogeography, Palaeoclimatology, Palaeoecology*, **115**, 61–89.

Kutzbach, J. E., Guetter, P. J., Ruddiman, W. F. and Prell, W. L. (1989). Sensitivity of climate to late Cenozoic uplift in southern Asia and the American west: numerical experiments. *Journal of Geophysical Research*, **94**, 18393–407.

Kutzbach, J. E., Prell, W. L. and Ruddiman, W. F. (1993). Sensitivity of Eurasian climate to surface uplift of the Tibetan Plateau. *The Journal of Geology*, **101**, 177–90.

Lauscher, F. (1976). Methoden zur Weltklimatologie der Hydrometeore. Der Anteil des festen Niederschlags am Gesamtneiderschlag. *Archives für Meteorolgie, Geophysik und Bioklimatologie*, **24**, 129–76.

Leopold, E. A. and MacGinitie, H. D. (1972). Development and affinities of Tertiary floras in the Rocky Mountains. In *Floristics and Paleofloras of Asia and Eastern North America*, ed. A. Graham, pp. 147–200. Amsterdam: Elsevier.

MacGinitie, H. D. (1969). *The Eocene Green River Flora of Northwestern Colorado and Northeastern Utah*. Berkeley, CA: University of California Press.

Manabe, S. and Broccoli, A. J. (1990). Mountains and arid climates of middle latitudes. *Science*, **247**, 192–5.

Manabe, S. and Terpstra, T. B. (1974). The effects of mountains on the general circulation of the atmosphere as identified by numerical experiments. *Journal of Atmospheric Sciences*, **31**, 3–41.

Manzi, A. O. and Planton, S. (1994). Implementation of the ISBA parameterization scheme for land surface processes in a GCM – an annual cycle experiment. *Journal of Hydrology*, **155**, 353–87.

Markwick, P. (1994). 'Equability', continentality, and Tertiary 'climate': the crocodilian perspective. *Geology*, **22**, 613–16.

Matthew, M. D. and Permutter, M. A. (1994). Global cyclostratigraphy: an application to the Eocene Green River Basin. *Special Publication of the International Association of Sedimentologists*, **19**, 459–82.

McGuffie, K., Henderson-Sellers, A., Zhang, H., Durbidge, T. B. and Pitman, A. J. (1995). Global climate sensitivity to tropical deforestation. *Global and Planetary Change*, **10**, 97–128.

Meehl, G. A. (1992). Effect of tropical topography on global climate. *Annual Review of Earth and Planetary Sciences*, **20**, 85–112.

Meyer, H. W. (1992). Lapse rates and other variables applied to estimating paleolatitudes from fossil floras. *Palaeogeography, Palaeoclimatology, Palaeoecology*, **99**, 71–99.

Moore, G. T., Sloan, L. C., Hayashida, D. N. and Umrigar, N. P. (1992). Paleoclimate of the Kimmeridgian/Tithonian (Late Jurassic) world: II. Sensitivity tests comparing three different paleotopographic settings. *Palaeogeography, Palaeoclimatology, Palaeoecology*, **95**, 229–52.

Moore, J. G. (1965). Petrology of deep-sea basalt near Hawaii. *American Journal of Science*, **263**, 40–52.

Moore, J. G. (1977). Petrology of deep-sea basalt near Hawaii. *American Journal of Science*, **263**, 40–52.

Norris, R. D., Jones, L. S., Corfield, R. M. and Cartlidge, J. E. (1996). Skiing in the Eocene Uinta mountains? Isotopic evidence for snow melt and large mountains in the Green River Formation. *Geology*, **24**, 403–6.

Otto-Bliesner, B. (1993). Tropical mountains and coal formation: a climate model study of the Westphalian (306 Ma). *Geophysical Research Letters*, **20**, 1947–50.

Povey, D. A. R., Spicer, R. A. and England, P. C. (1994). Palaeobotanical investigation of early Tertiary palaeoelevations in northeastern Nevada: initial results. *Review of Palaeobotany and Palynology*, **81**, 1–10.

Prell, W. L. and Kutzbach, J. E. (1992). Sensitivity of the Indian monsoon to forcing parameters and implications for its evolution. *Nature*, **360**, 647–52.

Prentice, I. C., Cramer, W., Harrison, S. P., Leemans, R., Monserud, R. A. and Soloman, A. M. (1992). A global biome model based on plant physiology and dominance, soil properties and climate. *Journal of Biogeography*, **19**, 117–34.

Raymo, M. E. (1991). Geochemical evidence supporting T. C. Chamberlin's theory of glaciation. *Geology*, **19**, 344–7.

Raymo, M. E., Ruddiman, W. F. and Froelich, P. N. (1988). Influence of late Cenozoic mountain building on ocean geochemical cycles. *Geology*, **16**, 649–53.

Roehler, H. W. (1991). Correlation and depositional analysis of oil shale and associated rocks in the Eocene Green River Formation, greater Green River Basin, southwest Wyoming. *United States Geological Survey Map*, No. I-2226.

Roehler, H. W. (1993). Eocene climates, depositional environments, and geography, greater Green River Basin, Wyoming, Utah, and Colorado. *United States Geological Survey Professional Paper*, **1506-F**, 1–74.

Rozanski, K., Araguás-Araguás, L. and Gonfaiantini, R. (1993). Isotopic patterns in modern global precipitation. In *Climate Change in Continental Isotopic Records*, Geophysical Monograph, vol. 78, eds. P. K. Swart, K. C. Lohmann, J. McKenzie and S. Savin, pp. 1–36. Washington: American Geophysical Union.

Ruddiman, W. F. and Kutzbach, J. E. (1990). Late Cenozoic plateau uplift and climate change. *Transactions of the Royal Society of Edinburgh*, **81**, 301–14.

Ruddiman, W. F., Sarnthein, M., Backman, J. *et al.* (1989). Late Miocene to Pleistocene evolution of climate in Africa and the low-latitude Atlantic: overview of Leg 108 results. In *Proceedings of the Ocean Drilling Program, Scientific Results*, vol. 108, eds. W. Ruddiman, M. Sarnthein *et al.*, pp. 463–84. College Station, TX: Ocean Drilling Program.

Sahagian, D. L. and Maus, J. E. (1994). Basalt vesicularity as a measure of atmospheric pressure and palaeoelevation. *Nature*, **372**, 449–51.

Seltzer, G. O. (1990). Recent glacial history and paleoclimate of the Peruvian–Bolivian Andes. *Quaternary Science Reviews*, **9**, 137–52.

Sloan, L. C. (1994). Equable climates during the early Eocene: significance of regional paleogeography for North American climate. *Geology*, **22**, 881–4.

Sloan, L. C. and Barron, E. J. (1992*a*). A comparison of Eocene climate model results to quantified paleoclimatic interpretations. *Palaeogeography, Palaeoclimatology, Palaeoecology*, **93**, 183–202.

Sloan, L. C. and Barron, E. J. (1992*b*). Eocene climate model results: quantitative comparison to paleoclimatic evidence. *Palaeogeography, Palaeoclimatology, Palaeoecology*, **93**, 183–202.

Sloan, L. C., Walker, J. C. G. and Moore, T. C., Jr. (1995). Possible role of oceanic heat transport in early Eocene climate. *Paleoceanography*, **10**, 347–56.

Smith, G. I., Freidman, I., Klieforth, H. and Hardcastle, K. (1979). Areal distribution of deuterium in eastern California precipitation. *Journal of Applied Meteorology*, **18**, 172–88.

Sullivan, R. (1985). Origin of lacustrine rocks of Wilkins Peak Member, Wyoming. *American Association of Petroleum Geologists Bulletin*, **69**, 913–22.

Surdam, R. C. and Wolfbauer, C.A. (1975). Green River Formation, Wyoming: a playa-lake complex. *Geological Society of America Bulletin*, **86**, 335–45.

Talbot, M. R. (1990). A review of the palaeohydrological interpretation of carbon and oxygen isotopic ratios in primary lacustrine carbonates. *Chemical Geology (Isotope Geoscience Section)*, **80**, 261–79.

Talbot, M. R. and Kelts, K. (1990). Palaeolimnological signatures from carbon and oxygen isotopic ratios in carbonates from organic-rich lacustrine sediments. In *Lacustrine exploration: Case Studies and Modern Analogs*, eds. B. J. Katz and B. R. Rosendahl, pp. 99–112. Volume AAPG Studies in Geology. Tulsa, OK: American Association of Petroleum Geologists.

Weber, J. N. (1964). Carbon-oxygen isotopic composition of Flagstaff carbonate rocks and its bearing on the history of Paleocene–Eocene Lake Flagstaff of central Utah. *Geochimica et Cosmochimica Acta*, **28**, 1219–42.

Williams, M. W., Losleben, M., Caine, N. and Greenland, D. (1996). Changes in climate and hydrochemical responses in a high-elevation catchment, Rocky Mountains, USA. *Limnology and Oceanography*, **41**, 939–46.

Wing, S. and Greenwood, D. L. (1993). Fossils and fossil climate: the case for equable continental interiors in the Eocene. *Philosophical Transaction of the Royal Society of London*, **341**, 243–52.

Wolfe, J. A. (1990). Palaeobotanical evidence for a marked temperature increase following the Cretaceous/Tertiary boundary. *Nature*, **343**, 153–6.

Wolfe, J. A. (1992). An analysis of present day terrestrial lapse rates in the western conterminous United States and their significance to paleoaltitudinal estimates. *United States Geological Survey Bulletin*, **1964**, 1–35.

Wolfe, J. A. (1994). Tertiary climatic changes at middle latitudes of western North America. *Palaeogeography, Palaeoclimatology, Palaeoecology*, **108**, 195–205.

Wolfe, J. A. (1995). Paleoclimatic estimates from Tertiary leaf assemblages. *Annual Review of Earth and Planetary Sciences*, **23**, 119–42.

Zachos, J. C., Stott, L. D. and Lohmann, K. C. (1994). Evolution of early Cenozoic marine temperatures. *Paleoceanography*, **9**, 353–87.

Ziegler, A. M., Hulver, M. L. and Rowley, D. B. (1997). Permian world topography and climate. In *Late Glacial and Postglacial Environmental Changes: Quaternary, Carboniferous–Permian and Proterozoic*, ed. I. P. Martini, pp. 111–46. New York: Oxford University Press.

Ziegler, A. M., Rowley, D. B., Lottes, A. L., Sahagian, D. L., Hulver, M. L. and Gierlowski, T. C. (1985). Paleogeographic interpretation: with an example from the mid-Cretaceous of North America. *Annual Reviews of Earth and Planetary Sciences*, **13**, 385–425.

7

An early Eocene cool period? Evidence for continental cooling during the warmest part of the Cenozoic

SCOTT L. WING, HUIMING BAO, AND PAUL L. KOCH

ABSTRACT

During the late Paleocene global temperatures began to rise toward their Cenozoic acme, which was reached during the early Eocene. Most detailed studies of temperature change during the Paleocene–Eocene time interval have been based on isotopic analyses of microfossils from marine sections. For this study we estimated changes in paleotemperature for continental environments in the Bighorn Basin, Wyoming, based on leaf margin analyses of fossil floras. The leaf margin analyses show that over the last 2 million years of the Paleocene mean annual temperature increased from 12.9($\pm$2.4) to over 15($\pm$2.4) °C, but that during the first million years of the Eocene temperatures dropped from 18.2($\pm$2.3) to 10.8($\pm$3.3) °C. The strong temperature decline was followed by a rapid increase to 15.8($\pm$2.2) °C then 22.2($\pm$2.0) °C in the lower part of Chron 24n. The absence of megafloras during the short Late Paleocene Thermal Maximum (LPTM) prevented us from making temperature estimates based on leaf margin analysis. Early Eocene cooling is corroborated through oxygen isotope analysis of authigenic minerals in paleosols. These minerals record changes in the $\delta^{18}O$ value of surface water, which may be related to changes in temperature and vapor transport to the region. The $\delta^{18}O$ value of surface water dropped 4‰ during the early part of the early Eocene, consistent in direction and magnitude with the cooling inferred from the flora. The $\delta^{18}O$ values of soil carbonates from the LPTM increase by ~1‰, consistent with warming above background Paleocene temperatures by ~4 °C. Few previous studies of this globally warm time interval have noted cooling during the early Eocene. Here we briefly summarize faunal and floral turnover during the Paleocene–Eocene transition and explore the implications of early Eocene cooling for climatic and biotic change.

INTRODUCTION

It is clear from geological and paleontological evidence that warm middle and high latitudes were the typical climatic state for the earth during most of the Phanerozoic. The most extreme and extended warm intervals in earth history occurred in the Mesozoic and Paleozoic, distant enough in the past so that the continental positions are arguable, few isotopic temperature estimates are available, fine resolution of geological time is difficult, and biotas are highly dissimilar to living faunas and floras. All of these factors make the climates of these remote times difficult to characterize both geographically and in terms of shorter-term fluctuations.

The early Cenozoic, especially early Eocene, is increasingly being recognized as a key period for understanding the climate and biogeography of globally warm intervals of earth history. The early Cenozoic is recent enough that continental configurations are not radically different from the present, fossils and sediments are widely distributed and amenable to relatively precise radiometric and paleomagnetic dating, and the organisms present can be understood relatively well by comparing them with the living biota. The early Eocene is a kind of outpost of warm Mesozoic climate surviving to within 50 million years of the present.

The pattern of temperature change during the late Paleocene and early Eocene has been defined largely through measurements of stable oxygen isotopes in foraminiferal tests taken from ocean floor drill cores (e.g., Miller *et al*., 1987; Zachos *et al*., 1994). Although individual records vary, the overall trend is that ocean waters warmed substantially during the last half of the Paleocene and the first 2–3 million years of the Eocene. Oxygen isotopic analyses of benthic foraminifera indicate that deep-sea temperatures warmed from 7–10 °C in the late Paleocene to 10–14 °C in the early Eocene; analyses of planktic foraminifera indicate that middle- to high-latitude sea-surface temperatures increased from 11–20 °C in the late Paleocene to 14–24 °C in the early Eocene (Zachos *et al*., 1994). During this strong middle- and high-latitude warming, tropical sea-surface temperatures apparently remained nearly constant in the 20–26 °C range (Schrag *et al*., 1992; Zachos *et al*., 1994). Warming around the Paleocene–Eocene transition appears to be largely a middle- and high-latitude phenomenon.

Middle- to high-latitude continental climates also appear to have warmed from the late Paleocene to the early Eocene. Terrestrial floras from southeastern coastal Alaska indicate a warmer climate during deposition of the early Eocene Tolstoi Formation than the late Paleocene Chickaloon Formation (Wolfe, 1966, 1972, 1977), and floristic evidence for warming has been noted in North Dakota and Wyoming (Leopold and MacGinitie, 1972; Hickey, 1977). The increasing species diversity and size range of ectothermic vertebrates also indicate warming from the Paleocene to the Eocene in the interior of western North America (Hutchison 1982; Markwick, 1994). Previous paleotemperature estimates based on floras from the Bighorn and Wind River basins in Wyoming show that mean annual temperature increased from 13 to 21 °C during the late Paleocene and early Eocene (Hickey, 1980; Wing *et al*., 1991). The early Eocene has also been noted as a time of exceptional warmth in East Asia (Guo, 1985), Siberia (Budantsev, 1992), Australia (Christophel and Greenwood, 1989), western Europe (Collinson and Hooker, 1987), and southern South America (Romero, 1986). Low-latitude floral data for this time are scarce, but the floristic composition of Eocene floras from Panama is consistent with low-latitude temperatures in the modern range (Graham, 1992).

High-resolution studies of the Paleocene–Eocene boundary interval have revealed a short period of intense middle- and high-latitude warmth in the latest

Paleocene, commonly referred to as the Late Paleocene Thermal Maximum, or LPTM (Kennett and Stott, 1991; Zachos *et al.*, 1994; Corfield and Norris, 1996). During this ~100 ka interval, isotopic temperature estimates for high-latitude ocean waters suggest an ~5 °C increase in sea-surface temperature and near elimination of vertical temperature gradients in high-latitude oceans. A coeval increase in kaolinite in nearshore marine sediments indicates more rapid chemical weathering and enhanced runoff from continental surfaces at middle and high latitudes (Gibson *et al.*, 1993; Robert and Kennett, 1994). Several major biotic turnover events also took place during the LPTM, including a major immigration of new mammalian lineages in North America; this immigration event marks the beginning of the Wasatchian land mammal age, and the distinctive earliest Wasatchian assemblage has been called the Wa0 fauna (Gingerich, 1989). The LPTM also coincides with a ~4‰ negative excursion in carbon isotopes that has been observed in both marine and continental sections (e.g., Kennett and Stott, 1991; Koch *et al.*, 1992). This isotopic marker has been used to establish the synchroneity of biotic and climatic events in the oceans and on the continents during the LPTM (e.g., Koch *et al.*, 1995). The cause of the carbon isotope excursion and LPTM is debated, but may relate to the release of large quantities of methane over a period of ~10 ka that resulted from the dissociation of sea-floor clathrates (Dickens *et al.*, 1997).

In this paper we present a high-resolution record of temperature across the Paleocene–Eocene boundary interval in the Bighorn Basin of northern Wyoming, based on a combination of leaf margin data and oxygen isotope data from hematite and calcite in paleosols. Our data show the expected warming across the Paleocene–Eocene interval, but also an interval of strong cooling during the early Eocene that lasted ~1 million years. Scattered data from other areas suggest that this cooling was at least continental to hemispheric in scale rather than strictly local. We also briefly compare the record of biotic changes in the Bighorn Basin with the record of temperature change, and find evidence that diversity and composition of floras and faunas were affected by the cooling. This phenomenon will have to be accounted for in models of climatic change during the early Cenozoic and in considering the biotic consequences of climatic change.

GEOLOGICAL CONTEXT

Lithostratigraphy

The Paleocene–Eocene transition in the Bighorn Basin is recorded by rocks of the upper Fort Union and lower Willwood formations (Bown, 1980; Rose, 1981; Wing and Bown, 1985). Stratigraphic sections have been measured across the Paleocene–Eocene interval in seven areas of the Bighorn Basin: North Butte, Worland East, Sand Creek Divide, Gould Butte, Antelope Creek/Elk Creek, Foster Gulch/McCullough Peaks, and Polecat Bench/Clarks Fork Basin (Fig. 7.1; Bown, 1979; Gingerich *et al.*, 1980; Schankler, 1980; Clyde *et al.*, 1994; Bown *et al.*, 1994a; Wing, 1998). Areas of exposed Paleogene rocks are separated from one another by areas of Quaternary alluvium, so that correlations among them are generally made through mammalian biostratigraphy, magnetostratigraphy, and

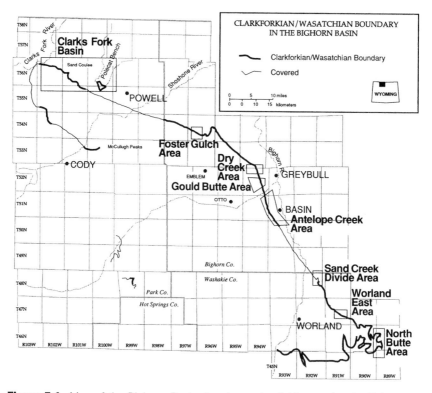

Figure 7.1. Map of the Bighorn Basin showing major field areas for the Paleocene-Eocene boundary interval.

carbon isotope values (Schankler, 1980; Gingerich, 1989; Clyde *et al.*, 1994; Tauxe *et al.*, 1994; Koch *et al.*, 1995). Within each field area local stratigraphic sections have been correlated by bed tracing and combined into composite sections that indicate the relative stratigraphic levels of hundreds of fossil localities. In all of the areas described in this work the sedimentary sequence across the Paleocene–Eocene transition lacks noticeable unconformities.

The dominant lithology in both the Fort Union and Willwood formations is mudstone, with variable lesser amounts of lenticular and sheet sandstones, carbonaceous shales, and in the Fort Union Formation, coal (Bown, 1980). Mudstones of the Willwood Formation are generally red or variegated in color, reflecting oxidation of iron compounds during pedogenesis in seasonally dry floodplain soils (Bown and Kraus, 1981*a*). Fort Union Formation mudstones are usually drab in color and higher in organic carbon content, and they probably represent wetter floodplain soils. The transition from drab Fort Union Formation mudstones to variegated Willwood Formation mudstones is coincident with the terminal Paleocene carbon isotope excursion and mammalian faunal change (the Clarkforkian/Wasatchian boundary) in many parts of the Bighorn Basin. However, in the Clarks Fork Basin the carbon isotope excursion and the beginning of the Wasatchian are more than 350 m above the beginning of Willwood deposition (Rose, 1981), whereas in the Gould Butte area the lowest Wasatchian faunas occur at least 100 m below the lowest Willwood Formation (Wing and Bown, 1985). This diachroneity strongly suggests that soil-forming conditions on the floodplains were influenced by local subsidence or tectonism as well as by regional and global climate change (Wing and Bown, 1985). Within the Willwood Formation, carbonaceous shales alternate with oxidized mudstones on a stratigraphic scale of tens of meters. Lithological variation at this scale may be the result of autogenic fluvial processes, but could also be influenced by high-frequency climatic change such as Milankovitch cyclicity (Kraus and Aslan, 1993).

Chronology of the Paleocene–Eocene boundary interval

The chronology of the Paleocene–Eocene transition has been reviewed extensively in recent publications (e.g., Berggren et al., 1995; Berggren and Aubry, 1996). Discussion here is limited to points that directly influence our age estimates for the Bighorn Basin. The two central problems that affect the calculation of ages in the Bighorn Basin are the calibration of the Geomagnetic Polarity Time Scale (GMPTS) of Cande and Kent (1992, 1995), and the calculation of an age for the carbon isotope excursion.

Cande and Kent (1992, 1995) used an age of 55 Ma for the Paleocene–Eocene boundary as one of the calibration points for their GMPTS. This age estimate was derived in the following fashion. The −17 ash in the North Sea, which has been dated at 54.5 Ma by Swisher and Knox (1991), occurs at 400 mbsf at Deep Sea Drilling Project (DSDP) Site 550. This level is 8 m above the NP9/NP10 boundary in the same hole (Aubry et al., 1996). The NP9/NP10 boundary, which was about two-thirds of the way down from the top of C24r, was taken as an approximation of the Paleocene–Eocene boundary, and this position (C24r.66) was assigned an age of 55 Ma. The adjacent calibration points for the GMPTS used by Cande and Kent (1992, 1995) were 65 Ma for the K/T boundary (C29r.3), and 46.8 Ma for C21n.33. The ages of reversals between these calibration points were interpolated by scaling the widths of the corresponding sea-floor magnetic stripes to time using a spline function forced through the calibration points. Errors in dating the calibration points would cause the ages of surrounding reversals to be wrong as well.

In the last few years many aspects of the Paleocene–Eocene record at DSDP 550 have been reinterpreted. The transition from C24r to C25n may lie several meters lower than originally thought (Ali and Hailwood, 1998); this would have made the −17 ash and the NP9/NP10 boundary appear to be too near the bottom of C24r in the original interpretation. Sediment accumulation rates may not have been constant at DSDP 550. Specifically, Aubry et al. (1996) suggested that there might be significant hiatuses and dissolution intervals in the lower part of C24r. If these hiatuses represent a substantial interval of non-deposition or erosion they too would make the −17 ash appear to be nearer the bottom of C24r than it really is. These problems suggest that 55 Ma may be closer to the middle of C24r than to a position two-thirds of the way down in C24r, as assumed by Cande and Kent (1992,

Table 7.1. *Calibration points for Bighorn Basin sections*

Calibration point	Section*	Meter level	Age model 1	Age model 2	References for age and stratigraphic level
Top Chron 24n.2n	MPS	1400	52.757	52.907	Clyde et al., 1994; Cande and Kent, 1995
Bentonitic tuff (^{40}Ar.^{39}Ar)	ECS	634	52.8±0.3	52.8±0.3	Wing et al., 1991
Top Chron 24n.3n	MPS	1300	52.903	53.072	Clyde et al., 1994; Cande and Kent, 1995
Base Chron 24n.3n	MPS	950	53.347	53.566	Clyde et al., 1994; Cande and Kent, 1995
Carbon isotope excursion	MPS	0	54.955	55.234	Koch et al., 1995; this paper
Carbon isotope excursion	HCS	0	54.955	55.234	Koch et al., 1995; this paper
Carbon isotope excursion	WES	0	54.995	55.234	Koch et al., 1995; this paper
Carbon isotope excursion	CFB	1520	54.995	55.234	Koch et al., 1995; this paper
Top of 25n	CFB	1070	55.904	56.215	Butler et al., 1981; Cande and Kent, 1995
Base Chron 25n	CFB	820	56.391	56.688	Butler et al., 1981; Cande and Kent, 1995
Top Chron 26n	CFB	500	57.554	57.791	Butler et al., 1981; Cande and Kent, 1995

*CFB, Polecat Bench/Clarks Fork Basin composite section; ECS, Antelope Creek/Elk Creek composite section; HCS, Honeycombs section; MPS, Foster Gulch/McCullough Peaks composite section; WES, Worland East section. Models 1 and 2 are explained in the text.

1995). If this is true, then reversals in this part of the GMPTS are up to 300 ka older than indicated in the timescale of Cande and Kent (1995).

A second problem in the chronology of the Paleocene–Eocene boundary interval is the age of the carbon isotope excursion and its relative position within C24r. Aubry *et al.* (1996) argued that all of the DSDP sections being studied for Paleocene–Eocene boundary questions have long hiatuses within C24r, and that these hiatuses cause the carbon isotope excursion to appear stratigraphically higher within C24r than would be expected based on its temporal position near the base of the chron. It has also been postulated that there may be more than one carbon isotope excursion within C24r (Aubry *et al.*, 1996), although multiple excursions have not been demonstrated in any single section.

Berggren *et al.* (1995) and Aubry *et al.* (1996) estimated an age of 55.5 Ma for the carbon isotope excursion based in part on its position 1 m below the NP9/NP10 boundary and 9 m below the −17 ash at DSDP 550. They argued that much of the million years between the ash and the isotopic excursion was represented in this core by a hiatus at the NP9/NP10 boundary. There remains, however, sub-stantial uncertainty in these interpretations. Aubry *et al.* (1996, p. 367) stated: 'We emphasize that the temporal framework used here is highly speculative because our reconstruction is based on the unverifiable assumption that the sedimentation rates were essentially the same at Sites 550 and 690.' There is no direct evidence for the length of the proposed NP9/NP10 hiatus at DSDP 550, and therefore we are only willing to conclude that the time between the excursion and the −17 ash must have been sufficient for the deposition of the 9 m of sediment between them at Site 550. At typical deep-sea sedimentation rates of $1–3$ cm ka^{-1}, the 9 m should represent 300–900 ka, which gives a range of minimum ages for the carbon isotope excursion of 54.8–55.4 Ma. This minimum estimate is consistent with the 55.5 Ma estimate made by Aubry *et al.* (1996), but ages as young as 54.8 Ma are equally possible. We also point out that if much of the time between the carbon isotope excursion and the −17 ash is represented by a hiatus at Site 550 (Aubry *et al.*, 1996), this requires high rates of sedimentation above the NP9/NP10 boundary.

Age estimates for the Bighorn Basin

In order to place the temperature and isotopic records from the Bighorn Basin in a global context, we have calculated age estimates by linear interpolation between tie points of known or estimated age. These tie points include six magnetic polarity reversals, one radiometrically dated tuff and the carbon isotope excursion (Table 7.1). We have developed two age models for the stratigraphic levels within the three major composite sections: the Foster Gulch/McCullough Peaks section, the Antelope Creek/Elk Creek section, and the Polecat Bench/Clarks Fork Basin section (Tables 7.2–7.4). Age model 1 uses the ages for the GMPTS published by Cande and Kent (1992, 1995). Age model 2 relies on recalculated ages for the relevant polarity reversals. This recalculation has been done by removing the 55 Ma calibration point at C24r.33, and replacing it with a calibration point of 52.8 ± 0.3 Ma at the base of 24n.1n based on magnetic stratigraphy and a radiometric date in the Antelope Creek/Elk Creek section in the Bighorn Basin (Wing *et al.*, 1991; Tauxe *et al.*,

Table 7.2. *Age calibration (in Ma) of Foster Gulch/McCullough Peaks Section*

Meter level	Age model 1	Age model 2	Sample/Event
1420	−52.728	−52.874	*Lambdotherium* FAD
1400	**−52.757**	**−52.907**	Top 24n.2n
1300	**−52.903**	**−53.072**	Top 24n.3n
1100	−53.157	−53.354	*Heptodon* FAD
950	**−53.347**	**−53.566**	Base 24n.3n
880	−53.436	−53.665	*Bunophorus* FAD
0	**−54.955**	**−55.234**	Wa0 – carbon isotope excursion
	0.685	0.561	Accumulation rate 0-880 m (m ka^{-1})
	0.788	0.709	Accumulation rate 880-1300 m (m ka^{-1})
	0.579	0.606	Accumulation rate 1300-1400 m (m ka^{-1})

Bold numbers are calibration points (age fixed); models 1 and 2 are explained in the text.

1994). Ages of the magnetic reversals were recalculated by linear interpolation (Table 7.5). Age model 2 does not rely on any marine calibration points in the late Paleocene or early Eocene.

Although the age estimates for the paleomagnetic reversals and the tuff are straightforward, the age estimate for the carbon isotope excursion requires explanation. Given the difficulties described above in using deep-sea marine sections to derive an age model for events during C24r, we do not accept the most recent published age estimate of 55.5 Ma for the carbon isotope excursion (Aubry *et al.*, 1996, Berggren and Aubry, 1996). Instead we calculate the age of the carbon isotope excursion based on its stratigraphic position within the Polecat Bench/Clarks Fork Basin section, where there is no evidence for lengthy hiatuses or large shifts in depositional rates within C24r. This requires a three-step process because there is no single Bighorn Basin section in which both the top and the bottom of C24r have been identified, and therefore there is no direct way to establish the relative position of the carbon isotope excursion within C24r. The first step is to estimate the age of the first appearance of the mammal *Bunophorus*, which is known from all three composite sections in the Bighorn Basin. This is done in the Foster Gulch/McCullough Peaks section, which we chose because of the stable magnetic behavior of the sediments there (Clyde *et al.*, 1994). The top and base of C24n.3 occur at 1300 and 950 m, respectively, in this section (Table 7.2; Clyde *et al.*, 1994). The ages for these magnetic boundaries are used to calculate the sedimentation rate, and this rate is extrapolated to the 880 m level to estimate the age of the First Appearance Datum (FAD) of *Bunophorus* in this section (53.436 Ma in model 1, 53.665 in model 2). In the second step the age for the FAD of *Bunophorus* is assigned to the 2240 m level of the Polecat Bench/Clarks Fork Basin section, the level of its FAD in this section (Gingerich, 1991). Finally, the age of the carbon isotope excursion (1520 m [Koch *et al.*, 1992]) is estimated by linear interpolation between the FAD of *Bunophorus* at

Table 7.3. *Age calibration (in Ma) of Elk Creek section*

Meter level	Age model 1	Age model 2	Sample/Event
634	**−52.800**	**−52.800**	^{39}Ar/^{40}Ar date
621	−52.831	−52.842	LMA-8
601	−52.878	−52.906	HOA-YPM33
591	−52.902	−52.938	*Lamdotherium* FAD
571	−52.949	−53.002	HOA-YPM1
546	−53.008	−53.083	HOA-D1467
485	−53.152	−53.279	HOA-D1531
468	−53.192	−53.334	LMA-7 top
438	−53.263	−53.456	HOA-D1398
430	−53.282	−53.430	*Heptodon* FAD
420	−53.306	−53.488	LMA-7 bottom
409	−53.332	−53.523	HOA-D1454
380	−53.400	−53.616	Biohorizon B
378	−53.405	−53.623	HOA-D1300
378	−53.405	−53.623	HOA−D1301
365	**−53.436**	**−53.665**	*Bunophorus* FAD (from MPS)
353	−53.486	−53.716	LMA-6 top
336	−53.556	−53.789	HOA-D1374
311	−53.661	−53.897	LMA-6 bottom
290	−53.748	−53.987	HOA-YPM350R
290	−53.748	−53.987	HOA-YPM350
210	−54.081	−54.331	HOA-YPM290
200	−54.122	−54.374	LMA-5 top
190	−54.164	−54.417	Biohorizon A
140	−54.372	−54.632	HOA-YPM104
110	−54.497	−54.761	LMA-5 bottom
105	−54.518	−54.783	LMA-4 top
100	−54.539	−54.804	HOA-YPM119
38	−54.797	−55.071	FAD *Platycarya* pollen
30	−54.830	−55.105	HOA-YPM115
10	−54.913	−55.191	LMA-4 bottom
0	**−54.955**	**−55.234**	Wa0 red beds (lithology only)
−7	−54.984	−55.264	LMA-3 top
−67	−55.234	−55.522	LMA-3 bottom
−172	−55.670	−55.974	LMA-2 bottom
−305	−56.224	−56.546	LMA-1 bottom
	0.240	0.233	Accumulation rate 0-365 m (m ka^{-1})
	0.423	0.311	Accumulation rate 365-634 m (m ka^{-1})

Bold numbers are calibration points (age fixed); models 1 and 2 are explained in the text.

2240 m and the bottom of C24r at 1070 m (Table 7.4; Butler *et al.*, 1981). Note that regardless of the age model, the carbon isotope excursion is likely to be only slightly less than two-thirds (63%) of the way down from the top of C24r, because the FAD of *Bunophorus* is only slightly older than the top of C24r. This stratigraphic position relative to the top and bottom of C24r is roughly the same one seen in DSDP Sites

Table 7.4. *Age calibration (in Ma) of Polecat Bench/Clarks Fork Basin Section*

Meter level	Age model 1	Age model 2	Sample/Event
2240	−53.436	−53.665	*Bunophorus* FAD date from MRS
1895	−54.164	−54.417	HOA-SC232
1860	−54.237	−54.493	Biohorizon A
1850	−54.259	−54.515	HOA-SC34
1780	−54.406	−54.667	HOA-SC313
1720	−54.533	−54.798	HOA-SC12
1620	−54.744	−55.016	HOA-SC18
1570	−54.849	−55.125	HOA-SC4
1570	−54.849	−55.125	HOA-SC123
1535	−54.923	−55.201	HOA-SC40
1520	−54.955	−55.234	Wa0 – carbon isotope excursion
1500	−54.997	−55.278	HOA-SC138
1455	−55.092	−55.376	HOA-SC90
1455	−55.092	−55.376	HOA-SC90R
1400	−55.208	−55.496	Youngest Cf2/Cf3
1355	−55.303	−55.594	HOA-SC127
1310	−55.398	−55.692	Oldest Cf2/Cf3
1250	−55.524	−55.823	LMA-2 midpoint
1210	−55.609	−55.910	HOA-SC92
1160	−55.714	−56.019	Tfu/Tw contact
1160	−55.714	−55.019	Cf1/Cf2
1090	−55.862	−56.171	HOA-SC171
1070	**−55.904**	**−56.215**	**Top of 25n**
1000	−56.040	−56.347	LMA-1 midpoint
950	−56.138	−56.442	Youngest base for Clarkforkian
850	−56.333	−56.631	Oldest base for Clarkforkian
820	**−56.391**	**−56.688**	**Base of 25n**
500	**−57.554**	**−57.791**	**Top of 26n**
	0.275	0.290	Accumulation rate 500–820 m (m ka^{-1})
	0.513	0.529	Accumulation rate 820–1070 m (m ka^{-1})
	0.474	0.459	Accumulation rate 1070–1520 m (m ka^{-1})

Bold numbers are calibration points (age fixed); models 1 and 2 explained in the text.

549 and 550. Age model 1 estimates the carbon isotope excursion to be at 54.995 Ma; model 2 yields an estimate of 55.234 Ma (Tables 7.1 and 7.4). These age estimates are within the range anticipated for the carbon isotope excursion based on DSDP 550 (see discussion above).

The age calibration of stratigraphic levels for each of the three major composite sections is shown in Tables 7.2–7.4. Below we briefly discuss the positions and correlation of the tie points in each section.

The lower three paleomagnetic tie points, the top of C26n, and the base and top of C25n, have been observed in the Clarks Fork Basin section along Polecat Bench (Butler *et al.*, 1981), where they occur in rocks that produce late Tiffanian or Clarkforkian mammals. Clarkforkian faunas in the eastern and southeastern

Table 7.5. *Recalibration of magnetic reversals*

Reversal	Cande and Kent (1995)	This paper (model 2)	Difference
C21n (0.33)	**46.800**	**46.800**	0.000
Top C22n	49.037	49.252	0.215
Base C22n	49.714	49.958	0.244
Top C23n.1n	50.778	51.028	0.250
Base C23n.1n	50.946	51.192	0.246
Top C23n.2n	51.047	51.291	0.244
Base C23n.2n	51.743	51.955	0.212
Top C24n.1n	52.364	52.529	0.165
Base C24n.1n	52.663	**52.800**	0.137
Top C24n.2n	52.757	52.907	0.150
Base C24n.2n	52.801	52.956	0.155
Top C24n.3n	52.903	53.072	0.169
Base C24n.3n	53.347	53.566	0.219
C24r (0.66)	**55.000**	55.315	0.315
Top C25n	55.904	56.215	0.311
Base C25n	56.391	56.688	0.297
Top C26n	57.554	57.791	0.237
Base C26n	57.911	58.125	0.214
Top C27n	60.920	60.903	−0.017
Base C27n	61.276	61.234	−0.042
Top C28n	62.499	62.385	−0.114
Base C28n	63.634	63.486	−0.148
Top C29n	63.976	63.827	−0.149
Base C29n	64.745	64.614	−0.131
C29r (0.3)	**65.000**	**65.000**	0.000

Bold numbers are age calibration points.

Bighorn Basin are very poorly known, and paleomagnetic analysis has not been done in the Fort Union Formation in this area, making correlation to the northern Bighorn Basin difficult

The carbon isotope excursion has been observed in the Clarks Fork Basin, Foster Gulch, Sand Point Divide, Worland East, and North Butte areas, outcropping over a distance of ~130 km (Koch *et al.*, 1992, 1995). In all regions, the negative carbon isotope values come from distinctive, thick, red paleosols that also produce the basal Wasatchian (Wa0) mammalian fauna first noted by Gingerich (1989). The Wa0 fauna has been collected from the thick red beds in one additional area on the west side of the basin (Hole-in-the-Ground), but this region has not been sampled for carbon isotopes. In the Antelope Creek/Elk Creek area in the eastern Bighorn Basin there are distinctive, thick, red beds at the base of the Willwood Formation that appear to correspond to the Wa0 red beds, but mammalian fossils and nodules suitable for stable carbon isotope analysis have not been found. Besides their stratigraphic position at the base of the Willwood Formation, there are three other reasons to think that the basal red beds on Antelope Creek represent the carbon isotope excursion and Wa0 interval. Firstly, early Wasatchian faunas (Wa1 or lower

Haplomylus–Ectocion zone) are known from a level 50 m above the basal red beds (Schankler, 1980). Secondly, late Paleocene palynomorphs and megafossils occur in Fort Union rocks less than 50 m beneath the basal red beds (Wing, 1984). Thirdly, the first occurrence of *Platycarya* pollen is 38 m above the basal red bed in this section (G. Harrington, personal communication, 1997; *contra* Wing *et al.*, 1991). In eastern North America the *Platycarya* FAD is in the uppermost decimeters of NP9 (Frederiksen, 1979, 1980), and the NP9/NP10 boundary occurs above the carbon isotope excursion, but within C24r (Gibson *et al.*, 1993). The inference that the thick red beds at the base of the Antelope Creek/Elk Creek section contain the carbon isotope excursion, as they do elsewhere in the basin, is consistent with the occurrence of the carbon isotope excursion within NP9 (Aubry *et al.*, 1996).

The base of C24n.3n has been putatively identified in two separate paleomagnetic studies in different parts of the Bighorn Basin. Tauxe *et al.* (1994), working in the Antelope Creek/Elk Creek section of the central Bighorn Basin, correlated a normal polarity zone (their N2) beginning at 312 m above the inferred Wa0 red beds with the base of C24n.3n. Clyde *et al.* (1994), working in the Foster Gulch/McCullough Peaks section in the northern Bighorn Basin, found a normal polarity zone (their C+) beginning 950 m above the Wa0 red beds. They interpreted this zone as equivalent to the beginning of C24n.3n. In both the Clyde *et al.* (1994) and Tauxe *et al.* (1994) studies, mammalian fossils were recovered from the rocks recording the polarity transition. In the Antelope Creek/Elk Creek section, these faunas lack the index fossil *Bunophorus*, which first occurs at 365 m, 53 m higher than the polarity transition (Bown *et al.*, 1994*a*). In the Foster Gulch/McCullough Peaks section, the first occurrence of *Bunophorus* is 50 m below the polarity transition. Thus the *Bunophorus* FAD and the magnetic polarity data are not consistent between the two sections. The Polecat Bench/Clarks Fork Basin section records the same relationship between the FAD *Bunophorus* and magnetic polarity events that is seen in the Foster Gulch/McCullough Peaks section, that is, there is no normal polarity interval below the FAD of *Bunophorus* but above the base of the Wasatchian (Butler *et al.*, 1981), although paleomagnetic sampling density is low in this part of the section.

Biostratigraphic relationships also are not consistent between the northern and central Bighorn Basin. The last occurrences of *Haplomylus* and *Ectocion* are above the first occurrence of *Bunophorus* in the Antelope Creek/Elk Creek section, whereas in both the Clarks Fork Basin and Foster Gulch/McCullough Peaks sections their ranges do not overlap *Bunophorus* (Bown *et al.*, 1994*a*; Clyde, 1997). This is probably the result of stratigraphic condensation in the Antelope Creek/Elk Creek section that results from local cut-and-fill deposits (Bown *et al.*, 1994*a*), although it could also result from lower sample density in the northern sections, or real differences in temporal range.

Because the Foster Gulch/McCullough Peaks section is much thicker than the Antelope Creek/Elk Creek section, displays more stable magnetic behavior (Clyde *et al.*, 1994), and agrees with the Polecat Bench/Clarks Fork Basin section, we think it is more likely that the reversal reported by Tauxe *et al.* (1994) at the 312 m level of the Antelope Creek/Elk Creek section is a spurious normal overprint than that Clyde *et al.* (1994) missed a normal event. Indeed Tauxe *et al.* (1994, p. 165) raised the possibility that some or all of the normal polarity intervals in the Antelope Creek/Elk Creek section are overprints. Consequently we have chosen to disregard the paleomagnetics of Tauxe *et al.* (1994) for the middle part of the section.

The youngest of the tie points in the Bighorn Basin is a radiometric age of 52.8 ± 0.3 Ma derived from a bentonitic tuff at the 634 m level of the Antelope Creek/Elk Creek section (Wing *et al.*, 1991). The date is based on $^{40}Ar/^{39}Ar$ analysis of sanidine. The ash is 53 m above the first occurrence of *Lambdotherium*, which indicates the base of the Lostcabinian land mammal assemblage zone (Bown *et al.*, 1994a). The FAD of *Lambdotherium* is also recorded at the 1420 m level of the Foster Gulch/McCullough Peaks section, where it occurs 20 m above the top of a normal polarity interval inferred to be 24n.2n (Clyde *et al.*, 1994). Given Cande and Kent's (1995) estimate of 52.76 Ma for the top of 24n.2n, the FAD of *Lambdotherium* is approximately synchronous in the northern and central Bigorn Basin.

Although the discussion above reveals several sources of uncertainty in our calibration points to a global geochronology, we note that the order of magnetic, geochemical, and biological events is robust. Varying interpretations of the magnetic stratigraphy affect the inferred ages of climatic and biotic changes by only a few hundred thousand years or less, which is unlikely to affect interpretation of climatic and biotic changes. For example, we have interpolated numerical ages for four biostratigraphic events (Biohorizon A, Biohorizon B, *Heptodon* FAD, *Lamdotherium* FAD) in two separate composite sections (Tables 7.2–7.4). For no event do the interpolated ages deviate from one another by more than 174 ka between the different sections (Biohorizon A – 73 ka, Biohorizon B – 120 ka, *Heptodon* FAD – 125 ka, *Lamdotherium* FAD – 174 ka), in spite of the potential for the age estimates to be affected by sampling intensity as well as by incorrect calibration of the sections.

Age calibration of the Polecat Bench/Clarks Fork Basin, Foster Gulch/McCullough Peaks, and Antelope Creek/Elk Creek sections reveals variation in rock accumulation rates within and among sections (Fig. 7.2, Tables 7.2–7.4). The more northerly sections have higher rates, probably because they are also further west and closer to the structural axis of the basin. Both the Antelope Creek/Elk Creek and Foster Gulch/McCullough Peaks sections show increasing rates of accumulation up section. The lower parts of both composite sections are closer to the margin of the basin, and their upper parts are closer to the center, so increases in accumulation rate within long composite sections may reflect position in the basin more than climatic or tectonic change through time (Clyde, 1997). Based on the relative thickness of mammalian biostratigraphic zones, Gingerich (1983) argued that maximum rates of basin subsidence occurred during the late Clarkforkian.

The Antelope Creek/Elk Creek section shows nearly a twofold increase in rock accumulation rate from the lower 365 m to the upper 270 m, which seems high in light of the continuity of mudstone and sandstone deposition throughout. However, there is independent evidence for increased sediment accumulation rates in the middle of the section. Bown and Kraus (1993) and Kraus and Bown (1993)

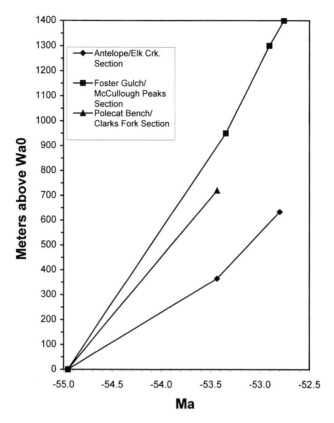

Figure 7.2. Rock accumulation rates in the three major stratigraphic sections across the Paleocene-Eocene transition in the Bighorn Basin. Rates calculated using age model 1 (see text). Calibrations for sections in Tables 7.2–7.4.

pointed out that there was a substantial decrease in paleosol maturity in the middle of the Antelope Creek/Elk Creek section, implying an increase in frequency and size of overbank depositional events. Our confidence in the age calibration of the Antelope Creek/Elk Creek section is also reinforced by the observation of similar-sized shifts in accumulation rates in the other composite sections (Fig. 7.2, Tables 7.2 and 7.4). Finally, our decision to disregard the paleomagnetics of Tauxe *et al.* (1994) for the middle part of the Antelope Creek/Elk Creek section decreases the difference in accumulation rates between the lower and upper half of the section. If we follow Tauxe *et al.* (1994) in placing the top of C24r at 312 m, this implies a threefold increase in accumulation rate within the Antelope Creek/Elk Creek section. In summary, rates of rock accummulation and comparisons between Bighorn Basin sections suggest that our age calibrations are robust and unlikely to change substantially unless ages for the paleomagnetic timescale or the carbon isotope excursion are modified.

PALEOCLIMATIC AND PALEOHYDROLOGIC RECONSTRUCTION

Leaf margin analysis

Leaf margin analysis is based on the observation made on living vegetation that the proportion of dicotyledonous species with entire-margined (untoothed) leaves is directly correlated with mean annual temperature (Wolfe, 1979). The functional basis of this correlation is not fully understood, but the primitive condition in dicots is probably to have entire-margined leaves, and teeth have evolved independently in many lineages (Hickey and Wolfe, 1975). Experimental studies suggest that teeth on leaves serve a variety of functions. In some temperate deciduous species teeth are photosynthetically active soon after bud-break, and long before the main part of the lamina (Brosh-Baker and Peet, 1997). Early photosynthetic activity in teeth may allow leaves to export photosynthate to the rest of the plant earlier in the year, a potential advantage in climates with short growing seasons and therefore a brief period during which the leaf can be a net source of energy (Baker-Brosh and Peet, 1997). Teeth have also been observed to be points of rapid transpirational water loss even after the leaf is fully expanded, presumably because tooth apices are associated with vascular tissue and emerge through the boundary layer of humid air that surrounds the lamina (Canny, 1990). Rapid loss of water through tooth apices suggests that teeth have a cost in terms of water loss as well as a benefit in terms of more rapid expansion and photosynthetic activity at the beginning of the growing season. The prevalence of teeth in a local flora probably represents a balance between the transpirational cost and the photosynthetic benefit that depends on a combination of local temperature, precipitation, and humidity. This hypothesis is consistent with the rarity of toothed leaves in wet and dry tropical and subtropical forests, where accelerated water loss under high temperatures could lead rapidly to wilting, and also with the rarity of teeth in cold, low-productivity environments where low nutrient levels have a similar effect to water stress (Givnish, 1979).

As with any correlational method, the use of leaf margin proportion (P) to infer mean annual temperature (MAT) makes the uniformitarian assumption that there was the same relationship between P and MAT in the past as has been observed in the present. In other words we assume that the leaves of plants in the early Cenozoic evolved morphological responses to climate in the same way that they do today. We think this assumption is valid for three reasons. Firstly, the evolution of teeth appears to hinge on basic ecophysiological relationships that affect all dicots: optimizing the rate of photosynthesis under a given temperature and precipitation regime. Thus the same factors that influence leaf morphology today operated in similar ways in the past. Secondly, the evolutionary change from entire to toothed leaves (and back) appears to be a relatively easy transition – both toothed and entire species are found within many living genera, and the variety of anatomical patterns in teeth implies that they have evolved many times within the dicotyledons (Hickey and Wolfe, 1975). The evolutionary lability of teeth makes it unlikely that the relationship between teeth and climate in the early Cenozoic would have been very different owing to a different set of lineages being present. There appears to be a high capacity for leaf form to respond to climate through evolution. Thirdly, the early Cenozoic floras analyzed here contain a substantial proportion of extant genera and families, many of which are in the East Asian and North American data sets used to calibrate the MAT/P relationship. Even if there were phylogenetic constraints on the development of teeth (e.g., if some dicot lineages do not evolve teeth no matter what climatic conditions they live under), these constraints should be similar for early Cenozoic and extant floras of the northern hemisphere because the floras of the two times are taxonomically similar.

The correlation between MAT and P has been quantified for a group of 34 sites in East Asia (Wolfe, 1979), for two separate groups of sites from the Americas (Wolfe, 1993; Wilf, 1997), and for one set of sites from Australia (Greenwood, 1992). The three northern hemisphere data sets demonstrate similar relationships between P and MAT (Wilf, 1997). The 34 sites in the East Asian data set show the strongest correlation of P with MAT ($r = 0.98$, $p < 0.001$), perhaps because they were drawn from areas with adequate growing season precipitation (Wolfe, 1979), and because of the large number of species per sample (Wilf, 1997). The correlation of P with MAT in the largely North American Climate–Leaf Analysis Multivariate Program (CLAMP) data set (106 sites; Wolfe, 1993) is less precise, although still highly significant ($r^2 = 0.76$, $p < 0.0005$). The lower correlation of P with MAT in the CLAMP data set may reflect the large number of samples from dry and extremely cold sites, and also the smaller number of species per site (Wilf, 1997). The leaf margin data set assembled by Wilf (1997) includes nine sites from North and South America and the Caribbean, and demonstrates a correlation between P and MAT that has a slope and intercept very similar to that seen in the East Asian data sets ($r^2 = 0.94$, $p < 0.0005$). The slope and intercept for the Australian data set are different from any of the northern hemisphere data (MAT $= 6.4176 + [17.63P]$), and the r^2 value is 0.78 (Greenwood, 1992; Greenwood, personal communication, 1997). The different relationship between P and MAT in Australian vegetation may reflect factors such as greater seasonality of rainfall and lower soil nutrient levels in Australian forests, or strong phylogenetic control of leaf morphology in some southern hemisphere lineages.

The estimates of MAT in this paper are based on the East Asian data set of Wolfe (1979) using the regression equation

$$MAT = 1.14 + (30.6P), \tag{7.1}$$

where P is the proportion of entire-margined species in the assemblage. We use the East Asian data set to make our temperature estimates because it is climatically the most analogous to the early Cenozoic of the Bighorn Basin in lacking extremely cold or dry sites. The standard error of the estimate of MAT based on Wolfe's (1979) data is 0.8 °C (Wing and Greenwood, 1993), but this is not the only component of error. Wilf (1997) has pointed out that there is also binomial error caused by uncertainty in estimating the true proportion of entire-margined species for the whole regional flora based on a sample with a finite number of species. This error (σ) is calculated in °C by

$$\sigma = 30.6\sqrt{P\,(1 - P)/r}, \tag{7.2}$$

where P is the proportion of entire-margined species in the sample, r is the total number of species in the sample, and the constant 30.6 is the regression coefficient for the East Asian data set (Wilf, 1997). For samples with 25–60 species and equal proportions of entire and non-entire forms, binomial sampling error creates a 2–3 °C uncertainty in estimating MAT (Wilf, 1997).

The accuracy of leaf margin analysis may also be affected by taphonomic processes that bias the representation of entire vs. toothed leaves, although no such effect has been demonstrated in actualistic studies (e.g., Greenwood, 1982). In addition, MacGinitie (1969) reported that toothed species are more common in stream-side vegetation than the regional vegetation in extant subtropical forests. This observation has not been quantified, but to the extent that it is true, fluvially deposited leaf assemblages may tend to underestimate MAT systematically.

Oxygen isotope geochemistry

In temperate and polar regions the mean oxygen isotope composition ($\delta^{18}O$ value*) of meteoric water (e.g., precipitation such as rain and snow) is strongly correlated with MAT, with the ^{18}O-depleted values in cold regions and ^{18}O-enriched values in warm regions (Dansgaard, 1964). The relationship between temperature and meteoric water $\delta^{18}O$ value results from the formation of vapor masses in warm, low-latitude regions, followed by rainout from these masses in colder, high-latitude regions, a phenomenon which has been successfully modeled as a Rayleigh distillation process (Dansgaard, 1964; Rozanski et al., 1993). In tropical areas, meteoric water $\delta^{18}O$ values are more strongly correlated to the amount of precipitation than to MAT. Complexities due to differences in vapor sources, orographic effects, and monsoonal climates can affect these relationships. However, the positive correlation between temperature and the $\delta^{18}O$ of meteoric water should have been present in temperate and polar regions in the geologic post. Changes in meteoric water $\delta^{18}O$ value provide a monitor of major, regional changes in the climatic and hydrologic cycle.

Our monitor of meteoric water $\delta^{18}O$ value is the isotopic composition of authigenic minerals in paleosols. The $\delta^{18}O$ of a mineral is determined by the fractionation of oxygen isotopes between the solid phase and the water from which it forms (which commonly varies as a function of temperature) and the $\delta^{18}O$ of the water. Thus minerals that form in near-surface environments record the isotopic composition of surface water, which will track the $\delta^{18}O$ of meteoric water to a greater or less extent. Pedogenic and vein-filling calcites have been used extensively as proxies for surface water $\delta^{18}O$ values (e.g., Quade et al., 1989; Winograd et al., 1992; Cerling and Quade, 1993). The oxygen isotope compositions of clay minerals and oxides in soils have been used increasingly to monitor the $\delta^{18}O$ of surface water (Savin and Epstein, 1970; Lawrence and Taylor, 1972; Yapp, 1987, 1993; Bird et al., 1992, 1993).

---

* $\delta^{18}O = ((R_{\text{sample}}/R_{\text{standard}}) - 1) \times 1000$, where R is the $^{18}O/^{16}O$ in the sample or standard. The standard is Vienna standard mean ocean water (V-SMOW). The units are in per mil (‰).

Vertebrate fossils encrusted with hematite (α-Fe_2O_3) are common in alluvial paleosols that developed on fluvial sediments that accumulated in the Bighorn Basin during the late Paleocene and early Eocene (Bown and Kraus, 1981a,b). Based on the number of paleosols, the timespan of sediment accumulation, and paleosol maturity, it has been estimated that individual paleosols formed in thousands to tens of thousand of years (Bown and Kraus, 1993; Kraus and Bown, 1993). A thorough examination of the timing of hematite mineralization within paleosols is beyond the scope of this paper (see Bao et al., 1998), but several lines of evidence provide some constraint. High concentrations of hematite are found only as encrustations on vertebrate fossils, not in the soil matrix or on nodules. Not all fossils have hematite coatings; encrustation is closely associated with fossils from hydromorphic paleosols, particularly 'class A' gray mudstones (Bown and Kraus, 1981b; Bown et al., 1994a). Petrographic observations of overgrowth relationships demonstrate that iron oxide and micrite were deposited synchronously, and that both were deposited prior to void-filling sparry calcite. Oxygen isotope analysis indicates that micrites are near-surface, pedogenic products, whereas sparry calcites are deep-burial products (Koch et al., 1995). These observations suggest that initial iron oxide formation occurred in near-surface environments. The inital precipitate was probably ferrihydrite ($5Fe_2O_3 \cdot 9H_2O$), a highly disordered and poorly crystalline precursor, that was subsequently altered by dehydration and internal rearrangement to hematite (Johnston and Lewis, 1983; Schwertmann and Murad, 1983; Schwertmann and Taylor, 1989), or insoluble Fe–organic complexes, which were transformed into hematite by oxidation (McKeague et al., 1986).

Several characteristics of iron oxides make them ideal for reconstruction of the $\delta^{18}O$ of the surface water. Iron oxides hold their original oxygen isotope compositions with high fidelity; even drastic chemical treatments do not affect the $\delta^{18}O$ of hematite and goethite (Becker and Clayton, 1976; Yapp, 1991). In addition, there is evidence that oxygen isotope fractionation between water and hematite is relatively insensitive to temperature change in the surface temperature range (Clayton and Epstein, 1961; Zheng, 1991), though the fractionation relationship determined by Yapp (1990a) shows greater temperature sensitivity. The source of this discrepancy is under investigation, but in any case the fractionation of oxygen isotopes between water and hematite is much less sensitive to temperature than the water/calcite fractionation. Consequently, hematite is an excellent indicator of the $\delta^{18}O$ of ancient surface water. Finally, recent papers by Yapp (1987, 1990a,b, 1991, 1993, 1997) have demonstrated the utility of oxygen isotope analysis of iron oxides and addressed many questions associated with their genesis and analysis.

MATERIALS AND METHODS

Leaf margin analysis

Fossil assemblages for leaf margin analysis come from > 250 sites in the Fort Union Formation and Willwood Formation of the Bighorn Basin that have been collected by field parties from the Smithsonian Institution and Yale University. Most assemblages come from small quarries of $\sim 3 \times 3$ m laterally and less than 0.5 m stratigraphic thickness. These sites occur in all parts of the Bighorn Basin, but are concentrated in the areas around North Butte, Elk Creek, and upper Fifteenmile Creek, and in the Clarks Fork Basin (Fig. 7.1; Hickey, 1980, Brown *et al.*, 1994a; Wing *et al.* 1995, Wing, 1998). Most of the localities are in measured sections that can be correlated lithostratigraphically to the carbon isotope excursion and LPTM.

Most of the leaf assemblages examined for this study were collected from carbonaceous shales that accumulated as the result of fine-grained sedimentation events on poorly drained distal floodplains (Davies-Vollum and Wing, 1998). Backswamp leaf assemblages are largely autochthonous, representing only a very small area of floodplain vegetation (Davies-Vollum and Wing, 1998). Grain size, bed forms, and thin-section features are consistent with low-energy deposition, and typically riparian genera such as *Platanus* are rare in the backswamp assemblages (Wing *et al.*, 1995). The fossil plant assemblage from any given backswamp site probably accumulated during one or a few sedimentation events spread over no more than a few years, and carbonaceous beds of 1–3 m thickness probably accumulated over hundreds to thousands of years (Davies-Vollum and Wing, 1998). Thus time averaging within or among sites along a single bed is insignificant for the purposes of this study.

A smaller number of floral assemblages come from abandoned channel and near-channel settings. Current energy in the near-channel environments was greater than in the backswamps or abandoned channels, but identifiable leaves still appear to be derived from local vegetation (Wing *et al.*, 1995). Temporal and spatial averaging of individual near-channel and abandoned channel floras is probably on the order of years and hundreds of meters. Many of these assemblages are dominated in numbers of specimens by typical riparian genera such as *Platanus*, but they contain a diversity of other, non-riparian species as well.

In order to increase the number of species for each leaf margin estimate of MAT, and hence decrease the effect of sampling error on the temperature estimate (Eq. 7.2), we have lumped floras from stratigraphic intervals of varying thickness for each estimate. This process introduces analytical time averaging on the order of hundreds of thousands of years, which means that temperature fluctuations of less than that duration cannot be discerned.

There are a total of eight leaf margin estimates of MAT derived from Bighorn Basin floras, designated Leaf Margin Analysis (LMA)-1 through LMA-8 (Table 7.6). All 14 of the early Clarkforkian sites that were included in LMA-1 were collected from the Fort Union Formation in the Clarks Fork Basin in northernmost Wyoming or southernmost Montana (Hickey, 1980). Some of these sites can be correlated with meter levels in the Polecat Bench/Clarks Fork Basin composite section, while the remainder can be assigned to the Cf1 faunal zone based on nearby mammalian faunal collections (Hickey, 1980). The most extensive collection from the early Clarkforkian is from the Bear Creek Mine and Foster Gulch (Montana) area. Although this locality is not in the Polecat Bench/Clarks Fork Basin composite section, it can be dated biostratigraphically because of the occurrence of early Clarkforkian mammals in the roof shales that are stratigraphically close to the flora (Hickey, 1980). The Bear Creek fauna may be somewhat earlier in the

Table 7.6. *Leaf margin estimates of MAT*

Sample	Sections with samples*	Meter level	Age model 1	Age model 2	Duration	MAT	MAT error	No. of entire spp.	No. of toothed spp.
LMA-8	ECS	621	−52.831	−52.842	0.01	22.2	2.0	33	15
LMA-7 (midpoint)	ECS	444	−53.249	−53.411	0.11	15.8	2.2	23	25
LMA-6 (midpoint)	ECS	332	−53.573	−53.806	0.17	10.8	3.3	6	13
LMA-5 (midpoint)	ECS	155	−54.310	−54.568	0.37	16.4	2.7	16	16
LMA-4 (midpoint)	ECS	57.5	−54.715	−54.987	0.40	18.2	2.3	24	19
LMA-3 (midpoint)	HCS, ECS, MPS	−37	−55.109	−55.393	0.25	15.7	2.4	20	22
LMA-2 (midpoint)	CFB, WES	−123	−55.467	−55.763	0.33	15.8	2.2	24	26
LMA-1 (midpoint)	CFB, Bear Creek	1000	−56.040	−56.347	0.62	12.9	2.4	15	24

*Section abbreviations as in Table 7.1; models 1 and 2 are explained in the text.

Clarkforkian than the Cf1 faunas collected near Polecat Bench (J. Alroy, personal communication, 1997), and therefore LMA-1 may include floras that span the entire early Clarkforkian, an interval we infer to be as much as 610 ka (Tables 7.4 and 7.5).

Floras collected at 30 sites were combined for LMA-2. Four of the sites are located in the Clarks Fork Basin, where they are associated with mammals of zone Cf2 and can be correlated to the Polecat Bench/Clarks Fork Basin section. The other 26 sites are in the Fort Union Formation in the North Butte, Worland East, Sand Creek Divide, and Antelope Creek areas of the eastern and southern Bighorn Basin. These sections lack age-diagnostic mammalian fossils or magnetostratigraphy, so floral sites were assigned to LMA-2 based on their stratigraphic level with respect to the Wa0 red bed. Sites that were 75–165 m below the Wa0 red bed were assigned to LMA-2. If we assume that accumulation rates in the upper Fort Union Formation of the eastern Bighorn Basin were the same as those in the lower Willwood Formation in the same area (0.23 m ka^{-1}; Table 7.2), then these meter levels correspond to the upper and lower bounds of Cf2 time as determined from the age calibration of the Polecat Bench/Clarks Fork Basin section (Table 7.4). Although we lack dated horizons in the Fort Union of the eastern Bighorn Basin, the assumption of constant rates of rock accumulation during the middle Clarkforkian through early Wasatchian in this area is consistent with other evidence. Firstly, the areas from which the floras were collected are all distibuted along strike near the margin of the basin, which should favor similar low rates of rock accumulation in all sections. Secondly, the Fort Union Formation in the eastern Bighorn Basin rests on a major erosional unconformity developed on the Late Cretaceous Lance Formation and Meeteetse Formation, but is gradational with the overlying Willwood Formation. This is consistent with Fort Union Formation deposition beginning in the late Paleocene and continuing without substantial hiatus into the early Eocene. Thirdly, the composition of the megafloras indicates a late Paleocene age (Wing, 1998). Fourthly, scattered mammalian fossils from near the base of the Fort Union Formation west of Greybull are Tiffanian or Clarkforkian (Gingerich, personal communication, 1990). Clearly, more paleomagnetic, radiometric, or biostratigraphic data are needed in the Fort Union Formation sections of the eastern Bighorn Basin, but in the absence of such data we think the stratigraphic grouping used here is likely to be a close approximation of Cf2 time. Sites grouped for LMA-2 span 480 ka (Tables 7.4 and 7.6).

Floras from 35 sites were combined for LMA-3. One of these sites is in the Clarks Fork Basin and can be correlated directly with the Polecat Bench/Clarks Fork Basin section and Cf3 mammals (Hickey, 1980). The remaining 34 sites are from the uppermost 75 m of the Fort Union Formation in the North Butte, Worland East, Gould Butte, and Antelope Creek areas. As with the sites used for LMA-2, rock accumulation rates for the lower part of the Antelope Creek/Elk Creek section were assumed to apply to the uppermost Fort Union Formation. Fifteen of the sites are 30 m or less below the Wa0 red bed in conformable sequence. The lower bound for LMA-3 sites was set at 75 m below the Wa0 red bed because this is the level that is age equivalent to the base of Cf3 (Tables 7.3 and 7.4) if a depositional rate of 0.23 m ka^{-1} is assumed. Sites grouped for LMA-3 span 300 ka.

LMA-4 is based on floras from 20 sites in the Antelope Creek, Gould Butte, and Foster Gulch areas. The sites in the Antelope Creek/Elk Creek section are all 10–100 m above the Wa0 red bed, the site from Foster Gulch is 18 m above the Wa0 red bed, and the sites from the Gould Butte area are from a stratigraphic interval 20–105 m above the first appearance of Eocene plant megafossils. Based on rock accumulation rates calculated from the lower part of the Antelope Creek/Elk Creek section (Table 7.3), sites grouped for LMA-4 span 410 ka.

The remaining leaf margin analyses (LMA-5 through LMA-8) are based almost exclusively on floras from the Antelope Creek/Elk Creek section. Most of the 31 sites from which floras were combined for analysis in LMA-5 have stratigraphic levels of 110–200 m in the Antelope Creek/Elk Creek section, except for one diverse site in the Gould Butte area that is about 200 m above the first occurrence of Eocene plants. Based on rock accumulation rates calculated from the lower part of the Antelope Creek/Elk Creek section (Table 7.3), sites grouped for LMA-5 span 390 ka.

The LMA-6 is based on floras from only six sites, all between 311 and 353 m in the Antelope Creek/Elk Creek section. Based on rock accumulation rates calculated from the lower part of the Antelope Creek/Elk Creek section (Table 7.3), sites grouped for LMA-6 span 180 ka.

The LMA-7 is also based on floras from six sites in a narrow stratigraphic interval, 420–468 m above the Wa0 red bed. Rock accumulation rates calculated for the upper part of the Antelope Creek/Elk Creek section (Table 7.3) indicate that the sites grouped for LMA-7 span 100 ka. The stratigraphically highest temperature estimate, LMA-8, is derived from the flora of a single, laterally extensive carbonaceous shale at the 621 m level of the Antelope Creek/Elk Creek section. This is only 13 m below the bentonitic tuff (52.8±0.3 Ma). There are 24 sites in this bed distributed across about 10 km of outcrop. The estimated maximum timespan for LMA-8, 10 ka, is based on depositional models discussed by Davies-Vollum and Wing (1998).

Oxygen isotope analysis

We obtained samples spanning the Clarkforkian to middle Wasatchian from fossil mammals collected in the southern and central Bighorn Basin (United States Geological Survey, Denver) and the Clarks Fork Basin (Museum of Paleontology, University of Michigan). Details of the isotopic analysis of hematite are fully explained in Bao *et al.* (1998). Briefly, hematite was collected from bones by drilling under a microscope. Samples were pretreated with 1 M HCl (24–36 h) and dilute NaHCl (12–24 h) at room temperature to remove calcite, apatite, clay minerals, and organic contaminants. To obtain $\delta^{18}O$ values for hematite, which is intimately mixed with silicate minerals, we used a modified version of the mass-balance approach of Bird *et al.* (1992). Firstly, samples were treated with 5 M NaOH at 95 °C (3–10 h). This treatment removes a large fraction of the silicates and concentrates hematite, but has no effect on the $\delta^{18}O$ of hematite (Yapp, 1991). The pretreated bulk sample was split, and one split was treated with 6 M HCl at 80 °C (<30 min), leaving a residue of silicate minerals, chiefly quartz. The $\delta^{18}O$ of pure hematite was calculated using the following relationship:

$$\delta^{18}O_{bulk} = \delta^{18}O_{hematite} * f_{Fe} + \delta^{18}O_{residue} * (1 - f_{Fe}), \qquad (7.3)$$

where $\delta^{18}O_{bulk}$ is the measured isotope composition of pretreated bulk powder, $\delta^{18}O_{residue}$ is the composition of the silicate residue, and f_{Fe} and $1 - f_{Fe}$ are the oxygen mole fractions of hematite and silicates, respectively, in pretreated bulk powder.

To calculate mole fractions, the concentrations of Fe, Si, and Al (the concentrations of other elements are trivial) were determined for the split of the pretreated bulk powder by inductively coupled plasma atomic emission spectroscopy (ICP-AES) using a Perkin Elmer 6000. The error associated with elemental analysis for each sample was determined by analyses of duplicates. The mean value for error was ± 0.0007 ($n = 29$) for the oxygen mole fraction in hematite (≈ 1 weight %). This analysis was conducted at the Department of Geosciences, Princeton University.

The $\delta^{18}O$ value of pretreated bulk powder and silicate residues was determined by laser fluorination. Approximately 1–2 mg of sample powder was loaded into a stainless steel holder, then pretreated overnight with BrF_5 at room temperature to remove adsorbed atmospheric moisture and gases. Oxygen was generated by fluorination with a 20 W CO_2 laser in a BrF_5 atmosphere (Valley *et al.*, 1995; Rumble *et al.*, 1994). The O_2 evolved by laser fluorination was collected on a molecular sieve at $-190\,°C$, then immediately analyzed as O_2 on a Finnigan MAT 252. The error associated with laser fluorination and isotopic analysis was assessed through analysis of duplicates. The mean value for error was $<0.15‰$. Laser fluorination and oxygen isotope analysis on a Finnigan MAT 252 were conducted at the Geophysical Laboratory, Carnegie Institution of Washington.

The error for each calculated value of $\delta^{18}O_{hematite}$ was estimated from the error in determination of weight % hematite. The mean value for the effect of this error on $\delta^{18}O_{hematite}$ was $\pm 0.26‰$ ($n = 29$). The average for the total analytical error, which also includes errors associated with isotopic analysis, was $\pm 0.7‰$.

To test the reliability of the method, whole process replicate analyses (performed on separate sample powders drilled from the same bone) were conducted on two samples: HOA-YPM350 and HOA-SC90. The differences in $\delta^{18}O_{hematite}$ were $0.11‰$ and $0.45‰$ for these whole process duplicates. Data consistency was tested through comparisons of values between localities at different lateral positions at the same meter level: SC4 and SC123 at 1570 m in the Polecat Bench/Clarks Fork Basin section, and D1300 and D1301 at 78 m in the Antelope Creek/Elk Creek section. Differences in $\delta^{18}O_{hematite}$ for different samples from the same meter level were $0.84‰$ and $1.2‰$.

PALEOCLIMATIC RESULTS

Mean annual temperature from leaf assemblages

The eight estimates of MAT (LMA-1 through LMA-8) are shown in Table 7.6 and plotted against time in Fig. 7.3. Durations associated with each estimate are

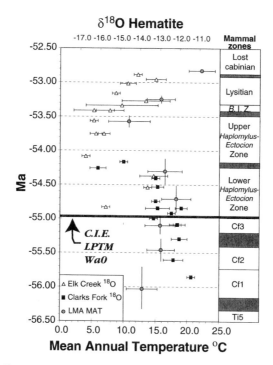

Figure 7.3. Temperature estimates from leaf margin analysis (LMA) and estimates of the $\delta^{18}O$ of precipitation from iron oxides plotted against time. C.I.E., carbon isotope excursion; LPTM, Latest Paleocene Thermal Maximum; *B.I.Z.*, *Bunophorus* Interval Zone.

discussed above. MAT estimates were calculated using Equation (7.1) and errors of the MAT estimates were calculated using Equation (7.2).

MAT estimates rise from 12.9 to 15.3 °C during the Clarkforkian. The earliest Wasatchian estimate of MAT is 18.2 °C, after which MAT declines to 16.4 °C and then 10.8 °C at an interpolated age of 53.56 Ma (uppermost C24r). The coldest MAT estimate has a relatively large uncertainty (±3.3 °C) because of the low diversity of the flora from this time period. This cold estimate of MAT is as low as the mid-Paleocene (Tiffanian) MAT estimates made by Hickey (1980). In the middle Wasatchian, MAT estimates rise first to 15.8 °C and finally to 22.2 °C by 52.83 Ma, near the beginning of the Lostcabinian and C24n.1n.

Although the small number of MAT estimates and the large amount of analytical time averaging for the Clarkforkian estimates do not make for a detailed record of temperature change through time, a strong pattern emerges from the data. MAT rose during the last million years of the Paleocene and into the early Eocene, declined to less than late Paleocene values by about 1.5 Ma following the carbon isotope excursion, and then warmed rapidly over the succeeding 700 ka to reach a long-term maximum in the middle to later early Eocene.

Table 7.7. $\delta^{18}O$ of hematite-encrusting bones

Sample	Section*	Meter level	Age model 1	Age model 2	$\delta^{18}O$ hematite	Error
HOA-YPM33	ECS	601	−52.878	−52.906	−14.1	0.2
HOA-YPM1	ECS	571	−52.949	−53.002	−13.2	0.5
HOA-D1467	ECS	546	−53.008	−53.083	−14.6	0.4
HOA-D1531	ECS	485	−53.152	−53.279	−15.2	0.2
HOA-D1398	ECS	438	−53.263	−53.430	−13.7	1.2
HOA-D1454	ECS	409	−53.332	−53.523	−14.9	1.8
HOA-D1300	ECS	378	−53.405	−53.623	−16.3	1.0
HOA-D1301	ECS	378	−53.405	−53.623	−15.5	0.7
HOA-D1374	ECS	336	−53.556	−53.789	−16.3	0.2
HOA-YPM350R	ECS	290	−53.748	−53.987	−16.2	0.3
HOA-YPM350	ECS	290	−53.748	−53.987	−15.8	0.3
HOA-YPM290	ECS	210	−54.081	−54.331	−16.7	0.2
HOA-YPM104	ECS	140	−54.372	−54.632	−13.3	0.7
HOA-YPM119	ECS	100	−54.539	−54.804	−13.6	0.2
HOA-YPM115	ECS	30	−54.830	−55.105	−15.7	0.2
HOA-SC232	CFB	1895	−54.164	−54.417	−14.8	0.2
HOA-SC34	CFB	1850	−54.259	−54.515	−16.1	0.4
HOA-SC313	CFB	1780	−54.406	−54.667	−13.2	0.2
HOA-SC12	CFB	1720	−54.533	−54.798	−13.1	0.3
HOA-SC18	CFB	1620	−54.744	−55.016	−13.2	0.2
HOA-SC4	CFB	1570	−54.849	−55.125	−11.9	0.3
HOA-SC123	CFB	1570	−54.849	−55.125	−13.1	0.6
HOA-SC40	CFB	1535	−54.923	−55.201	−12.4	0.2
HOA-SC138	CFB	1500	−54.997	−55.278	−13.3	0.2
HOA-SC90	CFB	1455	−55.092	−55.376	−12.1	0.2
HOA-SC90R	CFB	1455	−55.092	−55.376	−12.1	0.4
HOA-SC127	CFB	1355	−55.303	−55.594	−12.0	0.4
HOA-SC92	CFB	1210	−55.609	−55.910	−12.3	0.5
HOA-SC171	CFB	1090	−55.862	−56.171	−11.4	0.2

*Section abbreviations as in Table 7.1; models 1 and 2 are explained in the text.

Isotopic composition of hematite

Hematite $\delta^{18}O$ values are given in Table 7.7 for samples from the Clarks Fork Basin and the eastern and central Bighorn Basin. In Fig. 7.3, hematite $\delta^{18}O$ values are plotted vs. age for comparison with the MAT estimates derived from leaf margin analysis. In the Clarks Fork Basin, hematite $\delta^{18}O$ values are fairly consistent from 55.86 Ma (1090 m) to 54.43 Ma (1780 m), then drop $\sim$3‰ by 54.27 Ma. In the central Bighorn Basin, the $\delta^{18}O_{hematite}$ at 54.39 Ma (140 m) is nearly identical to that observed at 54.43 Ma in the Clarks Fork Basin. As in the north, $\delta^{18}O_{hematite}$ dropped by $\sim$3.5‰ by 54.09 Ma (210 m) in the Antelope Creek/Elk Creek section. Values remained low for over 500 ka before beginning to rise between 53.36 Ma (378 m) and 53.11 Ma (409 m).

Given independent evidence that hematite forms in the pedogenic environment (Bao et al., 1998), $\delta^{18}O_{hematite}$ can be used to infer relative changes in the $\delta^{18}O$

of surface water in the Bighorn Basin. Current uncertainty regarding the correct fractionation relationship for the hematite/water system precludes reconstruction of the absolute $\delta^{18}O$ value of surface water. However, all three published fractionation relationships exhibit low temperature sensitivity in the earth surface temperature range. Estimates of sensitivity range from $0.14‰\,°C^{-1}$ (Yapp, 1990a) to $0.08‰\,°C^{-1}$ (Zheng, 1991), to $0.04‰\,°C^{-1}$ (Clayton and Epstein, 1961). Ongoing experiments, as well as constraints supplied by the isotope chemistry of co-occurring phases, suggest that the fractionations of Zheng (1991) or Clayton and Epstein (1961) are most applicable to Bighorn Basin hematite (Bao *et al.*, 1998). If true, even climatically driven changes in temperature of mineral formation as large as $40\,°C$ would generate shifts in $\delta^{18}O_{hematite}$ of only 1.5–3‰. While we cannot presently exclude the possibility that a portion of the signal in $\delta^{18}O_{hematite}$ records the effect of temperature change on fractionation, for this discussion we assume that $\delta^{18}O_{hematite}$ dominantly records change in the $\delta^{18}O$ value of surface water.

Thus the record from hematite indicates that the $\delta^{18}O$ of Bighorn Basin surface water was essentially constant from mid-Clarkforkian through the first 700 ka of the Wasatchian, then dropped rapidly and remained low for over 500 ka before beginning to rise $\sim$53.5 Ma. This shift in the isotopic composition of ancient surface water can be related to paleoclimate by assuming that: (1) in this mid-latitude, mid-continental region, variations in the $\delta^{18}O$ of meteoric water (the ultimate source of surface water) were more strongly correlated to temperature than to amount of precipitation; (2) the slope of the modern relationship between mean annual meteoric water $\delta^{18}O$ value and MAT can be applied to the Paleogene of North America; and (3) the source of precipitation arriving in the region remained constant throughout the interval. We recognize that these assumptions may be in error. Careful treatment of the meteoric water $\delta^{18}O$/MAT relationship for the Paleocene–Eocene transition would require linking oxygen isotope fractionations in atmospheric processes to a climate simulation for the interval. This is beyond the scope of this paper. Instead our goal is to evaluate whether the mid-latitude $\delta^{18}O$/MAT slopes of 0.5–$0.6‰\,°C^{-1}$ observed in the Quaternary (Dansgaard, 1974; Boyle, 1997) supply plausible, consistent estimates of temperature change when applied to the record of change in the $\delta^{18}O$ of Bighorn Basin surface water. Given these assumptions, the drop in surface water $\delta^{18}O$ values of 3.25‰ in the earliest Eocene that we infer from $\delta^{18}O_{hematite}$ corresponds to a temperature decrease of 5.5–$6.5\,°C$.

No hematite-encrusted bones have been recovered from Wa0 beds, so we cannot evaluate changes in hematite $\delta^{18}O$ values during the LPTM. However, isotopic records of change in MAT and the $\delta^{18}O$ of surface water are available for the Bighorn Basin. Koch *et al.* (1995) demonstrated that both soil carbonate and bivalve aragonite from Wa0 beds are enriched in ^{18}O by $\sim$1‰ relative to material in underlying Clarkforkian or overlying early Wasatchian sediments. In order to assess changes in MAT from calcite or aragonite, we must consider the effects of temperature both on the $\delta^{18}O$ of meteoric (and surface) water and on the fractionation of oxygen isotopes between mineral and water. Assuming that the effect of temperature on the fractionation of oxygen isotopes between calcium carbonate and water is

~0.23‰ °C^{-1} (Friedman and O'Neil, 1977), and that meteoric water $\delta^{18}O$ increases by a range of +0.5 to 0.6‰ °C^{-1} with increases in MAT, an increase in MAT ranging from 2.7 to 3.7 °C is required to explain the 1‰ increase in Bighorn Basin soil carbonates and bivalve shells during the LPTM. This would be accomplished by a shift of +1.6 to +1.9‰ in the $\delta^{18}O$ of meteoric and surface water.

CONTINENTAL BIOTIC CHANGE IN THE PALEOCENE–EOCENE TRANSITION

Faunas

The late Paleocene–early Eocene was a period of major mammalian immigration into North America and western Europe. The beginning of the Clarkforkian in North America (during C25n) is recognized by the first appearances of rodents and coryphodontid pantodonts. There are no close relatives of these higher taxa in the earlier Paleocene of North America, implying that they immigrated into North America from another continent. The most significant mammalian immigration event of the Cenozoic occurred at the beginning of the Wasatchian, traditionally recognized as the beginning of the Eocene in continental deposits of North America (Krishtalka *et al.*, 1987). These immigrants also represent orders and families new to North America: primates, even-toed ungulates, odd-toed ungulates, and hyaenodontid creodonts. As at the beginning of the Clarkforkian, the lack of close relatives earlier in the Paleocene implies that these taxa dispersed into North America from another continent. Wasatchian immigrants are known from approximately coeval strata across much of North America, from Mississippi and Baja California in the south to Ellesmere Island in the north (Krishtalka *et al.*, 1987; Beard and Tabrum, 1991). The first appearance of Wasatchian immigrants is approximately synchronous in Europe and North America (Hooker, 1996), strengthening the hypothesis that these taxa dispersed to both continents across high-latitude land bridges. Relatively sketchy knowledge of African and Asian faunas from the Paleocene has made the origins of the Wasatchian immigrants a contentious issue. Recent discoveries in Asia may point to an Asian origin for many of these groups (Beard, 1997). If rodents, coryphodontids, primates, odd-toed and even-toed ungulates, and hyaenodontids all originated in Asia, then the opening of high-latitude land bridges between Asia, North America, and Europe probably had a major effect on the timing of immigration into the latter two continents (Beard, 1997).

The earliest Wasatchian (Wa0) faunas have been identified only in the Bighorn Basin (Gingerich, 1989), where they are contemporaneous with the carbon isotope excursion (Koch *et al.*, 1992, 1995). The Wa0 fauna has the familial composition of typical later Wasatchian assemblages, but many of the species are small compared with earlier and later congenerics (Gingerich, 1989; Clyde, 1997). Among living mammals, populations with small body size are associated with warmer climate (Searcy, 1980; Koch, 1986), supporting isotopic evidence for warmer climate during the LPTM. The Wa0 fauna also has the highest species richness and the most equitable distribution of individuals among species of any Paleocene or Eocene faunal interval in the Bighorn Basin sequence (Clyde, 1997), another characteristic of mammalian faunas from areas with warmer climates. The basal Wasatchian turnover event also modified the trophic structure of the mammalian fauna by increasing the proportion of herbivore/frugivore species (Clyde, 1997). Clyde (1997) has also shown that the taxa immigrating into North America at the beginning of the Wasatchian rapidly became numerically important or dominant members of the mammalian fauna, implying that the faunal transition was ecologically as well as taxonomically significant.

Detailed studies of faunal change during the early Eocene have been confined to the northern Rocky Mountains, therefore the generality of the observations is unknown. Faunal change within the early Eocene is at the generic or specific level. A small pulse of faunal change, called Biohorizon A, has been identified among mammalian faunas from the central Bighorn Basin (Schankler, 1980). Although the existence of Biohorizon A has been attributed to variations in sampling intensity (Badgley and Gingerich, 1988), more recent data tend to confirm the presence of accelerated compositional change at this time (Badgley, 1990; Bown *et al.*, 1994*a*), which we infer to be about 54.2 Ma. A small number of taxa have last appearances at Biohorizon A (Schankler, 1980, 1981), and three of them reappear synchronously, higher in the Antelope Creek/Elk Creek section near Biohorizon B. Bown *et al.* (1994*b*) identified several lineages that underwent size increases at Biohorizon A, which is consistent with cooler climates. Faunas from the Foster Gulch/McCullough Peaks section show a small increase in the mean size of individuals in the fauna from Wa2 to Wa3, approximately the same time as Biohorizon A (Clyde, 1997).

A third pulse of mammalian faunal change occurred during the middle of the Wasatchian in the Bighorn Basin, recognized as Biohorizons B and C by Schankler (1980). Bown *et al.* (1994*a*) argued that Biohorizons B and C were separated by only 40–60 m, and referred to this zone of accelerated turnover as Biohorizon B-C. During this short period, approximately contemporaneous with the beginning of C24n (see disussion above), there are several closely spaced first occurrences (*Bunophorus, Heptodon*) and last occurrences (*Haplomylus, Ectocion*) of mammalian genera (Bown *et al.*, 1994*a*). A similar pattern of faunal turnover has been observed in the northern part of the Bighorn Basin (Badgley and Gingerich, 1988; Clyde, 1997). During and just prior to Biohorizon B there was a large increase in morphological variability in many lineages of mammals, followed by a two- to threefold increase in the number of lineages at the species level in the subset of genera that have been studied thoroughly (K. D. Rose, personal communication, 1996). Size decreases in many lineages also occurred in this time interval, which is consistent with renewed climatic warming (Bown *et al.*, 1994*b*). Size decrease is less noticeable in faunas from the McCullough Peaks area, although the mean individual size for the whole fauna does decrease in the latter part of Wa4, approximately the same as Biohorizon B (Clyde, 1997). Some species-level faunal changes at Biohorizon B may represent *in situ* evolution or intracontinental migration, but intercontinental migration from Asia has been proposed as an explanation for the first appearances of the ungulate genera *Bunophorus* and *Heptodon*, and the slightly later first appearance of *Lambdotherium* (Beard, 1997).

In summary, there are four intervals of accelerated change in the composition of mammalian faunas during the Paleocene and early Eocene. The beginning

Clarkforkian and beginning Wasatchian events are at a high taxonomic level and probably record northern hemisphere-wide migration across high-latitude land bridges, quite possibly from Asia (Beard, 1997). In the Bighorn Basin the Wasatchian immigration is accompanied by increasing diversity and decreasing body size. These features are consistent with warming climates at middle and high latitudes. The species-level events at Biohorizon A have been observed only in the Bighorn Basin, and may be local, but the body size increases are consistent with paleobotanical and isotopic evidence for cooling in the same sections. The generic-level immigration and body size decrease at Biohorizon B has also been documented best in the Bighorn Basin, and could also be local, but again the changes are con-sistent with the paleobotanical and isotopic evidence for increasing temperature (Bown *et al.*, 1994*b*). The last three periods of faunal change are associated with inflections in the temperature curve. The basal Wasatchian immigration coincides with the LPTM (and with the carbon isotope excursion), Biohorizon A coincides with falling temperatures in the early Eocene, and Biohorizon B with rapidly rising temperatures near the beginning of C24n. Later early Eocene faunas are not well represented in the Bighorn Basin, but the transition to Lostcabinian mammalian faunas appears to follow Biohorizon B by less than 500 ka, and takes place during the same rapid warming period.

Floras

Floras of the Paleocene–Eocene interval have been far less studied than mammalian faunas. It has been recognized for some time that Eocene floras of North America have more lineages of modern tropical distribution than do Paleocene floras (e.g., Leopold and MacGinitie, 1972; Wolfe, 1972, 1977; Hickey, 1977). Similarity between the extant warm-temperate floras of North America and East Asia has long been taken as an indication of dispersal across Beringia during the early Cenozoic (e.g., Graham, 1972). Tropical and subtropical taxa shared among Eocene floras of western Europe and North America imply the existence of warm-climate migration corridors across the North Atlantic (Tiffney, 1985*a,b*). Substantial floral similarity existed between East Asia and North America as early as the late Paleocene (e.g., Brown, 1962; Hickey, 1977; Guo *et al.*, 1984).

Paleocene–Eocene floral change in the Bighorn Basin follows the overall pattern observed elsewhere in the northern hemisphere, but the greater stratigraphic resolution reveals more detail and complexity in the pattern. Change in floral com-position is concentrated during two stratigraphic intervals. The first interval begins just following the LPTM (or perhaps during it, since there are no samples from the LPTM interval) and extends through about 30 m of section (approximately 200 ka). The second interval of accelerated change begins between 380 and 420 m in the Antelope Creek/Elk Creek section (near the base of C24n.1), and is completed by 621 m (although it may be shorter because there are almost no plant fossils between 468 and 621 m). This corresponds approximately to the first 600–700 ka of C24n. These intervals of compositional change stand out despite the substantial variability in the floral composition of sites that results from the original heterogeneity in floodplain vegetation and differences in sedimentary environments (Fig. 7.4).

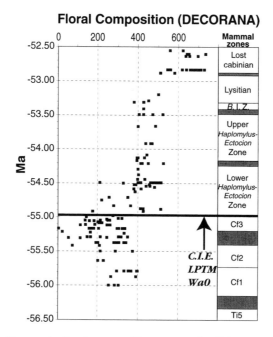

Figure 7.4. Change in floral composition across the Paleocene-Eocene boundary interval. Each point represents one site. Site ages were interpolated using age model 1 (see text). Scores on the *x*-axis are from the first axis of a detrended correspondence analysis (DECORANA; Hill, 1979) of a site by species matrix of presence/absence data. Abbreviations as in Fig. 7.3.

The first interval of floral turnover is largely the result of last occurrences in the last part of the Clarkforkian and the earliest part of the Wasatchian, combined with the first appearances of several typical Eocene species within 50 m above the Wa0 red beds. Last appearances predominated over first appearances in the late Clarkforkian (Cf3) and throughout the early part of the Wasatchian (50–350 m or 53.5–54.75 Ma), and plant diversity decreased as a consequence (Fig. 7.5). Several of the last occurrences just before the LPTM are of long-ranging Paleocene taxa such as 'Ficus' artocarpoides and Porosia verrucosa, but many common Paleocene taxa, such as 'Eucommia' serrata, Cornus, and Persites argutus, survived into the early Wasatchian. Although the majority of last appearances near the LPTM are among taxa that have modern relatives that are deciduous and temperate, the number is too small to demonstrate a clear preference in the extinction event.

In the absence of megafloral samples in the Wa0 red beds, it is not possible to tell if floral first occurrences are exactly concurrent with the Wa0 immigration event in mammals, or if they are spread out over a longer time interval. Although marine and terrestrial isotopic records indicate that the LPTM was an interval of rapid warming, the plant taxa making first appearances in the first 20 m above the Wa0 red beds are not thermophiles. Earliest Wasatchian plant immigrants include the deciduous tree *Alnus*, the tree fern *Cnemidaria*, the scrambling fern *Lygodium*,

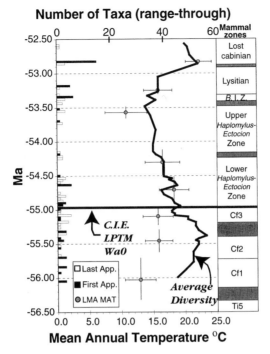

Figure 7.5. Floral richness and turnover across the Paleocene–Eocene boundary interval. Bars indicate the number of first and last occurrences at each stratigraphic level. The line is a two-point running average of the number of species. All calculations were made based on standard range-through assumptions (i.e., taxa were assumed to occur at all levels between their first and last appearance). Abbreviations as in Fig. 7.3.

and the aquatic fern *Salvinia*. *Cnemidaria* and *Salvinia* are today neotropical montane and cosmopolitan subtropical to tropical genera, respectively, but *Alnus* and *Lygodium* are both widely distributed in temperate climates as well as in montane tropical and subtropical areas. If these immigrants share an ecological characteristic, it is the ease with which they are dispersed. The ferns disperse via small, wind-blown spores, and *Alnus* has small, winged seeds that can be carried by wind or water. Although studies of tree migration rates during the Holocene have shown that population movement can potentially be very fast (100–500 m year^{-1}, Delcourt and Delcourt, 1987), these rates reflect dispersal into recently deglaciated landscape. Tree populations spreading across high-latitude land bridges during the LPTM would presumably have encountered pre-existing forest vegetation, which might have depressed rates of successful establishment following dispersal. This would have greatly reduced migration rates, and might have prevented the spread of all but the most easily dispersed plant species to North America during the 100 ka of the LPTM. Slow rates of plant migration might also explain why there is so much mammalian immigration and so little plant immigration at the base of the Wasatchian.

The second interval of accelerated floral turnover recorded in the Wasatchian is dominated by first occurrences, including many taxa that are ubiquitous in Lostcabinian and early Bridgerian floras in the northern Rocky Mountains. The sharp increase in first occurrences at 621 m (≈ 52.8 Ma) is likely the result of limited sampling in the 200 m below, but the high rates of sediment accumulation in this part of the section imply that the immigration event was rapid, taking place in 500 ka or less (Fig. 7.5). The taxa that appear in the upper part of the Antelope Creek/Elk Creek section tend to have subtropical to tropical distributions today (e.g., Elaeocarpaceae, Apocynaceae, Lauraceae, Myrtaceae, *Machaerium* [Leguminosae]), and the thickness of the compressions and preservation of the cuticle are consistent with their being broad-leaved evergreen trees or shrubs. The influx of new taxa in the interval between 53.3 and 52.8 Ma results in an increase in diversity. Both the diversity increase and the climatic preferences of living relatives of the immigrants are consistent with increasing temperatures at this time in the Bighorn Basin.

As is true with the mammalian fauna, intervals of accelerated floral turnover in the Bighorn Basin appear to correspond to times of more rapid climatic change. The basal Wasatchian plant immigrants are as nearly contemporaneous with the LPTM as sampling allows. The long decline in plant diversity in the early Wasatchian is concurrent with the period of lower temperatures indicated by both isotopic and paleobotanical temperature estimates. The sharp increase in first appearances and in diversity occurs in the same part of the section as the rapid increase in temperature.

Paleobotanical data from outside the Bighorn Basin suggest that the trends described above are at least regional to continental in scope. Changes in floral composition similar to those around the LPTM are seen during the Paleocene–Eocene transition in the Williston Basin of North Dakota (Hickey, 1977), and the increasing floral diversity and change in composition observed in the upper part of the Antelope Creek/Elk Creek section is also observed in association with Lostcabinian mammalian faunas in several parts of southern and central Wyoming (Wing, 1987; Wilf, 1999). The pattern of low latest Paleocene and earliest Eocene diversity followed by rapidly increasing mid–early Eocene diversity is also found in palynofloras from the Gulf Coast of the USA (Frederiksen, 1994). Finally, palynofloras from England also indicate an early Eocene period of lowered diversity (Jolley, 1996). Together these disparate floral data sets provide preliminary evidence that the climatic events associated with late Paleocene and early Eocene floral change in the Bighorn Basin were continental to hemispheric in extent.

CLIMATIC IMPLICATIONS

Two independent lines of evidence confirm the existence of a period of substantial cooling during the early Eocene in the Bighorn Basin, probably beginning ~ 700 ka after the carbon isotope excursion and LPTM, and lasting for approximately the same amount of time. Following this decline, temperatures increased rapidly to their Cenozoic maximum within the first ~ 600 ka after the beginning of C24n. The amplitude of the drop in MAT estimated from leaf margin analysis is

$\sim 7.4\,°C$. Oxygen isotope analysis of hematite, which overlaps the early Eocene part of the floral record, corroborates a change in temperature of $\sim 6\,°C$. Between 53.5 and 52.8 Ma, leaf margin estimates indicate an $\sim 10\,°C$ increase in MAT. Although oxygen isotopic measurements of hematite are not available for rocks younger than 53 Ma, the initial part of the rise in MAT is accompanied by an $\sim 2‰$ increase in the inferred $\delta^{18}O$ of surface water, consistent with a temperature change of 3.5–4 °C.

Because the isotopic and plant-based estimates of MAT come from different sediment, which may represent different parts of short-term climatic cycles and different edaphic conditions on the floodplain, some consistent offset might be expected in temperature estimates from the two different environments. The alternation of oxidized and organic-rich floodplain sediments also dictates that the isotopic and paleobotanical temperature estimates do not represent precisely the same times. All of these factors (different amounts of time averaging, different local conditions, and offsets in the ages of samples) should tend to create differences in MAT records derived from isotopic and paleobotanical sources. Yet both methods indicate an ~ 700 ka long cool period during the earliest part of the Eocene.

The consistency of floral and oxygen isotope results demonstrates that the observed temperature changes are basin-wide phenomena affecting a range of environments through the region. The thermally driven perturbations in vapor transport that are recorded by the oxygen isotope data are region-to-continental scale features. Furthermore, the faunal and floral events that are coeval with temperature changes in the Bighorn Basin are also known to occur at roughly the same times in other parts of North America. If, as seems likely, the biotic correlates of temperature change are responses rather than coincidences, then the occurrence of the biotic changes over much of North America implies that the climatic driving forces had effects at a continental scope.

Changes in sea level could have caused climatic fluctuations in the late Paleocene and early Eocene of the Bighorn Basin because regressions might be expected to correspond with increased seasonality and colder winter temperatures in continental interiors. During the late Paleocene and early Eocene sea level rose and fell at frequencies roughly similar to those of the temperature changes that we have observed in the Bighorn Basin. The bulk of the Clarkforkian has been correlated to cycle TA 2.1, a time of increasing sea level during the generally high sea levels of the early Paleogene (Haq et al., 1988; Woodburne and Swisher, 1995). A Type 1 sequence boundary (TA 2.1/TA 2.2) during which much of the continental shelf is inferred to have been exposed, has been correlated with the late Clarkforkian (Woodburne and Swisher, 1995). Sea level rose briefly across the Clarkforkian/Wasatchian boundary, but a second Type 1 sequence boundary (TA 2.2/TA 2.3) has been correlated with the earliest Wasatchian (Woodburne and Swisher, 1995). A third Type 1 boundary (TA 2.3/TA 2.4) occurred within the first million years of the Wasatchian, followed by two more minor drops in sea level before the beginning of the Lostcabinian (Woodburne and Swisher, 1995).

Although the Paleocene–Eocene interval is characterized by fairly large transgressive–regressive cycles, the Bighorn Basin temperature records do not show a good temporal correspondence with sea-level change. The TA 2.1/TA 2.2

regression occurs during the warm late Clarkforkian, and the TA 2.2/TA 2.3 regression occurs during the warm earliest Wasatchian. Only the TA 2.3/TA 2.4 regression migth correspond to the onset of the cooling we have documented ~700 ka following the carbon isotope excursion. The lack of correspondence between sea level and temperature casts doubt on this possible mechanism, althoug the lack of correspondence could also reflect miscorrelation of marine and continental sequences.

We currently have no good hypotheses to explain early Eocene cooling, but if future observation confirmed this perturbation it considerably complicates the climatic history of the late Paleocene and early Eocene. It appears that over the period from about 58 to 53 Ma, climate in central North America warmed, cooled, then warmed again, with major cooling and warming phases lasting on the order of 1–2 million years. The durations of these fluctuations, whether global or continental in scale, demonstrate climatic variability during a period of early history characterized by globally warm climate and low latitudinal temperature gradients.

CONCLUSIONS

1. High-resolution paleoclimate records can be obtained from continental sections, such as those in the Bighorn Basin, where high sedimentation rates make up for short hiatuses in deposition.
2. Better correlation of continental and marine rocks, based on chemostratigraphy, magnetostratigraphy, and direct dating of strata, will make such records useful in developing a global understanding of paleoclimatic change, as will the use of multiple isotopic and paleontological proxies for climate change.
3. Greater temporal resolution reveals a more complex pattern of climate change than was previously expected, including a substantial cool interval in the early Eocene of the Bighorn Basin.
4. The similarity of floral change through the Paleocene–Eocene transition in different parts of North America suggests that the early Eocene cooling affected at least the whole continent.
5. The causes of cool–warm–cool–warm fluctuation during the late Paleocene and early Eocene are at this point unknown. The fluctuation is too slow to be related to Milankovitch orbital processes, and is not correlated in an obvious way with seal-level changes.
6. The consequences of the cool period show up in patterns of turnover and diversity of terrestrial plants and mammals, and in the body size distribution of the mammals.
7. The duration as well as the magnitude of warming (and cooling?) events may turn out to be an important determinant of their biotic effects.

ACKNOWLEDGEMENTS

We thank Karen Bice and Will Clyde for careful reviews of an earlier draft of this paper, and Jim Zachos and Lisa Sloan for valuable conversations on the stratigraphy and paleoclimate of the Paleocene–Eocene boundary interval. S.L.W. was supported by grants from the Scholarly Studies program of the Smithsonian

Institution and the Evolution of Terrestrial Ecosystems program. P.L.K. was supported by NSF grant EAR 9627953. This is ETE publication no. 65.

REFERENCES

Ali, J. R. and Hailwood, E. A. (1998). Resolving possible problems associated with magneto-stratigraphy of the Paleocene/Eocene boundary in holes 549, 550 (Goban Spur) and 690 B (Maud Rise). *Strata*, **9**, 16–20.

Aubry, M.-P., Berggren, W. A., Stott, L. and Sinha, A. (1996). The upper Paleocene–lower Eocene stratigraphic record and the Paleocene–Eocene boundary carbon isotope excursion: implications for geochronology. In *Correlation of the Early Paleogene in Northwest Europe*, Geological Society Special Publication No. 101, eds. R. W. O'B. Knox, R. M. Corfield and R. E. Dunay, pp. 353–80. London: Geological Society.

Badgley, C. (1990). A statistical assessment of last appearances in the Eocene record of mammals. In *Dawn of the Age of Mammals in the Northern Part of the Rocky Mountain Interior, North America*, Geological Society of America Special Paper No. 243, eds. T. M. Bown and K. D. Rose, pp. 153–68. Boulder, CO: Geological Society of America.

Badgley, C. and Gingerich, P. D. (1988). Sampling and faunal turnover in early Eocene mammals. *Palaeogeography, Paleoclimatology, Palaeoecology*, **63**, 141–57.

Baker-Brosh, K. F. and Peet, R. K. (1997). The ecological significance of lobed and toothed leaves in temperate forest trees. *Ecology*, **78**, 1250–5.

Bao, H., Koch, P. L. and Hepple, R. P. (1998). Hematite and calcite coatings on fossil vertebrates. *Journal of Sedimentary Research*, A, **68**, 727–38.

Beard, K. C. (1997). East of Eden: Asia as an important center of taxonomic origination in mammalian evolution. In Dawn of the Age of Mammals in Asia, *Bulletin of the Carnegie Museum of Natural History*, eds. K. C. Beard and M. R. Dawson, **34**, 5–39.

Beard, K. C. and Tabrum, A. R. (1991). The first early Eocene mammal from eastern North America: an omomyid primate from the Bashi Formation, Lauderdale County, Mississippi. *Mississippi Geology*, **11**, 1–6.

Becker, R. H. and Clayton. R. N. (1976). Oxygen isotope study of a Precambrian banded iron-formation, Hamersley Range, Western Australia. *Geochimica et Cosmochimica Acta*, **40**, 1153–65.

Berggren, W. A. and Aubry, M. P. (1996). A late Paleocene–early Eocene NW European and North Sea magnetobiochronological correlation network. In *Correlation of the Early Paleogene in Northwest Europe*, Geological Society Special Publication No. 101, eds. R. W. O'B Knox, R. M. Corfield and R. E. Dunay, pp. 309–52. London: Geological Society.

Berggren, W. A., Kent, D. V., Swisher, C. C., III and Aubry, M. P. (1995). A revised Cenozoic geochronology and chronostratigraphy. In *Geochronology, Time Scales and Global Stratigraphic Correlation*, SEPM Special Publication No. 54, eds. W. A. Berggren, D. V. Kent, M. P. Aubry and J. Hardenbol, pp. 129–212. Tulsa, OK: SEPM.

Bird, M. I., Longstaffe, F. J., Fyfe, W. S. and Bildgen, P. (1992). Oxygen-isotope systematics in a multiphase weathering system in Haiti. *Geochimica et Cosmochimica Acta*, **56**, 2831–8.

Bird, M. I., Longstaffe, F. J., Fyfe, W. S., Kronberg, B. I. and Kishida, A. (1993). An oxygen-isotope study of weathering in the eastern Amazon Basin, Brazil. In Climate Change in Continental Isotopic Records, eds. P. K. Swart *et al.*, *Geophysical Monograph*, **78**, 295–307.

Bown, T. M. (1979). Geology and mammalian paleontology of the Sand Creek facies, lower Willwood Formation (lower Eocene), Washakie County, Wyoming. *Geological Survey of Wyoming Memoir*, **2**, 1–151.

Bown, T. M. (1980). Summary of latest Cretaceous and Cenozoic sedimentary, tectonic, and erosional events, Bighorn Basin, Wyoming. In *Early Cenozoic Paleontology and Stratigraphy of the Bighorn Basin, Wyoming*, University of Michigan Papers on Paleontology No. 24, ed. P. D. Gingerich, pp. 25–32. Ann Arbor, MI: University of Michigan.

Bown, T. M., Holroyd, P. A. and Rose, K. D. (1994*b*). Mammal extinctions, body size, and paleotemperature. *Proceedings of the National Academy of Science*, **91**, 10403–6.

Bown, T. M. and Kraus, M. J. (1981*a*). Lower Eocene alluvial paleosols (Willwood Formation, northwestern Wyoming, USA) and their significance for paleoecology, paleoclimatology, and basin analysis. *Palaeogeography, Palaeoclimatology, Palaeoecology*, **34**, 1–30.

Bown, T. M. and Kraus, M. J. (1981*b*). Vertebrate fossil-bearing paleosol units (Willwood Formation, lower Eocene, northwest Wyoming USA: implications for taphonomy, biostratigraphy, and assemblage analysis. *Palaeogeography, Palaeoclimatology, Palaeoecology*, **34**, 31–56.

Bown, T. M. and Kraus, M. J. (1993). Time-stratigraphic reconstruction and integration of paleopedologic, sedimentologic, and biotic events (Willwood Formation, lower Eocene, northwest Wyoming, USA). *Palaios*, **8**, 68–80.

Bown, T. M., Rose, K. D., Simons, E. L. and Wing, S. L. (1994*a*). Distribution and stratigraphic correlation of upper Paleocene and lower Eocene fossil mammal and plant localities of the Fort Union, Willwood, and Tatman Formations, southern Bighorn Basin, Wyoming. *US Geological Survey Professional Paper*, **1540**, 1–269.

Boyle, E. A. (1997). Cool tropical temperatures shift the global $\delta^{18}O$–T relationship: an explanation for the ice core $\delta^{18}O$-borehole thermometry conflict? *Geophysical Research Letters*, **24**, 273–6.

Brown, R. W. (1962). Paleocene flora of the Rocky Mountains and Great Plains. *US Geological Survey Professional Paper*, **375**, 1–119.

Budantsev, L. Y. (1992). Early stages of formation and dispersal of the temperate flora in the Boral reigon. *The Botanical Review*, **58**, 1–48.

Butler, R. F., Gingerich, P. D. and Lindsay, E. H. (1981). Magnetic polarity stratigraphy and biostratigraphy of Paleocene and lower Eocene continental deposits, Clark's Fork Basin, Wyoming: *Journal of Geology*, **89**, 299–316.

Cande, S. C. and Kent, D. V. (1992). A new geomagnetic polarity time scale for the Late Cretaceous and Cenozoic. *Journal of Geophysical Research*, **97**, 13917–51.

Cande, S. C. and Kent, D. V. (1995). Revised calibration of the geomagnetic polarity time scale for the Late Cretaceous and Cenozoic. *Journal of Geophysical Research*, **100**, 6093–5.

Canny, M. J. (1990). What becomes of the transpiration stream? *New Phytologist*, **114**, 341–68.

Cerling, T. E. and Quade, J. (1993). Stable carbon and oxygen isotope in soil carbonates. In Climate Change in Continental Isotopic Records, eds. P. K. Swart *et al.*, *Geophysical Monograph*, **78**, 217–31.

Christophel, D. C. and Greenwood, D. R. (1989). Changes in climate and vegetation in Australia during the Tertiary. *Review of Palaeobotany and Palynology*, **58**, 95–109.

Clayton, R. N. and Epstein, S. (1961). The use of oxygen isotopes in high temperature geological thermometry. *Journal of Geology*, **69**, 447–52.

Clyde, W. C. (1997). *Stratigraphy and Mammalian Paleontology of the McCullough Peaks, Northern Bighorn Basin, Wyoming: Implications for Biochronology, Basin*

Development, and Community Reorganization across the Paleocene–Eocene Boundary. PhD thesis, Department of Geology, University of Michigan, Ann Arbor, MI.

Clyde, W. C., Stamatakos, J. and Gingerich, P. D. (1994). Chronology of the Wasatchian land-mammal age (early Eocene): magnetostratigraphic results from the McCullough Peaks section, northern Bighorn Basin, Wyoming. *Journal of Geology*, **102**, 367–77.

Collinson, M. E. and Hooker, J. J. (1987). Vegetational and mammalina faunal changes in the Early Tertiary of southern England. In *The Origins of Angiosperms and their Biological Consequences*, eds. E. M. Friis, W. G. Chaloner and P. R. Crane, pp. 259–304. New York: Cambridge University Press.

Corfield, R. M. and Norris, R. D. (1996). Deep water circulation in the Paleocene Ocean. In *Correlation of the Early Paleogene in Northwest Europe*, Geological Society Special Publication No. 101, eds. R. W. O'B. Knox, R. M. Corfield and R. E. Dunay, pp. 443–56. London: Geological Society.

Dansgaard, W. (1964). Stable isotopes in precipitation. *Tellus*, **16**, 436–68.

Davies-Vollum, S. K. and Wing, S. L. (1998). Sedimentological, taphonomic, and climatic aspects of Eocene swamp deposits (Willwood Formation, Bighorn Basin, Wyoming). *Palaios*, **13**, 28–40.

Delcourt, P. A. and Delcourt, H. R. (1987). *Long-Term Forest Dynamics of the Temperate Zone*. New York: Springer-Verlag.

Dickens, G. R., Castillo, M. M. and Walker, J. C. G. (1997). A blast of gas in the latest Paleocene: simulating first-order effects of massive dissociation of oceanic methane hydrate. *Geology*, **25**, p. 259-62.

Frederiksen, N. O. (1979). Paleogene sporomorph biostratigraphy, northeastern Virginia. *Palynology*, **3**, 129–67.

Frederiksen, N. O. (1980). Paleogene sporomorphs from South Carolina and quantitative correlations with the Gulf Coast. *Palynology*, **4**, 125–79.

Frederiksen, N. O. (1994). Paleocene floral diversities and turnover events in eastern North America and their relation to diversity models. *Review of Palaeobotany and Palynology*, **82**, 225–38.

Friedman, I. and O'Neil, J. R. (1977). Compilation of stable isotope fractionation factors of geochemical interest. *US Geological Survey Professional Paper*, **440**, 1–12.

Gibson, T. G., Bybell, L. M. and Owens, J. P. (1993). Latest Paleocene lithologic and biotic events in neritic deposits of southwestern New Jersey. *Paleoceanography*, **8**, 495–514.

Gingerich, P. D., (1983). Paleocene–Eocene faunal zones and a preliminary analysis of Laramide structural deformation in the Clark's Fork Basin, Wyoming. *Wyoming Geological Association Annual Field Conference Guidebook*. **34**, 185–95.

Gingerich, P. D. (1989). New earliest Wasatchian mammalian fauna from the Eocene of northwestern Wyoming: Composition and diversity in a rarely sampled high-flood-plain assemblage. *University of Michigan Papers on Paleontology*, **28**, 1–97.

Gingerich, P. D. (1991). Systematics and evolution of Early Eocene Perissodactyla (Mammalia) in the Clarks Fork Basin, Wyoming. *Contribution from the Museum of Paleontology, University of Michigan*, **28**, 181–213.

Gingerich, P. D., Rose, K. D. and Krause, D. W. (1980). Cenozoic mammalian faunas of the Clark's Fork Basin–Polecat Bench area, northwestern Wyoming. In *Early Cenozoic Paleontology and Stratigraphy of the Bighorn Basin, Wyoming*, University of Michigan Papers on Paleontology No. 24, ed. P. D. Gingerich, pp. 51–64. Ann Arbor, MI: University of Michigan.

Givnish, T. J. (1979). On the adaptive significance of leaf form. In *Topics in Plant Population Biology*, eds. O. T. Solbrig, S. Jain, G. B. Johnson and P. H. Raven, pp. 375–407. New York: Columbia University Press.

Graham, A. (1972). Outline of the origin and historical recognition of floristic affinities between Asia and eastern North America. In *Floristics and Paleofloristics of Asia and Eastern North America*, ed. A. Graham, pp. 1–18. Amsterdam: Elsevier.

Graham, A. (1992). Neotropical Paleogene coastal floras $^{18}O/^{16}O$-estimated warmer vs. cooler equatorial waters. *American Journal of Botany*, **79** (supplement), 102.

Greenwood, D. R. (1992). Taphonomic constraints on foliar physiognomic interpretations of Late Cretaceous and Tertiary palaeoclimates. *Review of Palaeobotany and Palynology*, **71**, 149–90.

Guo, S-X. (1985). Preliminary interpretation of Tertiary climate by using megafossil floras in China. *Palaeontologia Cathayana*, **2**, 169–75.

Guo, S-X., Sun, Z-H., Li, H-M. and Dou, Y-W. (1984). Paleocene megafossil flora from Altai of Xinjiang. *Bulletin of Nanjing Institute of Geology and Palaeontology, Academia Sinica*, **8**, 119–46.

Haq, B. U., Hardenbol, J. and Vail, P. R. (1988). Mesozoic and Cenozoic chronostratigraphy and cycles of sea-level change. In *Sea-Level Changes: An Integrated Approach*, SEPM Special Publication No. 42, eds. C. K. Wilgus, B. S. Hastings, C. A. Ross, H. Posamentier, J. Van Wagoner and C. G. St. C. Kendall, pp. 71–108. Tulsa, OK: SEPM.

Hickey, L. J. (1977). Stratigraphy and paleobotany of the Golden Valley Formation (early Tertiary) of western North Dakota. *Geological Society of America Memoir*, **150**, 1–181.

Hickey, L. J. (1980). Paleocene stratigraphy and flora of the Clark's Fork Basin. In *Early Cenozoic Paleontology and Stratigraphy of the Bighorn Basin, Wyoming*, University of Michigan Papers on Paleontology No. 24, ed. P. D. Gingerich, pp. 33–49. Ann Arbor, MI: University of Michigan.

Hickey, L.J. and Wolfe, J.A. (1975). The bases of angiosperm phylogeny: vegetative morphology. *Annals of the Missouri Botanical Garden*, **62**, 538–89.

Hill, M. O. (1979). *DECORANA, a FORTRAN program for Detrended Correspondence Analysis and Reciprocal Averaging*. Ithaca, New York: Microcomputer Power.

Hooker, J. J. (1996). Mammalian biostratigraphy across the Paleocene–Eocene boundary in the Paris, London and Belgian basins. In *Correlation of the Early Paleogene in Northwest Europe*, Geological Society Special Publication No. 101, eds. R. W. O'B. Knox, R. M. Corfield and R. E. Dunay, pp. 205–18. London: Geological Society.

Hutchison, J. H. (1982). Turtle, crocodilian, and champsosaur diversity changes in the Cenozoic of the north-central region of western United States. *Palaeogeography, Palaeoclimatology, Palaeoecology*, **37**, 149–64.

Johnston, J. H. and Lewis, D. G. (1983). A detailed study of the transformation of ferrihydrite to hematite in an aqueous medium at 92 °C. *Geochimica et Cosmochimica Acta*, **47**, 1823–31.

Jolley, D. W. (1996). The earliest Eocene sediments of eastern England: and ultra-high resolution palynological correlation. In *Correlation of the Early Paleogene in Northwest Europe*, Geological Society Special Publication No. 101, eds. R. W. O'B. Knox, R. M. Corfield and R. E. Dunay, pp. 219–54. London: Geological Society.

Kennett, J. P. and Stott, L. D. (1991). Abrupt deep-sea warming, palaeoceanographic changes and benthic extinctions at the end of the Palaeocene. *Nature*, **353**, 225–9.

Koch, P. L. (1986). Clinal geographic variation in mammals: implications for the study of chronoclines. *Paleobiology*, **12**, 269–81.

Koch, P. L., Zachos, J. C. and Dettmann, D. L. (1995). Stable isotope stratigraphy and paleoclimatology of the Paleogene Bighorn Basin (Wyoming, U.S.A.). *Palaeogeography, Palaeoclimatology, Palaeoecology*, **115**, 61–89.

Koch, P. L., Zachos, J. C. and Gingerich, P. D. (1992). Correlation between isotope records in marine and continental carbon reservoirs near the Palaeocene/Eocene boundary. *Nature*, **358**, 319–22.

Kraus, M. J. and Aslan, A. (1993). Eocene hydromorphic paleosols: significance for interpreting ancient floodplain processes. *Journal of Sedimentary Petrology*, **63**, 453–63.

Kraus, M. J. and Bown, T. M. (1993). Short-term sediment accumulation rates determined from Eocene alluvial paleosols. *Geology*, **21**, 743–6.

Krishtalka, L., West, R. M., Black, C. C. *et al.* (1987). Eocene (Wasatchian through Duchesnean) biochronology of North America. In *Cenozoic Mammals of North America Geochronology and Biostratigraphy*, ed. M. O. Woodburne, pp. 77–117. Berkeley, CA: University of California Press.

Lawrence, J. R. and Taylor, H. P. Jr. (1972). Hydrogen and oxygen isotope systematics in weathering profiles. *Geochimica et Cosmochimica Acta*, **36**, 1377–93.

Leopold, E. B. and MacGinitie, H. D. (1972). Development and affinities of Tertiary floras in the Rocky Mountains. In *Floristics and Paleofloristics of Asia and Eastern North America*, ed. A. Graham, pp. 147–200. Amsterdam: Elsevier.

MacGinitie, H. D. (1969). The Eocene Green River flora of northwestern Colorado and northeastern Utah. *University of California Publications in Geological Sciences*, **83**, 1–202.

Markwick, P. J. (1994). 'Equability,' continentality and Tertiary 'climate': the crocodilian perspective. *Geology*, **22**, 613–6.

McKeague, J. A., Cheshire, M. V., Andreux, F. and Berthelin, J. (1986). Organo-mineral complexes in relation to pedogenesis. In *Interactions of Soil Minerals with Natural Organics and Microbes*, Soil Science Society of America Special Publication No. 17, eds. P. M. Huang and M. Schnitzer, pp. 549–92. Madison, WI: Soil Science Society of America.

Miller, K. G., Fairbanks, R. G. and Mountain, G. S. (1987). Tertiary oxygen isotope synthesis, sea level history, and continental margin erosion. *Paleoceanography*, **2**, 1–19.

Quade, J., Cerling, T. E. and Bowman, J. R. (1989). Systematic variation in the carbon and oxygen isotopic composition of pedogenic carbonate along elevation transects in the southern Great Basin, United States. *Geological Society of America Bulletin*, **101**, 464–75.

Robert, C. and Kennett, J. P. (1994). Antarctic subtropical humid episode at the Paleocene–Eocene boundary: clay-mineral evidence. *Geology*, **22**, 211–4.

Romero, E. J. (1986). Paleogene phytogeography and paleoclimatology of South America. *Annals of the Missouri Botanical Garden*, **73**, 449–61.

Rose, K. D. (1981). The Clarkforkian land-mammal age and mammalian faunal composition across the Paleocene–Eocene boundary. *University of Michigan Papers on Paleontology*, **26**, 1–197.

Rozanski, K., Araguas-Araguas, L. and Gonfiantini, R. (1993). Isotopic patterns in modern global precipitation. In Climate Change in Continental Isotopic Records, eds. P. K. Swart *et al.*, *Geophysical Monograph*, **78**, 1–36.

Savin, S. M. and Epstein, S. (1970). The oxygen and hydrogen isotope geochemistry of clay minerals. *Geochimica et Cosmochimica Acta*, **34**, 25–42.

Schankler, D. M. (1980). Faunal zonation of the Willwood Formation in the central Bighorn Basin, Wyoming. In *Early Cenozoic Paleontology and Stratigraphy of the Bighorn Basin, Wyoming*, University of Michigan Papers on Paleontology No. 24, ed. P. D. Gingerich, pp. 99–114. Ann Arbor, MI: University of Michigan.

Schankler, D. M. (1981). Local extinction and ecological re-entry of early Eocene mammals. *Nature*, **293**, 135–8.

Schrag, D., DePaolo, D. J. and Richter, F. M. (1992). Oxygen isotope exchange in a two-layer model of oceanic crust. *Earth and Planetary Science Letters*, **111**, 305–17.

Schwertmann, U. and Murad, E. (1983). Effect of pH on the formation of goethite and hematite from ferrihydrite. *Clays and Clay Minerals*, **31**, 277–84.

Schwertmann, U. and Taylor, R. M. (1989). Iron oxides. In *Minerals in Soil Environments*, SSSA Book Series 1, eds. J. B. Dixon and S. B. Weed, pp. 379–438. Madison, WI: Soil Science Society of America.

Searcy, W. A. (1980). Optimum body sizes at different temperatures: an energetics explanation of Bergmann's rule. *Journal of Theoretical Biology*, **83**, 579–93.

Swisher, C. C., III and Knox, R. W. O'B. (1991). The age of the Paleocene/Eocene boundary: 40Ar/39Ar dating of the lower part of NP10, North Sea Basin and Denmark. In *IGCP Project 308 (Paleocene/Eocene boundary events)*, International Annual Meeting and Field Conference, Brussels, 2–6 December 1991, Abstracts with Program, p. 16.

Tauxe, L., Gee, J., Gallet, Y., Pick, T. and Bown, T. M. (1994). Magnetostratigraphy of the Willwood Formation, Bighorn Basin, Wyoming: new constraints on the location of the Paleocene/Eocene boundary. *Earth and Planetary Science Letters*, **125**, 159–72.

Tiffney, B. H. (1985a). Perspectives on the origin of the floristic similarity between eastern Asia and eastern North America. *Journal of the Arnold Arboretum*, **66**, 73–94.

Tiffney, B. H. (1985b). The Eocene North Atlantic land bridge: its importance in Tertiary and modern phytogeography of the northern hemisphere. *Journal of the Arnold Arboretum*, **66**, 243–73.

Valley, J. W., Kitchen, N., Kohn, M. J., Niendorf, C. R. and Spicuzza, M. J. (1995). UWG-2, a garnet standard for oxygen isotope ratios: strategies for high precision and accuracy with laser heating. *Geochimica et Cosmochimica Acta*, **59**, 5223–31.

Wilf, P. (1997). When are leaves good thermometers? A new case for Leaf Margin Analysis. *Paleobiology*, **23**, 373–90.

Wilf, P. D. (1999). Paleobotanical analysis of late Paleocene–early Eocene climate changes in the greater Green River Basin of southwestern Wyoming. *GSA Bulletin* (in press).

Wing, S. L. (1984). A new basis for recognizing the Paleocene/Eocene boundary in western interior North America. *Science*, **226**, 439–41.

Wing, S. L. (1987). Eocene and Oligocene floras and vegetation of the northern Rocky Mountains. *Annals of the Missouri Botanical Garden*, **74**, 748–84.

Wing, S. L. (1998). Late Paleocene–early Eocene floral and climatic change in the Bighorn Basin, Wyoming. In *Late Paleocene–Early Eocene Biotic and Climatic Events*, eds. W. Berggren, M. P. Aubry and S. Lucas. New York: Columbia University Press.

Wing, S. L., Alroy, J. and Hickey, L. J. (1995). Plant and mammal diversity in the Paleocene to early Eocene of the Bighorn Basin. *Palaeogeography, Palaeoclimatology, Palaeoecology*, **115**, 117–56.

Wing, S. L. and Bown, T. M. (1985). Fine scale reconstruction of late Paleocene–early Eocene paleogeography in the Bighorn Basin of northern Wyoming. In *Cenozoic Paleogeography of West-Central United States*, eds. R. M. Flores and S. S. Kaplan, pp. 93–105. Denver, CO: SEPM, Rocky Mountain Section.

Wing, S. L., Bown, T. M. and Obradovich, J. D. (1991). Early Eocene biotic and climatic change in interior western North America. *Geology*, **19**, 1189–92.

Wing, S. L. and Greenwood, D. R. (1993). Fossils and fossil climate: the case for equable continental interiors in the Eocene. In *Palaeoclimates and their Modeling with Special Reference to the Mesozoic Era*, eds. J. R. L. Allen, B. J. Hoskins, B. W. Sellwood and R. A. Spicer. *Philosophical Transactions of the Royal Society, London B, Biological Sciences*, **341**, 243–52.

Winograd, I. J., Coplen, T. B, Landwehr, J. M. *et al.* (1992). Continuous 500,000-year climate record from vein calcite in Devils Hole, Nevada. *Science*, **258**, 255–60.

Wolfe, J. A. (1966). Tertiary plants from the Cook Inlet Region, Alaska. *US Geological Survey Professional Paper*, **398-B**, 1–32.

Wolfe, J. A. (1972). An interpretation of Alaskan Tertiary floras. In *Floristics and Paleofloristics of Asian and Eastern North America*, ed. A. Graham, pp. 201–33. Amsterdam: Elsevier.

Wolfe, J. A. (1977). Paleogene floras from the Gulf of Alaska region. *US Geological Survey Professional Paper*, **997**, 1–108.

Wolfe, J. A. (1979). Temperature parameters of humid to mesic forests of eastern Asia and relation to forests of other regions of the northern hemisphere and Australasia. *US Geological Survey Professional Paper*, **1106**, 1–37.

Wolfe, J. A. (1993). A method of obtaining climatic parameters from leaf assemblages. *US Geological Survey Bulletin*, **2040**, 1–71.

Woodburne, M. O. and Swisher, C. C. (1995). Land mammal high-resolution geochronology, intercontinental overland dispersals, sea level, climate, and vicariance. In *Geochronology, Time Scales and Global Stratigraphic Correlation*, SEPM Special Publication No. 54, eds. W. A. Berggren, D. V. Kent, M. P. Aubry and J. Hardenbol, pp. 335–64. Tulsa, OK: SEPM.

Yapp, C. J. (1987). Oxygen and hydrogen isotope variations among goethites (α-FeOOH) and the determination of paleotemperatures. *Geochimica et Cosmochimica Acta*, **51**, 355–64.

Yapp, C. J. (1990a). Oxygen isotopes in iron (III) oxides: 1, mineral-water fractionation factors. *Chemical Geology*, **85**, 329–35.

Yapp, C. J. (1990b). Oxygen isotope effects associated with the solid-state α-FeOOH to α-Fe$_2$O$_3$ phase transformation. *Geochimica et Cosmochimica Acta*, **54**, 229–36.

Yapp, C. J. (1991). Oxygen isotopes in an oolitic ironstone and the determination of goethite δ^{18}O values by selective dissolution of impurities: The 5 M NaOH method. *Geochimica et Cosmochimica Acta*, **55**, 2627–34.

Yapp, C. J. (1993). The stable isotope geochemistry of low temperature Fe(III) and Al "oxides" with implications for continental paleoclimates. In Climate Change in Continental Isotopic Records, eds. P. K. Swart *et al.*, *Geophysical Monograph*, **78**, 285–94.

Yapp, C. J. (1997). An assessment of isotopic equilibrium in goethites from a bog iron and a lateritic regolith. *Chemical Geology*, **135**, 159–71.

Zachos, J. C., Stott, L. D. and Lohmann, K. C. (1994). Evolution of early Cenozoic marine temperatures. *Paleoceanography*, **9**, 353–87.

Zheng, Y.-F. (1991). Calculation of oxygen isotope fractionation in metal oxides. *Geochimica et Cosmochimica Acta*, **55**, 2299–307.

III

Case studies: Mesozoic

8

Paleontological and geochemical constraints on the deep ocean during the Cretaceous greenhouse interval

KENNETH G. MACLEOD, BRIAN T. HUBER, AND MY LE DUCHARME

ABSTRACT

For intervals of global warmth, poleward heat transport by warm, saline water masses may be a critical variable in the global climate equation and is widely discussed as a potential mechanism to reconcile differences between computer models and empirical data. However, ocean structure during greenhouse times is poorly constrained, especially on a global scale. The distribution of fossil organisms provides one way to map deep ocean conditions. Inoceramid bivalves reached the acme of their 200 million year range during the Late Cretaceous and then virtually disappeared ~67 million years ago in an event associated with the deterioration of the Cretaceous greenhouse climate (e.g., MacLeod *et al.*, 1996). In 1448 samples (73 sites) representing bathyal paleodepths across the last ~45 million years of the Cretaceous, inoceramid abundance is remarkably constant in time (extinction interval excluded) but not in space. In general, inoceramids were common to abundant throughout the Atlantic and Indian oceans but relatively rare in the Pacific. Geochemical studies of two Indian Ocean sites revealed changes in the $\delta^{18}O$ values of benthic and deep dwelling planktic foraminifers suggesting that bottom waters became cooler and less saline at the time of the inoceramid extinction (MacLeod and Huber, 1996*a*); new data from a peri-Tethyan site in the western North Atlantic suggest that surface waters became warmer and/or less saline at approximately the same time. The correlation between these changes and the end of the Cretaceous greenhouse climate supports the proposition that warm, saline water masses have a role in creating and/or maintaining greenhouse climate conditions. Further, the paleobiogeography of inoceramids may indicate that intermediate and deep waters sourced in low latitudes were more prevalent in the Cretaceous Atlantic and Indian oceans than in the Pacific.

Results from climate simulations coupled with ocean circulation models include implicit predictions regarding regional and bathymetric differences in water mass properties (e.g., Bice *et al.* Chapter 4, this volume; DeConto *et al.*, Chapter 9, this volume). Geologic data concerning conditions in the deep ocean (e.g., the distribution of Cretaceous inoceramids) provide important tests of these predictions, help infer paleoceanographic processes, and increase the likelihood that improved agreement between model results and geologic data reflects improved understanding of the climate system.

INTRODUCTION

Conditions in the deep oceans during intervals of global warmth may have been fundamentally different from those at present. The modern deep ocean is dominated by cold, polar water masses. The only significant warm, saline intermediate or deep water mass is sourced in the Mediterranean, sinks to a depth of ~ 1200 m in the North Atlantic, and represents a small fraction of the world's oceans. During greenhouse times, though, the majority of downwelling may have occurred in mid- to low latitudes rather than high latitudes, and warm, saline waters may have partially to largely filled the deep ocean (e.g., Chamberlin, 1906; Brass *et al.*, 1982; Hay, 1988; Kennett and Stott, 1991). Such changes in deep ocean circulation patterns should have affected (paleo)ecological conditions at the seafloor and could have affected latitudinal gradients in surface temperature (e.g., Barron *et al.*, 1995; Sloan *et al.*, 1995; Schmidt and Mysak, 1996; Lyle, 1997).

Latitudinal temperature gradients are the most problematic aspect of greenhouse intervals. Global warmth is well documented for intervals of the geologic past and seems to correspond with elevated atmospheric concentrations of CO_2 and other greenhouse gases (e.g., Crowley, 1993; Compton and Mallinson, 1996). In addition, fairly good agreement exists between average temperatures estimated from geologic data and simple energy balance models, and these energy balance models specify pCO_2 values that are consistent with those independently estimated from geochemical models for the same time intervals (Crowley and North, 1991; Berner, 1994). Agreement breaks down, though, when the global distribution of temperatures (rather than the global average) is considered. Geologic data for greenhouse climates indicate warmer polar regions and cooler tropics, i.e., shallower latitudinal temperature gradients, than general circulation models (GCMs) reproduce (e.g., Barron *et al.*, 1995; Greenwood and Wing, 1995; Huber *et al.*, 1995; Sloan *et al.*, 1995; Crowley and Zachos, Chapter 3, this volume). This disagreement demonstrates shortcomings in our understanding of greenhouse climates and undermines confidence in predictions regarding present warming trends.

Variation in ocean heat transport is a leading candidate to reconcile differences between empirical and modeling results (other mechanisms that could warm the poles are discussed in Sloan *et al.* (1995), Schmidt and Mysak (1996), and DeConto *et al.* (Chapter 9, this volume)). As intuitively expected, forcing model oceans to transport more heat from the equator to the poles decreases simulated latitudinal temperature gradients (e.g., Barron *et al.*, 1993, 1995), but it is difficult to evaluate the importance of these simulated changes because of limitations in the way the ocean has been treated in the models. While warm, saline water masses are commonly proposed as a means of effecting increased oceanic heat transport, most simulations have not attempted to resolve oceanic processes at this scale. Rather, they have parameterized oceanic influences on climate using control values that best reproduce modern conditions. For example, in the GENESIS GCM with a mixed layer ocean, the control value (Q) for oceanic heat transport is only 15–30%

of measured values (Bryden and Hall, 1980; Roemmich, 1980; Hastenrath, 1982; Carissimo et al., 1985; Barron et al., 1993), allowing several divergent interpretations of similar experiments. Barron et al. (1995) considered the heat flux necessary to double (2Q) or quadruple (4Q) oceanic heat transport in the model modest relative to measured values and argued, because simulated latitudinal temperature gradients with a 4Q mid-Cretaceous ocean are as shallow as the steepest gradients allowed by geologic data, that deep ocean circulation is a potentially important climatic variable. Alternatively, Sloan et al. (1995) assumed the control value for oceanic heat transport accurately scales modern oceanic processes and calculated that doubling ocean transport (using a 2Q value improved agreement between an Eocene simulation and paleotemperature estimates) with warm, saline water masses would require a mechanistically prohibitive poleward flow equivalent to the outflow of 80 Mediterraneans. Finally, DeConto et al. (Chapter 9, this volume) also discounted the potential importance of increased oceanic heat transport for mechanistic reasons but specified an oceanic heat flux of 4Q in their favored simulation which they justified as being a value comparable to present measured values. Studies focused specifically on oceanic processes have not clarified the issue (Schmidt and Mysak, 1996; Lyle, 1997). Both experiments found that changes in the rate and pattern of deep ocean circulation could cause greenhouse oceans to transport more heat than the modern oceans but were divided on the effect, prevalence, and relative importance of deep water masses sourced in different regions (i.e., the role of warm, saline water masses).

Like latitudinal temperature gradients, vertical temperature gradients in greenhouse oceans are reduced relative to the present, but warm benthic temperatures could reflect elevated temperatures in source regions at high latitudes as well as downwelling of saline waters at mid- to low latitudes (e.g., Hay, 1988; Bice et al., Chapter 4, this volume). Distinguishing between these two possibilities is difficult because increases in temperature and increases in salinity (the principal variables affecting seawater density and thereby downwelling) have opposing effects on the $\delta^{18}O$ of biogenic calcite (e.g., Crowley and Zachos, Chapter 3, this volume). However, reversed $\delta^{18}O$ depth gradients observed in fossils from the Middle Ordovician Taconic Basin (Railsback et al., 1989) and the Late Cretaceous South Atlantic (Saltzman and Barron, 1982) seem to imply warmer water at depth and, thus, require saline bottom water masses. In addition, biological and geochemical changes in benthic microfossil assemblages have been well documented at shifts to and from greenhouse states (e.g., warming during the late Paleocene–early Eocene (e.g., Kennett and Stott, 1991; Pak and Miller, 1992; Zachos et al., 1993; Thomas and Shackleton, 1996) and cooling during the Maastrichtian (e.g., Barrera, 1994; MacLeod and Huber, 1996a)). The strongest indications that benthic changes occurring at climatic transitions could be related to reorganization of deep circulation are the rapid, within-site increase in $\delta^{18}O$ values of benthic foraminiferal tests at the Late Paleocene Thermal Maximum (e.g., Stott and Kennett, 1990; Thomas and Shackleton, 1996) and a mid-Maastrichtian switch in the rank order of benthic and some planktic foraminiferal $\delta^{18}O$ ratios at a mid- and a high paleolatitude site (MacLeod and Huber, 1996a).

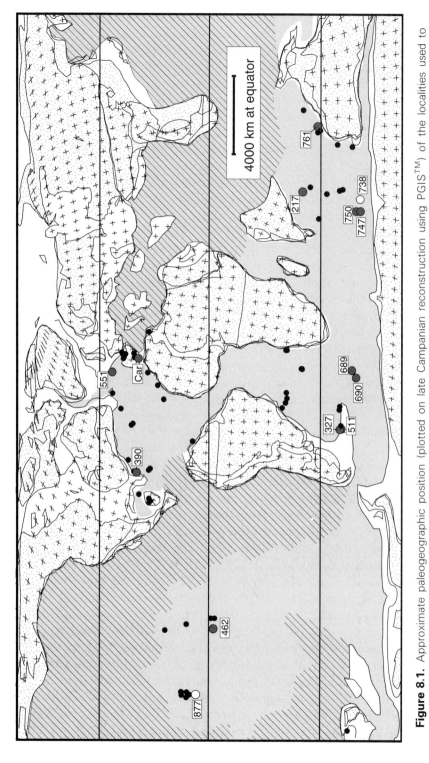

Figure 8.1. Approximate paleogeographic position (plotted on late Campanian reconstruction using PGIS™) of the localities used to constrain the paleobiogeography of inoceramids living at bathyal paleodepths during the Late Cretaceous. Labeled localities are specifically mentioned in the text; filled circles are sites included in the compilation of inoceramid occurrences whereas hollow circles represent sites discussed for which inoceramid data were either completely excluded (738) or not available (877). Diagonal ruling indicates areas of ocean crust subducted since the Cretaceous.

Existing geologic data have limited spatial/temporal coverage, and finding coincidence between changes in the deep ocean and climatic shifts is a long way from demonstrating cause and effect, but, if heat transported by deep waters is an important factor in greenhouse climates, these coincidences are required. This study is aimed at examining these correlations on the spatial and temporal scale of the Cretaceous greenhouse world. To test more directly whether deep ocean structure and circulation had a role in creating and/or maintaining greenhouse conditions, we have examined (1) paleobiogeographic patterns of Late Cretaceous deep sea inoceramid bivalves, (2) δ^{13}C and δ^{18}O analyses of planktic and benthic microfossils, and (3) counts of planktic foraminiferal assemblages through the Maastrichtian at two sites.

MATERIALS AND METHODS

Inoceramid paleobiogeography

The distribution of organisms provides one way to map paleoecological conditions and, for the Cretaceous deep ocean, inoceramid bivalves are an excellent group on which to focus. Inoceramids first appeared in the Permian, reached their acme during the Late Cretaceous (e.g., Dhondt, 1992; Voigt, 1995) when they are common to abundant at bathyal paleodepths (e.g., Saltzman and Barron, 1982; Barron et al., 1984; MacLeod et al., 1996), and, with the exception of one genus, went extinct ~2 million years before the end of the Cretaceous. Inoceramid diversity seems to broadly parallel changes in Cretaceous temperature and pCO$_2$ (e.g., Berner, 1994; Huber et al., 1995) and they suffered a major pulse of extinction during Maastrichtian cooling (e.g., MacLeod and Huber, 1996a; MacLeod et al., 1996) suggesting they may be sensitive indicators of benthic conditions associated with greenhouse climates. Further, inoceramid shells commonly disaggregated into characteristic prismatic shell fragments that are easily recognized in washed residues (e.g., MacLeod and Orr, 1993). Thus changes in inoceramid paleobiogeography can be efficiently constrained on a global scale using Ocean Drilling Program/Deep Sea Drilling Project (ODP/DSDP) samples.

We have generated data on the distribution of inoceramid shell fragments in 1448 Albian–Maastrichtian samples from 73 sites (Fig. 8.1). In rare cases generic assignment of shell fragments seems possible (MacLeod and Orr, 1993; MacLeod et al., 1996), but identification at lower taxonomic levels was not possible for the shell fragments discussed herein. The samples included in our analysis span outer neritic to abyssal paleodepths (e.g., MacLeod et al., 1996) but dominantly represent bathyal environments. New results include presence/absence data from 344 samples (Albian–Maastrichtian) from the DSDP/ODP Micropaleontological Reference Center (MRC) foraminiferal collection at the Smithsonian Institution's Museum of Natural History; relative abundance data for 133 of the samples (Cenomanian–Maastrichtian) from DSDP holes 327A and 511 discussed in Huber et al. (1995), as well as absolute abundance data for 56 samples (Cenomanian–Maastrichtian) from Caravaca, Spain and 30 samples (Campanian–Maastrichtian) from DSDP Hole 390A. For counting methodology see MacLeod et al. (1996). In addition, we compiled presence/absence data for 77 samples (Santonian–Maastrichtian) from

DSDP holes 462 and 462A (Premoli Silva and Sliter, 1981), 31 samples (Maastrichtian) from DSDP Site 217 (Pessagno and Michael, 1974), and 777 samples (Campanian–Maastrichtian) with a global distribution (MacLeod *et al.*, 1996). It should be noted that the latter group of samples was initially used to determine details of the timing and pattern of inoceramid disappearance during the Maastrichtian. Many sections were chosen because of their paleogeographic location without prior knowledge of inoceramid occurrence, but other sections were chosen specifically because previous work suggested they preserved a good record of the extinction event. Sampling density was highest in the mid-Maastrichtian and, in inoceramid-bearing sections, was particularly intense below the extinction horizon (to better estimate the pace of inoceramid decline).

To interpret the occurrence of inoceramid shell fragments, we follow the taphonomic model of MacLeod and Orr (1993) where an inoceramid shell that was not preserved intact is assumed to have become millions of individual foraminiferal-sized prismatic fragments. These shell fragments were then dispersed by burrowing organisms in a fashion that can be described using a diffusion equation (MacLeod, 1994). In this model the abundance of shell fragments is expected to be high near the original position of a shell but to decrease exponentially over tens of centimeters to meters, the length scale of biological mixing (e.g., Berger and Killingley, 1982; Officer and Lynch, 1983; Wheatcroft, 1991); the presence of even a few shell fragments in a sample is indicative of a local (tens of centimeters to meters) occurrence of at least one inoceramid, but it is possible to collect a sample rich in shell fragments in intervals where inoceramids were rare or to collect a sample with low abundance of shell fragments in intervals where inoceramids were fairly common (see fig. 9 in MacLeod and Orr, 1993). Therefore the trend in shell fragment abundance across many samples in a section is more important than the absolute abundance of shell fragments in any one sample.

Physical reworking and non-preservation of the local fauna invalidates interpretations based on the taphonomic model, so we excluded the following: all samples from ODP Hole 738C (MacLeod *et al.*, 1996) because there is evidence for large scale reworking throughout the section (e.g., MacLeod and Huber, 1996*b*); two Maastrichtian samples from DSDP Hole 390A because they contained specimens of the lower upper Campanian planktic foraminifer *Radotruncana calcarata*; and ~30 samples from the MRC collection, Site 511, and Caravaca because of evidence for strong dissolution or lack of identifiable microfossils (i.e., poor preservation). We also excluded Albian and older samples from DSDP Site 511 because the fossil assemblages suggest deposition in relatively shallow water (Fassell and Bralower, 1999), indicating these samples are not representative of the environment of interest. Including any or all of these excluded samples in the analyses does not affect our conclusions.

Stable isotopic data

The use of stable isotopic analyses of calcareous microfossils in paleoceanographic studies is common. $^{13}C/^{12}C$ ratios are typically used to estimate changes in the intensity of productivity or shifts of carbon between organic and inorganic reservoirs. $^{18}O/^{16}O$ ratios are typically used to estimate paleotemperature but also vary with changes in salinity. Diagenetic alteration is a major confounding factor that can be subtle and difficult to rigorously discount (e.g., Schrag et al., 1992, 1995; Huber et al., 1995), and a number of other systematic and random errors inherent in stable isotopic analyses can also introduce artifacts (Crowley and Zachos, Chapter 3, this volume). To maximize the probability that our data were not compromised by these types of error, we have focused on results from well-preserved microfossils; where possible we have examined results from multiple taxa; and we emphasize relative changes within and between sections and are not particularly concerned with absolute comparisons among sites or estimated paleotemperatures/paleosalinities.

We measured $\delta^{13}C$ and $\delta^{18}O$ values of various microfossils from DSDP holes 390A (Blake Nose: 30° 89′ N, 76° 7′ W; water depth 2665 m) and 551 (Goban Spur: 48° 55′ N, 13° 30′ W; water depth 3887 m) to test for changes in water column structure (e.g., MacLeod and Huber 1996a) as well as to augment existing data concerning latitudinal temperature gradients during the Cretaceous (e.g., Huber et al., 1995). Well-preserved, size-specific separates of planktic foraminifers representing different surface habitats as well as benthic foraminifers and inoceramid prisms were picked from washed residues. Separates were heated under vacuum to remove organic contaminants and analyzed on a VG Prism series II mass spectrometer at the University of Florida. Results are reported in standard δ-notation relative to the Pee Dee Belemnite (PDB) standard. Analytical error is $<0.1‰$ (2σ) for both $\delta^{13}C$ and $\delta^{18}O$.

Population dynamics among planktic foraminifers

Changes in planktic foraminiferal assemblages could help guide paleoceanographic interpretations. For samples from DSDP Hole 390A, we divided the $>125\,\mu m$ fraction using a sediment microsplitter until a separate containing approximately 300–600 foraminifers remained. These individuals were identified to the lowest taxonomic level possible, and the abundances of the various taxa were used to estimate species richness, $H(S)$ diversity, and equitability (or evenness) through the section. Species richness (S) is the total number of species observed in any sample without normalizing for sample size. $H(S)$ diversity was calculated using the Information Function (Shannon, 1948)

$$H(S) = \Sigma p_i \ln p_i$$

for $i = 1$ to the number of species observed where p_i is the proportion of the total sample represented by the ith species. This function characterizes a population, taking into account both the number of species and their relative abundances, while making no assumptions about an underlying distribution (Hayek and Buzas, 1997). Equitability was calculated as

$$E = e^{H(S)}/S$$

and measures how evenly the individuals are distributed among the species present in any given sample. These three metrics are interrelated but are not redundant. Species

richness could reflect the diversity of available habitats (e.g., water column stratification, seasonality) as well as evolution (including extinction) or migration. $H(S)$
diversity varies with species richness, but, for any given number of species (S), $H(S)$
has a maximum value when all taxa are equally abundant and a minimum value when
most of the individuals in a sample represent a single taxon. High $H(S)$ values for
planktic foraminiferal assemblages are expected for samples from stable or oligotrophic environments, whereas low values are expected for assemblages from extreme,
stressed, or eutrophic environments. Equitability varies from 0.0 to 1.0 as the individuals present become more evenly distributed among the species present, and is
independent of the number of species in the sample. This measure is difficult to
interpret uniquely, but high E values are found in samples representing oligotrophic
conditions (Nederbragt, 1991).

To see if patterns observed at relatively low paleolatitudes (Hole 390A) are
similar to those for high latitudes we calculated the same population metrics for
ODP Hole 690C using the counts reported in Huber (1990). The 690C counts were
based on the $>150\,\mu m$ size fraction, but this difference should not invalidate a
comparison of trends between the two sites. We have placed the Campanian/
Maastrichtian boundary in Hole 690C at the top of Chron C32N (after Gradstein
et al., 1994) rather than upper C33N as reported by Huber (1990).

Age assignments

Comparison of trends among sites and regions depends on accurate temporal correlation. We used the time scale of Gradstein *et al.* (1994) and have generally followed age assignments reported in the referenced publications or relevant
DSDP/ODP volumes. In instances where age boundaries were poorly constrained,
we interpolated between well-dated samples. Because the placement of the
Campanian/Maastrichtian boundary has recently been modified, we have updated
Campanian and Maastrichtian age determinations so that all results are consistent
with the zonal scheme of Premoli Silva and Sliter (1994). From youngest to oldest
the relevant biostratigraphic zones are: (1) *Abathomphalus mayaroensis* Zone (late
Maastrichtian), (2) *Racemiguembelina fructicosa* Zone (early Maastrichtian), (3)
Gansserina gansseri Zone (early Maastrichtian to late Campanian), (4)
Globotruncana falsostuarti Zone (*sensu* Salaj and Samuel (1966); late Campanian),
and (5) *Radotruncana calcarata* Zone (late Campanian). Although the base of the *A.
mayaroensis* Zone has been demonstrated to be diachronous with the nominate
taxon first appearing in upper C31R in austral latitudes and lower C31N in
Tethyan sections (Huber, 1992; Huber and Watkins, 1992), it has been consistently
recognized on a global scale and we use it to approximate the early/late
Maastrichtian boundary.

RESULTS

Inoceramid biostratigraphy and paleobiogeography

Inoceramids were a common member of the bathyal community for most of
the Late Cretaceous and then disappeared from the deep sea during the late

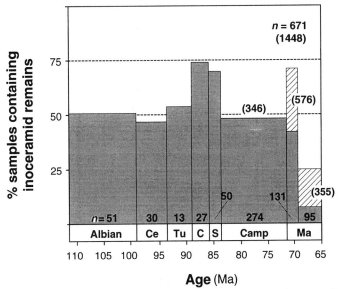

Figure 8.2. Proportion of samples containing inoceramid remains as a function of time (Table 8.1). Inoceramids are common from the Albian through the early Maastrichtian and then decline in the late Maastrichtian. Late Maastrichtian samples containing inoceramids are concentrated near the early/late Maastrichtian boundary consistent with a mid-Maastrichtian inoceramid extinction event (e.g., MacLeod *et al.*, 1996). Horizontal axis is scaled to the time scale of Gradstein *et al.* (1994). The number of samples (*n*) is listed for each subdivision. For the Campanian and Maastrichtian, percentages shown by the diagonal ruling are calculated with the data of MacLeod *et al.* (1996) included (*n* shown in parentheses), whereas the shaded bars exclude those data. See text for further discussion.

Maastrichtian. Inoceramid shell fragments occur in nearly 50% to almost 75% of Albian–lower Maastrichtian samples (Fig. 8.2). Inoceramid remains occur in only ~25% of upper Maastrichtian samples and the inoceramid-bearing samples are concentrated near the lower/upper Maastrichtian boundary. In addition to these temporal trends, inoceramid distributions suggest geographic differences among Cretaceous oceans (Fig. 8.3). Inoceramids occur most frequently in samples from the Atlantic and Indian oceans (~60%) and least frequently in samples from the Pacific (~15%). When the temporal data are divided by ocean basin (Table 8.1) the same patterns seem to hold; there is considerable scatter in the results which is not surprising as some times and regions either have no data or are represented by few samples from a single locality. Finally, in the spatial and temporal plots (Figs. 8.2 and 8.3), exclusion of the data from MacLeod *et al.* (1996) shifts the proportion of samples containing inoceramids in a non-random fashion, but both the mid-Maastrichtian event and the general scarcity of inoceramid remains in Pacific samples are still apparent.

Samples from the Caravaca section (southern Spain), DSDP Hole 390A (western Atlantic), and DSDP holes 327A and 511 (South Atlantic) provide a greater

Table 8.1. Compilation of the occurrence of inoceramid remains in 1448 Late Cretaceous samples. Under each category n is the number of samples examined, p is the number of samples which contained inoceramid remains, and % is the percent of samples containing inoceramid remains. Time is resolved to the age level with the exception of the Maastrichtian which is subdivided into two parts. Age determinations for the Caravaca section and Hole 390A are based on our observations of the foraminiferal assemblage. All other age determinations are based on published reports (typically the relevant ODP/DSDP volume); see text for details. The values in italics exclude the data of MacLeod et al. (1996); see text for discussion. Bold is for emphasis. Spaces indicate no data

	Atlantic Ocean			Indian Ocean			Pacific Ocean			Southern Ocean			Total		
	n	p	%	n	p	%	n	p	%	n	p	%	n	p	%
Late Maastrichtian	185	53	**29**	94	32	**34**	13	1	**8**	63	2	**3**	355	88	**25**
	49	*5*	*10*	*29*	*1*	*3*	*4*	*0*	*0*	*13*	*0*	*0*	*95*	*7*	*7*
Early Maastrichtian	376	320	**85**	70	54	**77**	72	6	**8**	58	29	**50**	576	409	**71**
	61	*40*	*66*	*22*	*8*	*36*	*33*	*5*	*15*	*15*	*6*	*40*	*131*	*55*	*42*
Campanian	183	118	**64**	27	25	**93**	79	9	**11**	57	13	**23**	346	165	**48**
	170	*105*	*62*	*14*	*12*	*86*				*11*	*6*	*55*	*274*	*132*	*48*
Santonian	39	28	**72**	3	3	**100**	7	3	**43**	1	1	**100**	50	35	**70**
Coniacian	20	17	**85**	5	3	**60**	2	0	**0**				27	20	**74**
Turonian	8	5	**63**	2	1	**50**	3	1	**33**				13	7	**54**
Cenomanian	27	12	**44**	2	2	**100**	1	0	**0**				30	14	**47**
Albian	40	22	**55**	10	4	**40**	1	0	**0**				51	26	**51**
Total	878	575	**65**	213	124	**58**	178	20	**11**	179	45	**25**	1448	764	**53**
	414	*234*	*57*	*87*	*34*	*39*	*51*	*9*	*18*	*40*	*13*	*33*	*671*	*296*	*44*

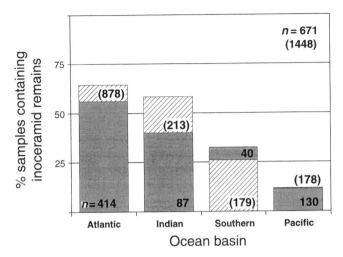

Figure 8.3. Proportion of samples containing inoceramid remains in different ocean basins (Table 8.1). Shell fragments are most common in samples from the Atlantic and Indian oceans whereas they are relatively rare in samples from the Pacific. The number of samples (*n*) is listed for each subdivision. Percentages shown by the diagonal ruling are calculated with the data of MacLeod *et al.* (1996) included (*n* shown in parentheses), whereas the shaded bars exclude those data. See text for further discussion.

temporal resolution than the stage-level data and also show a general constancy of inoceramid occurrence across millions of years followed by a terminal decline during the Maastrichtian. At Caravaca, several samples from the Cenomanian lack inoceramids and inoceramid abundance is low in a few upper Campanian samples (Fig. 8.4), but preservation is poor in both of these intervals (the samples did contain identifiable, calcitic microfossils and therefore were not excluded from the analysis). The disappearance of inoceramid remains in the mid-Maastrichtian, on the other hand, is documented by numerous samples and occurs within an interval of fair to good preservation. The Maastrichtian extinction event was also observed in DSDP holes 390A (Fig. 8.5) and 327A (Fig. 8.6). A comparison of results from Caravaca and Hole 327A supports the conclusion that inoceramids disappeared earlier at high southern latitudes than at low latitudes (MacLeod *et al.*, 1996). Finally, across more than 100 m (represented by 66 samples) of the lower Campanian in Site 511, inoceramids are consistently absent to rare and this observation does not seem to be a preservational artifact.

Stable isotopic data

New stable isotopic results are reported in Tables 8.2 and 8.3. In general samples from DSDP Hole 390A have good to very good preservation (Table 8.4; see also D'Hondt and Arthur (1995)), but we analyzed only individuals with very good preservation. There is an ~0.7‰ negative excursion in $\delta^{13}C$ associated with the disappearance of inoceramid remains, and late Maastrichtian planktic foraminifers

Table 8.2. *Oxygen and carbon isotopic values of selected microfossils from DSDP Hole 390A expressed in standard δ-notation relative to the PDB standard. Spaces indicate no data*

Core	Section	Interval (cm)	Depth (mbsf)	Globotruncana arca				Rugoglobigerina spp.				Globigerinelloides subcarinatus				Heterohelix globulosa				Benthic		
				n	Mass	δ¹⁸O	δ¹³C	n	Mass	δ¹⁸O	δ¹³C	n	Mass	δ¹⁸O	δ¹³C	n	Mass	δ¹⁸O	δ¹³C	Mass	δ¹⁸O	δ¹³C
12	3	61	118.06	5	132	−1.10	1.62	7	66	−1.38	2.22	25	52	−0.86	1.62	28	92	−0.90	1.67	80	0.29	1.72
12	4	37	119.32	5	144	−0.81	2.12									32	110	−0.78	1.99			
12	5	111	121.56	5	136	−0.88	1.73	9	144	−1.32	2.41	27	56	−0.64	1.85	32	98	−1.08	1.70	72	0.44	1.56
12	6	112	123.07	5	118	−0.88	2.02	10	136	−1.14	2.82	34	74	−0.98	1.85	32	76	−1.13	1.77	38	0.46	1.65
13	1	47	124.37	5	122	−0.91	2.08	10	142	−1.44	2.60	12	30	0.00	2.17	32	122	−0.77	1.61	96	0.49	1.60
13	1	111	125.01	5	124	−0.78	1.90	10	144	−1.43	2.50	32	54	−0.61	1.94	32	94	−0.93	1.90	150	0.66	1.66
13	2	96	126.36	5	134	−0.65	2.12					34	56	−0.26	2.05	32	78	−0.44	2.13	52	0.31	0.74
13	4	113	129.53	5	142	−0.51	2.24	10	100	−0.81	2.13	34	54	−0.30	2.08	32	84	−0.41	1.99	42	0.56	1.91
13	6	35	131.75	5	118	−0.55	2.00	2	34	−0.58	1.97	34	60	−0.44	2.03	30	62	−0.58	1.90	106	0.21	1.57
14	1	47	133.47	5	150	−0.65	1.83	6	52	−0.39	2.54	34	44	0.03	1.99	32	74	−0.46	1.89	70	0.78	1.39
14	2	65	135.15	5	133	−0.60	1.81	10	100	−0.72	2.03	34	52	−0.08	1.83	32	86	−0.51	1.67	74	0.56	1.08
14	3	42	136.42	5	136	−0.47	2.08	8	80	−0.80	2.12	34	47	0.06	1.87	32	72	−0.52	1.61	62	0.60	1.28
14	4	24	137.74	5	142	−0.63	1.65	11	68	−1.10	2.06	30	46	0.26	2.07	28	74	−0.66	1.55	72	0.64	1.33
14	5	65	139.65	5	122	−0.50	1.62	9	80	−0.48	2.10	34	64	−0.31	2.02	32	90	−0.41	1.88	54	0.41	1.38

Table 8.3. *Oxygen and carbon isotopic values of selected microfossils from DSDP Hole 551 expressed in standard δ-notation relative to the PDB standard. Abbreviations under preservation are as follows: G, good; M, moderate; P, poor. Spaces indicate no data*

Core	Section	Interval (cm)	Depth (mbsf)	Preservation	Whiteinella aprica				Dicarinella hagni				Rotalipora cushmani				Lenticulina sp.				Planulina sp.				Gavelinella sp.			
					n	Mass	$\delta^{18}O$	$\delta^{13}C$	n	Mass	$\delta^{18}O$	$\delta^{13}C$	n	Mass	$\delta^{18}O$	$\delta^{13}C$	n	Mass	$\delta^{18}O$	$\delta^{13}C$	n	Mass	$\delta^{18}O$	$\delta^{13}C$	n	Mass	$\delta^{18}O$	$\delta^{13}C$
5	1	16	132.66	M-G	31	142	-2.37	2.78	17	138	-1.95	2.55					16	132	-1.55	1.07	8	150	-1.21	1.87				
5	1	91	133.41	M	12	120	-2.84	2.50	12	150	-2.11	2.25																
5	1	14	134.14	M-P	8	110	-1.93	2.50	14	150	-1.98	2.45					5	150	-1.74	0.86								
5	2	30	134.30	M	15	148	-2.23	2.71	11	150	-2.00	2.52					7	136	-1.51	1.55								
5	2	26	138.76	G	22	142	-2.25	2.46					13	140	-1.84	2.13	7	146	-1.18	-0.21								
6	1	61	139.11	M	36	132	-2.19	2.56					10	148	-1.85	2.09	8	140	-1.24	-0.13					10	108	-0.63	1.13
6	2	12	140.12	M	12	140	-2.30	2.37					8	148	-2.01	2.13												
6	2	117	141.17	M	10	132	-1.81	2.38					9	136	-1.80	2.13	7	146	-1.16	0.79	9	136	-0.62	1.37	8	144	-0.78	1.02
6	3	12	141.62	M-G	22	148	-1.73	2.32					9	140	-1.96	1.94												
6	3	64	142.14	M-G	19	146	-2.04	2.29					10	142	-1.98	1.88	7	142	-1.15	0.44								

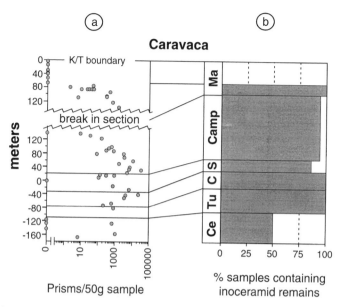

Figure 8.4. The distribution of inoceramid shell fragments through ~450 m of Late Cretaceous section exposed near Caravaca, Spain. (a) Absolute abundance of shell fragments normalized to raw sample weight (MacLeod and Orr, 1993) plotted against stratigraphic level. (b) Data in (a) plotted in a fashion analogous to Fig. 8.2, i.e., proportion of samples/time interval containing inoceramid remains. Both plots indicate that inoceramids were abundant throughout most of the Late Cretaceous but disappeared in the mid-Maastrichtian. Age determinations are based on our observations of planktic foraminifers.

have $\delta^{18}O$ values ~1‰ lower than the same taxa from the early Maastrichtian. The sharpest change in $\delta^{18}O$ occurs at the level of the inoceramid disappearance (Fig. 8.5), but there is also an unconformity at this level (see below).

Data from DSDP Hole 551 (Table 8.3) are incorporated into plots of latitudinal temperature gradients (after Huber *et al.*, 1995), but note generally moderate preservation at this site.

Hole 390A: completeness of the Maastrichtian

While the *A. mayaroensis* Zone was thought to be absent from Hole 390A based on shipboard studies (Benson *et al.*, 1978), *A. mayaroensis* was subsequently reported for a sample collected approximately 10 m below the Cretaceous/Tertiary boundary (D'Hondt and Arthur, 1995). We confirm that ~10 m of *A. mayaroensis* Zone strata occur in Hole 390A and found that the nominate species is rare but occurs in most samples throughout the 10 m (Table 8.4). Below the first occurrence (FO) of *A. mayaroensis*, though, an unconformity is recognized. Between the sample containing the FO of *A. mayaroensis* (390A-13-2, 37 cm) and the subjacent sample (390A-13-2, 96 cm) there is a distinct color change (Benson *et al.*, 1978). Markers for the *R. fructicosa* Zone (*Contusotruncana contusa* and *R. fructicosa*) occur throughout

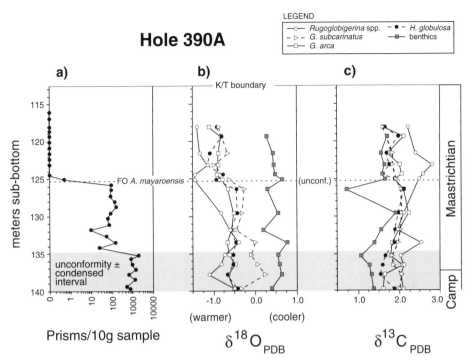

Figure 8.5. Trends in inoceramid abundance and stable isotopic ratios through the upper Campanian and Maastrichtian portions of DSDP Hole 390A. (a) Absolute abundance of shell fragments normalized to raw sample weight (MacLeod and Orr, 1993). (b) $\delta^{18}O$ values of selected foraminifers and inoceramid remains. (c) $\delta^{13}C$ values of selected foraminifers and inoceramid remains. Planktic foraminifers show an apparent warming trend across the entire interval but especially at the level where inoceramids disappear; benthic foraminifers show a ~0.7‰ negative $\delta^{13}C$ excursion at the level of the inoceramid extinction. These correlations, however, may result from a hiatus at this level in the section (see text).

the *A. mayaroensis* Zone but have not been found in any sample below the FO of *A. mayaroensis,* and there are numerous other first occurrences coincident with the FO of *A. mayaroensis.* Estimating the duration of the hiatus is problematic. It could span the entire *G. gansseri* and *R. fructicosa* zones as well as portions of the sub- and superjacent zones. However, *G. gansseri* was not found in any samples from Hole 390A (including those from the *A. mayaroensis* Zone) so our failure to observe it could indicate either absence of *G. gansseri* Zone strata or local rarity/absence of the nominate taxon. No complementary nannofossil data are available. In short, placement of the Campanian/Maastrichtian boundary is uncertain in Hole 390A and lower Maastrichtian strata could be absent.

Population dynamics among planktic foraminifers

Planktic foraminifers are abundant throughout the upper Campanian–Maastrichtian interval of Hole 390A. Preservation ranges from good in the

Table 8.4. *Numerical abundance of planktic foraminifers from the Campanian–Maastrichtian of DSDP Hole 390A. Abbreviations are as follows: Maas, Maastrichtian; Camp, Campanian; A. may., A. mayaroensis; G. fals., Globotruncana falsostuarti; R. calc., Radotruncana calcarata; VG, very good; G, good; M, moderate; P, present. Spaces indicate no data*

Sample metadata:

#	Age	Zone	Sample ID	Top depth	Preservation
S1	Maas.	A. may.	12-2, 12-14	116.07	VG
S2	Maas.	A. may.	12-2, 91-93	116.86	VG
S3	Maas.	A. may.	12-3, 61-63	118.06	VG
S4	Maas.	A. may.	12-4, 37-39	119.32	VG
S5	Maas.	A. may.	12-5, 37-39	120.82	VG
S6	Maas.	A. may.	12-6, 36-38	122.31	VG
S7	Maas.	A. may.	13-1, 47-49	124.37	VG
S8	Maas.	A. may.	13-2, 37-39	125.77	VG
S9	Camp.	G. fals.	13-2, 96-98	126.36	G
S10	Camp.	G. fals.	13-2, 37-39	127.27	VG
S11	Camp.	G. fals.	13-3, 113-11	128.03	M-G
S12	Camp.	G. fals.	13-4, 113-11	129.53	VG
S13	Camp.	G. fals.	13-5, 34-36	130.24	VG
S14	Camp.	G. fals.	13-6, 35-37	131.75	G
S15	Camp.	G. fals.	13-6, 121-12	132.61	VG
S16	Camp.	G. fals.	14-1, 47-49	133.47	G
S17	Camp.	G. fals.	14-1, 112-11	134.12	G
S18	Camp.	G. fals.	14-2, 115-11	135.65	G
S19	Camp.	G. fals.	14-3, 42-44	136.42	M-G
S20	Camp.	G. fals.	14-4, 24-26	137.74	G
S21	Camp.	G. fals.	14-5, 32-34	139.32	G
S22	Camp.	R. calc.	14-5, 65-67	139.65	G

Species abundances (columns S1–S22 correspond to the samples above):

Species	S1	S2	S3	S4	S5	S6	S7	S8	S9	S10	S11	S12	S13	S14	S15	S16	S17	S18	S19	S20	S21	S22
Globotruncanella petaloidea		4	5	8	21	6	13	10	14	9	25	9	6	13	12	3	25	11	20	1		3
Globotruncanella havanensis	1	2					2	1				10							1	3	13	3
Globotruncana subcircumnodifer																		38	30	45	23	38
Contusotruncana plummerae																			4	1	1	
Archaeoglobigerina sp.	3			10	2	2	38	15	33	16	52	25	12	14	27	28	26	27	47	3		31
Trinitella scotti	4	6	1			3				16					2	30	4	10	61	5	60	20
Rugotruncana sp.							2	4	17	7					1	8	14		33	2	17	15
Rugoglobigerina rugosa	23	12	8	16	1		3	11	13	20	8	1	20	1	18	8	17	9	31	20	8	17
Rugoglobigerina hexacamerata	2	3		37		5		2		5		1	6	7		13			8	28		3
Radotruncana calcarata																						
Pseudotextularia nuttalli	9	6	3	8	18	4	4	4	3	4	12	8	2	1	7		6	1		2		3
Pseudoguembelina costulata										3			5	5	13		5		3		7	2
Planoglobulina multicamerata	2	1		3	10	2	1	3	1	3	15	1	3	1	1	8	8	1	2		1	
Planoglobulina acervulinoides	4	2	1	2	4		3	3		2	3		2	1	1		5		0			2
Laevihetrohelix glabrans	2	3	2	5	5		2	7	2		3	3	3		2	24	15	21		3		3
Heterohelix planata	18	19	22	26	19	9	14	4	7	39	19	23	22	24	12	15	6	33		6		3
Heterohelix navarroensis	16	8	10	6	8	23	33	31	29	19	18	13	9		1	7	17	9				2
Heterohelix globulosa	156	129	121	100	94	61	128	161	171	109	151	110	77	92	113	134	120	90	222	126	102	67
Gublerina acuta	3	1	1		1	4		1			1	4				3	3		1			4
Globotruncanita stuartiformis	P	1	2				12	3	19	9	7	1	1	2	2	3	7	2	6	6	3	12
Globotruncanita cf. angulata																						1
Globotruncana ventricosa						1			2	5	5	2			P	P	2	3	4		P	2
Globotruncana orientalis	19		15	12	12	22	13	37		40	40	36	35		44					14		4
Globotruncana linneiana			25	2		7	3	4	8	4	9	6	2	1	8	8		12	23	3	5	42
Heterohelix sp.			2	5	7	4	21		12	18	14		4			14	3		8	3		6
Globotruncana falsostuarti	1	1									1	1	1						2	1	1	2
Globotruncana bulloides											3								4	5	5	3
Globotruncana arca	8	9	9	30	42	24	31	34	46	22	20	47	26	23	11	5	45	70	20	35	29	
Globigerinelloides subcarinatus	7	11	13	17	16	8	18	13	19	9	26	16	9	15	12	15	17	14	55	25	13	13
Globigerinelloides prairiehillensis	3	1		8	2	5	6	1	6	7	8	6	4	12	6	19	24	34	83	29	37	9
Globigerinelloides alvarezi	7	9	7	4		1	10	9	9	10	19	13	3	10	12	22	10	10	45	11	8	8
Contusotruncana patelliformis								2	4	4	2	1	5	5	4	2	4	2	2	3	5	
Contusotruncana fornicata							12	7	22	26	35	19	102	13	29	15	39	29	49	9	12	42

Continued

Table 8.4. (cont.)

Age	Zone	Sample ID	Top depth	Preservation	Hedbergella monmouthensis	Archaeoglobigerina blowi	Globotruncana esnehensis	Globotruncanella sp.	Hedbergella holmdelensis	Heterohelix striata	Rugoglobigerina reicheli	Schackoina multispinata	Globotruncanella minuta	Pseudoguembelina palpebra	Globotruncana petersi	Pseudotextularia intermedia	Pseudoguembelina excolata	Heterohelix punctulata	Rugoglobigerina rotundata	Gumbelitria cretacea	Globotruncana rosetta	Contusotruncana walfischensis	Globotruncana dupleubei	Contusotruncana contusa	Racemiguembelina fructicosa	Abathomphalus mayaroensis	Gansserina wiedenmayeri	Globotruncanita stuarti	Pseudoguembelina kempensis	Racemiguembelina powelli	Pseudotextularia elegans	Globotruncana aegyptiaca	Contusotruncana plicata	Globigerinelloides ultramicrus	Globotruncanita insignis	Hedbergella sp.	Total
Maas.	A. may.	12-2, 12-14	116.07	VG	6		9			9	9		9	13	5		7		1		P	1	2	P	13	1		1			10	7		3	1	2	389
Maas.	A. may.	12-2, 91-93	116.86	VG	2		12			13	1		10	12			2	1			1	P	2	P	2			3	4		6	1			15		322
Maas.	A. may.	12-3, 61-63	118.06	VG	8		9			28	2		14	11			16				1	2	3	1	10	1		10	7	3		1			14	1	370
Maas.	A. may.	12-4, 37-39	119.32	VG	3		44			9	11			8			5						1	2	10	4	P	19			6				12		457
Maas.	A. may.	12-5, 37-39	120.82	VG	3		30			22	5		11	5			20				3	2	2	2	20	P		7	2	13	3	1					433
Maas.	A. may.	12-6, 36-38	122.31	VG	1		12			24			2	2			15				2	2	1	P	13		P	5	10		4	P	P	2			260
Maas.	A. may.	13-1, 47-49	124.37	VG	3	2	38			21	2			3	1		4				2	P	P	1	17	P	P	6			3	1					498
Maas.	A. may.	13-2, 37-39	125.77	VG	5		63			24	3		8	9			10	6	1		3		1	2	18	1	P	2	4	2	6						510
Camp.	G. fals.	13-2, 96-98	126.36	G	4		7		7	7	2			2	1		2				2		1	1													470
Camp.	G. fals.	13-2, 37-39	127.27	VG	6		5			27				13			1																				445
Camp.	G. fals.	13-3, 113-11	128.03	M-G	8	17			5	23				3			1		7	1	1	1															530
Camp.	G. fals.	13-4, 113-11	129.53	VG	8		37		2	6				1																							351
Camp.	G. fals.	13-5, 34-36	130.24	VG	5	11			2	18				3				1																			443
Camp.	G. fals.	13-6, 35-37	131.75	G	5		37			5		1		2		1																					341
Camp.	G. fals.	13-6, 121-12	132.61	VG	4		7			2				1	1	1	3																				391
Camp.	G. fals.	14-1, 47-49	133.47	G	18	8					3		2			1																					425
Camp.	G. fals.	14-1, 112-11	134.12	G	12		4									2																					417
Camp.	G. fals.	14-2, 115-11	135.65	G	14	1			1							3																					368
Camp.	G. fals.	14-3, 42-44	136.42	M-G	4		13		2	2	3																										937
Camp.	G. fals.	14-4, 24-26	137.74	G	2	2	12		3			1																									389
Camp.	G. fals.	14-5, 32-34	139.32	G	4	1																															427
Camp.	R. calc.	14-5, 65-67	139.65	G				11																													321

Holes 327A & 511

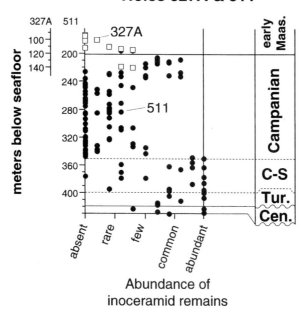

Figure 8.6. The distribution of inoceramid remains through the ~50 m of upper Campanian and lower Maastrichtian sediment from DSDP Hole 327A and ~300 m of Cenomanian–Maastrichtian sediment from DSDP Hole 511 plotted on a relative abundance scale (after MacLeod *et al.*, 1996). Inoceramid remains are rare across ~100 m of lower Campanian strata in Hole 511 and the highest samples in Hole 327A seem to record the Maastrichtian inoceramid extinction event. Age determinations follow Huber *et al.* (1995).

Campanian–lower Maastrichtian to very good in the upper Maastrichtian (Table 8.4). The assemblages in Hole 390A are dominated by *Heterohelix globulosa* which comprises between 20 and 40% of the total assemblage in every sample examined. Various species within *Heterohelix* have been characterized as opportunistic taxa (e.g., Hart, 1980; Caron and Homewood, 1983; Leckie, 1987) and stable isotopic data suggest that *H. globulosa* was often a relatively deep dwelling species. A number of other heterohelicids (both biserial and multiserial) are consistently present but in much lower numbers. Species of *Archaeoglobigerina*, *Globigerinelloides*, and *Rugoglobigerina* are similarly consistent, but numerically subordinate, components of the fauna. Various double-keeled taxa (e.g., *Globotruncana* spp., *Contusotruncana* spp.) are quite common, but single-keeled species (e.g., *Radotruncana calcarata*, *Globotruncanita* spp., *Gansserina weidenmayeri*) are relatively rare.

The low latitude fauna from Hole 390A has a higher species richness and *H*(*S*) diversity, but similar or slightly lower equitability, than the high latitude fauna of Hole 690C (Fig. 8.7). At both sites there seems to be an increase of 5–10 taxa near the base of the *A. mayaroensis* Zone, but it is important to remember that the FO of *A. mayaroensis* is diachronous (Huber, 1992; Huber and Watkins, 1992) and that

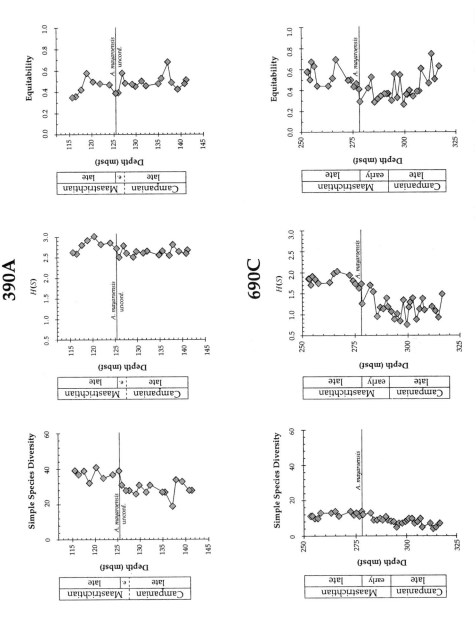

Figure 8.7. Stratigraphic trends in species richness (total number of taxonomic units), $H(S)$ diversity (diversity normalized for different species abundances), and equitability (species evenness) for subtropical DSDP Hole 390A and high latitude ODP Hole 690C. See text for explanation.

there is likely a significant unconformity at this level in Hole 390A. Thus the increase should not be considered a correlative event although the possibility is not excluded by our data. In Hole 390A there is a decrease in equitability and $H(S)$ diversity in the top few samples which reflects an increase in the relative abundance of *H. globulosa*; otherwise, these metrics are relatively stable across the low latitude sample set. In Hole 690C, on the other hand, there is a trend of decreasing equitability through the Campanian and early Maastrichtian and an increase in the late Maastrichtian. Greater equitability and higher species richness both contribute to higher $H(S)$ diversity estimates for the younger samples in Hole 690C.

DISCUSSION

Inoceramid biostratigraphy and paleobiogeography

The rise and fall of Cretaceous inoceramids seems to co-vary with Cretaceous climate (Fig. 8.8). Taxonomically, inoceramids diversified during the Aptian–Albian (e.g., Dhondt, 1992; Voigt, 1995) near the start of the Cretaceous greenhouse interval (Douglas and Savin 1973, 1975; Leckie, 1987; Bralower *et al.*, 1993; Huber *et al.*, 1995; Fassell and Bralower, 1999). Our data do not include the initiation of greenhouse conditions, but, for the balance of the Late Cretaceous, inoceramids seem to have flourished at bathyal depths in the Atlantic and Indian oceans (Figs. 8.2 and 8.3) and then disappeared in a global pulse of extinction during the deterioration of greenhouse conditions (e.g., MacLeod and Huber 1996*a*; MacLeod *et al.*, 1996).

Inoceramids were a phenomenally abundant and successful group in the deep sea for most of the Late Cretaceous. In Albian through lower Maastrichtian strata inoceramids were consistently found in $\sim 50\%$ of the samples examined (Fig. 8.2). Based on the taphonomic model employed (MacLeod and Orr, 1993) this frequency indicates that about half of Late Cretaceous deep sea samples containing calcareous microfossils were deposited within a meter or so of the life position of an inoceramid. Shell fragments occur in a higher proportion of samples from the Coniacian, Santonian, and early Maastrichtian than for other time intervals. For the Coniacian and Santonian, sample size is too small to determine if inoceramids were actually more common at these times, but for the early Maastrichtian the elevated value is likely a sampling artifact. If samples from MacLeod *et al.* (1996) are excluded, the early Maastrichtian value decreases to $\sim 50\%$ (and the late Maastrichtian value to $< 10\%$). As noted above, a major goal of MacLeod *et al.* (1996) was to document precisely the last occurrence (LO) of inoceramids on a global scale. To accomplish this goal, relatively high density sampling was undertaken in inoceramid-bearing sections across the interval where inoceramids decline and disappear (near the FO of *A. mayaroensis*). Many of these mid-Maastrichtian samples contain relatively few shell fragments compared with older samples (MacLeod *et al.*, 1996), but preferentially sampling from within the known range of inoceramids in inoceramid-bearing sections created a bias toward samples containing inoceramids.

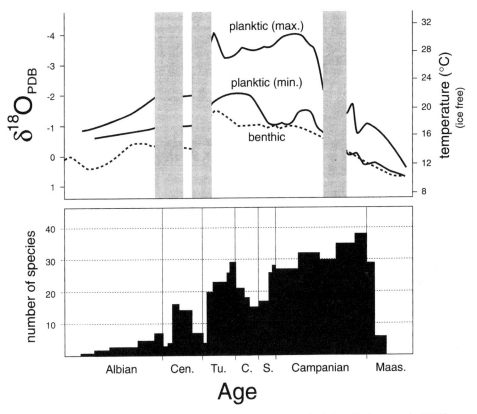

Figure 8.8. Generalized austral paleotemperature trends (after Huber *et al.*, 1995) and inoceramid diversity trends (after Voigt, 1995) during the Late Cretaceous. Gray bars on the paleotemperature plot indicate significant gaps in the high latitude record. Inoceramid diversity and high latitude warmth (a characteristic of greenhouse climates) seem to be positively correlated, and on a regional scale correlation between diversity and temperature might be tighter than the figure suggests as inoceramids disappeared earlier at austral paleolatitudes than elsewhere (MacLeod *et al.*, 1996). For the diversity plot we adjusted the position of the Campanian/Maastrichtian boundary and for both plots we rescaled the data so that the horizontal axis approximates the Gradstein *et al.* (1994) time scale. Differences in the apparent distribution of Maastrichtian inoceramids between Figs. 8.2 and 8.8 result from differences in the stratigraphic subdivisions recognized and in the stratigraphic resolution of the studies summarized; upper Maastrichtian inoceramid occurrences reported in Fig. 8.2 are restricted to the lower upper Maastrichtian and therefore support the suggested correlation between inoceramids and temperature/greenhouse conditions.

The stability of the inoceramid record does not seem to be an artifact of poor temporal resolution caused by pooling data at the stage level. Long records from Caravaca (Fig. 8.4) and DSDP holes 327A and 511 (Fig. 8.6) have a high temporal resolution and are characterized by high inoceramid abundances in all samples across intervals representing tens of millions of years. All three high resolution records show variation, but these variations do not contradict the generalization. The interval of low inoceramid abundance in the lower Campanian at Site 511 does not correspond to any recognized global events, but is associated with local changes in the structure of the water column (see below). The upper Campanian decline in shell fragment abundance in Caravaca corresponds to an interval of poor preservation and may be an artifact. The absence of inoceramids from several upper Cenomanian samples at Caravaca may also be an artifact of poor preservation or it may be related to events at the Cenomanian/Turonian boundary.

Inoceramid occurrences through the Cenomanian, Turonian, and subsequent stages are perhaps the most interesting aspect of the global compilation. Using the same temporal resolution as employed in Fig. 8.2, Johnson *et al.* (1996) showed widespread rudist reefs during the Albian, a major restriction in the latitudinal extent of reefs during the Cenomanian–Coniacian ages, and expansion during the Santonian. In the inoceramid data Albian, Cenomanian, and Turonian abundances are similar, and, if anything, inoceramid abundance was elevated during both the Coniacian and the Santonian. The Cenomanian/Turonian boundary event could have significantly affected bathyal communities (including inoceramids) and have been of too short a duration to be detected in our global data. However, if the decline of rudist reefs during the Cenomanian–Coniacian was caused by tropical cooling brought about by the export of heat by warm, saline water masses as hypothesized in Johnson *et al.* (1996), these water masses either did not impinge widely on the bottom at bathyal depths or the associated changes in benthic conditions did not affect inoceramids in a simple fashion.

The relatively low frequency of inoceramids in the Pacific suggests that bathyal environments in the Atlantic and Indian oceans were more hospitable to inoceramids than in the Pacific Ocean (Fig. 8.3). For the same reasons discussed above, the proportion of Atlantic and Indian ocean samples containing inoceramids decreases when the samples from MacLeod *et al.* (1996) are excluded. For the Southern Ocean, excluding these samples has the opposite effect (estimated abundance increases), but this difference also can be explained as a sampling bias. Inoceramids disappeared earlier in austral latitudes than elsewhere (MacLeod *et al.*, 1996), and because MacLeod *et al.* (1996) focused on Maastrichtian samples, their Southern Ocean material contained a relatively large number of samples from above the regional LO of inoceramids. Regardless of the data set used, the Pacific exhibits the lowest proportion of samples containing inoceramid shell fragments.

Stable isotopic data

In DSDP Hole 390A there is an $\sim 0.7‰$ negative $\delta^{13}C$ excursion in benthic foraminifers associated with the LO of inoceramids (Fig. 8.5); a similar pattern has been observed in ODP holes 750A and 761B (Fig. 8.9). The benthic foraminiferal carbon record is unlikely to represent changes in primary productivity or storage of organic carbon on a global scale because parallel shifts are not observed among planktic taxa in either Hole 761B or Hole 390A. MacLeod and Huber (1996*a*) proposed that the shift might be a characteristic inherited from a new source region for intermediate to deep waters or that it may have been forced by a change in the benthic cycling of carbon caused by a difference in bottom water properties (e.g.,

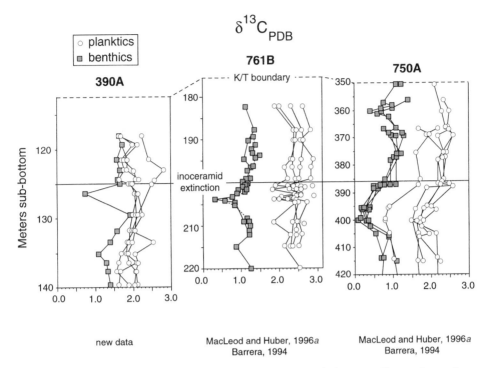

new data

MacLeod and Huber, 1996a
Barrera, 1994

MacLeod and Huber, 1996a
Barrera, 1994

Figure 8.9. Carbon isotopic trends in foraminifers through the upper Campanian and Maastrichtian portions of DSDP Hole 390A and ODP holes 750A and 761B. At all three sites there is an ~0.7‰ negative $\delta^{13}C$ excursion associated with the last occurrence of inoceramid remains despite the fact that the sites are separated by thousands of kilometers and the disappearance of inoceramids is diachronous (MacLeod *et al.*, 1996).

increased concentration of dissolved oxygen). A third possibility suggested by culturing experiments of planktic foraminifers (Spero *et al.*, 1997) is that variations in bottom water alkalinity caused shifts in the $\delta^{13}C$ composition of benthic foraminifers. However, it is difficult to understand how any of these hypotheses could yield a consistent $\delta^{13}C$ excursion over relevant spatial or temporal scales. Assuming the carbon excursions at the three sites are related, the distance between Site 390 and the other two sites (Fig. 8.1) and the diachroneity in the LO of inoceramids (MacLeod *et al.*, 1996) exacerbate this temporal/spatial problem that was already troubling when only sites 750 and 761 were considered (MacLeod and Huber, 1996a). The peak of the benthic excursion occurs below the LO of inoceramid remains (MacLeod and Huber, 1996a; MacLeod *et al.*, 1996), and in Hole 390A there is a biostratigraphically and lithologically recognized unconformity immediately above the excursion. Thus while our new data from Hole 390A support a Maastrichtian benthic $\delta^{13}C$ excursion in the Atlantic, better understanding of the benthic $\delta^{13}C$ record will require additional data from stratigraphically complete sections.

Oxygen isotopic data suggest a relationship between inoceramids and bottom waters sourced in mid- to low latitudes. At sites 750 and 761 increases in $\delta^{18}O$ values of benthic foraminifers (apparent cooling of bottom waters) are coincident with the decline of inoceramids (MacLeod and Huber, 1996*a*). More suggestive in terms of ocean circulation, deep dwelling planktic foraminifers (particularly *H. globulosa*) at these sites have $\delta^{18}O$ values as high as or higher than values in co-occurring benthic foraminifers where inoceramids are abundant, but lower than benthic values where inoceramids are absent (MacLeod and Huber, 1996*a*). These data indicate that in the inoceramid-bearing intervals, bottom waters were warmer than some overlying waters. At constant salinity, warm water is less dense than cool water; so, because water density should increase with depth, warmth at depth must be accompanied by elevated salinity. High salinity water masses are enriched in ^{18}O, and $\delta^{18}O$ values of tests secreted in ^{18}O-enriched water masses are similarly enriched. Therefore bottom water temperatures in the inoceramid-bearing intervals were greater than would be estimated if observed $\delta^{18}O$ differences between benthic and planktic taxa were attributed solely to temperature (see fig. 3 in MacLeod and Huber, 1996*a*). Such warm, saline bottom waters must have been sourced in mid- to low latitudes; the switch in the rank order of $\delta^{18}O$ was taken as evidence for a reversal in deep ocean circulation patterns when inoceramids disappeared (MacLeod and Huber, 1996*a*). A similar shift in $\delta^{18}O$ between *Heterohelix* and benthic foraminifers occurs at the level of inoceramid disappearance in Hole 327A (Huber *et al*., 1995). No carbon excursion was observed at Site 327, but there are large coring gaps and sampling density was low.

The lower Campanian of Site 511 (South Atlantic) provides a natural experiment of the relationship between $\delta^{18}O$ values and inoceramid abundance. The scarcity of inoceramid shell fragments across more than 100 m of lower Campanian strata at this site is not associated with changes in lithology or style of preservation and seems to represent a temporary decline in the local inoceramid population (Fig. 8.10). In this interval, $\delta^{18}O$ values of *H. globulosa* are among the lowest observed and plot far from co-occurring benthic taxa; in the bracketing inoceramid-rich strata, $\delta^{18}O$ values of *H. globulosa* are similar to those of benthic taxa. Temporary scarcity of inoceramids at Site 511 is thus associated with a divergence between $\delta^{18}O$ values of deep dwelling planktic (*H. globulosa*) and benthic foraminifers similar to the pattern observed at the levels where inoceramids disappear in holes 750A and 761B. At Site 511, though, there are no obvious changes in benthic $\delta^{18}O$ values correlative with changes in inoceramid abundance. Inoceramids are well known from shallow, normal marine paleoenvironments during the Late Cretaceous and lack of benthic $\delta^{18}O$ shifts associated with changes in inoceramid abundance at Site 511 (and 390; see below) indicate that bathyal inoceramids were not obligate residents of warm, saline water masses. Even in the absence of evidence for changes in bottom water masses, though, inoceramid abundance is correlated with planktic isotopic shifts, suggesting that changes in the water column structure, and presumably circulation, affected conditions on the seafloor below. Evidence for saline water masses in the Campanian/Maastrichtian South Atlantic is provided by Saltzman and Barron (1982), who reported that $\delta^{18}O$ values of 72–70 Ma inoceramid

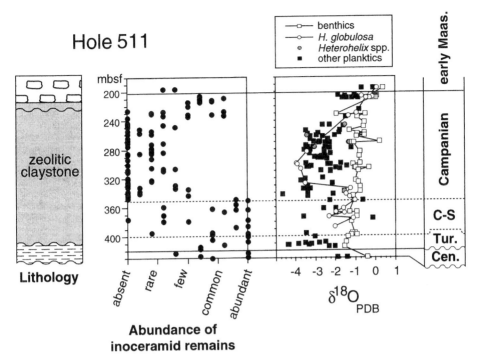

Figure 8.10. The distribution of inoceramids in DSDP Hole 511 plotted with a schematic depiction of lithology and $\delta^{18}O$ analyses of foraminifers (after Huber *et al.*, 1995). The lower Campanian interval during which inoceramids were rare does not correspond with changes in lithology or preservation but does correspond to a shift in $\delta^{18}O$ of *H. globulosa*. *H. globulosa* $\delta^{18}O$ values are close to benthic values where inoceramids are common, a pattern similar to the one observed in ODP holes 750A and 761B (MacLeod and Huber, 1996*a*).

shells from a 4000 m paleodepth site were lower (warmer) than values observed at a 3000 m and a 500 m paleodepth site.

In Hole 390A the gradient between planktic and benthic taxa is greater in the upper Maastrichtian (above the disappearance of inoceramids) than it is in the Campanian/lower Maastrichtian (where inoceramids are present), but there is no reversal in $\delta^{18}O$ values with depth (Fig. 8.5). In addition, like the record from Site 511, the increase in gradient results from shifts in the $\delta^{18}O$ values of planktic foraminifers rather than benthic foraminifers. The decrease by $\sim 1\%_{0}$ in planktic $\delta^{18}O$ values could reflect an increase in surface water temperature and/or a decrease in surface water salinity. D'Hondt and Arthur (1996) addressed this problem from the perspective of latitudinal gradients in $\delta^{18}O$ analyses of late Maastrichtian planktic foraminifers including samples from Site 390. They proposed that observed $\delta^{18}O$ values could indicate either (1) cool tropics, (2) saline tropics, or (3) globally elevated $\delta^{18}O_{seawater}$ (all relative to the present). They concluded that Maastrichtian tropical waters were significantly cooler (~ 20–$21\,^{\circ}C$) but had similar salinity to modern tropical waters. Their use of GCM results to discount significant changes in tropical sea surface salinity, though, seems arbitrary as these same GCM experiments also argue against the cool sea surface temperatures they support. An increase in $\delta^{18}O_{seawater}$, regardless of its likelihood, would shift the problem from apparently cool tropics to apparently very warm poles. Wilson and Opdyke (1996) subsequently estimated Maastrichtian tropical sea surface temperatures of $\sim$27–32 °C (at least as warm as the present) at ODP Site 877 (equatorial Pacific) based on analyses of exceptionally preserved rudists and calcitic cements. The presence of rudists suggests an early Maastrichtian age for the Pacific samples, but the $\delta^{18}O$ values we observed in lower Maastrichtian/Campanian samples from Hole 390A accentuate the difference between paleotemperature estimates between the results of D'Hondt and Arthur (1996) and Wilson and Opdyke (1996). Thus the discrepancy is unlikely to result from secular temperature variation (Crowley and Zachos, Chapter 3, this volume). We feel that diagenetic artifacts and/or salinity effects may compromise the absolute paleotemperature estimates of D'Hondt and Arthur (1996). Since we have focused on relative changes through Hole 390A, though, absolute paleotemperature/paleosalinity estimates are of secondary concern in our analysis. There is declining preservation downward in Hole 390A, but we analyzed only well-preserved individuals. Further, increasing alteration with depth would be expected to result in lower $\delta^{18}O$ in older samples whereas we observed the opposite.

The Maastrichtian is generally characterized as a time of global cooling (e.g., Douglas and Savin, 1973, 1975; Boersma and Shackleton, 1981; Barrera *et al.*, 1987; Huber *et al.*, 1995), which seems to favor interpreting lower $\delta^{18}O$ values in late Maastrichtian planktic foraminifers from Site 390 as resulting from a decrease in surface water salinity rather than warming. On the other hand, Wolfe and Upchurch (1987) proposed warming during the mid-Maastrichtian based on leaf physiognomy of collections from the southeastern USA, the most proximate region to Hole 390A for which independent paleotemperature estimates have been generated. Increasing temperature in low latitude surface waters at the same time as high latitude cooling would suggest that some process(es) act(s) to cool the tropics (as well as to warm the poles) during greenhouse times. Alternatively, decreasing sea surface salinity in low latitudes could have caused a reduction or cessation of downwelling in these regions consistent with the deep ocean circulation model suggested by mid- and high latitude $\delta^{18}O$ results. These scenarios are not mutually exclusive, but deconvolving the influence of temperature and salinity on $\delta^{18}O$ values without additional information is impossible.

Population dynamics among planktic foraminifers

We hoped that population data for planktic foraminifers would constrain interpretation of the Maastrichtian $\delta^{18}O$ data. An increase in planktic foraminiferal species richness occurs in the Maastrichtian at holes 390A and 690C (Fig. 8.7). Immigration of mid- to low latitude species largely accounts for the increased species richness at Site 690 (e.g., Huber and Watkins, 1992). These authors noted that poleward expansion of low latitude taxa is unexpected during the Maastrichtian high latitude cooling (e.g., Barrera *et al.*, 1987; Huber *et al.*, 1995). Based on paleobiogeographic evidence they postulated that emergence of an isthmus between South

America and Antarctica may have led to increased north–south surface water communication and diminished convective mixing of the surface waters in the southern South Atlantic region. The large relative increase in species richness in Hole 690C seems to be responsible for the increase in $H(S)$ diversity as there are no apparent correlative changes in equitability. Population data at Site 690 are thus consistent with (but not diagnostic of) a more stratified water column during the late Maastrichtian without major shifts in trophic structure.

Unlike Hole 690C, the increase in species richness from the Campanian/early Maastrichtian to the late Maastrichtian in Hole 390A reflects the evolution of new taxa more than the immigration of taxa from other regions. Some authors (e.g., Douglas and Savin, 1978; Hart, 1980; Caron and Homewood, 1983; Hodell and Vayavananda, 1993) have suggested that planktic foraminiferal speciation may be largely driven by changes in the vertical structure of the water column. Increased surface water stratification or increased depth of the thermocline could lead to the opening of new niche space that is subsequently occupied by new taxa. The increased gradient in $\delta^{18}O$ values between planktic and benthic taxa and the divergence of $\delta^{18}O$ and $\delta^{13}C$ values in *Rugoglobigerina* spp. from values in other planktic taxa could be cited as evidence of increased stratification. On the other hand, the decrease of $\delta^{13}C$ gradients among these other planktic taxa and the benthic taxa as well as the convergence of $\delta^{18}O$ values for *Globigerinelloides subcarinatus* and the other planktics do not support this conclusion (Fig. 8.5). Further, Norris *et al.* (1993) concluded that evolution of new morphospecies does not necessarily reflect changes in depth ecology. Across the FO of *A. mayaroensis* at Site 390, equitability and $H(S)$ diversity (after correcting for changes in species richness) values are relatively stable at Site 390 and the decreases in both measures in the top few samples are not correlative with the isotopic or species richness shifts. At Site 390, then, there may have been an increase in surface water stratification during the Maastrichtian, but diversity and dominance estimates do not help us distinguish between increasing temperatures or decreasing salinity (Fassell and Bralower, 1999), and the nature of paleoceanographic changes remains ambiguous.

Latitudinal temperature gradients

Because shallow latitudinal temperature gradients seem to be characteristic of greenhouse climates, a final test of correlation between changes in the deep ocean and greenhouse climates is to compare the distribution of inoceramids with a compilation of available paleotemperature estimates (Fig. 8.11). Many uncertainties are associated with such reconstructions (e.g., Huber and Hodell, 1996; Price *et al.*, 1996) and they are best viewed as suggestive rather than indicative. Differences in the vital effects among species analyzed, in diagenetic artifacts, in temporal sampling density at each site (i.e., the number and spread of data that were averaged to generate the points plotted), and in the spatial coverage for each time slice are all sources of error.

These caveats notwithstanding, the reconstructed gradients show several interesting features. As is typical for warm intervals, reconstructed surface temperature gradients are shallower than those observed at present. Maastrichtian data yield

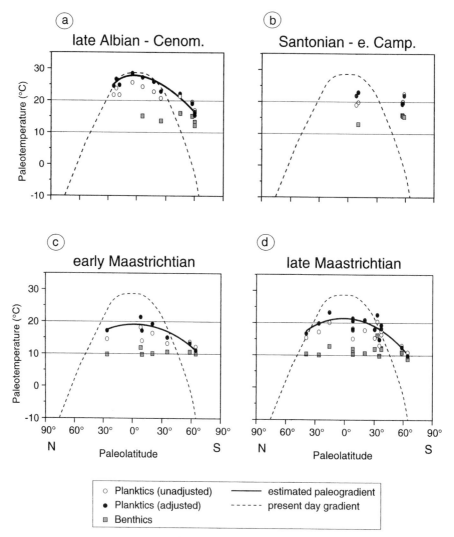

Figure 8.11. Latitudinal temperature gradients estimated from $\delta^{18}O$ analyses of planktic foraminifers (after Huber *et al.*, 1995; Huber and Hodell, 1996) for four different time intervals in the Cretaceous: (a) Albian-Cenomanian, (b) Santonian-Campanian, (c) early Maastrichtian, and (d) late Maastrichtian. Benthic (gray squares) and planktic (hollow circles) points show the average of analyses from a given site and time interval whereas filled circles represent planktic points adjusted assuming modern surface water salinity (and $\delta^{18}O_{seawater}$) gradients (Zachos *et al.*, 1994). Paleotemperature estimates are based on the equation of Erez and Luz (1983). In each plot the dashed line shows the modern latitudinal temperature gradient and the solid line represents a best fit parabola through the data adjusted for salinity. Estimated gradients are all shallower than the present. The Santonian-Campanian estimate is almost flat and the late Maastrichtian estimate is steeper than the early Maastrichtian estimate. The Santonian-Campanian plot contains the fewest data points, and the cool tropical temperatures in this time slice, as well as in both Maastrichtian time slices, are based on analyses of specimens that are slightly recrystallized (see Huber *et al.*, (1995) and Huber and Hodell (1996) for references). The two Maastrichtian plots are relatively data rich and contain many sites in common.

cooler paleotemperature estimates overall (both planktic and benthic data) than earlier time slices, suggesting either whole ocean cooling at the end of the Cretaceous or a systematic temporal bias in the isotopic data (e.g., Schrag et al., 1992). Benthic latitudinal gradients are flat for all time slices, and there are no apparent systematic differences in benthic values between the two Maastrichtian time slices. If there was a mid-Maastrichtian reversal in ocean circulation associated with the extinction of inoceramids, the lack of apparent change in $\delta^{18}O$ values places significant restrictions on interpretation of the Maastrichtian benthic data. Possible explanations include: (1) errors in benthic $\delta^{18}O$ values (see Crowley and Zachos, Chapter 3, this volume) prevent meaningful comparisons among sites; (2) decreases in $\delta^{18}O$ due to cooling were balanced by increases in $\delta^{18}O$ due to lower salinity; (3) the duration of the mid-Maastrichtian event was short and is obscured by averaging on longer time scales; or (4) saline water masses did not fill the deep ocean and $\delta^{18}O$ shifts in localities where saline water masses did impinge on the bottom (e.g., MacLeod and Huber, 1996a) are obscured by averaging in the global compilation. On the other hand, surface temperature gradients do seem to have changed correlative with the inoceramid extinction event, and the late Maastrichtian plot has the steepest reconstructed surface water temperature gradient of the four time slices. Disagreement regarding tropical temperatures during the Maastrichtian (D'Hondt and Arthur, 1996; Wilson and Opdyke, 1996) illustrates large continuing uncertainties in generalization about paleotemperature. Still, apparent warming in Hole 390A (Fig. 8.5) and apparent cooling in Hole 690C (fig. 4 in Barrera and Huber, 1990) support the conclusion of increasing latitudinal temperature gradients based on observations from single sections and argues against some, but not all, potential artifacts in the latitudinal plots.

IMPLICATIONS FOR GREENHOUSE CLIMATES

Significant questions remain concerning paleoceanographic changes in different regions during the Cretaceous. However, the data presented show that the correlations between Maastrichtian faunal and geochemical changes that have been used to implicate changes in the deep ocean as a significant variable in the greenhouse climate equation (MacLeod and Huber, 1996a; MacLeod et al., 1996) also hold on a global scale across the Late Cretaceous. The distribution of inoceramid bivalves suggests that benthic conditions (as perceived by inoceramids) at bathyal paleodepths were remarkably constant and hospitable in the Atlantic and Indian oceans during most of the Late Cretaceous but were less favorable in the Pacific over the same interval. Stable isotopic data suggest that the occurrence of inoceramids in the deep sea is often associated with evidence for warm, saline water masses and their extinction may have been caused by changes in the source region of intermediate and/or deep water masses (MacLeod and Huber, 1996a). Even in instances where there are not $\delta^{18}O$ or $\delta^{13}C$ changes among benthic taxa, the disappearance of inoceramids (either terminal or temporary) is correlated with $\delta^{18}O$ shifts in planktic taxa that are indicative of changes in water column structure. Thus in total our data suggest that warm, saline water masses may have been prevalent at bathyal depths in the Atlantic and Indian oceans during most of the Late Cretaceous and that these water masses seem to have disappeared during the Maastrichtian interval of global cooling.

ACKNOWLEDGEMENTS

We thank T. Bralower, M. Lyle, and Y. Maestas for critical reviews of this manuscript and the Ocean Drilling Program for providing samples from ODP/DSDP cores. Funding was provided by the Smithsonian Institution.

REFERENCES

Barrera, E. (1994). Global environmental changes preceding the Cretaceous–Tertiary boundary: Early–late Maastrichtian transition. *Geology*, **22**, 877–80.

Barrera, E. and Huber, B. T. (1990). Evolution of Antarctic waters during the Maestrichtian: foraminifer oxygen and carbon isotope ratios, ODP Leg 113. In *Proceedings of the Ocean Drilling Program, Scientific Results*, vol. 113, eds. P. F. Barker, J. P. Kennett *et al.*, pp. 813–23. College Station, TX: Ocean Drilling Program.

Barrera, E., Huber, B. T., Savin, S. M. and Webb, P. N. (1987). Antarctic marine temperatures: late Campanian through early Paleocene. *Paleoceanography*, **2**, 21–47.

Barron, E. J., Fawcett, P. J., Petersen, W. H., Pollard, D. and Thompson, S. L. (1995). A "simulation" of mid-Cretaceous climate. *Paleoceanography*, **10**, 953–62.

Barron, E. J., Peterson, W. H., Pollard, D. and Thompson, S. (1993). Past climate and the role of ocean heat transport: model simulations for the Cretaceous. *Paleoceanography*, **8**, 785–98.

Barron, E. J., Saltzman, E. and Price, D. A. (1984). Occurrence of *Inoceramus* in the South Atlantic and oxygen isotopic paleotemperatures. In *Initial Reports of the Deep Sea Drilling Project*, vol. 75, eds. W. W. Hay, J. C. Sibuet *et al.*, pp. 893–904. Washington, DC: Government Printing Office.

Benson, W. E., Sheridan, R. E. *et al.* (1978). *Initial Reports of the Deep Sea Drilling Project.* Washington, DC: US Government Printing Office.

Berger, W. H. and Killingley, J. S. (1982). Box cores from the equatorial Pacific; ^{14}C sedimentation rates and benthic mixing. *Marine Geology*, **45**, 93–125.

Berner, R. A. (1994). GEOCARB II: A revised model of atmospheric CO_2 over Phanerozoic time. *American Journal of Science*, **294**, 56–91.

Boersma, A. and Shackleton, N. J. (1981). Oxygen and carbon isotope variations and planktonic foraminiferal depth habitats: Late Cretaceous to Paleocene, Central Pacific, DSDP Sites 463 and 465, Leg 65. In *Initial Reports of the Deep Sea Drilling Project*, eds. J. Thiede, T. L. Vallier *et al.*, pp. 513–26. Washington, DC: US Government Printing Office.

Bralower, T. J., Sliter, W. V., Arthur, M. A., Leckie, R. M., Allard, D. and Schlanger, S. O. (1993). Dysoxic/anoxic episodes in the Aptian–Albian (Early Cretaceous). In The Mesozoic Pacific: Geology, Tectonics, and Volcanism, eds. M. S. Pringle, W. W. Sager, W. V. Sliter and S. Stein. *American Geophysical Union Monograph*, **77**, 5–37.

Brass, G. W., Southam, J. R. and Peterson, W. H. (1982). Warm saline bottom water in the ancient ocean. *Nature*, **296**, 620–3.

Bryden, H. L. and Hall, M. M. (1980). Heat transport by ocean currents across 25° N latitude in the Atlantic. *Science*, **207**, 884–6.

Carissimo, B. C., Oort, A. H. and Voner Haar, T. H. (1985). Estimating the meridional energy transports in the atmosphere and ocean. *Journal of Physical Oceanography*, **15**, 82–91.

Caron, M. and Homewood, P. (1983). Evolution of early planktic foraminifera. *Marine Micropaleontology*, **7**, 453–62.

Chamberlin, T. C. (1906). On a possible reversal of deep-sea circulation and its influence on geologic climates. *Journal of Geology*, **14**, 363–73.

Compton, J. S. and Mallinson, D. J. (1996). Geochemical consequences of increased late Cenozoic weathering rates and the global CO_2 balance since 100 Ma. *Paleoceanography*, **11**, 431–46.

Crowley, T. J. (1993). Geological assessment of the greenhouse effect. *Bulletin of the American Meteorological Society*, **74**, 2363–73.

Crowley, T. J. and North, G. R. (1991). *Paleoclimatology*. Oxford: Oxford University Press.

Dhondt, A. V. (1992). Cretaceous inoceramid biogeography: a review. *Palaeogeography, Palaeoclimatology, Palaeoecology*, **92**, 217–32.

D'Hondt, S. and Arthur, M. A. (1995). Interspecific variation in stable isotopic signals of Maastrichtian planktic foraminifera. *Paleoceanography*, **10**, 123–35.

D'Hondt, S. and Arthur, M. A. (1996). Late Cretaceous ocean and the cool tropic paradox. *Science*, **271**, 1838–41.

Douglas, R. G. and Savin, S. M. (1973). Oxygen and carbon isotope analyses of Cretaceous and Tertiary foraminifera from the central North Pacific. In *Initial Reports of the Deep Sea Drilling Project*, eds. E. L. Winterer, J. I. Ewing *et al.*, pp. 591–607. Washington, DC: US Government Printing Office.

Douglas, R. G. and Savin, S. M. (1975). Oxygen and carbon isotope analyses of Tertiary and Cretaceous microfossils from the Shatsky Rise and other sites in the North Pacific Ocean. In *Initial Reports of the Deep Sea Drilling Project*, eds. R. L. Larson, R. Moberly *et al.*, pp. 509–20. Washington, DC: US Government Printing Office.

Douglas, R. G. and Savin, S. M. (1978). Oxygen isotopic evidence for the depth stratification of Tertiary and Cretaceous foraminifera. *Palaeogeography, Palaeoclimatology, Palaeoecology*, **3**, 175–96.

Erez, J. and Luz, B. (1983). Experimental paleotemperature equation for planktonic foraminifera. *Geochimica et Cosmochimica Acta*, **47**, 1025–31.

Fassell, M. L. and Bralower, T. J. (1999). A warm, equable middle Cretaceous: stable isotope evidence. In *The Evolution of Cretaceous Ocean/Climate Systems*, eds. E. Barrera and C. Johnson (in press).

Gradstein, F. M., Agterberg, F. P., Ogg, J. G. *et al.* (1994). A Mesozoic time scale. *Journal of Geophysical Research*, **99**, 24051–74.

Greenwood, D. R. and Wing, S. L. (1995). Eocene continental climates and latitudinal temperature gradients. *Geology*, **23**, 1044–8.

Hart, M. B. (1980). A water depth model for the evolution of the planktonic Foraminiferida. *Nature*, **286**, 252–4.

Hastenrath, S. (1982). On meridional heat transports in the world's ocean. *Journal of Physical Oceanography*, **12**, 922–7.

Hay, W. W. (1988). Paleoceanography: A review for the GSA Centennial. *Geological Society of America Bulletin*, **100**, 1934–56.

Hayek, L.-A. C. and Buzas, M. A. (1997). *Surveying Natural Populations*. New York: Columbia University Press.

Hodell, D. A. and Vayavananda, A. (1993). Middle Miocene paleoceanography of the western equatorial Pacific (DSDP Site 289) and the evolution of *Globorotalia (Fohsella)*. *Marine Micropaleontology*, **22**, 279–310.

Huber, B. T. (1990). Maastrichtian planktonic foraminifer biostratigraphy of the Maud Rise (Weddell Sea, Antarctica): ODP Leg 113 Holes 689B and 690C. In *Proceedings of the Ocean Drilling Program, Scientific Results*, vol. 113, eds. P. F. Barker and J. P. Kennett, pp. 489–513. College Station, TX: Ocean Drilling Program.

Huber, B. T. (1992). Paleobiogeography of Campanian–Maastrichtian foraminifers in the southern high latitudes. *Palaeogeography, Palaeoclimatology, Palaeoecology*, **92**, 325–60.

Huber, B. T. and Hodell, D. A. (1996). Middle–Late Cretaceous climate of the southern high latitudes: Stable isotopic evidence for minimal equator-to-pole thermal gradients: Discussion and reply. *Geological Society of America Bulletin*, **108**, 1192–6.

Huber, B. T., Hodell, D. A. and Hamilton, C. P. (1995). Mid- to Late Cretaceous climate of the southern high latitudes: Stable isotopic evidence for minimal equator-to-pole thermal gradients. *Geological Society of America Bulletin*, **107**, 1164–91.

Huber, B. T. and Watkins, D. K. (1992). Biogeography of Campanian–Maastrichtian calcareous plankton in the region of the Southern Ocean: paleogeographic and paleoclimatic implications. In *The Antarctic Paleoenvironment: A Perspective on Global Change*, eds. J. P. Kennett and D. A. Warnke, pp. 31–60. Antarctic Research Series. Washington, DC: American Geophysical Union.

Johnson, C. C., Barron, E. J., Kauffman, E. G., Arthur, M. A., Fawcett, P. J. and Yasuda, M. K. (1996). Middle Cretaceous reef collapse linked to ocean heat transport. *Geology*, **24**, 376–80.

Kennett, J. P. and Stott, L. D. (1991). Abrupt deep-sea warming, palaeoceanographic changes and benthic extinctions at the end of the Palaeocene. *Nature*, **353**, 225–9.

Leckie, R. M. (1987). Paleoecology of mid-Cretaceous planktonic foraminifera: a comparison of open ocean and epicontinental sea assemblages. *Micropaleontology*, **33**, 164–76.

Lyle, M. (1997). Could early Cenozoic thermohaline circulation have warmed the poles? *Paleoceanography*, **12**, 161–7.

MacLeod, K. G. (1994). Bioturbation, inoceramid extinction, and mid-Maastrichtian ecological change. *Geology*, **22**, 139–42.

MacLeod, K. G. and Huber, B. T. (1996a). Reorganization of deep ocean circulation accompanying a Late Cretaceous extinction event. *Nature*, **380**, 422–5.

MacLeod, K. G. and Huber, B. T. (1996b). Strontium isotopic evidence for extensive reworking in sediments spanning the Cretaceous/Tertiary boundary at ODP Site 738. *Geology*, **24**, 463–6.

MacLeod, K. G., Huber, B. T. and Ward, P. D. (1996). The biostratigraphy and paleobiogeography of Maastrichtian inoceramids. In *The Cretaceous–Tertiary Event and Other Catastrophes in Earth History*, Geological Society of America Special Paper No. 307, eds. G. Ryder, D. Fastovsky and S. Gartner, pp. 361–73. Boulder, CO: Geological Society of America.

MacLeod, K. G. and Orr, W. N. (1993). The taphonomy of Maastrichtian inoceramids in the Basque region of France and Spain and the pattern of their decline and disappearance. *Paleobiology*, **19**, 235–50.

Nederbragt, A. J. (1991). Late Cretaceous biostratigraphy and development of Heterohelicidae (planktic foraminifera). *Micropaleontology*, **37**, 329–72.

Norris, R. D., Corfield, R. M. and Cartlidge, J. E. (1993). Evolution of depth ecology in the planktic foraminifera lineage *Globorotalia* (*Fohsella*). *Geology*, **21**, 975–8.

Officer, C. B. and Lynch, D. R. (1983). Determination of mixing parameters from tracer distributions in deep-sea sediment cores. *Marine Geology*, **52**, 59–74.

Pak, D. K. and Miller, K. G. (1992). Paleocene to Eocene benthic foraminiferal isotopes and assemblages: implications for deepwater circulation. *Paleoceanography*, **7**, 405–22.

Pessagno, E. A. and Michael, F. Y. (1974). Mesozoic foraminifera, leg 22, site 217. In *Initial Reports of the Deep Sea Drilling Project*, eds. C. C. von der Borch and J. G. Sclater, pp. 629–34. Washington, DC: US Government Printing Office.

Premoli Silva, I. and Sliter, W. V. (1981). Cretaceous planktonic foraminifers from the Nauru Basin, Leg 61, Site 462, Western Equatorial Pacific. In *Initial Reports of the Deep Sea*

Drilling Project, eds. R. L. Larson, S. O. Schlanger *et al.*, pp. 423–37. Washington, DC: US Government Printing Office.

Premoli Silva, I. and Sliter, W. V. (1994). Cretaceous planktonic foraminiferal biostratigraphy and evolutionary trends from the Bottaccione section, Gubbio, Italy. *Palaeontographica Italica*, **82**, 1–89.

Price, G. D., Sellwood, B. W. and Pirrie, D. (1996). Middle–Late Cretaceous climate of the southern high latitudes: Stable isotopic evidence for minimal equator-to-pole thermal gradients: Discussion. *Geological Society of America Bulletin*, **108**, 1192–3.

Railsback, L. B., Anderson, T. F., Ackerly, S. C. and Cisne, J. L. (1989). Paleoceanographic modeling of temperature–salinity profiles from stable isotopic data. *Paleoceanography*, **4**, 585–91.

Roemmich, D. (1980). Estimation of meridional heat flux in the North Atlantic by inverse methods. *Journal of Physical Oceanography*, **10**, 1792–3.

Salaj, J. and Samuel, O. (1966). *Foraminifera der West karpaten-Kreide (Slowakei)*. Bratislava: Geologicky stav Dionyzastra.

Saltzman, E. S. and Barron, E. J. (1982). Deep circulation in the Late Cretaceous: oxygen isotope paleotemperatures from *Inoceramus* remains in DSDP cores. *Palaeogeography, Palaeoclimatology, Palaeoecology*, **40**, 167–81.

Schmidt, G. A. and Mysak, L. A. (1996). Can increased poleward oceanic heat flux explain the warm Cretaceous? *Paleoceanography*, **11**, 579–93.

Schrag, D. P., DePaolo, D. J. and Richter, F. M. (1992). Oxygen isotope exchange in a two-layer model of oceanic crust. *Earth and Planetary Science Letters*, **111**, 305–17.

Schrag, D P., DePaolo, D. J. and Richter, F. M. (1995). Reconstructing past sea surface temperatures: correcting for diagenesis of bulk marine carbonate. *Geochimica et Cosmochimica Acta*, **59**, 2265–78.

Shannon, C. E. (1948). A mathematical theory of communication. *Bell System Technical Journal*, **27**, 379–423, 623–56.

Sloan, L. C., Walker, J. C. G. and Moore, T. C. Jr (1995). Possible role of oceanic heat transport in the early Eocene climate. *Paleoceanography*, **10**, 347–56.

Spero, H. J., Bijma, J., Lea, D. W. and Bemis, B. E. (1997). Effect of seawater carbonate concentration on foraminiferal carbon and oxygen isotopes. *Nature*, **390**, 497–500.

Stott, L. D. and Kennett, J. P. (1990). Antarctic Paleogene planktonic foraminifer biostratigraphy: ODP Leg 113 sites 689 and 690. In *Proceedings of the Ocean Drilling Program, Scientific Results*, vol. 113, eds. P. F. Barker, J. P. Kennett *et al.*, pp. 549–69. College Station, TX: Ocean Drilling Program.

Thomas, E. and Shackleton, N. J. (1996). The Palaeocene–Eocene benthic foraminiferal extinction and stable isotope anomalies. In *Correlation of the early Paleogene in Northwest Europe*, eds. R. W. O'B. Knox, R. Corfield and R. E. Dunay, pp. 401–41. London: Geological Society.

Voigt, S. (1995). Palaeobiogeography of early Late Cretaceous inoceramids in the context of a new global palaeogeography. *Cretaceous Research*, **16**, 343–56.

Wheatcroft, R. A. (1991). Conservative tracer study of horizontal sediment mixing rates in a bathyal basin, California borderland. *Journal of Marine Research*, **49**, 565–88.

Wilson, P. A. and Opdyke, B. N. (1996). Equatorial sea-surface temperatures for the Maastrichtian revealed through remarkable preservation of metastable carbonate. *Geology*, **24**, 555–8.

Wolfe, J. A. and Upchurch, G. R. Jr (1987). North American nonmarine climates and vegetation during the Late Cretaceous. *Palaeogeography, Palaeoclimatology, Palaeoecology*, **61**, 33–77.

Zachos, J. C., Lohmann, K. C., Walker, J. C. G. and Wise, S. W. (1993). Abrupt climate change and transient climates during the Paleogene: A marine perspective. *Journal of Geology*, **101**, 191–213.

Zachos, J. C., Stott, L. D. and Lohmann, K. C. (1994). Evolution of early Cenozoic marine temperatures. *Paleoceanography*, **9**, 353–87.

9

Late Cretaceous climate, vegetation, and ocean interactions

ROBERT M. DECONTO, ESTHER C. BRADY, JON BERGENGREN, AND
WILLIAM W. HAY

ABSTRACT

General Circulation Models (GCMs) have had difficulty simulating the low meridional thermal gradients and warm winter continents characteristic of the Late Cretaceous and other warm paleoclimatic periods, forcing both the validity of the models and interpretations of the proxy climate data to be questioned. In a new approach to pre-Quaternary paleoclimate modeling, Campanian (80 Ma) climate and vegetation have been simulated using a GCM, interactively coupled to a predictive vegetation model (DeConto *et al.*, 1998, 1999*a,b*). This climate–vegetation simulation reproduced the overall warmth, warm polar temperatures, and warm winter continental interiors characteristic of the Late Cretaceous geologic record. High latitude forests played an important role in the maintenance of polar warmth and equable continental interiors both directly and indirectly. The model required 1500 ppm atmospheric CO_2 (about 4.4 times present-day value) and 2×10^{15} W poleward ocean heat transport to maintain high-polar latitude forests. Smaller values of atmospheric CO_2 and poleward ocean heat transport predicted replacement of forest with tundra in high latitude continental interiors and cold (especially winter and spring) temperatures. To determine if the Late Cretaceous ocean could maintain low meridional thermal gradients an Ocean General Circulation Model (OGCM) was forced by the Campanian climate simulated by the climate–vegetation model (Brady *et al.*, 1998). The OGCM predicts that most of the Campanian ocean heat transport is through meridional overturning, with most deep water formation occurring at high southern latitudes in a process similar to that found in today's North Atlantic. The OGCM produces 10–12 °C deep water and contributes about the same poleward heat transport as control simulations of the present-day ocean circulation, without relying on marginal seas in regions of high net evaporation to produce salinities high enough for deep convection at low latitudes.

INTRODUCTION

The geologic record suggests Cretaceous climate was much warmer than today (Frakes, 1979; Frakes *et al.*, 1992). The evidence for a generally warm Cretaceous climate includes the latitudinal expansion of thermophilic organisms (e.g., Kauffman, 1973; Lloyd, 1982; Huber and Watkins, 1992), the expansion of dinosaurs into polar latitudes of both hemispheres (Colbert, 1973; Olivero *et al.*, 1991; Crame, 1992), and the poleward migration of vegetational provinces (Vakhrameev, 1991). Meridional thermal gradients were low in the oceans (Barrera *et al.*, 1987, Huber *et al.*, 1995; MacLeod *et al.*, Chapter 8, this volume) and on land (Wolfe and Upchurch, 1987; Parrish and Spicer, 1988).

The Campanian Stage of the Late Cretaceous was not as warm as the mid-Cretaceous thermal maximum. However, global mean annual surface temperatures, reconstructed from climate proxies, were still ~10 °C warmer than today (DeConto, 1996; DeConto *et al.*, 1998, 1999*a*). Unlike the Early Cretaceous (Frakes and Francis, 1988; Stoll and Schrag, 1996), there is no evidence of significant ice during the Late Cretaceous. The early Campanian is the warm end member of a Late Cretaceous cooling trend, recognized in both the terrestrial (Wolfe and Upchurch, 1987; Parrish and Spicer, 1988) and marine realms (Douglas and Savin, 1975; Boersma and Shackleton, 1981; Barrera *et al.*, 1987; Kauffman and Johnson, 1988; Pirrie and Marshall, 1990; Jenkyns *et al.*, 1994).

High concentrations of atmospheric CO_2, an important 'greenhouse' gas affecting the earth's radiation energy balance, have been offered as the primary contributor to Cretaceous warmth (Budyko and Ronov, 1979; Fischer, 1982; Barron and Washington, 1984, 1985; Schneider *et al.*, 1985; Barron *et al.*, 1993*a*). Estimates of Late Cretaceous atmospheric CO_2 fall between 1.5 and 9 times larger than present-day values (Cerling, 1991; Freeman and Hayes, 1992; Berner, 1994; Andrews *et al.*, 1995). Early General Circulation Model (GCM) studies of Cretaceous climate (e.g., Barron and Washington, 1984) suggested that an increase in atmospheric CO_2 of four times present-day values provides a plausible explanation for Cretaceous warmth. However, most modeling studies have shown that increases in atmospheric CO_2 tend to increase surface temperatures at all latitudes. Increased levels of atmospheric CO_2 do not account for the very low meridional thermal gradients recognized in the geologic record (see Crowley and Zachos, Chapter 3, this volume).

The Campanian physical environment was very different from that of today. The distribution of continents reflected the recent break-up of Pangea and sea level was high (Haq *et al.*, 1987), flooding the continental interiors and coastal zones. Total land area was about 20% less than today, significantly reducing continentality in the northern hemisphere. The distribution of land and sea has been cited as an explanation for Cretaceous warmth (e.g., Donn and Shaw, 1977) because of the different thermal characteristics of land and ocean. This idea was supported by early GCM experiments (e.g., Barron and Washington, 1984). However, these simulations used mean annual climate models with simplistic energy balance oceans that did not account for seasonal changes in insolation, limiting the seasonal cycle over the high latitude land masses. Subsequent modeling efforts (e.g., Barron *et al.*, 1993*a*), using GCMs with seasonally varying insolation and a 50 m slab ocean model that captures the seasonal thermal cycle of the ocean's mixed layer, showed that the role of geography is small compared with forcing from increased atmospheric CO_2. However, a reduction in northern hemisphere continentality during the Cretaceous and Eocene contributed to warmer land areas adjacent to epicontinental seas (Valdes *et al.*, 1996) and lakes (Sloan, 1994), emphasizing the importance of regional paleogeography to paleoclimate model results.

Today, the oceans play an important role in regulating the planetary energy budget by contributing approximately as much poleward heat transport as the atmosphere (Carissimo *et al.*, 1985; Peixoto and Oort, 1992; Trenberth and Solomon, 1994). Increased modeled ocean heat transport has often been cited as a means of maintaining low meridional thermal gradients during warm paleoclimate modes (Barron, 1983; Covey and Barron, 1988; Covey and Thompson, 1989; Rind and Chandler, 1991; Barron *et al.*, 1993*b*, 1995). However, a viable mechanism for increasing ocean heat transport beyond present-day values in a world with substantially reduced thermal gradients has not been adequately explained (Sloan *et al.*, 1995). In addition, meridional thermal gradients as low as those indicated by interpretations of the Late Cretaceous geologic record (e.g., Huber *et al.*, 1995; D'Hondt and Arthur, 1996; MacLeod *et al.*, Chapter 8, this volume) may have weakened the zonal component of atmospheric circulation (Schneider, 1984), limiting the transport of heat from the oceans to the continental interiors. Prior simulations of Cretaceous climate using elevated values of both atmospheric CO_2 and ocean heat transport (e.g., Barron *et al.*, 1993*b*, 1995), still produced subfreezing ($< -20\,°C$) temperatures in the winter hemisphere continental interiors, contradicting evidence of tropical and para-tropical vegetation in the fossil plant record of North America (Upchurch and Wolfe, 1987; Wolfe and Upchurch, 1987), frost-sensitive palms in the Asian interior (Vakhrameev, 1991), Mongolian crocodiles (Lefield, 1971), and non-marine temperature estimates determined from foliar analysis (e.g., Upchurch and Wolfe, 1987; Wolfe and Upchurch, 1987; Herman and Spicer, 1997). This discrepancy between climate models and geological data for greenhouse times has become a classic problem of paleoclimatology (see Sloan and Barron, 1990; Valdes *et al.*, 1996).

GENESIS V. 2.0 AND EVE

Climate models test the sensitivity of climate to external forcing, such as variations in insolation, and internal forcing from changes intrinsic to the earth's surface and atmosphere (see Valdes, Chapter 1, this volume). However, climate is the result of complex interactions between the atmosphere, hydrosphere, cryosphere, biosphere, and solid earth, with each component operating on a different timescale (see DeConto *et al.*, Chapter 2, this volume). The GENESIS (Global ENvironmental Ecological Simulation of Interactive Systems) global climate model (v. 2.0) (Thompson and Pollard, 1997) is an earth system model, designed to simulate climate by allowing interaction between individual models of climate system components (see Fig. 2.1, this volume).

GENESIS (see DeConto *et al.*, Chapter 2, this volume), uses an Atmospheric General Circulation Model (AGCM) as its core component, coupled to a non-dynamical slab ocean model and multilayer models of soil, snow, and sea ice. The slab ocean model captures the seasonal thermal capacity of the ocean's mixed layer and transports heat as a linear diffusion down the local ocean temperature gradient as a function of latitude and the zonal fraction of land vs. sea. A Land-

Surface-Transfer Scheme (LSX) with a horizontal resolution of $2° \times 2°$ calculates fluxes of heat, radiation, moisture, and momentum between the ground (soil and snow), vegetation, and atmosphere. In LSX, vegetation is represented by two canopy layers ('grass' and 'trees'), the structure and seasonal phenology of which depend on the type of vegetation specified in each $2° \times 2°$ grid cell. A new predictive Equilibrium Vegetation Ecology (EVE) model (Bergengren and Thompson, 1999) provides a description of vegetation for LSX and allows realistic feedbacks between the atmosphere and the vegetation.

EVE predicts the equilibrium state of plant community structure as a function of climate and fundamental ecological principles and has been incorporated as an interactive component of the Campanian climate simulation. The basic vegetation units in EVE are called 'lifeforms.' The lifeforms define the individual components of a plant community, like the individual and species levels of vegetation, but are based on their physiognomic and ecological characteristics at the biome level (Box, 1981). EVE is driven by seven ecoclimatic predictors derived from 12 monthly mean values of temperature, precipitation, and relative humidity provided by the GENESIS GCM. EVE predicts ecosystem structure by excluding potential lifeforms based on competition for light and succession after disturbance. Biome distributions (forest, shrubland, grassland, tundra, or desert) are defined by the fractional cover of the lifeforms present in each land surface grid cell.

CAMPANIAN (80 MA) BOUNDARY CONDITIONS

Boundary conditions for paleoclimate simulations using the GENESIS GCM include atmospheric chemistry, the solar constant, orbital parameters, and solid-earth boundary conditions describing the distribution of land and sea, elevation, orographic roughness, soil texture, and vegetation. Vegetation is dynamic and predicted, not prescribed and fixed as in prior simulations of Cretaceous climate (e.g., Barron *et al.*, 1993*a*, 1995; Otto-Bliesner and Upchurch, 1997). Two Campanian (80 Ma) simulations (CASE 1 and CASE 2) were performed using identical boundary conditions, except atmospheric CO_2 and ocean heat transport which were increased in CASE 2.

Campanian paleogeography

Because highly resolved, accurate paleogeographic reconstructions are critical to the simulation of realistic paleoclimates (Hay, 1996; Crowley, 1998), a new high-resolution Late Cretaceous paleogeography was constructed on a $2° \times 2°$ grid (DeConto, 1996; Fig. 9.1, see color plate). The Campanian paleogeography consists of a global plate tectonic reconstruction, with shoreline locations and elevations superposed on the tectonic model. The plate tectonic reconstruction (Hay *et al.*, 1999), includes 290 tectonic blocks and fragments restored to their Campanian positions. The position of paleoshorelines, representing a Late Cretaceous high-stand of sea level, were reconstructed from the *Atlas of Lithological–Paleogeographical Maps of the World* (Ronov *et al.*, 1989), modified to include newly recognized land areas around the Kerguelen Plateau. The flooding of conti-

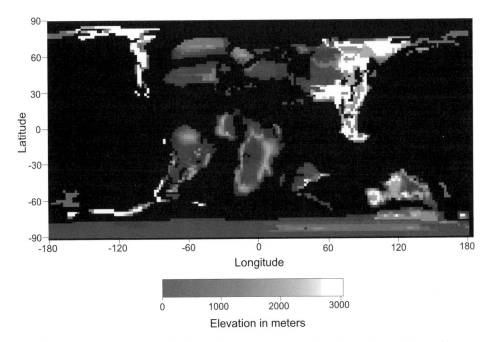

Figure 9.1. Campanian (80 Ma) paleogeography providing the solid-earth boundary conditions for the paleoclimate simulations. The paleogeography includes global reconstructions of Campanian shorelines and topography, superposed on a new global plate tectonic model of the Late Cretaceous (Hay *et al.*, 1999).

nental crust in the Campanian results in a global land area about 20% smaller than today, with a substantial reduction in northern hemisphere continentality.

Global topography is another important boundary condition for paleoclimate simulations. Prescribed topography can affect the zonal mean circulation, monsoonal circulation, storm track position, precipitation patterns, and snowline. Like the shoreline data, Campanian elevations were constructed largely from Ronov *et al.* (1989). Average elevations were applied to orogenic zones (1000–3000 m), intermontane basins (500–1000 m), and areas of continental erosion (200–500 m) and deposition (0–200 m). The data were superposed on the tectonic model and contoured with the paleoshorelines providing zero elevation. The resulting Campanian paleogeography is shown in Fig. 9.1.

Orographic roughness, the standard deviation of the small-scale orography in each grid cell ($3.75° \times 3.75°$), is used as a source term in the gravity-wave parameterization in the GENESIS AGCM. Orographically induced gravity waves play an important role in the large-scale flow in the troposphere and the lower stratosphere. Because small-scale topography in the Late Cretaceous is unknown, a single value of 157 m was prescribed globally. This value is the average orographic roughness of the present-day ice-free areas of the earth. Ice-covered regions (Greenland and Antarctica) are relatively smooth and are probably inappropriate for estimating the roughness of an ice-free Late Cretaceous earth.

In the soil model component of GENESIS, the thermal, hydraulic, and radiative properties of soils are determined from their empirical dependence on texture. A single soil texture value (the fraction of sand, silt, and clay) was prescribed for each of the six layers in the soil model component of GENESIS (Table 9.1) because, as with orographic roughness, adequate empirical constraints on Campanian soils are lacking. The values used are an average of the present-day global soil texture data set (Webb *et al.*, 1993) for land areas between 40° N and 40° S.

Solar constant and orbital parameters

The present-day solar constant, defined as the incident solar flux at the top of the atmosphere, is 1365 W m^{-2}. However, according to models of solar evolution, the sun's luminosity was smaller in the Cretaceous, having risen steadily since the beginning of the Main-sequence about 4.7×10^9 years ago, when solar luminosity was about 30% less than today (Crowley, Chapter 14, this volume). Campanian solar luminosity was calculated from a standard model of solar evolution (Gough, 1981), resulting in a value of 1355.7 W m^{-2} (0.632% less than present-day value). Eccentricity was prescribed as zero and obliquity as 23.5°; with an eccentricity of zero, seasonal precession has no climatological effect.

Atmospheric CO$_2$

Long-wave radiation emitted by the warm surface of the earth is partially absorbed and then re-emitted by water vapor (the biggest contributor), naturally occurring trace gases (mainly carbon dioxide, methane, nitrous oxide, and ozone) and aerosols in the troposphere and stratosphere. Of the trace gases, CO$_2$ is the largest contributor to the 'greenhouse effect' (see Valdes, Chapter 1, this volume). When prescribing atmospheric CO$_2$ for paleoclimate simulations, model biases due to the several-decade lag between the climate to which the GCM was tuned and the current CO$_2$ level should be taken into account. The GENESIS GCM was tuned to 1960s and 1970s observed climatology, but with atmospheric CO$_2$ set to its 1985 level (345 ppm). The 1960–70s climatology to which the GCM was tuned would most

Table 9.1. *Prescribed Campanian soil textures in the six layers of the soil model component of GENESIS*

Soil model layer (thickness)	Sand (%)	Silt (%)	Clay (%)
1 (0.05 m)	55	25	20
2 (0.10 m)	55	26	19
3 (0.20 m)	50	29	21
4 (0.40 m)	48	32	20
5 (1.00 m)	49	31	20
6 (2.50 m)	49	30	21

closely correspond to ~1940s CO_2, if the world was allowed to equilibrate to the several-decade thermal lag of the upper ocean. Thus the GENESIS GCM physics contain a cooling bias corresponding to the difference in radiative forcing between ~1985 and ~1945 CO_2 levels, 345 and 308 ppm respectively. To compensate for this often ignored bias, prescribed values of atmospheric CO_2 should be higher than the actual values by a factor of 1.12 (345/308). In the first of two simulations of Campanian climate (CASE 1), atmospheric CO_2 was set at 1380 ppm, representing an effective concentration of ~1230 ppm (about 3.6 times present-day value) when the model bias mentioned above is taken into account. In the second simulation (CASE 2), atmospheric CO_2 was increased to 1680 ppm, an effective concentration of 1500 ppm (about 4.4 times present-day value). These values are within the range of estimated values of 1.5 and 9 times present for the Late Cretaceous. Freeman and Hayes (1992) used the fractionation of carbon isotopes during photosynthesis and the concentration of dissolved CO_2 in seawater and estimated Late Cretaceous pCO_2 to be <1000 ppm, near the low end of estimated values. Estimates based on a long-term geochemical carbon cycle model (GEOCARB II; Berner, 1994) range between 1.5 and 5 times present. Cerling (1991) calculated a Barremian to mid-Albian (Early Cretaceous) atmospheric CO_2 of 1500–3000 ppm, or about 4.4–9 times present, from the carbon isotopic composition of soil carbonate in paleosols. Andrews et al. (1995) applied Cerling's method to Maastrichtian paleosols from India, resulting in an estimate of 1300 (±500) ppm.

Ocean heat transport

The slab ocean model component of GENESIS is a simple, non-dynamical ocean analog designed to capture the seasonal thermal capacity of the ocean mixed layer and to balance net surface–atmosphere heat exchange with changes in temperature. Poleward heat transport in the slab ocean model is parameterized as a linear diffusion down the local temperature gradient as a function of latitude and the zonal fraction of land and sea. Because there is no explicit circulation, the efficiency of poleward heat transport is adjusted by varying the diffusion coefficient (described herein as the multiplicative of the diffusion coefficient used in the modern control simulation). Because of the low meridional thermal gradients already in place at the start of these ice-free, Late Cretaceous climate simulations, the diffusion coefficient had to be multiplied by 2 (CASE 1) to maintain poleward ocean heat transport values close to those in the modern GENESIS v. 2.0 control simulation and 4 (CASE 2) to maintain converged ocean heat transport values close to present-day observed values (about 0.7×10^{15} W larger than the GENESIS v. 2.0 control simulation; Fig. 9.2).

Coupled climate–vegetation (GENESIS–EVE) simulations

Atmospheric inputs (energy and water) affect the structure and function of terrestrial ecosystems, and the characteristics of terrestrial ecosystems (canopy roughness, leaf area index, and seasonality of leaf display) influence global climate by altering fluxes of momentum, radiation, heat, and water vapor at or near the ground surface. Terrestrial ecosystems exhibit a wide range of structures and phy-

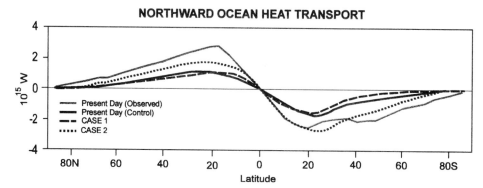

Figure 9.2. Converged poleward oceanic heat transport in the CASE 1 and CASE 2 simulations, calculated from mean annual surface fluxes in GENESIS (using a non-dynamical slab ocean model). In CASE 1, the multiplicative of the heat diffusion coefficient in the slab ocean heat transport scheme was set at 2. In CASE 2, the multiplicative was set at 4. Converged poleward oceanic heat transport in the GENESIS v. 2.0 control simulation (multiplicative = 1) and an estimate of present-day poleward oceanic heat flux (Trenberth and Solomon, 1994) are shown for comparison.

siologies; however, the global distribution of Campanian vegetation is difficult to reconstruct at the $2° \times 2°$ resolution of the land surface model in GENESIS. Modeling efforts using vegetation distributions reconstructed from the spatially fragmented fossil record may be forcing climate with inappropriate land surface boundary conditions in locations where the prescribed vegetation type cannot be supported by the simulated climate. To preclude such biases, and to assess the interactive nature of the coupled climate–vegetation system, a predictive vegetation model (EVE) was applied as an interactive component of the climate simulation (DeConto *et al.*, 1998, 1999a,b). EVE provides the seasonal phenology and physical attributes of the two canopy layers ('grass' and 'trees') in each $2° \times 2°$ grid cell and updates predicted vegetation once each year, according to the previous year's climatology provided by GENESIS. In this way, the climate simulated by GENESIS is affected by the vegetation predicted by EVE and vice versa, allowing realistic feedbacks between the atmosphere and the land surface.

EVE includes 110 physiognomically defined lifeforms. Examples include tropical rainforest broadleaf trees, temperate summergreen broadleaf trees, temperate rainforest evergreen needleleaf trees, and succulent forbs. However, not all the lifeforms found in today's terrestrial biosphere existed 80 million years ago, requiring each lifeform to be compared with the Late Cretaceous fossil record of vegetation. Only those lifeforms with known Late Cretaceous physiognomic analogs were included in the Campanian simulations shown here. The most notable exclusions were the graminoids, narrowleaf herbs with well-developed root stocks including grasses (Gramineae) and sedges (Cyperaceae). Gramineae may have existed in the Late Cretaceous (Crepet and Feldman, 1991). However, grasslands as we know them today did not exist until the mid-Tertiary. In sensitivity tests using a predictive vegetation ecology model forced by a modern climatology and no graminoids (grasses removed), forbs (broadleaf herbaceous angiosperms), ferns, and shrubs provide the dominant fractional cover in the climate niche occupied by grassland today (DeConto, 1996). The term 'forb–fern prairie' is applied to this Campanian biome.

CAMPANIAN CLIMATE–VEGETATION SIMULATIONS

In an attempt to achieve the Late Cretaceous 'target' temperatures, meridional thermal gradients, and equable continental interiors reconstructed from the geologic record (e.g., Barron, 1983; Wolfe and Upchurch, 1987; Horrell, 1991; Spicer and Corfield, 1992; Herman and Spicer, 1997), two Campanian climate–vegetation simulations were performed that differed only in prescribed atmospheric CO_2 and ocean heat transport. These simulations are not sensitivity tests, but are attempts to simulate a realistic Campanian climate that agrees with both marine and non-marine proxies from the geologic record.

CASE 1

In the first of two Campanian climate–vegetation simulations (CASE 1), GENESIS and EVE were run in coupled mode, with the Campanian solid-earth boundary conditions described above, an atmospheric CO_2 concentration of 1380 ppm (an effective concentration of 1230 ppm) and the multiplicative of the ocean heat transport diffusion coefficient set at 2 (Fig. 9.2).

Although vegetation was predicted and not fixed, a vegetation distribution was prescribed to reduce 'spin-up' time and to preclude biases caused by bare soil (bare soil becomes either desiccated or saturated as a function of regional evaporation minus precipitation, the gradient of water vapor between the soil surface and the overlying air, and net radiative heating) before vegetation can become established. The Campanian vegetation was initialized with a global forest, with the fractional cover of forest lifeforms prescribed according to a zonally symmetric meridional temperature gradient derived from a compilation of Senonian proxy climate data (DeConto, 1996).

CASE 1 results: a tundra-cooling feedback?

After the first 5 model years of climate–vegetation interaction, shrubland and forb–fern prairie became firmly established in South America, Africa, North America, and Asia. Tundra began to invade high latitude North America and the Antarctic Peninsula. After 10 model years, deserts began to form in the low-lying continental interiors of South America and Africa, centered at a latitude of 20° S. In Antarctica, tundra spread into coastal regions and the high elevations of the continental interior, mainly at the expense of needleleaf evergreen forest. Needleleaf forest is tall, with a high leaf area index and a small seasonal cycle that masks the high albedo of snow cover in winter. By contrast, tundra has a low canopy height and is easily covered by deep winter snow accumulations caused by the high water vapor content of the relatively warm high latitude atmosphere. The spread of tundra into high latitude continental interiors enhanced regional cooling from an increase in surface albedo, less absorbed solar flux, and a reduction in atmospheric moisture from decreased transpiration. Surface air temperatures over areas first occupied by evergreen-dominated forest and then replaced by tundra dropped by 7 °C in the annual mean and by as much as 15 °C in the early spring, prior to snow melt. Tundra-enhanced continental cooling contributed to decreased high latitude sea surface temperatures, manifested throughout the year by the seasonal thermal lag of the ocean's mixed layer. This effect is consistent with prior non-interactive modeling studies of the effects of deforestation on energy balance and climate (Bonan *et al.*, 1992, 1995; Thomas and Rowntree, 1992; Dutton and Barron, 1996; Gallimore and Kutzbach, 1996; Otto-Bliesner and Upchurch, 1997) and with modern observations of radiative temperature over forest and tundra (Rouse, 1984). The interactive modeling scheme showed that the spread of tundra was aided by the increasingly cool conditions caused by the tundra itself, resulting in a positive tundra-cooling feedback leading toward the ultimate collapse of the remaining forest. The CASE 1 simulation was integrated for 25 model years, but a climate–vegetation equilibrium was never reached as the spread of tundra and the runaway tundra-cooling feedback continued. The application of interactive modeling schemes can obscure interpretations as to the actual forcing mechanisms responsible for a specific result, such as the spread of tundra demonstrated here. However, the dramatic cooling over locations that were covered by forest in one model year and tundra the next, and the failure of the simulation to reach equilibrium after 25 model years of integration, suggest that the continued cooling was indeed enhanced by the spread of tundra, which was spreading, in large part from the cooling.

Most of the boreal and Antarctic tundra predicted by GENESIS–EVE (Fig. 9.3a, see color plate) is rich in both mesic and peat-forming bryophyte lifeforms. Saturated soil columns and standing water are predicted by the soil model component of GENESIS over the same areas, indicating that much of the tundra predicted over high latitudes would be analogous to peat-forming wetland. According to modern estimates of the carbon storage potential of peatland (112 kg C m^{-2} for peatland vs. 15 kg C m^{-2} for forest; Gorham, 1991), the spread of tundra at the expense of forest can sequester significant masses of carbon (Klinger *et al.*, 1996). The 11 250 000 km^2 of tundra predicted by CASE 1 could account for the fixation of ~1100 Gt C. If high latitude forest–tundra transitions like that simulated by CASE 1 occurred during the Cretaceous or other mostly ice-free paleoclimates, a significant transient reduction in atmospheric CO_2 (buffered on millennial timescales by exchange with the deep ocean) could have provided additional cooling, additive to the albedo and surface radiation balance feedbacks described above (DeConto *et al.*, 1999*b*).

CASE 2

The values of atmospheric CO_2 and ocean heat transport prescribed in CASE 1 could not maintain high latitude coastal and continental climates warm enough to maintain a vegetation distribution compatible with Campanian fossil assemblages in the high latitudes of either hemisphere (e.g., Parrish and Spicer, 1988; Askin, 1992; Herman and Spicer, 1997). A second simulation of Campanian climate (CASE 2) was performed using the same boundary conditions described above, except atmospheric CO_2 was increased to 1680 ppm (an effective concentration of 1500 ppm), and the multiplicative of the diffusion coefficient in the heat transport term in the slab ocean model was increased from 2 to 4. The CASE 2 vegetation was initialized with the tundra-dominated vegetation distribution predicted by CASE 1 at model year 15 (Fig. 9.3a).

CASE 2 results

After 5 model years forest began replacing tundra over the high latitude continental interiors. By model year 10, the desert biome had reached equilibrium in South America and Africa, and almost all the tundra in continental Antarctica, North America, and Asia had been replaced by forest. The Antarctic Peninsula, which had been dominated by tundra, polar desert, and shrubland in CASE 1 (Fig. 9.3a), was now completely forested by evergreen linear-leaf trees (*Auracaria* and *Podocarpus*) rainforest (Fig. 9.3b, see color plate), exhibiting better compatibility with fossil vegetation assemblages (e.g., Askin, 1992). Tundra was eliminated from high northern latitudes, replaced by boreal forest and a fringe of forb–fern prairie along the northern continental margins. Climate statistics showed that the simulation had reached an equilibrium state at model year 12. The simulation was run until model year 20 to provide 8 years of equilibrium climate data to be averaged for evaluation and to drive an OGCM in a study of Campanian ocean circulation discussed below.

Figure 9.4 (see color plate) shows the mean annual surface air temperature (MAT) simulated by GENESIS–EVE (CASE 2) and the zonally averaged MAT, relative to the present day. The simulated global MAT is 24.1 °C (about 9 °C warmer than today). Simulated tropical ocean surface temperatures are 30–33 °C, warmer than today and than predicted by oxygen isotopic studies (Crowley and Zachos, Chapter 3, this volume; MacLeod *et al.*, Chapter 8, this volume), but in agreement with the concept of 'Supertethys' (Kauffman and Johnson, 1988). Campanian high latitudes exhibit above-freezing MATs extending to both poles. North polar ocean surface MATs are 9 °C and no sea ice is predicted. The predicted south-polar MAT is 2 °C, much different from the −50 °C observed today. The warm polar latitudes result in an overall surface meridional thermal gradient less than half of that today, but greater than meridional gradients based on isotopic values (e.g., Huber *et al.*, 1995; MacLeod *et al.*, Chapter 8, this volume). The area of subzero mean annual temperatures is reduced from prior simulations of Cretaceous climate (e.g., Barron *et al.*, 1993a,b, 1995; Valdes *et al.*, 1996). Coldest monthly mean temperatures (−12 °C) occur in small regions of the higher elevations in northeast Asia and East Antarctica. In general, non-marine surface temperatures are in very close agreement with MATs determined from foliar analysis (e.g., Upchurch and Wolfe, 1987; Parrish and Spicer, 1988; Herman and Spicer, 1997). Summer and winter temperatures (Fig. 9.5, see color plate) exhibit a small seasonal cycle, in better agreement with the concept of Cretaceous equability. Asian interior temperatures have been interpreted as very warm based on the occurrence of Mongolian crocodiles (Lefield, 1971). The most cold-resistant modern crocodiles coincide with a cold monthly mean temperature of 4.4 °C (Hutchison, 1982), within the range of temperatures (-2 to 5 °C) predicted over Mongolia.

The CASE 2 climate can be described as wet (as well as warm) with a global mean annual precipitation of 3.9 mm day^{-1} (about 25% wetter than today). Precipitation was seasonally steady, resulting in large areas of saturated soil. The simulated precipitable water content of the Campanian atmosphere was 42 mm, twice that of the present-day atmosphere, as a result of the larger moisture-holding capacity of the warmer Campanian atmosphere (the saturation vapor pressure of air increases exponentially with increasing temperature). As expected, high latitude precipitation was much greater during the Campanian than today, with average values of between 1.0 and 3.0 mm day^{-1}. Despite substantial predicted snowfall at high latitudes, no snow survived through the summer months, implying that no ice sheets would have formed.

As a result of increased atmospheric water vapor and an enhanced hydrological cycle in the CASE 2 Campanian simulation, poleward atmospheric heat transport is dominated by latent heat flux. Simulated poleward latent heat transports at 40° latitude (in both hemispheres) are 0.5×10^{15} W larger in the CASE 2 simulation than in the present-day control, despite an overall reduction in total atmospheric heat transport of 0.7×10^{15} W (provided by a reduction in dry static poleward heat flux). Total atmosphere and ocean meridional heat transports are about the same in both cases.

The reconstructed Campanian topography exerts a strong influence on the distribution of precipitation (Fig. 9.6, see color plate). The dry deserts of Africa, South America, and Australia occur in low-lying intracontinental basins surrounded by higher elevations. This relief either deflects zonal winds or induces high precipitation where the zonal flow brings oceanic moisture to the windward side of mountain ranges and highlands. Downwind of the highlands, rain shadows develop, because most moisture is lost on the upwind side. Monsoonal precipitation patterns are limited to tropical Southeast Asia, where summer heating of continental Asia causes low-level convergence that allows moisture from the eastern Tethys to penetrate the Southeast Asian peninsula.

Despite a simulated surface meridional thermal gradient 60% smaller than today (Fig. 9.4b), winds in the lowest (surface) level of the AGCM are only 16% weaker in the Campanian simulation than in simulations of the present day. Large vertically integrated meridional thermal gradients are maintained, because the upper tropospheric meridional thermal gradient rises in response to increased CO_2 (Barron and Washington, 1982; Manabe and Bryan, 1985).

The role of vegetation in Campanian climate

The simulated CASE 2 vegetation (Fig. 9.3b) predicts cool temperate boreal evergreen needleleaf trees and summergreen broadleaf trees to be the dominant upper canopy lifeforms in the high latitude continental interiors. In the model, high latitude maritime regions are dominated by temperate summergreen broadleaf trees, in agreement with Herman and Spicer (1997) who suggest deciduousness as the most likely wintering strategy for Late Cretaceous Alaskan floras. With Late

Cretaceous obliquity assumed to be close to the present day (23.5°), there would have been negligible solar insolation during the winter months at high latitudes. Those high latitude and polar taxa that were not deciduous must have been able to enter dormancy (Spicer and Parrish, 1987; Spicer, 1990). Extant boreal needleleaf trees (mostly short-needled conifers) have the ability to withstand long periods of cold and dark, with almost no respiration. Boreal type forests tend to have dense canopies and high leaf area indices. In the model, evergreens dominate the fractional cover (up to 80% in some locations of interior Antarctica and Asia) of the high latitude continental interiors. Although some elements of high-polar latitude forests could have been evergreen (Herman and Spicer, 1997), there is little evidence to support or refute the predicted vegetation distributions, especially over continental Antarctica.

The maintenance of forest over most of Antarctica, Australia, and high northern latitudes simulated by CASE 2 maintains a relatively low albedo during the late winter and early spring, despite a large fractional snow cover. The influence of the distribution of terrestrial ecosystems can be seen in a comparison between September and March surface albedos (Fig. 9.7, see color plate) of the tundra-dominated CASE 1 simulation (Fig. 9.3a) and the forest-dominated CASE 2 simulation (Fig. 9.3b). Low continental albedos (<0.15) are maintained over the CASE 2 evergreen forest-dominated regions, even at latitudes covered by winter and spring snow (i.e., most of Antarctica and north-central Asia). Over the same regions, the tundra-dominated CASE 1 simulation has much higher spring albedos, approaching 0.8, resulting in a decrease in solar energy absorbed at the surface and decreased net radiation.

Thus the distribution of predicted biomes played an important role in the maintenance of low meridional thermal gradients and warm continental interiors, as suggested by prior, non-interactive sensitivity tests designed to examine the role of vegetation distribution in climate change (e.g., Bonan *et al.*, 1992; Dutton and Barron, 1996; Otto-Bliesner and Upchurch, 1997). High latitude needleleaf forests in continental Antarctica, North America, and Asia masked the high albedo of snow cover in the winter and spring, prior to snow melt (Fig. 9.7). Increased net radiation over forested land areas heated the overlying atmosphere, especially during late winter and spring. Spring temperatures in the forest-dominated high latitudes (CASE 2) were ~15 °C warmer than the same latitudes in the tundra-dominated simulation (CASE 1), and ~7 °C warmer in the annual mean. Increased high latitude warmth, especially in winter and spring, moderated seasonal cooling of the high latitude oceans. This, in turn, reduced summer continental cooling from the influence of the seasonal thermal lag of the oceans and limited sea-ice formation. The radiative warming effect of the high latitude forests caused by reduced albedo and increased transpiration increased surface to atmosphere latent heat flux by ~25 W m^{-2}. The adiabatic transfer of heat from near the surface to higher tropospheric levels reduced surface pressure, increasing net convergence over the continents and the advection of warm moist air from the oceans into the continental interiors. Warm high latitude and polar summer temperatures maintained healthy forest and limited tundra development, helping to maintain polar warmth in a self-perpetuating system. Increased atmospheric moisture, mainly from an increase in transpiration over forest, enhanced the latent heat transport potential by the atmosphere and added to the water vapor feedback component of the 'greenhouse' effect.

CAMPANIAN OCEAN CIRCULATION

The poleward ocean heat transports required to achieve the CASE 2 climate and vegetation distribution, although similar to present-day observed values, are larger (by $\sim 0.7 \times 10^{15}\,\text{W}$) than those required by present-day simulations (Fig. 9.2), suggesting that significant ocean heat transport is required for an accurate Late Cretaceous climate reconstruction. Proxy temperature data from oxygen isotope analysis suggest the Late Cretaceous deep oceans were as warm as $15\,°C$ (e.g., Savin, 1977; Saltzman and Barron, 1982). A low latitude source of deep water formation has been proposed for the Cretaceous, in which Warm Salty Deep Water (WSDW) is produced in marginal seas with restricted circulation, in regions of high evaporation minus precipitation (Chamberlin, 1906; Brass *et al.*, 1982; Saltzman and Barron, 1982; MacLeod and Huber, 1996; MacLeod *et al.*, Chapter 8, this volume). The formation of WSDW at low latitudes and its eventual upwelling at high latitudes has been proposed as a possible mechanism for enhanced poleward ocean heat transport during the Cretaceous (e.g., Hay, 1995; Huber *et al.*, 1995). Some paleontological and geochemical evidence supports this mode of circulation for the Cretaceous (Saltzman and Barron, 1982; MacLeod and Huber, 1996; MacLeod *et al.*, Chapter 8, this volume). To examine the role of the oceans in maintaining a warm Late Cretaceous climate with low equator-to-pole thermal gradients, Brady *et al.* (1998) applied an Ocean General Circulation Model (OGCM) to the surface forcing provided by the CASE 2 GENESIS–EVE simulation described above. Limitations of this asynchronous coupling approach are explained in DeConto *et al.* (Chapter 2, this volume).

The OGCM used by Brady *et al.* (1998) is a modified version of Semtner and Chervin's (1992) rigid lid, Boussinesq, hydrostatic primitive equation ocean model, with modifications used to accelerate the deep ocean into equilibrium (Bryan, 1984). A deep ocean acceleration factor of 10 results in ten times more deep ocean years than surface years for the same amount of integration. The model has an equator-to-pole resolution of $2° \times 2°$ and 20 vertical layers, providing adequate resolution to simulate the major current systems. In present-day control simulations, model transports in the wind-driven circulation are comparable to observations.

Surface forcing used to drive the ocean model includes wind stress, surface heat flux (calculated from traditional bulk formula), and evaporation minus precipitation (corrected globally so that the mean volume integral of salinity does not drift). In this case, the surface forcing was provided by monthly mean quantities averaged over 8 years of the CASE 2 GENESIS–EVE simulation.

Initial conditions for OGCMs include specifications of the initial three-dimensional velocity field (u, v, w), temperature (T), and salinity (S). Solid-earth boundary conditions include the position of continents and bathymetry. In this case, the velocity field was initialized as zero. Initial sea surface temperatures

(SSTs) were obtained from zonally averaged SSTs provided by the CASE 2 GENESIS simulation. Temperatures were decreased linearly from the surface to 2000 m, where the deep ocean was prescribed as 7 °C. Because OGCM simulations have been shown to be sensitive to prescribed bathymetry (Bice *et al.*, 1998), Campanian bathymetry was reconstructed with a $2° \times 2°$ resolution, based on the plate tectonic and shoreline reconstructions shown in Fig. 9.1, published magnetic lineation data (Cande *et al.*, 1989), an age–depth relationship for ocean crust (Stein and Stein, 1992), and a correction applied to account for the accumulation of sediment (Crough, 1983).

As expected, the OGCM simulation of the Campanian predicts an equator-to-pole SST thermal gradient much reduced from the present day (Brady *et al.*, 1998). Mean annual equatorial SSTs are about 32 °C, several degrees warmer than today, and polar SSTs are about 8 °C (Fig. 9.8a, see color plate). The vigor of the Campanian surface circulation is only slightly reduced from simulations of the modern ocean, with well-organized subtropical gyres and equatorial current systems in the Pacific (Fig. 9.8b). An Antarctic Circumpolar Current is precluded by the lack of a deep passage between Antarctica and South America. Flow through the Tethys is east–west along the entire south margin, with a weaker west–east return flow along the northern Tethyan margin. The enhanced Cretaceous hydrological cycle (increased poleward freshwater flux and latent heat transport) drives large open ocean surface salinity differences (Fig. 9.8b, see color plate).

Deep water formation

In the Campanian OGCM simulation (Brady *et al.*, 1998), most deep water is formed in high southern latitudes, not low latitudes as has been suggested (e.g., Brass *et al.*, 1982). Deep water is formed via the cooling of a saline water mass that originates in the low latitude South Atlantic. In the Angola Basin region, high net evaporation minus precipitation values of 6 mm day^{-1} drive open ocean salinities to 40‰. This warm (26 °C) saline water is advected southward in the western boundary current of the South Atlantic, and moves into the South Indian Ocean where it sinks after cooling in a process similar to deep water formation in the present-day North Atlantic. Although some convection occurs in the eastern Tethys and in the eastern North Pacific (also areas of high net evaporation), the convection is too shallow (~750 m) to significantly contribute to deep water formation. The deep water formed in high southern latitudes is 10–12 °C (Fig. 9.9a, see color plate), much warmer than today's deep water, and has salinities of ~35‰ (Fig. 9.9b, see color plate). The net transport associated with this convection produces a strong meridional overturning cell in the southern hemisphere (~38.5 Sv; Fig. 9.9c), about twice as large as present-day meridional overturning associated with deep water formation in the North Atlantic, but similar in strength to overturning in the North Atlantic ocean in the present-day control simulation using the same OGCM (Brady *et al.*, 1998). Deep water formation in high northern latitudes is limited by very low surface salinities (Figs. 9.8b and 9.9b).

Ocean heat transport

The simulated deep water circulation pattern described in Brady *et al*. (1998) results in ocean heat transport about the same as OGCM simulations of the present day (Fig. 9.10). Prior coupled atmosphere–ocean model CO_2 sensitivity tests also showed little change in total ocean heat transport between $1\times$ and $8\times$ present-day CO_2 (Manabe and Bryan, 1985). The main difference between the Campanian OGCM simulation and present-day OGCM experiments is in the partitioning of heat transport provided by the wind-driven and thermohaline circulation. In the Campanian simulation, ocean heat transport is dominated by strong meridional overturning, with a much smaller contribution from the wind-driven, gyre circulation (Fig. 9.10). Meridional sea surface density gradients are increased, despite reduced sea surface meridional thermal gradients, because the thermal expansion coefficient of seawater increases with temperature (Manabe and Bryan, 1985).

Brady *et al*.'s (1998) OGCM simulation shows that significant poleward ocean heat transport can be maintained despite reduced surface meridional thermal gradients, via enhanced thermohaline circulation characterized by deep water formation at high latitudes. This result is supported by prior sensitivity tests using simplified ocean modeling schemes (Schmidt and Mysak, 1996), suggesting that more poleward ocean heat flux can be maintained with deep water formation at high vs. low latitudes. In this case, the maintenance of relatively warm deep ocean temperatures (despite high latitude source areas) can be attributed to the warm high latitude surface air temperatures simulated by the original GENESIS CASE 2 climate simulation.

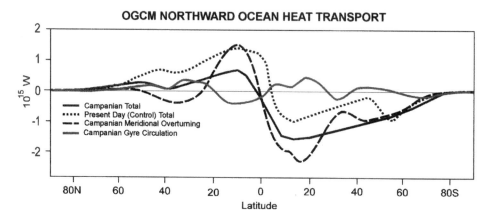

Figure 9.10. Zonally integrated northward ocean heat transport simulated by the OGCM. Total northward ocean heat transport (Campanian) is broken down into contributions from meridional overturning and the gyre circulation (horizontal plane). Heat transport from parameterized eddy diffusion (not shown) provides the residual. Total ocean heat transport, calculated from a control simulation of the present-day ocean circulation using the same OGCM (dotted line), is shown for comparison (after Brady *et al*., 1998).

SUMMARY

Late Cretaceous warmth, low meridional thermal gradients, and warm continental interiors, long recognized by geologists, are now being better simulated by numerical models. In the Late Cretaceous, the combination of increased atmospheric CO_2, high latitude forests, and ocean heat transport equal to or greater than today provided by vigorous meridional overturning may have contributed to low meridional thermal gradients and warm continental interiors.

On the continents, Antarctic and northern high latitude forests reduced surface albedo, especially in the late winter and early spring prior to snow melt, increasing net radiation and latent heat flux. The vegetation-induced warming moderated winter cooling of high latitude sea surface temperatures and increased atmospheric moisture at high latitudes, adding to the water vapor feedback component of the greenhouse effect, increasing the latent heat transport potential of the atmosphere, and aiding the advection of heat into the continental interiors (DeConto *et al.*, 1998, 1999a,b). As a result of the interactive climate–vegetation modeling approach applied to the Late Cretaceous, important feedbacks in the coupled climate–vegetation system have been recognized, especially over ice-free high latitudes. High-polar latitude terrestrial ecosystems have been shown to be sensitive to changes in atmospheric CO_2 and poleward oceanic heat transport. Shifts in these ecosystems (e.g., a transition from forest to tundra) could be an important feedback mechanism amplifying high latitude responses to climatic forcing. An OGCM simulation of the Campanian (Brady *et al.*, 1998) produces warm ($\sim$10–12 °C), saline ($\sim$35‰) deep water, via convection at high southern latitudes, contrary to the long-standing hypothesis of Warm Salty Deep Water formation in low latitude, marginal seas. This mode of ocean circulation is sufficient to produce significant poleward ocean heat transport, similar to present-day values. However, fully coupled atmosphere–ocean–vegetation modeling schemes (see DeConto *et al.*, Chapter 2, this volume) will be required to determine if the mode of ocean circulation demonstrated by Brady *et al.* (1998) is sufficient to maintain an accurate Cretaceous climatology and vegetation distribution, like that demonstrated by CASE 2.

Recent advances in paleoclimate modeling, including interactive vegetation schemes, better treatment of land surface–atmosphere processes and feedbacks, and more detailed reconstructions of paleogeography, are helping to narrow the gap between model simulations and reconstructions of pre-Quaternary climates, especially over mid–high latitudes. However, model simulations using elevated values of atmospheric CO_2 still result in tropical sea surface temperatures about 3–5 °C warmer than today, despite ocean heat transport increased beyond present-day values. If interpretations of isotopic data suggesting that tropical sea surface temperatures were cooler in the Cretaceous than today (Sellwood *et al.*, 1994; D'Hondt and Arthur, 1996; Crowley and Zachos, Chapter 3, this volume) are correct, then the models may be too sensitive to increases in atmospheric CO_2 over tropical latitudes and therefore do not explicitly account for some feedback(s) that moderate tropical surface temperatures (Manabe and Bryan, 1985; Brady *et al.*, 1998). Future advances in climate modeling (DeConto *et al.*, Chapter 2, this volume), including the inclusion of more explicit schemes to account for cloud feedbacks, will either reduce model–

data inconsistencies at tropical latitudes (Crowley and Zachos, Chapter 3, this volume) or lend support to the possibility that isotopic estimates of tropical sea surface temperatures during warm climate periods are too cold.

REFERENCES

Andrews, J. E., Tandon, S. K. and Dennis, P. E. (1995). Concentration of carbon dioxide in the Late Cretaceous atmosphere. *Journal of the Geological Society of London*, **152**, 1–3.

Askin, R. (1992). Late Cretaceous–Early Tertiary vegetation and climate. In *The Antarctic Paleoenvironment: A Perspective on Global Change*, eds. J. P. Kennett and D. A. Warnke, pp. 61–75, Washington, DC: American Geophysical Union.

Barrera, E., Huber, B. T., Savin, S. M. and Webb, P. N. (1987). Antarctic marine temperatures: Late Campanian through early Paleocene. *Paleoceanography*, **2**, 21–47.

Barron, E. J. (1983). A warm equable Cretaceous: The nature of the problem. *Earth Science Reviews*, **19**, 305–38.

Barron, E. J., Fawcett, P. J., Peterson, W. H., Pollard, D. and Thompson, S. L. (1995). A "simulation" of mid-Cretaceous climate. *Paleoceanography*, **10**, 953–62.

Barron, E. J., Fawcett, P. J., Pollard, D. and Thompson, S. L. (1993*a*). Model simulations of Cretaceous climates: the role of geography and carbon dioxide. *Philosophical Transactions of the Royal Society of London*, **341**, 307–16.

Barron, E. J., Peterson, W. H., Pollard, D. and Thompson, S. L. (1993*b*). Past climate and the role of ocean heat transport: Model simulations for the Cretaceous. *Paleoceanography*, **8**, 785–98.

Barron, E. J. and Washington, W. M. (1982). Atmospheric circulation during warm geologic periods: is the equator-to-pole surface-temperature gradient the controlling factor? *Geology*, **10**, 633–6.

Barron, E. J. and Washington, W. M. (1984). The role of geographic variables in explaining paleoclimates: results from Cretaceous climate model sensitivity studies. *Journal of Geophysical Research*, **89**, 1267–79.

Barron, E. J. and Washington, W. M. (1985). Warm Cretaceous climates: high atmospheric CO_2 as a plausible mechanism. In *The Carbon Cycle and Atmospheric CO_2: Natural Variations Archaen to Present*, eds. E. T. Sundquist and W. S. Broecker, pp. 546–53. Washington, DC: American Geophysical Union.

Bergengren, J. C. and Thompson, S. L. (1999). Modeling the effects of global climate change on natural vegetation, Part I. The equilibrium vegetation ecology model. *Climatic Change* (submitted).

Berner, R. A. (1994). GEOCARB II: A revised model of atmospheric CO_2 over Phanerozoic time. *American Journal of Science*, **294**, 56–91.

Bice, K. L., Barron, E. J. and Peterson, W. H. (1998). Reconstruction of realistic early Eocene paleobathymetry and ocean GCM sensitivity to specified basin configuration. In *Tectonic Boundary Conditions for Climate Reconstructions*, eds. T. J. Crowley and K. Burke, pp. 227–47. Oxford Monographs on Geology and Geophysics. New York: Oxford University Press.

Boersma, A. and Shackleton, N. J. (1981). Oxygen and carbon isotope variations and planktonic foraminifera depth habitats, Late Cretaceous to Paleocene. In *Initial Reports of the Deep Sea Drilling Project*, vol. 62, eds. J. Thiede, T. Vallier *et al.*, pp. 513–26. Washington, DC: US Government Printing Office.

Bonan, G. B., Chapin, F. S., III and Thompson, S. L. (1995). Boreal forest and tundra ecosystems as components of the climate system. *Climatic Change*, **29**, 145–67.

Bonan, G. B., Pollard, D. and Thompson, S. L. (1992). Effects of boreal forest vegetation on global climate. *Nature*, **359**, 716–8.

Box, E. O. (1981). *Macroclimate and Plant Forms: An Introduction to Predictive Modeling in Phytogeography*. The Hague: Dr W. Junk Publishers.

Brady, E. C., DeConto, R. M. and Thompson, S. L. (1998). Deep water formation and Poleward ocean heat transport in the warm climate extreme of the Cretaceous (80 Ma). *Geophysical Research Letters*, **25**, 4205–8.

Brass, G. W., Southam, J. R. and Peterson, W. W. (1982). Warm saline bottom water in the ancient ocean. *Nature*, **296**, 620–3.

Bryan, K. (1984). Accelerating the convergence to equilibrium of ocean climate models. *Journal of Physical Oceanography*, **16**, 927–33.

Budyko, M. and Ronov, A. (1979). Chemical evolution of the atmosphere in the Phanerozoic. *Geochemistry International*, **16**, 1–9.

Cande, S. C., LaBrecque, J. L., Larson, R. L., Pitman, W. C., III, Golovchenko, X. and Haxby, W. F. (1989). *Magnetic Lineations of the World's Ocean Basins*. Tulsa, OK: American Association of Petroleum Geologists.

Carissimo, B. C., Oort, A. H. and Vonder Harr, T. H. (1985). Estimating the meridional energy transports in the atmosphere and oceans. *Journal of Physical Oceanography*, **15**, 82–91.

Cerling, T. E. (1991). Carbon dioxide in the atmosphere: Evidence from Cenozoic and Mesozoic paleosols. *American Journal of Science*, **291**, 377–400.

Chamberlin, T. J. (1906). On a possible reversal of deep sea circulation and its influence on geologic climates. *Journal of Geology*, **14**, 363–73.

Colbert, E. H. (1973). Continental drift and the distribution of fossil reptiles. In *Implications of Continental Drift to the Earth Sciences*, eds. D. H. Tarling and S. K. Runcorn, pp. 395–412. New York: Academic Press.

Covey, C. and Barron. E. J. (1988). The role of ocean heat transport in climatic change. *Earth Science Reviews*, **24**, 429–45.

Covey, C. and Thompson, S. L. (1989). Testing the effects of ocean heat transport on climate. *Palaeogeography, Palaeoclimatology, Palaeoecology*, **75**, 331–41.

Crame, J. A. (1992). Review: Late Cretaceous paleoenvironments and biotas, an Antarctic perspective. *Antarctic Science*, **4**, 371–82.

Crepet, W. L. and Feldman, G. D. (1991). The earliest remains of grasses in the fossil record. *American Journal of Botany*, **78**, 1010–13.

Crough, S. T. ·(1983). The correction for sediment loading on the seafloor. *Journal of Geophysical Research*, **88**, 6449–54.

Crowley, T. J. (1998). Significance of tectonic boundary conditions for paleoclimate simulations. In *Tectonic Boundary Conditions for Climate Reconstructions*, eds. T. J. Crowley and K. Burke, pp. 3–20. Oxford Monographs on Geology and Geophysics. New York: Oxford University Press.

DeConto, R. M. (1996). *Late Cretaceous Climate, Vegetation, and Ocean Interactions: An Earth System Approach to Modeling an Extreme Climate*. PhD thesis, University of Colorado, Boulder, CO.

DeConto, R. M., Bergengren, J. C. and Hay, W. W. (1998). Modeling Late Cretaceous climate and vegetation. *Zentralblatt für Geologie und Palaeontologie, Teil I, 1996*, **11/12**, 1433–44.

DeConto, R. M., Hay, W. W., Thompson, S. L. and Bergengren, J. C. (1999a). Late Cretaceous climate and vegetation interactions: Cold continental paradox. In *The Evolution of Cretaceous Ocean/Climate Systems*, eds. E. Barrera and C. Johnson, pp. 391–432. Boulder, CO: Geological Society of America Special Paper 332.

DeConto, R. M., Pollard, D., Bergengren, J. C. and Hay, W. W. (1999*b*). The role of terrestrial ecosystems in maintaining "greenhouse" paleoclimates. *Journal of Climate*, (submitted).

D'Hondt, S. and Arthur, M. A. (1996). Late Cretaceous oceans and the cool tropical paradox. *Science*, **271**, 1838–41.

Donn, W. L. and Shaw, D. M. (1977). Model of climate evolution based on continental drift and polar wandering. *Geological Society of America Bulletin*, **88**, 390–6.

Douglas, R. G. and Savin, S. M. (1975). Oxygen and carbon isotope analysis of Tertiary and Cretaceous microfossils from Shatsky Rise and other sites in the North Pacific Ocean. In *Initial Reports of the Deep Sea Drilling Project*, vol. 32, eds. R. Larson, R. Moberly *et al.*, pp. 509–20. Washington, DC: US Government Printing Office.

Dutton, J. F. and Barron, E. J. (1996). GENESIS sensitivity to changes in past vegetation. *Paleoclimates*, **1**, 325–54.

Fischer, A. G. (1982). Long-term climatic oscillations recorded in stratigraphy. In *Climate in Earth History*, eds. W. H. Berger and J. C. Crowell, pp. 97–104. Washington, DC: National Academy Press.

Frakes, L. A. (1979). *Climates Throughout Geologic Time*. New York: Elsevier.

Frakes, L. A. and Francis, J. E. (1988). A guide to cold polar climates from high-latitude ice-rafting in the Cretaceous. *Nature*, **339**, 547–9.

Frakes, L. A., Francis, J. E. and Syktus, J. I. (1992). *Climate Modes of the Phanerozoic: the History of the Earth's Climate Over the Last 600 Million Years*. Cambridge: Cambridge University Press.

Freeman, K. H. and Hayes, J. M. (1992). Fractionation of carbon isotopes by phytoplankton and estimates of ancient CO_2 levels. *Global Biogeochemical Cycles*, **6**, 185–98.

Gallimore, R. G. and Kutzbach, J. E. (1996). Role of orbitally induced changes in tundra area in the onset of glaciation. *Nature*, **381**, 503–5.

Gorham, E. (1991). Northern peatlands: Role in the carbon cycle and probable responses to climatic warming. *Ecological Applications*, **1**, 182–95.

Gough, D. O. (1981). Solar interior structure and luminosity variations. *Solar Physics*, **74**, 21–34.

Haq, B. U., Hardenbol, J. and Vail, P. R. (1987). Chronology of fluctuating sea-levels since the Triassic. *Science*, **235**, 1156–66.

Hay, W. W. (1995). Cretaceous paleoceanography. *Geologica Carpathica*, **46**, 257–66.

Hay, W. W. (1996). Tectonics and climate. *Geologishe Rundschau*, **85**, 409–37.

Hay, W. W., DeConto, R. M., Wold, C. N. *et al.* (1999). An alternative global Cretaceous paleogeography. In *The Evolution of Cretaceous Ocean/Climate Systems*, eds. E. Barrera and C. Johnson. Boulder, CO: Geological Society of America (in press).

Herman, A. B. and Spicer R. A. (1997). New quantitative palaeoclimate data for the Late Cretaceous Arctic: Evidence for a warm polar ocean. *Palaeogeography, Palaeoclimatology, Palaeoecology*, **128**, 227–51.

Horrell, M. A. (1991). Phytogeography and paleoclimatic interpretation of the Maastrichtian. *Palaeogeography, Palaeoclimatology, Palaeoecology*, **86**, 87–138.

Huber, B. T., Hodell, D. A. and Hamilton, C. P. (1995). Middle–Late Cretaceous climate of the high southern latitudes: Stable isotope evidence for minimal equator to pole thermal gradients. *Geological Society of America Bulletin*, **107**, 1164–91.

Huber, B. T. and Watkins, D. K. (1992). Biogeography of Campanian–Maastrichtian calcareous plankton in the region of the southern ocean: Paleogeographic and paleoclimatic implications. In *The Antarctic Paleoenvironment: A Perspective on Global Change*, eds. J. P. Kennett and D. A. Warnke, pp. 31–60. Washington, DC: American Geophysical Union.

Hutchison, J. H. (1982). Turtle, crocodilian, champosaur diversity changes in the Cenozoic of the north-central region of western United States. *Palaeogeography, Palaeoclimatology, Palaeoecology*, **37**, 149–64.

Jenkyns, H. C., Gale, A. S. and Corfield, R. M. (1994). Carbon- and oxygen-isotope stratigraphy of the English chalk in Italian Scaglia and its paleoclimatic significance. *Geological Magazine*, **131**, 1–34.

Kauffman, E. G. (1973). Cretaceous bivalvia. In *Atlas of Paleobiogeography*, ed. A. Hallam, pp. 353–83. Amsterdam: Elsevier.

Kauffman, E. G. and Johnson, C. C. (1988). The morphological and ecological evolution of middle and Upper Cretaceous reef-building rudistids. *Palaios*, **3**, 194–216.

Klinger, L. F., Taylor, J. A. and Franzen, L. G. (1996). The potential role of peatland dynamics in ice-age initiation. *Quaternary Research*, **45**, 89–92.

Lefield, J. (1971). Geology of the Djadokhta Formation at Bayn Dzak (Mongolia). *Palaeontologia Polonica*, **25**, 101–27.

Lloyd, C. R. (1982). The Mid-Cretaceous Earth: Paleogeography, ocean circulation, temperature, and atmospheric circulation. *Journal of Geology*, **90**, 393–413.

MacLeod, K. G. and Huber, B. T. (1996). Reorganization of deep ocean circulation accompanying a Late Cretaceous extinction event. *Nature*, **380**, 422–5.

Manabe, S. and Bryan, K. (1985). CO_2-induced change in a coupled ocean atmosphere model and its paleoclimatic implications. *Journal of Geophysical Research*, **90**, 11689–707.

Olivero, E. B., Gasparini, Z., Rinaldi, C. A. and Scasso, R. (1991). First record of dinosaurs in Antarctica (Upper Cretaceous, James Ross Island): Paleogeographic implications. In *Geological Evolution of Antarctica*, eds. M. R. A. Thomson, J. A. Crame and J. W. Thomson, pp. 617–22. London: Cambridge University Press.

Otto-Bliesner, B. L. and Upchurch, G. R. (1997). Vegetation induced warming of high-latitude regions during the Late Cretaceous period. *Nature*, **385**, 804–7.

Parrish, J. T. and Spicer, R. A. (1988). Late Cretaceous terrestrial vegetation: A near polar temperature curve. *Geology*, **16**, 22–5.

Peixoto, J. P. and Oort, A. H. (1992). *Physics of Climate*. New York: American Institute of Physics.

Pirrie, D. and Marshall, J. D. (1990). High-paleolatitude Late Cretaceous paleotemperatures: new data from James Ross Island, Antarctica. *Geology*, **18**, 31–4.

Rind, D. and Chandler, M. (1991). Increased ocean heat transports and warmer climate. *Journal of Geophysical Research*, **96**, 7437–61.

Ronov, A., Khain, V. and Balukhovsky, A. (1989). *Atlas of Lithological-Paleogeographical Maps of the World: Mesozoic and Cenozoic of Continents and Oceans*. Leningrad: Ministry of Geology, USSR.

Rouse, W. R. (1984). Microclimate at arctic treeline 1. Radiation balance of tundra and forest. *Water Resources Research*, **20**, 57–66.

Saltzman, E. S. and Barron, E. J. (1982). Deep circulation in the Late Cretaceous: oxygen isotope paleotemperatures from *Inoceramus* remains in DSDP cores. *Palaeogeography, Palaeoclimatology, Palaeoecology*, **40**, 167–81.

Savin, S. M. (1977). The history of the Earth's surface temperature during the last 100 million years. *Annual Review of Earth and Planetary Sciences*, **5**, 319–56.

Schmidt, G. A. and Mysak, L. A. (1996). Can increased poleward oceanic heat flux explain the warm Cretaceous climate? *Paleoceanography*, **11**, 579–93.

Schneider, S. H. (1984). Response of the annual and zonal mean winds and temperatures to variations in the heat and momentum sources. *Journal of the Atmospheric Sciences*, **41**, 487–510.

Schneider, S. H., Thompson, S. L. and Barron, E. J. (1985). Are high CO_2 concentrations needed to simulate above freezing winter conditions? In *The Carbon Cycle and*

Atmospheric CO₂: Natural Variations Archean to Present, eds. E. T. Sundquist and W. S. Broecker, pp. 554–60. Washington, DC: American Geophysical Union.

Sellwood, B. W. Price, G. D. and Valdes, P. J. (1994). Cooler estimates of Cretaceous temperatures. *Nature*, **370**, 453–5.

Semtner, A. J., Jr. and Chervin, R. M. (1992). Ocean general circulation from a global eddy-resolving simulation. *Journal of Geophysical Research*, **97**, 5493–550.

Sloan, L. C. (1994). Equable climate during the early Eocene: Significance of regional paleogeography for North American climate. *Geology*, **22**, 881–4.

Sloan, L. C. and Barron, E. J. (1990). "Equable" climate during Earth history? *Geology*, **18**, 489–92.

Sloan, L. C., James, C. G. and Moore, T. C., Jr (1995). Possible role of ocean heat transport in Early Eocene climate. *Paleoceanography*, **10**, 347–56.

Spicer, R. A. (1990). Reconstructing high latitude Cretaceous vegetation and climate: Arctic and Antarctic compared. In *Antarctic Paleobiology and Its Role in the Reconstruction of Gondwana*, eds. T. N. Taylor and E. L. Taylor, pp. 15–26. New York: Springer-Verlag.

Spicer, R. A. and Corfield, R. M. (1992). A review of terrestrial and marine climates in the Cretaceous with implications for modelling the 'Greenhouse Earth'. *Geological Magazine*, **129**, 169–80.

Spicer, R. A. and Parrish, J. T. (1987). Plant megafossils, vertebrate remains, and paleoclimate of the Kogosukruk Tongue (Late Cretaceous), North Slope Alaska. *United States Geological Survey Circular*, **998**, 47–8.

Stein, C. and Stein, S. (1992). A model for the global variation in oceanic depth and heat flow with lithospheric age. *Nature*, **359**, 123–9.

Stoll, H. M. and Schrag, D. P. (1996). Evidence for glacial control of rapid sea level changes in the Early Cretaceous. *Science*, **272**, 1772–74.

Thomas, G. and Rowntree, P. R. (1992). The boreal forests and climate. *Quarterly Journal of the Royal Meteorological Society*, **118**, 469–97.

Thompson, S. L. and Pollard, D. (1997). Greenland and Antarctic mass balances for present and doubled atmospheric CO₂ from the GENESIS Version-2 Global Climate Model. *Journal of Climate*, **10**, 158–87.

Trenberth, K. E. and Solomon, A. (1994). The global heat balance: Heat transports in the atmosphere and ocean. *Climate Dynamics*, **10**, 107–4.

Upchurch, G. R. and Wolfe, J. A. (1987). Mid-Cretaceous to early Tertiary vegetation and climate: evidence from fossil leaves and woods. In *The Origin of Angiosperms and Their Biological Consequences*, eds. E. M. Friis, W. G. Chaloner and P. R. Crane, pp. 75–103. Cambridge: Cambridge University Press.

Vakhrameev, V. A. (1991). *Jurassic and Cretaceous Floras and Climates of the Earth*. Cambridge: Cambridge University Press.

Valdes, P. J., Sellwood, B. W. and Price, G. D. (1996). Evaluating concepts of Cretaceous equability. *Paleoclimates*, **2**, 139–58.

Webb, R. T., Rosenzweig, C. E. and Levine, E. R. (1993). Specifying land-surface characteristics in general circulation models: Soil profile data set and derived water holding capacities. *Global Biogeochemical Cycles*, **7**, 97–108.

Wolfe, J. A. and Upchurch, G. R. (1987). North American nonmarine climates and vegetation during the Late Cretaceous. *Palaeogeography, Palaeoclimatology, Palaeoecology*, **61**, 33–77.

10

Jurassic phytogeography and climates: new data and model comparisons

PETER MCA. REES, ALFRED M. ZIEGLER, AND PAUL J. VALDES

ABSTRACT

Leaves are a plant's direct means of interacting with the atmosphere, and their morphology is often attuned to and reflects prevailing environmental conditions. Although better understood and documented for angiosperms (or 'flowering plants'), non-angiosperms also exhibit a phytogeographic pattern linked most strongly to the evaporation/precipitation ratio, a relationship often reflected in their foliar morphologies. We have used this to interpret Jurassic terrestrial climate conditions along a spectrum defined by climate-sensitive lithological end-members such as evaporites and coals. Global climate zones, or biomes, were determined by exploring the foliar morphology/climate relationship using multivariate statistical analysis.

Jurassic plant productivity and maximum diversity were concentrated at mid-latitudes, where forests were dominated by a mixture of ferns, cycadophytes, sphenophytes, pteridosperms, and conifers. Low-latitude vegetation tended to be xeromorphic and only patchily forested, represented by small-leafed forms of conifers and cycadophytes. Polar vegetation was dominated by large-leafed conifers and ginkgophytes which were apparently deciduous. Tropical everwet vegetation was, if present at all, highly restricted. Five main biomes are recognized from the data: seasonally dry (summerwet or subtropical), desert, seasonally dry (winterwet), warm temperate, and cool temperate. Their boundaries remained at near-constant paleolatitudes while the continents moved through them (south, in the case of Asia, and north, in the case of North America). Net global climate change throughout the Jurassic appears to have been minimal.

The data-derived results are compared here with a new climate simulation for the Late Jurassic. The use of more detailed paleogeography and paleotopography has improved the overall data/model comparisons. Major discrepancies persist at high latitudes, however, where the model predicts cold temperate conditions far beyond the tolerance limits indicated by the plants. Nevertheless, our approach at least enables *direct* comparison of global paleoclimate interpretations from both a data and a model perspective, and indicates future research directions.

INTRODUCTION

We can determine climate zones or biomes for intervals in the geological past by studying the morphology and distributional patterns of fossil leaves. These provide a 'paleoclimate spectrum' between extreme end-member lithological indicators of climate such as coals and evaporites. There are three main reasons for believing this paleobotanical approach to be valid: (1) global distributional patterns of modern vegetation show a strong relationship with climate, especially temperature and precipitation, and the way these parameters are distributed through the annual cycle (e.g., Walter, 1985; Prentice *et al.*, 1992; Neilson, 1995); (2) leaves are a plant's direct means of interacting with the atmosphere, and their morphology is often attuned to and reflects prevailing environmental conditions; and (3) these relationships appear to have remained fairly constant since terrestrial vascular plants became established (e.g., Meyen, 1973). Fossil leaf genera and species are typically delimited taxonomically on the basis of relatively coarse characters such as size and shape, so are usually defined by morphological and not necessarily true biological criteria. Although such uncertainties may hinder evolutionary studies, these coarse subdivisions can be useful as paleoclimatic tools if one accepts that leaf morphologies represent environmental adaptations.

The relationship between modern angiosperm leaf physiognomy and climate is particularly well documented, from the early leaf margin analyses of Bailey and Sinnott (1915) to the advances made by J. A. Wolfe culminating in CLAMP (Climate Leaf Analysis Multivariate Program; see Wolfe (1993) for methodology). This technique has been applied to Tertiary and, more recently, Late Cretaceous floras (e.g., Herman and Spicer, 1996), resulting in quantitative estimates of climate parameters such as mean annual temperature and precipitation for these geological intervals. The results provide an important means of accurately determining paleoclimate signals and testing general circulation models (GCMs), with their capacity to predict future climates. However, such studies have been limited to approximately the last 120 million years (from the time when angiosperms became significant components of vegetation). Derived interpretations have also been compromised by 'patchiness' of the observations.

Paleoclimate interpretations using non-angiosperm leaf taxa (e.g., for the Jurassic: Barnard, 1973; Vakhrameev, 1991 and references therein) have been limited to the demarcation of floristic provinces based upon the distributions of only a few genera (e.g., *Dictyophyllum*, *Otozamites*, *Frenelopsis*) or orders (e.g., the Czekanowskiales). This was partly due to uncertainty regarding the climate tolerances of non-angiosperm fossils, given that most are now either extinct or at best 'relictual' in their distributions, having been marginalized by flowering plants. Another limitation was the absence of a method to arrange and analyze all of the available fossil plant data.

One means of compensating for the inferior non-angiosperm climate signal is to use a global 'whole-flora' approach to data collection, combined with statistical ordinations of the data. This is the approach we have used to interpret the Jurassic 'greenhouse' world. One purpose of this contribution is to show that we can use broad non-angiosperm leaf morphological characters combined with our 'whole-flora' approach to determine paleoclimates for pre-angiosperm times, and that it is possible to unravel the multiple effects (or constraints) of floral provinciality and evolution, taxonomic, and taphonomic bias, continental motion, and true global climate change. This is illustrated here with examples of results from multivariate statistical analyses of Jurassic floras, as well as climate modeling for the Late Jurassic. Our statistical approach enables us to interpret paleobotanical evidence of past climates in a more rigorous and repeatable manner. It also provides geologists and paleoclimate modelers with a means of utilizing directly and easily the detailed work of the paleobotanical community.

BACKGROUND

Broad-scale patterns of Jurassic climates have already been described, based upon various lines of sedimentological and paleobotanical evidence (e.g., Krassilov, 1972; Barnard, 1973; Hallam, 1984, 1994; Vakhrameev, 1991; Frakes *et al.*, 1992; Ziegler *et al.*, 1994, 1996; Price *et al.*, 1995, including previous authors referred to in each of these publications). Before describing our new approach we highlight some of the limitations, both of data and of technique, of previous studies.

Jurassic vegetation is commonly represented by leaf remains preserved as impressions or coalified compressions which, when no features (e.g., cuticle or fertile structures) are preserved to demonstrate that they are different, often leads to the creation of poorly defined and probably artificial groupings of leaf species and genera (e.g., *Cladophlebis, Sphenopteris, Taeniopteris*). This in turn has often given the impression of an apparent global homogeneity of Jurassic floras (see comments by Wesley, 1973). Other problems occur, such as taphonomic and collection bias, taxonomic inconsistency, poor stratigraphic control, uncertainty concerning the relation of foliar organs to each other and to other plant organs, poor preservation of most material, and morphological variability of different species of the same genus. In spite of these problems, strong patterns have emerged indicating that certain morphologies/taxa consistently co-occur, and that there is a correlation between the kind of foliage preserved and its paleogeographic distribution. However, previous studies using Jurassic plants to interpret global climates relied upon the distributional patterns of only a few individual genera or groups. Such an approach enables major differences to be highlighted, but does not show gradations in vegetation and climate. Since these are so obvious today, it seems reasonable to assume that they also existed in the geological past. Another disadvantage of the 'single-genus' approach (e.g., Krassilov, 1972) is that the framework collapses when individual genera become extinct. Hence, this method does not provide a consistent and repeatable means of interpreting climates through time.

The Russian paleobotanist V. A. Vakhrameev recognized the limitations of this approach, in a series of papers culminating in his posthumously published synthesis (Vakhrameev, 1991). He commented (1991, p. 4) that 'names for phytochoria [territorial floristic units] should be geographical rather than based on the terms of predominant plants since the areas of the latter undergo substantial changes with geological time.' His interpretations of Jurassic climates are based, nonetheless, upon the distributions of only a few plant genera or groups to delineate 'phytochoria' (i.e., proxies for phytogeographic and climate zones). The use of such terminology (phytochoria, as well as province, region, and area) results in uncertainty when a given region has moved tectonically with respect to latitude, particularly when the geographical term is retained for floras that have changed in response to climate by migration rather than evolution (see Ziegler, 1990). Hence, terms such as 'Euro-Sinian' and 'Siberian–Canadian' are not used here. We prefer to use descriptors derived from observations of modern vegetation and climate, independent of continent positions through time.

Vakhrameev's floristic and climatic interpretations for the Jurassic are further compromised since, of the plant taxa shown, only one fern genus (*Dictyophyllum*), two cycadophyte genera (*Otozamites*, *Ptilophyllum*), and one order (the Czekanowskiales) appear on each of his Jurassic maps. Studies of Present vegetation and climates do not use such limited data. Clearly, such a paleobotanical approach does not provide a reliable means of comparing global climates through Jurassic intervals which span some 60 million years. In spite of this, the Jurassic climate zones recognized by Vakhrameev and other paleobotanists agree, at least in the very broadest sense, with previous climate model results (e.g., Moore *et al.*, 1992*a,b*; Valdes and Sellwood, 1992; Valdes, 1994). However, any attempt to advance the role of fossil plants as paleoclimate indicators should be based upon as much as we know of the original plant assemblage and the global occurrences of such assemblages. Although imperfect, due to factors such as taphonomic, taxonomic, and collection bias, this approach provides our best proxy for original vegetation and therefore climate regimes in the geological past. We may not understand past vegetation and climates as completely as today's, but we can improve the paleoclimate data and model techniques to derive a closer approximation to real conditions.

NEW DATA AND METHODS

Fossil leaf genera, multivariate analysis, and 'floral gradients'

Our study uses multivariate analysis, specifically correspondence analysis, to assess co-occurrences and distributional patterns of fossil leaves at the genus level. This has ensured a more standardized approach to identifications and comparisons; fossil plant specimens are far more likely to have been assigned to the 'correct' genus or form genus than species.

For the Jurassic, we assembled a literature-derived floral database comprising 782 plant localities and 8175 genus occurrences worldwide. Lists of taxa from 644 northern hemisphere localities were submitted to correspondence analysis, those with only one or two leaf genera or spanning more than one Jurassic interval being omitted. Our analyses of these data represent an attempt to quantify gradations between floral provinces that have been identified previously by Russian authors (e.g., Vakhrameev *et al.*, 1978; Krassilov, 1981; Dobruskina, 1982; Vakhrameev, 1991). All have delineated climate zones using the floras; our aim is to explore the character of what is, in reality, a smooth gradient, but in a more rigorous and repeatable manner. For many years, Chinese and 'former Soviet

Union' (FSU) geologists and paleobotanists have collated paleofloristic data in the form of comprehensive taxonomic lists from specified sites covering a wide geographical area. The Chinese reports contain detailed information about sedimentary sequences from all regions of China, with lists of plant fossil species given for each fossiliferous stratigraphic unit, and the FSU taxon lists are of comparable quality (see Ziegler *et al.* (1994, 1996) for references). Central coordination ensured a taxonomic uniformity and 'whole-flora' approach seldom matched in the West; Asian data therefore provide, at present, the most obvious starting point for determining Jurassic phytogeography and climates. All lists and references are contained in databases of the Paleogeographic Atlas Project, University of Chicago. Using correspondence analysis these floral lists can be used to derive phytogeographic patterns and, because at the generic level the taxonomy of fossil vegetative organs reflects physiognomy, these patterns may be expected to have some correlation with climate.

We should point out that the use of paleobotanical genera for determining Mesozoic vegetational units or biomes is not a nearest living relative (NLR) approach because these taxa are not being compared directly with living forms. As with all fossils the name primarily denotes a morphotype that in many instances conveys the physiognomy of a particular organ. Indeed, it is sometimes necessary to 'side-step' traditional paleobotanical taxonomy, which is often hindered by political and regional biases (ensuring a highly specialized local but limited global view), as well as stratigraphic biases (with what is effectively the 'same' fossil plant type being assigned to a different genus or species depending upon its age). The morphotype approach can work in the paleoclimatologist's favor, once the taxonomic nomenclature is understood in terms of basic morphological characters, phytogeographic distributions, and likely paleoclimatic regimes.

We use correspondence analysis to arrange the fossil leaf genera from our 644 Jurassic northern hemisphere localities according to their degree of association with one another. Three separate analyses were carried out on data summed over relatively long time intervals (i.e., Early, Middle, and Late Jurassic), due to limitations in stratigraphic correlation and dating. A presence/absence matrix was constructed for each interval from locality taxonomic lists assembled throughout the northern hemisphere. These were then submitted to correspondence analysis (CA), which is a method used commonly in studies of modern ecology and vegetational succession. The advantages of CA are that it provides the same scaling of sample (locality) and character (taxa) plots, enabling direct comparison, and can accommodate 'incomplete' data matrices where some information is missing (Hill, 1979; Gauch, 1982). The version used is one of the programs in the CANOnical Community Ordination (CANOCO) package compiled by Ter Braak (1992), an extension of the Cornell Ecology Program DECORANA of Hill (1979). The general procedure has been described by Shi (1993): 'Geometrically, ordination involves rotation and transformation of the original multidimensional co-ordinate system and reduction of high dimensionality so that major directions of variation within the data set can be found and more readily comprehended than by looking at the original data alone.' Thus, we use CA as a means of arranging all of the elements (whether genera or localities) relative to axes in multidimensional space according to their similarity to each other. Most of the variation occurs on the first axis, with other axes accounting for progressively less, so two-dimensional plots (one for genera and the other for localities) were produced showing the variance within data sets on the two principal axes. Genera that frequently co-occur plot closest together on axis 1, whilst those that rarely co-occur are furthest apart. The same applies to the localities plot; those which share many floral elements plot closest to one another, whilst those with little in common plot furthest apart.

Ordination results for 57 Early Jurassic leaf genera from 196 northern hemisphere localities are shown in Fig. 10.1. Microphyllous (i.e., small-leafed) cycadophytes and microphyllous conifers plot to the left of axis 1, with macrophyllous (large-leafed) conifers and ginkgophytes toward the right. Thus, the plants of these two groups rarely co-occur, which makes sense if we consider their leaf morphologies in terms of climate (small-leafed forms of cycadophytes and conifers, often with thick cuticles, adapted to hot dry environments vs. large and presumably deciduous leaves of conifers and ginkgophytes adapted to seasonally cool and/or dark conditions). Other plant groups such as sphenophytes, ferns, and macrophyllous cycadophytes occupy the central and right-hand portions of axis 1, presumably since few of them were tolerant of water stress. It should be emphasized however that the symbols on the generic plot indicate only the centroids of the various floral elements and individual ones may have wide ranges. The corresponding Early Jurassic locality plot (Fig. 10.2) shows a broad correlation between axis 1 score

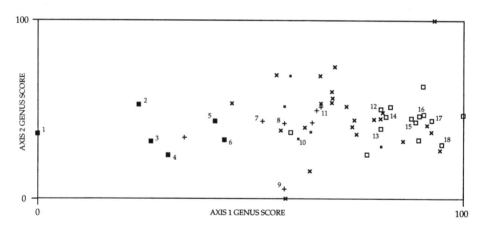

Figure 10.1. Correspondence analysis (CA) axis 1/axis 2 plot for 57 Early Jurassic leaf genera from northern hemisphere localities. The genera have been assigned to the following broad morphological categories: microphyllous cycadophytes, microphyllous conifers, and *Pachypteris* (large solid squares); macrophyllous cycadophytes (vertical crosses); ferns, sphenophytes, and lycophytes (diagonal crosses); 'unassigned' conifers (small solid squares); macrophyllous conifers and ginkgophytes (open squares). Numbers refer to the following leaf genera: 1, *Zamites*; 2, *Otozamites*; 3, *Brachyphyllum*; 4, *Pachypteris*; 5, *Ptilophyllum*; 6, *Pagiophyllum*; 7, *Pterophyllum*; 8, *Taeniopteris*; 9, *Nilssonia*; 10, *Elatocladus*; 11, *Ctenis*; 12, *Podozamites*; 13, *Baiera*; 14, *Ginkgo*; 15, *Pityophyllum*; 16, *Sphenobaiera*; 17, *Czekanowskia*; 18, *Desmiophyllum*.

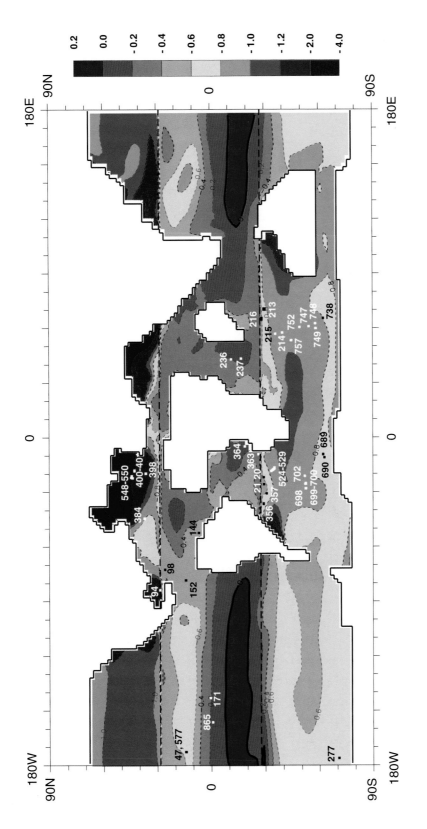

Figure 4.24. Map of early Eocene surface isotopic composition ($\delta^{18}O$, SMOW) calculated using the model-predicted surface salinity and the salinity–$\delta^{18}O$ relationship of Broecker (1989); poleward of 28° latitude and the western equatorial Atlantic equation of Fairbanks et al. (1992) equatorward of 28° latitude.

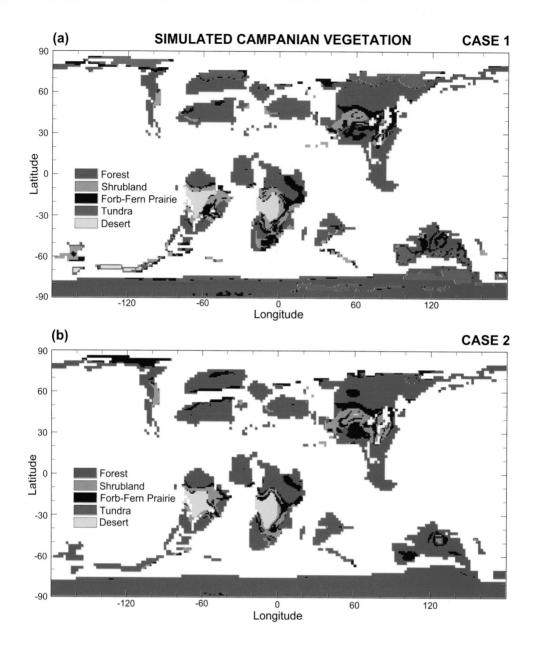

Figure 9.3. Campanian biomes predicted by EVE in the CASE 1 (a) and CASE 2 (b) simulations of climate and vegetation after 15 years of integration.

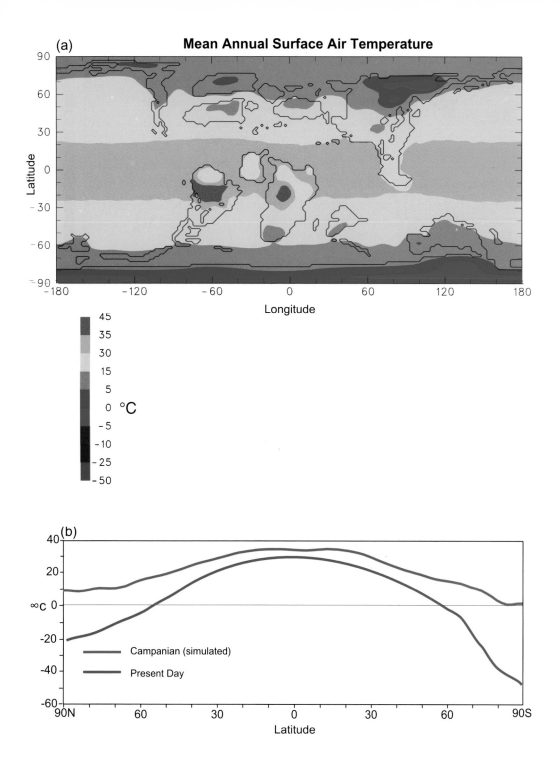

Figure 9.4. Mean annual surface (2 m) air temperature (MAT) simulated in the CASE 2 climate–vegetation simulation (a) and zonally averaged (meridional) MAT (b).

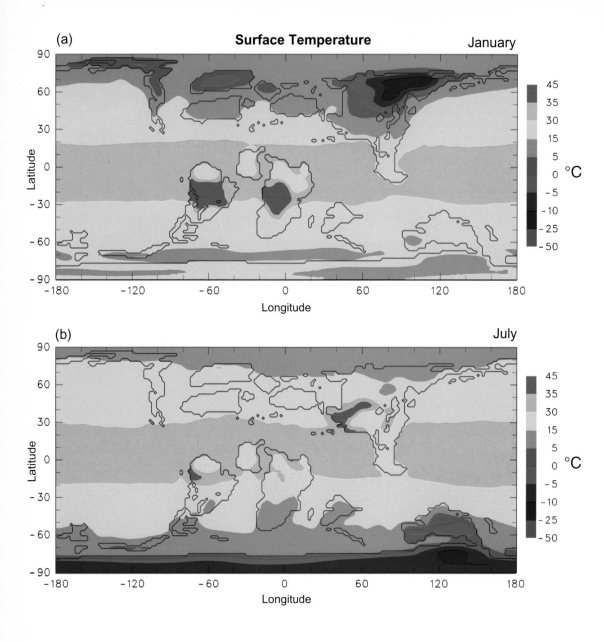

Figure 9.5. January (a) and July (b) monthly mean surface (2 m) air temperatures.

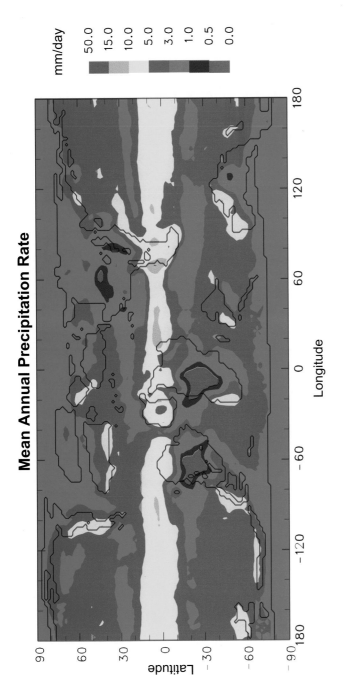

Figure 9.6. Mean annual precipitation in mm day^{-1} simulated by the forest-dominated CASE 2.

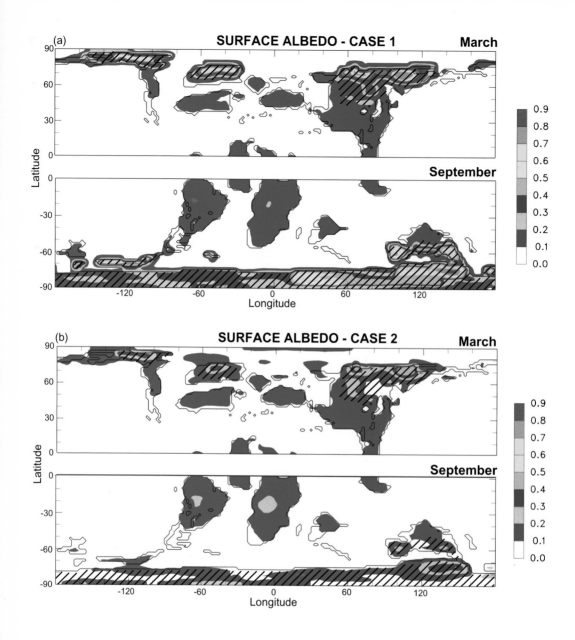

Figure 9.7. Spring surface albedo for northern and southern hemispheres shown for CASE 1 (a) and CASE 2 (b). The hatching shows the area covered by >50% fractional snow cover.

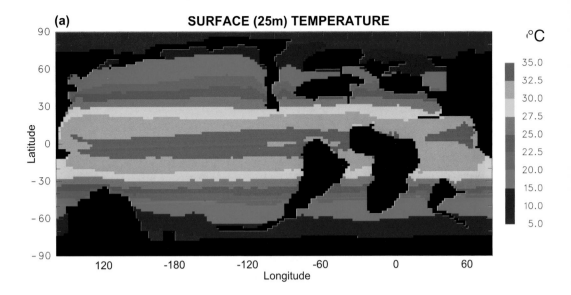

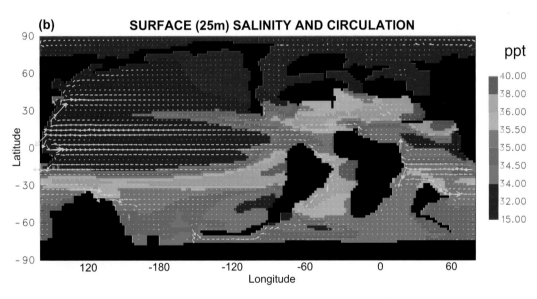

Figure 9.8. Temperature (a) and salinity (b) simulated by the OGCM forced with the CASE 2 surface climatology and integrated for 450 surface years (4500 deep ocean years). The maps are centred on a longitude that shows the entire Pacific basin. Velocity vectors showing the surface (25 m) circulation are superposed on surface salinity (b). The vector scale is 0.24° per cm 5^{-1}.

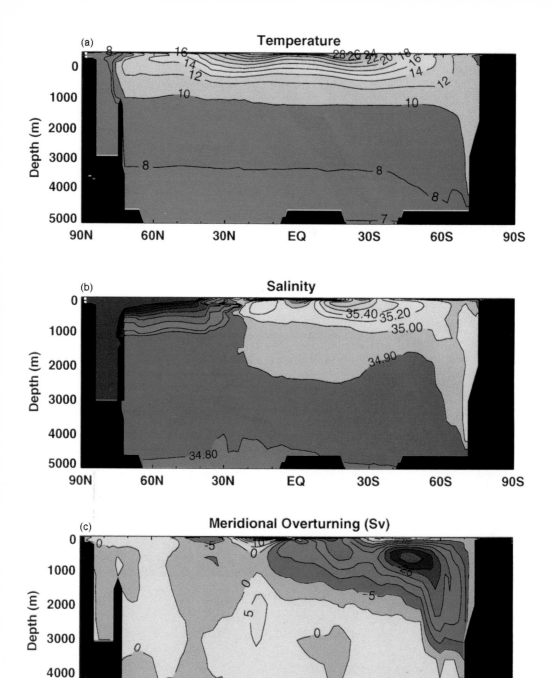

Figure 9.9. Globally averaged zonal profiles of temperature (a) in °C, at contour intervals of 2 °C, salinity (b) in ppt, at contour intervals of 0.2 ppt except between 34.8 and 35.0 ppt where the interval is 0.1 ppt, and meridional overturning stream function (c) in units of $10^6 \, m^3 \, s^{-1}$ (Sv) at contour intervals of 5 Sv.

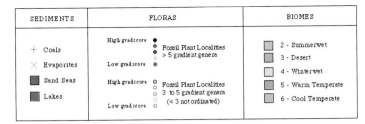

SEDIMENTS	FLORAS	BIOMES

SEDIMENTS

+ Coals
× Evaporites
■ Sand Seas
■ Lakes

FLORAS

High gradscore ●
● Fossil Plant Localities
● > 5 gradient genera
Low gradscore ●

High gradscore ○
○ Fossil Plant Localities
○ 3 to 5 gradient genera
○ (< 3 not ordinated)
Low gradscore ○

BIOMES

2 - Summerwet
3 - Desert
4 - Winterwet
5 - Warm Temperate
6 - Cool Temperate

Figure 10.7. New Early Jurassic (a), Middle Jurassic (b), and Late Jurassic (c) paleogeographic maps, showing floral and lithological data as well as inferred biomes.

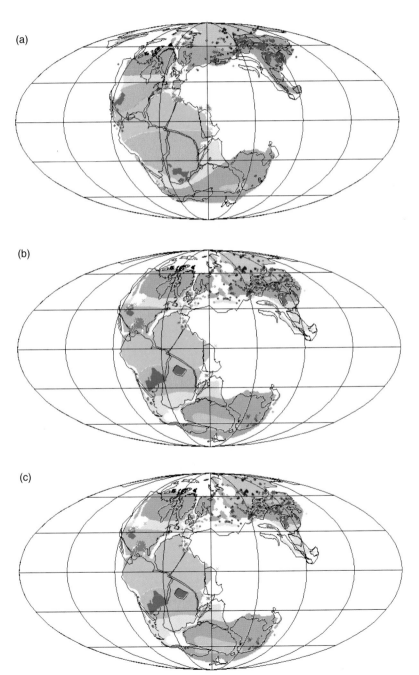

(a)

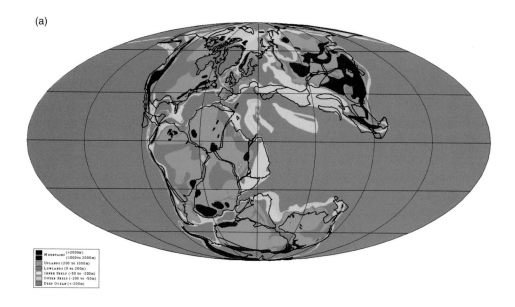

Mountains (>2000m)
(1000 to 2000m)
Uplands (200 to 1000m)
Lowlands (0 to 200m)
Inner Shelf (-50 to -200m)
Outer Shelf (-200 to -50m)
Deep Ocean (<-200m)

(b)

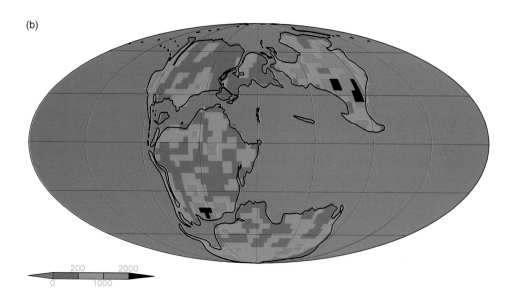

200   2000
0   1000

Figure 10.8. Comparison of (a) data-derived and (b) modeled paleogeography
(including topography).

(a)

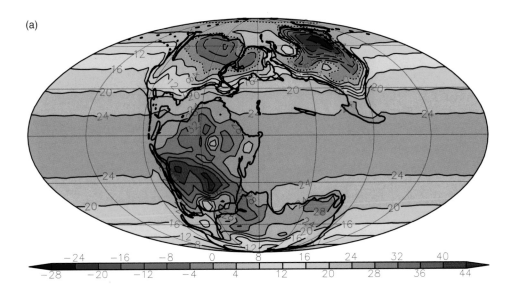

(b)

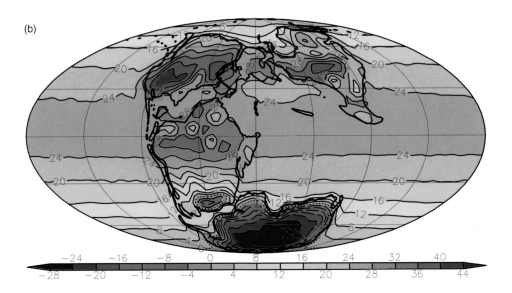

Figure 10.9. Modeled Late Jurassic surface air temperatures for (a) Dec–Jan–Feb seasonal average (i.e., northern hemisphere winter and southern hemisphere summer), and (b) June–Jul–Aug (i.e., northern hemisphere summer and southern hemisphere winter). The contour interval is 4 ˚C.

(a)

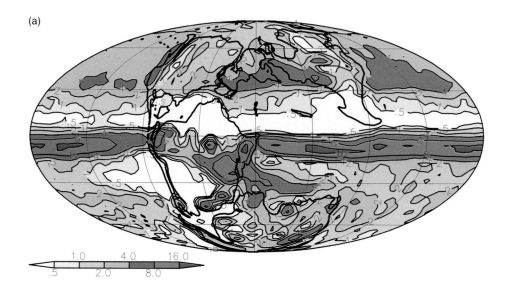

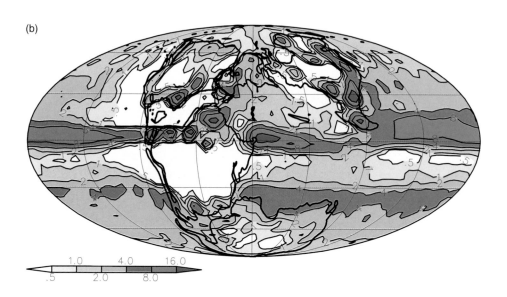

(b)

Figure 10.10. Modeled seasonal precipitation, as in Fig. 10.9. The contours are 0.5, 1, 2, 4, 8 and 16 mm day^{-1}.

(a)

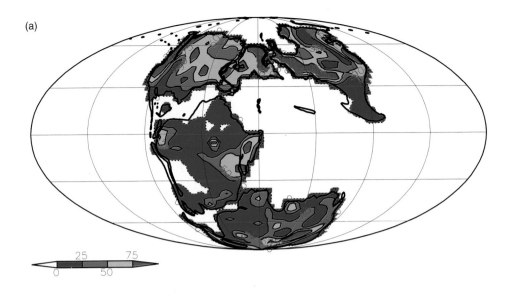

(b)

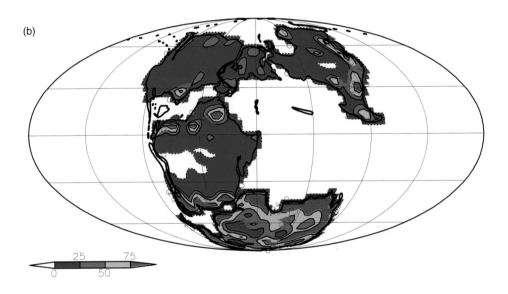

Figure 10.11. Modeled surface soil moisture (in the first 7.5 cm of the soil), as in Fig. 10.9. Contours are every 25% of total soil water capacity.

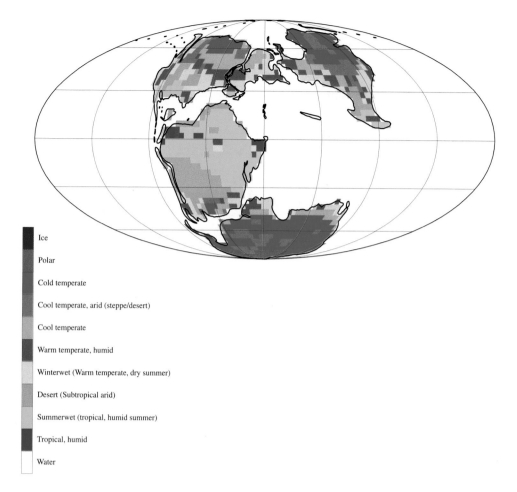

	Ice
	Polar
	Cold temperate
	Cool temperate, arid (steppe/desert)
	Cool temperate
	Warm temperate, humid
	Winterwet (Warm temperate, dry summer)
	Desert (Subtropical arid)
	Summerwet (tropical, humid summer)
	Tropical, humid
	Water

Figure 10.12. Predicted Walter biome/climate zones for the Late Jurassic (compare with the data-derived results in Fig. 10.7 c, below).

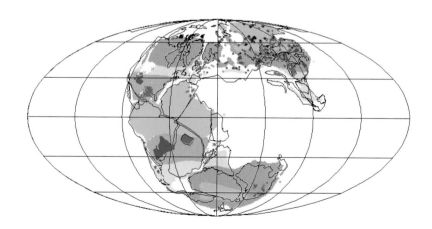

Figure 10.7(c). Late Jurassic paleogeographic map.

June - July - August (JJA)
Average Surface Temperature

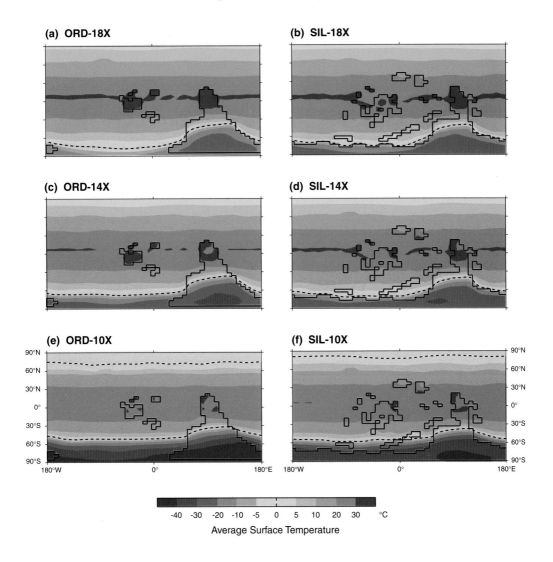

Figure 13.7. June–July–August (JJA) and December–January–February (DJF) average surface temperatures. (Columns: change due to CO_2 only; rows: change due to paleogeography only.)

December - January - February (DJF)
Average Surface Temperature

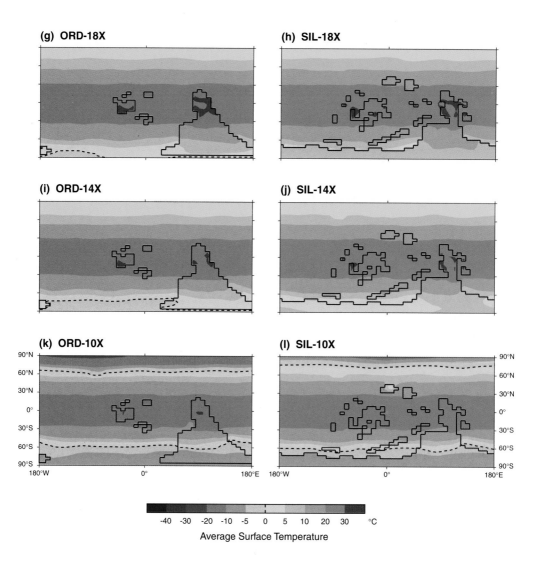

Figure 13.7. (*cont*).

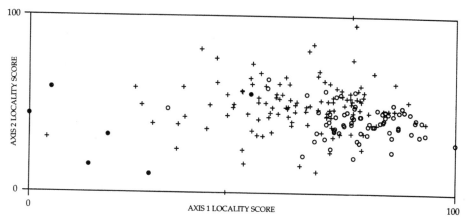

Figure 10.2. Early Jurassic CA axis 1/axis 2 plot for 196 northern hemisphere plant localities. Localities are coded according to paleolatitude: 0–40° N (solid circles), 40–60° N (vertical crosses), 60–90° N (open circles).

and paleolatitude. This becomes better defined when the localities are plotted on paleogeographic maps, where it is clear that factors such as longitudinal east vs. west, continental interior vs. maritime, and even topographic variations also contribute, just as in the modern world.

We can therefore interpret Early Jurassic phytogeographic patterns based on the axis 1 scores of individual leaf genera and corresponding plant localities, due to their relative degrees of association. We can then understand these climatically in terms of the basic morphological characteristics of individual leaf genera and the paleogeographic distribution of plant localities. However, this is an improvement on previous work only in that we have adopted a 'whole-flora' approach and applied statistical rules to arrange the data. How can we compare vegetation and climate signals from successive time intervals in a more rigorous and repeatable manner? By repeating the exercise for Middle and Late Jurassic floras, we arrive at three separate ordinations (for J1, J2, and J3 intervals, comprising 196, 288, and 160 northern hemisphere plant localities respectively). The ordering of leaf genera along axis 1 remains fairly constant in each interval, since overall floristic change was minimal throughout the Jurassic. By averaging the scaled (0 to 100) axis 1 scores of the 32 genera common to all three intervals, we derive a Jurassic 'floral gradient' (Fig. 10.3). This shows the gradient score for each genus, which is based upon its averaged axis 1 position relative to all other genera throughout the Jurassic. Microphyllous conifers and microphyllous cycadophytes have low scores, whereas macrophyllous conifers and ginkgophytes have high ones. Ferns (e.g., *Todites*, *Cladophlebis*, *Coniopteris*) and macrophyllous cycadophytes occupy the central portion of the gradient, along with sphenophyte genera such as *Equisetites* (fossil 'horsetails' or 'scouring rushes'). Using the floral gradient, we can assign a value to any Jurassic plant locality (whether J1, J2, or J3) simply by averaging the scores of its constituent leaf genera. Indeed, anyone can place a new locality list on the gradient by averaging the score

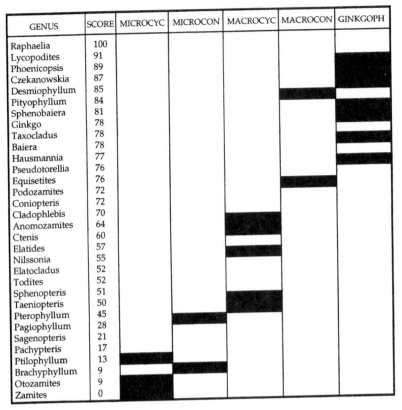

GENUS	SCORE	MICROCYC	MICROCON	MACROCYC	MACROCON	GINKGOPH
Raphaelia	100					
Lycopodites	91					
Phoenicopsis	89					
Czekanowskia	87					
Desmiophyllum	85					
Pityophyllum	84					
Sphenobaiera	81					
Ginkgo	78					
Taxocladus	78					
Baiera	78					
Hausmannia	77					
Pseudotorellia	76					
Equisetites	76					
Podozamites	72					
Coniopteris	72					
Cladophlebis	70					
Anomozamites	64					
Ctenis	60					
Elatides	57					
Nilssonia	55					
Elatocladus	52					
Todites	52					
Sphenopteris	51					
Taeniopteris	50					
Pterophyllum	45					
Pagiophyllum	28					
Sagenopteris	21					
Pachypteris	17					
Ptilophyllum	13					
Brachyphyllum	9					
Otozamites	9					
Zamites	0					

Figure 10.3. Jurassic floral gradient, derived from the averaged axis 1 scores of genera common to J1, J2, and J3 floras. Five broad morphological categories ('morphocats') and their constituent genera are highlighted, showing the gradation from microphyllous forms to macrophyllous conifers and ginkgophytes.

for each of the 32 genera represented. It should be stressed that the score of each genus on the gradient represents its 'centroid' in the northern hemisphere latitudinal spectrum; most genera appear in at least some lists throughout this range. This is due in part to the general uniformity of Mesozoic floras and in part to time-averaging of the taxonomic lists. The gradients are subtle and may be determined only by reference to the entire assemblage. Such a method at least provides the best available proxy for original vegetation and prevailing climate conditions at a given locality.

Using the floral gradient, we can compare Early, Middle, and Late Jurassic plant localities objectively, observe any spatial and temporal changes, and interpret these in terms of floral provinciality, continental motion, and global climate change. Figure 10.4 shows the correlation between floral gradient score and paleolatitude for Early Jurassic localities *worldwide*. Although based upon northern hemisphere ordinations (given the relative paucity and poor stratigraphic control of southern hemisphere/Gondwanan Jurassic floras), the derived floral gradient can also be used to assign scores to these Gondwanan localities. As with the Early Jurassic axis 1 ordi-

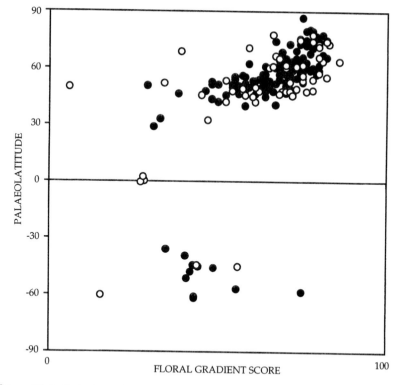

Figure 10.4. Early Jurassic locality gradient score vs. paleolatitude for northern and southern hemispheres. Locality scores are derived from the scores of the constituent genera shown in Fig. 10.3. Open circles represent small samples with three, four, or five genera present on the floral gradient; filled circles are larger samples with six or more gradient genera.

nation for the northern hemisphere (Fig. 10.2), a broad correlation of gradient score with paleolatitude is still evident, indicating a symmetry of Jurassic climate zones about the equator, consistent with coal, evaporite, and eolian sand distributions.

Lithological indicators of climates

The study of fossil leaf morphologies, associations, and distributional patterns provides a spectrum of climate information between end-member climatically sensitive lithologies such as coals (precipitation > evaporation) and evaporites (evaporation > precipitation). The global distributions of these lithologies provide another important data source. Occurrences of coal, lacustrine, evaporite, and eolian sand deposits were compiled from the literature and are recorded in Paleogeographic Atlas Project databases. There is an apparently symmetrical arrangement of Jurassic evaporites and coals about the equator, shown in Fig. 10.5. These lithological patterns provide information about climate extremes in terms of net evaporation and precipitation, and support the new paleobotanical interpretations. They also indicate that global application of the northern hemisphere floral gradient (e.g., Fig. 10.4) and direct northern and southern hemisphere floristic comparisons are valid. Although there are as yet no data from East Antarctica (i.e., the continental interior of southern Gondwana), we believe it reasonable to infer that fossil leaf genera such as *Phoenicopsis*, *Czekanowskia*, and *Podozamites* (or at least large-leafed and deciduous morphological equivalents) would have been present there during the Jurassic, similar to contemporaneous forms growing at high northern latitudes.

Our global climate interpretations are based upon the occurrences and ordinations of leaf genera, as well as selected lithological indicators. More localized distributional patterns of fossil palynomorph taxa such as *Classopollis*, indicative of dry conditions (e.g., Volkheimer, 1970; Vakhrameev, 1991 and references therein), as well as information from paleosols (e.g., Sellwood and Price, 1994; Singer *et al.*, 1994), support our new results for the low-latitude regions.

Leaf morphological categories

We assigned all Jurassic leaf genera, whether or not they were included in the correspondence analysis, to 10 coarser morphological categories: sphenophytes, ferns, pteridosperms, microphyllous cycadophytes, unassigned (intermediate or morphologically variable) cycadophytes, macrophyllous cycadophytes, ginkgophytes, microphyllous conifers, unassigned (intermediate or morphologically variable) conifers, and macrophyllous conifers ('sphenoph', 'fern', 'fern2', 'microcyc', 'cyc', 'macrocyc', 'ginkgoph', 'microcon', 'con', and 'macrocon', respectively). The paleolatitudinal distribution of these morphological categories ('morphocats') is shown in Fig. 10.6. Individual points represent the total numbers of morphocats within each 10° interval, whereas the curve represents an average of northern and southern hemisphere data. Shaded areas represent the approximate latitudinal extent of

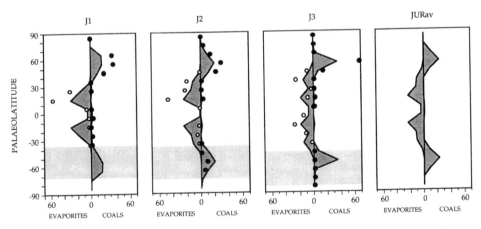

Figure 10.5. Distribution of Jurassic coals and evaporites by paleolatitude. Occurrences of coals within a 10° latitudinal interval were calculated as a percentage of total coal occurrences. The same calculation was used for evaporite occurrences. Curves represent averages of northern and southern hemisphere data, and shaded rectangles show the approximate latitudinal extent of East Antarctica, for which lithological data are absent.

East Antarctica. As with the lithological distributions shown in Fig. 10.5, the curve was calculated because the paucity of southern hemisphere data, particularly the 'datahole' caused by present-day Antarctic ice cover, otherwise renders such broad inter-hemisphere comparisons problematical. Highest numbers occur at mid-latitudes, decreasing pole- and equatorwards in each hemisphere, although with a small peak about the equator. The patterns are of interest if we accept them as a crude proxy for vegetational diversity: highest at mid-latitudes, decreasing pole- and equatorwards, but with some possible tropical diversity.

Determination of biomes

Our approach enables *direct* comparison of vegetation patterns throughout the Jurassic and means that we can determine biomes or climate zones in a more rigorous fashion. Although multivariate analysis serves to identify the degree of variance in the data it cannot, of course, specify the sources of variance. It is the physiognomy implicit in the names of individual fossil leaf genera that ultimately enables the determination of global paleoclimates. Ordinations of fossil leaf genera and localities are derived from the primary literature and these, combined with lithological data, enable climate zones (biomes) to be drawn on paleogeographic maps. An individual leaf genus is defined by basic morphological characters such as size and shape that can be interpreted in terms of prevailing environmental conditions. The relative position of each genus on a generic ordination plot is

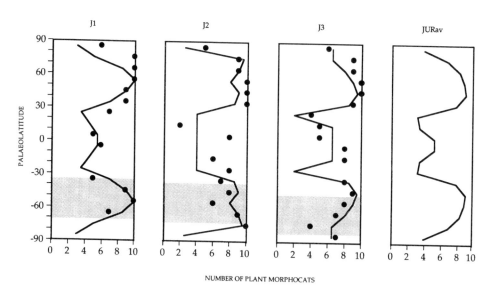

Figure 10.6. Occurrences of Jurassic plant morphocats by paleolatitude. Individual points represent total numbers of morphocats within each 10° interval, whereas the curve represents an average of northern and southern hemisphere data. Shaded areas represent the approximate latitudinal extent of East Antarctica, for which paleobotanical data are absent.

defined by its degree of association with other leaf genera. The relative position of each floral locality on a corresponding locality plot is defined by its constituent leaf genera. Correspondence analysis simply provides an objective assessment of variance in the original data matrix. The generic and locality plots both show that the data arrays are gradational rather than disjunct, indicating that climate influenced the patterns and not geographic barriers. This poses a problem in classification because there are no natural breaks in the distributions, but this is also true of the Present.

We use the biome scheme developed by Walter (1985), in which he reduced the macroclimate of the present-day land surface to nine major biomes. He compiled 'ecological climate diagrams' showing monthly temperature, precipitation, and other statistics for some 8000 meteorological ground stations worldwide and combined these with details of the corresponding vegetation. In effect, his choices of climatic boundary conditions were influenced by natural transitions in the vegetation. One attractive aspect of Walter's scheme is that it is simple and therefore applicable to the geological past, thus overcoming deficiencies inherent in the fossil record. The scheme (as modified by Ziegler (1990); see comments therein for further details) also retains information on seasonality of temperature and precipitation, factors that are of fundamental importance to controlling vegetation patterns and our derived interpretations of Jurassic climates. It is significant that fossil leaf morphologies are similar at a given paleolatitude (whether northern or southern hemisphere), often regardless of conventional taxonomic status. This would appear to indicate that plants developed similar strategies to maximize their efficiency in a given environment, regardless of their botanical affinities. Indeed, it should be emphasized that few fossil plant taxa are, in fact, true biological ones, being instead an approximation based upon morphological similarity rather than biological compatibility.

There are two extremes of vegetation type in our Jurassic example: localities comprising wholly microphyllous forms (of conifers and cycadophytes) and localities comprising wholly macrophyllous conifers and ginkgophytes. Based on leaf morphologies, these can be interpreted as plants adapted to hot dry, and seasonally cool and/or dark conditions respectively. These plant types do occasionally co-occur but it is relatively easy to define end-member biomes based on the locality scores defined by these leaf genera. Microphyllous plant localities occur at low paleolatitudes and can be assigned to a seasonally dry biome, this being consistent with evaporite and eolian sand distributions. The macrophyllous conifer/ginkgophyte localities occur at high paleolatitudes and can be assigned to a cool temperate biome, based upon the deciduous nature of the foliage. It is harder to define the latitudinal boundaries between these and the intermediate warm temperate biome. However, the occurrence of macrophyllous cycadophytes and ferns, changes in their relative abundance with respect to microphyllous and macrophyllous/ginkgophyte forms, and coal distributions, all enable subdivisions of the floral and climate spectrum. The patterns become even clearer when individual localities are plotted on the new paleogeographic maps shown in Fig. 10.7 (see color plates).

JURASSIC CLIMATE INTERPRETATIONS

Our paleogeographic maps (Fig. 10.7) illustrate the latitudinal consistency of biomes (or climate zones) throughout the Jurassic. Care must be taken to differentiate between real global climate change and observed effects due to continents passing beneath climate zones (see Ziegler *et al.*, 1996). For example, clockwise motion about an axis in Europe resulted in a polewards motion of North America and the reverse in eastern Asia. The latitudinal transitions between floras remained fairly constant, so we maintain that no net global change occurred in the area occupied by the individual biomes. Thus, while the climate of Asia became warmer and drier through the Middle and Late Jurassic, the reverse was true of North America. Climate changes that have been described for Eurasia (e.g., Vakhrameev, 1991; Hallam, 1994) are in our view the effect of the continent moving with respect to the climate zones, rather than the reverse (see Ziegler *et al.*, 1994, 1996). While Asia moved southwards, North American Lower and Middle Jurassic deserts represented by the Navajo and Entrada formations (Parrish and Peterson, 1988) gave way to seasonally dry Upper Jurassic climates of the Morrison Formation (Dodson *et al.*, 1980), followed by temperate coal swamps in the Cretaceous (Horrell, 1991) as that continent continued moving polewards.

Clearly, the effects of continental motion can be pronounced and must be considered when interpreting paleoclimates. Differences observed vertically within a sequence preserved at a particular locality may be due to local variations influenced by sedimentological and taphonomic biases, the effects of latitudinal motion through climate zones, genuine global climate change, or combinations of these factors. Further detailed studies and correlations between sequences are required in order to understand whether such locally observed variations are caused principally by regional or by global climate changes. For the Jurassic, we subscribe to the former view, given the evidence for marked latitudinal migrations of the North American and Asian continents. We also maintain that interpretations of global paleoclimate conditions based upon global data collection are more robust than those derived from selective observations (e.g., Hallam, 1984, 1994; Vakhrameev, 1991; Frakes *et al.*, 1992).

We have no geological evidence in the Jurassic for tropical everwet and, at the other extreme, tundra or glacial biomes. The equatorial regions were markedly drier than today, with large continental interiors. Admittedly, since large areas of the equatorial zone are devoid of Jurassic deposits, the lack of rainforests could be a preservational effect. Nonetheless, a large continent on the equator in the form of the combined Africa and South America must have experienced effects detrimental to everwet conditions. Much of this land mass would have been remote from moisture sources and the latitudinal excursion of the Intertropical Convergence Zone (ITCZ) would have been more extreme (see Ziegler *et al.*, 1987). Large excursions away from the equator, allowed by weak polar highs in a warm world, would have led to little or no constantly wet zones, merely seasonally wet areas. So, the limited moisture available to the system is theorized to have been spread over a wider area. The zone of peak productivity therefore shifts from low to higher latitudes at times of global warmth.

The fact that we do not recognize tundra vegetation at high latitudes in the Jurassic does not mean that the biome approach is invalid. Even if it were argued that modern vegetation and climate regimes cannot be used as a basis for assigning strictly similar biomes in the geological past, the lack of evidence for persistent polar ice (from either data or models) during this period of globally warm conditions means that a similar absence of tundra and permafrost is unsurprising. Of course, such vegetation may have existed during other geological intervals but may be hard to recognize as such, given that small-stature plants belonging to low-diversity floras would have had a correspondingly reduced preservation potential.

Based on the global distributions of lithological climate indicators such as coals (precipitation > evaporation), as well as evaporites and eolian sands (evaporation > precipitation), a general symmetry of climate zones about the (paleo)equator is seen throughout the Jurassic (Fig. 10.7). Such indicators provide useful information about extreme climate conditions (e.g., Lottes and Ziegler, 1994; Price *et al.*, 1995), but it is fossil plant data which enable the entire terrestrial climate spectrum to be determined, particularly when analyzed using our more rigorous 'whole-flora' approach. A problem arises with Jurassic floral sites from Gondwana; data are relatively sparse and stratigraphic control is often less certain, which means there are only 50 usable localities (cf. 644 in the northern hemisphere). In addition, key high-latitude information is typically missing, potential localities being covered today by the Antarctic ice sheet. All of this prevents an intra-Gondwanan floral gradient from being compiled and so the northern hemisphere one is used. Nevertheless, relative spatial and temporal variations of floral gradient score and inferred climate signal calculated for these Gondwanan localities are consistent with lithological indicators, paleoclimate models, and the paleogeography of the continent (Fig. 10.7). Most plant localities in southern Gondwana have similar gradient scores to those in the northern hemisphere interpreted as belonging to the warm temperate biome. One exception is the Argentine floras, most of which have low gradient scores indicative of seasonally dry conditions. Palynological data from Argentina (e.g., Volkheimer, 1970) agree with our biome designation for this region. These differences in floral gradient score across southern Gondwana are also consistent with the distributional patterns of lithological indicators such as coals (which are common in Australia) and eolian sandstones (preserved in southern Africa and South America). There is no direct geological evidence for a cool temperate biome in Gondwana similar to that interpreted from high latitude sites in Eurasia. However, given the otherwise-symmetrical arrangement of biomes about the equator, it is reasonable to infer such a biome for areas distant from maritime influence in the interior of southern Gondwana.

From the combined floral and lithological data, the Jurassic world is interpreted as essentially one in which low latitudes were seasonally dry (summerwet or subtropical), succeeded polewards in both hemispheres by desert, seasonally dry (winterwet), warm temperate, and cool temperate biomes (Fig. 10.7). Despite the caveats outlined above (primarily due to patchiness of the geological record) the biome concept provides, at present, the most rigorous method for interpreting climate signals from pre-angiosperm floras. Furthermore, the standard scale presented here (Fig. 10.3; modified from Ziegler *et al.*, 1996) at least enables direct comparison of such fossil plant assemblages worldwide. Comparisons of our new approach with results of paleoclimate modeling are of particular interest, enabling us to understand Jurassic climates from both a biological and a physical perspective.

COMPARISON OF LATE JURASSIC DATA AND MODEL RESULTS

New model results for the Late Jurassic can be compared with climate predictions based on the GCM simulations of Moore *et al.* (1992*a,b*), Valdes and Sellwood (1992), and Valdes (1994). These model simulations showed broad agreement but there were differences at high latitudes, where Moore *et al.* (1992*a*) produced results significantly colder than those of Valdes and Sellwood (1992). However, these simulations used different paleogeographic reconstructions and hence it is difficult to accurately compare these model simulations to the data. For this reason, we have performed a new simulation so that the GCM uses exactly the same paleogeography as the data. The previous model was based on Kimmeridgian paleogeography, whereas the new one uses our more refined Volgian reconstruction. In all other aspects the GCM is identical, including prescribed CO_2 concentrations of 4× present day, to that used by Valdes (1994; see that paper for a detailed description of physical parameterizations). The model is typical of most current GCMs; it includes a detailed radiation scheme with both seasonal and diurnal variations, an interactive, relative humidity-based cloud scheme, the Betts–Miller convection scheme, and a simple three-layer surface parameterization.

The model has 19 levels in the vertical and the simulation described here uses a spectral representation with a triangular truncation at total wavenumber 31. All physical processes are considered on a grid approximately $4° \times 4°$, although the true resolution of the model is somewhat poorer. The effect of this relatively coarse resolution can be seen in Fig. 10.8 (see color plates), which compares the detailed (data-derived) paleogeography with that used in the model. The GCM provides a good approximation, despite its relative coarseness; all of the major mountain ranges are represented, although there is a tendency to broaden the relatively narrow features on the western side of the Americas. The model has been run for a total of 7 years and the last 5 years have been averaged to give the predicted climatology. It should be noted that this simulation used the same prescribed sea surface temperature as in Valdes (1994). This means that the model reaches a dynamic equilibrium relatively rapidly and an extended length of run is not essential.

The major changes between this reconstruction and that used by Valdes and Sellwood (1992) are seen in the southern hemisphere. The new reconstruction indicates a major seaway extending across Gondwana, so that South America and Africa are now separated from Antarctica, India, and Australia. Owing to the ameliorating effects of this large water mass, the regions adjacent to it are likely to have had a smaller seasonal range of temperatures, but would have been generally moister. One other important modification is that there is considerably larger orography on the Antarctic peninsula and much-reduced orography in the center of the African/South American continent. This is likely to have had important effects on precipitation and temperature patterns. Figure 10.9 (see color plate) shows the predicted seasonal patterns of surface air temperature. The model is predicting cold temperatures during the winters in northern America, northern Eurasia, and Antarctica. In this last region, the temperatures decrease to $-24\,^{\circ}\text{C}$ or less over a substantial part of the continent. This is somewhat colder than shown in the model produced by Valdes (1994), which used the previous (restricted-seaway) paleogeographic reconstruction. It appears to be due to two different factors. Firstly, the general elevation of the region is greater and hence there is a simple effect of lapse rate. This can contribute quite significantly, since a typical change of temperature with altitude is approximately $6.5\,^{\circ}\text{C}$ per 1000 m. In addition, it appears that the enhanced orography on the western side of Antarctica effectively blocked the flow of relatively warm oceanic air in winter, resulting in a cooling over a large part of the western region of the continent. Moore *et al*. (1992*b*) also showed important differences in the simulations depending on the paleotopography which was used, although their choice of paleo-topography is not the same as ours.

In the vicinity of the southern Gondwanan seaway this simulation is indeed less extreme than that of Valdes (1994). This is particularly noticeable for the summer since temperatures do not become anywhere near as high as in the previous simulation. In winter, the differences are more subtle since the effect of high orography over South Africa is important. However, a difference map (not shown) does reveal a warming of up to $10\,^{\circ}\text{C}$ in coastal regions. In other regions, the simulation is broadly similar to that of Valdes (1994). The tropical regions can reach temperatures as high as $40\,^{\circ}\text{C}$ during parts of the summer months. The cool winter temperatures in North America are somewhat more severe than previously and, again, this can largely be explained by enhanced orography in this region.

Temperature is not the only aspect of the simulation that is important when making comparisons with geological data; moisture availability is also vital. Figure 10.10 (see color plate) shows the simulated seasonal precipitation. There are heavy bands of rainfall over the tropics, as well as in mid-latitudes; these are very similar to those described by Valdes (1994). During the southern hemisphere summer, there is a very strong monsoonal-type precipitation extending over the whole South American/ African continent. However, in the corresponding winter season, the precipitation has completely disappeared and is replaced by marked drought conditions. The model predicts rainfall amounts near to zero. In mid-latitudes, the rainfall belts correlate closely with the model's simulation of storm tracks (not shown). In the winter season, there is a clear band of precipitation at approximately 40° S and 40° N. This corresponds well with locations of the winterwet biome. There is a substantial change compared with the previous simulation due to the southern seaway. This region is much wetter in the new simulation and therefore agrees better with the observed biomes. In addition to precipitation, another important indicator of hydrological conditions is the surface soil moisture content. This shows (Fig. 10.11, see color plate) the amount of moisture in the soil, expressed as a percentage of the maximum allowed. It is effectively a balance between the total precipitation and the total evaporation, which is largely controlled by the temperature and moisture availability. Although the patterns of soil moisture mimic the precipitation   patterns, there is a considerable amount of small-scale variability.

This is in part the result of orography; elevated surfaces are cooler and hence there is less evaporation.

The maps of temperature, precipitation, and soil moisture give a clear idea as to the type of biomes that are expected. However, for a more rigorous comparison between the model and data, it is useful to compute the Walter biomes based on the monthly mean temperatures and monthly mean precipitation values. Our procedure is identical to that shown in Kutzbach and Ziegler (1994). The resulting prediction of Walter biome/climate zones for the Late Jurassic is shown in Fig. 10.12 (see color plate), and can be compared directly with the data-derived biome map shown in Fig. 10.7c.

Overall comparison between the data and model is encouraging, maintaining the broad pattern of summerwet equatorial regions, succeeded polewards by desert then warm and cool temperate biomes. In the tropics, there are a few grid points predicting tropical rainforest-type biomes, but there are no data near any of these grid points and so the model could be correct. The rest of the tropics is predicted to be summerwet, which is generally in good agreement with the data. The model predicts that this summerwet region (which is part of the tropical summer monsoon) extends into the Arabian region and up to 30° S. This is in direct conflict with the evaporite and eolian sand data; the reason for this discrepancy is not clear, although the lithological distributions may represent too broad a time interval (i.e., correlations may not be sufficiently precise). Small changes in orography and sea surface temperature could influence the latitudinal extent but the disagreement is greater than 20° of latitude, which is more than could simply be accounted for by the presently available data.

In mid- to high latitudes, the most striking feature of the northern hemisphere is that the model predicts more extensive regions of winterwet climates and more restricted regions of warm temperate climates. In some senses, the distinction between these climates is relatively subtle and so perhaps the model error is correspondingly small here. Nonetheless, the model is somewhat too cool at these latitudes. The tendency of the model to be too cool is more striking in the southern hemisphere. The data and model results agree on the southern side of the African/South American continent, and indicate a mixture of winterwet and warm temperate-type climates. One exception is that the model predicts much cooler conditions over the orography, which is as one would expect. We performed an additional calculation by converting the surface temperatures into mean sea level temperatures, by assuming a uniform change of temperature with height of 6.5 °C per 1000 m of elevation. This converted the cool and cold temperate climates into a mixture of warm and winterwet climates. Nevertheless, overall data and model agreement at high latitudes on the Gondwanan continent is poor. On the coast itself, the agreement is reasonable and it can be argued that this suggests that our choice of sea surface temperatures is acceptable. However, in the southern interior of the continent the model is predicting a mixture of cool and cold temperate climates and very few areas of warm temperate climates. As with northern high latitudes, the model is clearly predicting temperatures which are too cold. Using mean sea level temperatures does reduce the regions of cold temperate climates, but does not help greatly since the temperatures remain substantially colder than indicated by the geological data.

DISCUSSION

Despite the relatively subtle temperature gradients apparent in the Jurassic, floral variation does occur and can be employed to measure spatial and temporal climate changes. Individual leaf genera can occur in almost any floral list, but the climate signal is lodged in the sum total of the floral elements; our statistical approach at least provides an objective scale to compare the floras. It also highlights the scope for developing a more quantitative analysis of pre-angiosperm floras and paleoclimate interpretation. Our approach has already been applied to Mesozoic floras from Eurasia (Spicer *et al.*, 1994; Ziegler *et al.*, 1994, 1996) and shows the potential of pre-angiosperm floras as at least semiquantitative indicators of paleoclimate. Initial results from Cretaceous floral ordinations (Spicer *et al.*, 1994) have demonstrated that the appearance of angiosperms did not significantly alter the relationship between non-angiosperm distributions, physiognomy, and climate signals. It should therefore be possible to calibrate the climate signals derived from non-angiosperms against the quantitative signals obtained through studies of angiosperm leaf physiognomy (e.g., Wolfe, 1993).

It is the overall climate signal lodged in individual and collective floral assemblages which enables larger-scale interpretations of past climates. It should be emphasized that this is possible only if *all* the plants in a fossil assemblage are described or listed, instead of just the well-preserved or biologically interesting ones. Without this basic information the full potential of the plant fossil record to reveal past environmental and climate change cannot be exploited. Of course, this must be based primarily on accurate plant determinations, an appreciation of the limitations of the fossil plant data, and the critical use, whenever possible, of independent age evidence. Despite the reasonable agreement between our data-derived climate map (Fig. 10.7c) and the modeled ones (Figs. 10.8–10.12), our interpretations should still be regarded as preliminary. It is only by combining our data with that from other paleobotanical (e.g., palynology and fossil wood), paleosol, and lithological climate indicators, in addition to marine paleontological and isotopic data, that we can understand Jurassic climates based upon all of the available geological evidence.

Previous Late Jurassic (Kimmeridgian) GCMs (e.g., Moore *et al.*, 1992a; Valdes, 1994) match reasonably the global climate results from the new floral and lithological data, particularly for the northern hemisphere. There are some problems, however, particularly in southern Gondwana, where the models predict a cold temperate central core for the continent, precisely where geological data are absent. It should be noted that agreement between the new (Volgian) model and data on the southern edge of the African/South American continent is much better than with the original simulation of Valdes (1994). Thus, as well as providing an important test of the model, it could be argued that this provides further confirmation of the existence of a seaway in this region.

We suggest that, in some regions, the discrepancies between data and model results could be reduced if mean sea level temperatures were used instead of surface temperatures (which could be greater than 2000 m above sea level and hence more than 13 °C cooler). This raises the issue of one potential preservational bias on fossil plants. It is more likely that a fossil plant assemblage will be preserved at the bottom of a valley than at the crest of a ridge or mountain. Therefore the assemblage may tend to indicate the valley-bottom temperatures, which would be relatively warm. One means of reducing (though not overcoming) such bias is to use our 'whole-flora' approach to data collection, so that even locally rare elements growing more distally to the depositional sites are included in the floral lists. Despite this, the paleogeographic data and the model cannot really resolve valleys. Hence elevation in the model is considerably higher than the valley bottom and consequently the temperatures will be cooler. It can be argued that the correction to mean sea level provides a more realistic comparison between model and data. It is worth noting that such corrections are used routinely in weather forecasting: Salzburg, Austria, for example, is in a valley at approximately 300 m whereas many weather forecast models use a mean orography of more than 1000 m since they cannot resolve the valley. Hence the raw weather forecasts from the models are much too cool, and post-processing is applied to correct for the elevation difference.

The data/model comparison clearly suggests that the model is too cold at high latitudes in both hemispheres. Changing CO_2 levels and sea surface temperatures could make some difference, but is unlikely to completely resolve the problem. It appears to be another example of the 'equable climates' issue noted for the Cretaceous and Eocene (e.g., Sloan and Barron, 1990; Barron et al., 1994). Recent work (Dutton and Barron, 1996; DeConto et al., Chapter 9, this volume) has suggested that some of the disagreement can be reconciled by including feedbacks between the climate and the vegetation. This effect is also likely to be important for the Jurassic and we are in the process of incorporating this in our simulations.

Although we apply terms such as cool temperate to describe high-latitude Jurassic climates, it should be borne in mind that we are referring primarily to the length of plant growing season (4–6 months for the cool temperate biome). The main limitation on growing season length at such latitudes today is temperature, whereas in the Jurassic it was most probably light. So there is a difference, although many of the 'cool temperate' plants appear to respond in a similar fashion by shedding their leaves seasonally. The important point is that we are applying the biome label to introduce some consistency to the interpretation of past vegetation and climates and that this represents an initial attempt to interpret global patterns. It is the relative occurrence, abundance, and degree of association of different fossil leaf genera that help define these patterns. By conducting the exercise on a global scale and by applying statistical methods to arrange the plant data, we can produce 'biome/ paleobiome/climate/pattern' maps for intervals in the geological past which can be compared directly with the climate model results.

The terminology we have adopted is arguably less important than the fact that our approach is more comprehensive and repeatable than any other that uses pre-angiosperm plant data. Ultimately, if all of the 'patterns' match between the data and model results, then we can use the parameters of the model to accurately define global patterns of temperature, precipitation, and soil moisture. However, we can only be reasonably certain that the model is correct if we have a robust means of testing it initially. Our contribution represents just one example of ongoing studies aimed at testing, refining, and improving feedbacks between the data and models.

ACKNOWLEDGEMENTS

D. B. Rowley and M. L. Hulver provided invaluable help throughout, not least in facilitating production of the new paleogeographic maps illustrated here. We thank R. A. Spicer for discussion during initial stages of this work and for subsequent comments. We are also grateful to L. D. Boucher, L. A. Frakes, and S. L. Wing for their helpful reviews.

REFERENCES

Bailey, I. W. and Sinnot, E. W. (1915). A botanical index of Cretaceous and Tertiary climates, *Science*, **41**, 831–4.

Barnard, P. D. W. (1973). Mesozoic floras. *Special Papers in Palaeontology*, **12**, 175–87.

Barron, E. J., Fawcett, P. J., Pollard, D. and Thompson, S. L. (1994). Model simulations of Cretaceous climates: the role of geography and carbon dioxide. In *Palaeoclimates and their Modelling: With Special Reference to the Mesozoic Era*, eds. J. R. L. Allen, B. J. Hoskins, B. W. Sellwood, R. A. Spicer and P. J. Valdes, pp. 99–108. London: Chapman & Hall.

Dobruskina, I. A. (1982). Triassic flora of Eurasia. *Trudy Akademii Nauk SSSR, Seria geologicheskaya*, **365**, 1–195 (in Russian).

Dodson, P., Behrensmeyer, A. K., Baker, R. T. and McIntosh, J. S. (1980). Taphonomy and paleoecology of the dinosaur beds of the Jurassic Morrison Formation. *Paleobiology*, **6**, 208–32.

Dutton, J. F. and Barron, E. J. (1996). Genesis sensitivity to changes in past vegetation. *Palaeoclimates*, **1**, 325–54.

Frakes, L. A., Francis, J. E. and Syktus, J. I. (1992). *Climate Modes of the Phanerozoic: The History of the Earth's Climate over the Past 600 Million Years*. Cambridge: Cambridge University Press.

Gauch, H. G., Jr. (1982). Multivariate analysis in community ecology. In *Cambridge Studies in Ecology*, eds. E. Beck, H. J. B. Birks and E. F. Connor, pp. 1–298. New York: Cambridge University Press.

Hallam, A. (1984). Continental humid and arid zones during the Jurassic and Cretaceous. *Palaeogeography, Palaeoclimatology, Palaeoecology*, **47**, 195–223.

Hallam, A. (1994). Jurassic climates as inferred from the sedimentary and fossil record. In *Palaeoclimates and their Modelling: With Special Reference to the Mesozoic Era*, eds. J. R. L. Allen, B. J. Hoskins, B. W. Sellwood, R. A. Spicer and P. J. Valdes, pp. 79–88. London: Chapman & Hall.

Herman, A. B. and Spicer, R. A. (1996). Palaeobotanical evidence for a warm Cretaceous Arctic Ocean. *Nature*, **380**, 330–3.

Hill, M. O. (1979). Correspondence analysis: a neglected multivariate method. *Applied Statistics*, **23**, 340–54.

Horrell, M. A. (1991). Phytogeography and paleoclimatic interpretation of the Maestrichtian. *Palaeogeography, Palaeoclimatology, Palaeoecology*, **86**, 87–138.

Krassilov, V. A. (1972). Phytogeographical classification of Mesozoic floras and their bearing on continental drift. *Nature*, **237**, 49–50.

Krassilov, V. A. (1981). Changes of Mesozoic vegetation and the extinction of dinosaurs. *Palaeogeography, Palaeoclimatology, Palaeoecology*, **34**, 207–24.

Kutzbach, J. E. and Ziegler, A. M. (1994). Simulation of Late Permian climate and biomes with an ocean–atmosphere model: comparisons with observations. In *Palaeoclimates and their Modelling: With Special Reference to the Mesozoic Era*, eds. J. R. L. Allen, B. J. Hoskins, B. W. Sellwood, R. A. Spicer and P. J. Valdes, pp. 119–32. London: Chapman & Hall.

Lottes, A. L. and Ziegler, A. M. (1994). World peat occurrence and the seasonality of climate and vegetation. *Palaeogeography, Palaeoclimatology, Palaeoecology*, **106**, 23–37.

Meyen, S. V. (1973). Plant morphology and its nomothetical aspects. *Botanical Review*, **39**, 205–60.

Moore, G. T., Hayashida, D. N., Ross, C. A. and Jacobsen, S. R. (1992*a*). Paleoclimate of the Kimmeridgian/Tithonian (Late Jurassic) world: I. Results using a general circulation model. *Palaeogeography, Palaeoclimatology, Palaeoecology*, **93**, 113–50.

Moore, G. T., Sloan, L. C., Hayashida, D. N. and Umrigar, N. P. (1992*b*). Paleoclimate of the Kimmeridgian/Tithonian (Late Jurassic) world: II. Sensitivity tests comparing three different paleotopographic settings. *Palaeogeography, Palaeoclimatology, Palaeoecology*, **95**, 229–52.

Neilson, R. P. (1995). A model for predicting continental-scale vegetation distribution and water balance. *Ecological Applications*, **5**, 362–85.

Parrish, J. T. and Peterson, F. (1988). Wind directions predicted from global circulation models and wind directions determined from eolian sandstones of the western United States – a comparison. *Sedimentary Geology*, **56**, 261–82.

Prentice, I. C., Cramer, W., Harrison, S. P., Leemans, R., Monserud, R. A. and Solomon, A. M. (1992). A global biome model based on plant physiology and dominance, soil properties and climate. *Journal of Biogeography*, **19**, 117–34.

Price, G. D., Sellwood, B. W. and Valdes, P. J. (1995). Sedimentological evaluation of general circulation model simulations for the 'greenhouse' Earth: Cretaceous and Jurassic case studies. Sedimentary Geology, **100**, 159–80.

Sellwood, B. W. and Price, G. D. (1994). Sedimentary facies as indicators of Mesozoic palaeoclimate. In *Palaeoclimates and their Modelling: With Special Reference to the Mesozoic Era*, eds. J. R. L. Allen, B. J. Hoskins, B. W. Sellwood, R. A. Spicer and P. J. Valdes, pp. 17–26. London: Chapman & Hall.

Shi, G. R. (1993). Multivariate data analysis in paleoecology and paleobiogeography – a review. *Palaeogeography, Palaeoclimatology, Palaeoecology*, **105**, 199–234.

Singer, A., Wieder, M. and Gvirtzman, G. (1994). Paleoclimate deduced from some Early Jurassic basalt-derived paleosols from northern Israel. *Palaeogeography, Palaeoclimatology, Palaeoecology*, **111**, 73–82.

Sloan, L. C. and Barron, E. J. (1990). "Equable" climates during the Earth history? *Geology*, **18**, 489–92.

Spicer, R. A., Rees, P. M. and Chapman, J. L. (1994). Cretaceous phytogeography and climate signals. In *Palaeoclimates and their Modelling: With Special Reference to the Mesozoic Era*, eds. J. R. L. Allen, B. J. Hoskins, B. W. Sellwood, R. A. Spicer and P. J. Valdes, pp. 69–78. London: Chapman & Hall.

Ter Braak, C. J. F. (1992). *CANOCO – a FORTRAN Program for Canonical Community Ordination*. Ithaca, NY: Microcomputer Power (plus software v. 3.11, November 1990).

Vakhrameev, V. A. (1991). *Jurassic and Cretaceous Floras and Climates of the Earth*. Cambridge: Cambridge University Press.

Vakhrameev, V. A., Dobruskina, I. A., Meyen, S. V. and Zaklinskaya, E. D. (1978). *Paläozoische und Mesozoische Floren Eurasiens und die Phytogeographie dieser Zeit*. Jena: VEB Gustav Fischer Verlag.

Valdes, P. J. (1994). Atmospheric general circulation models of the Jurassic. In *Palaeoclimates and their Modelling: With Special Reference to the Mesozoic Era*, eds. J. R. L. Allen, B. J. Hoskins, B. W. Sellwood, R. A. Spicer and P. J. Valdes, pp. 109–18. London: Chapman & Hall.

Valdes, P. J. and Sellwood, B. W. (1992). A palaeoclimate model for the Kimmeridgian. *Palaeogeography, Palaeoclimatology, Palaeoecology*, **95**, 47–72.

Volkheimer, W. (1970). Jurassic microfloras and paleoclimates in Argentina. *Proceedings, Second Gondwana Symposium, South Africa*, pp. 543–9.

Walter, H. (1985). *Vegetation of the Earth and Ecological Systems of the Geo-Biosphere*, 3rd edn. New York: Springer-Verlag.

Wesley, A. (1973). Jurassic plants. In *Atlas of Palaeobiogeography*, ed. A. Hallam, pp. 329–38. Amsterdam: Elsevier.

Wolfe, J. A. (1993). A method of obtaining climatic parameters from leaf assemblages. *US Geological Survey Bulletin*, **2040**, 1–73.

Ziegler, A. M. (1990). Phytogeographic patterns and continental configurations during the Permian Period. In *Palaeozoic Palaeogeography and Biogeography*, Geological Society Memoir No. 12, eds. W. S. McKerrow and C. R. Scotese, pp. 363–79. London: Geological Society.

Ziegler, A. M., Parrish, J. M., Yao, J. P. *et al.* (1994). Early Mesozoic phytogeography and climate. In *Palaeoclimates and their Modelling: With Special Reference to the Mesozoic Era*, eds. J. R. L. Allen, B. J. Hoskins, B. W. Sellwood, R. A. Spicer and P. J. Valdes, pp. 89–97. London: Chapman & Hall.

Ziegler, A. M., Raymond, A. L., Gierlowski, T. C., Horrell, M. A., Rowley, D. B. and Lottes, A. L. (1987). Coal, climate and terrestrial productivity: the Present and Early Cretaceous compared. In *Coal and Coal-Bearing Strata: Recent Advances*, Geological Society Special Publication No. 32, pp. 25–49. London: Geological Society.

Ziegler, A. M., Rees, P. M., Rowley, D. B., Bekker, A., Qing Li and Hulver, M. L. (1996). Mesozoic assembly of Asia: constraints from fossil floras, tectonics and paleomagnetism. In *The Tectonic Evolution of Asia*, eds. A. Yin and M. Harrison, pp. 371–400. Cambridge: Cambridge University Press.

IV

Case studies: Paleozoic

Permian and Triassic high latitude paleoclimates: evidence from fossil biotas

EDITH L. TAYLOR, THOMAS N. TAYLOR, AND N. RUBÉN CÚNEO

ABSTRACT

High latitude fossil floras provide an important source of data on past climates, since the plants were living in a strongly seasonal light regime and often existed at the limit of their environmental tolerances. Permian and Triassic rocks from Antarctica have been a rich source of biological information because of the large number of sites that have yielded exquisitely preserved fossil plants. Anatomically preserved plants from several sites in the central Transantarctic Mountains and southern Victoria Land provide a unique source of fossil tree ring data. Samples of wood have come from a variety of sites including fluvial settings, permineralized peat deposits, and forest sites where the trees are preserved *in situ* (in growth position). The wood exhibits distinct growth rings which have been analyzed for paleoclimate signals. Rings from both Permian and Triassic woods are large, ranging from a few millimeters to several centimeters in width, and represent growth rates that are 1–2 orders of magnitude larger than those at high latitudes today. In addition, the structure of the individual rings differs from that seen in modern temperate rings. The presence of a considerable amount of earlywood and only 1–2 cells of latewood suggests that light may have been the limiting factor in the growth of these forest trees. Tree ring data, as well as the level of diversity in the Antarctic Permian and Triassic floras are at variance with the majority of paleoclimate models that have been produced for the region. The biological data suggest that these time periods were far warmer than predictions based on physical data alone.

INTRODUCTION

Interest in modern global climate change, coupled with increased computer power and efficiency, have been two of the major driving forces in the surge of interest relating to the fossil climatic history of the earth. Until relatively recently most of the paleoclimate data focused on events and climate proxy records from the recent ice ages, measured in tens of thousands of years. Today interest in paleoclimate extends well back into the Precambrian ($\sim$0.6–4.0 billion years ago), and can be documented by an increasing amount of both physical and biological evidence. Modern numerical climate modeling has contributed a substantial amount of data that has been used in association with continental configurations and, more recently, paleotopographic reconstructions to suggest climatic conditions during various periods of geologic time. The disciplines that encompass understanding paleoclimates are at an interesting crossroads in the evolution of this highly synergistic science. While the models may be accurate based on extrapolations of modern physical parameters, the crucial question now is how they will compare when measured against biological data from the fossil record of the same time periods.

Plants have played an increasingly important role as proxy indicators of the earth's paleoclimatic history. In general there have been three basic approaches that have utilized fossil plants to interpret climate. One of these is the nearest living relative (NLR) method, which infers the climatic tolerance of fossil plants based on the habitat of one or more extant relatives. This approach has the greatest reliability when there is a strong correlation between the modern and the fossil species, and thus is of limited value the further one goes back in geologic time. A second method of obtaining climate signals from fossil plants depends upon certain characteristics of leaf morphology (physiognomy). It assumes that leaves have a particular morphology relative to the climate in which they live regardless of their genetic history. This foliar physiognomic technique has been used extensively with flowering plants and relies on a number of features (e.g., margin, leaf tip morphology, cuticle thickness, size) applied over an entire flora to determine climate. The third method used to interpret paleoclimate is the analysis of fossil tree rings from wood samples. This procedure requires cellular preservation (permineralization) and relies on the fact that climatic changes affect the physiology of the plant, and thus become translated into stages of periodic growth, and are manifested as growth rings. The limitation of this method is the difficulty in determining whether fossil growth rings represent annual rings (i.e., one produced each year) or periodic growth phases caused by environmental fluctuations (e.g., seasonal drought or shifting water table). Based on the paleolatitude alone (80–85° S for the Late Permian and 70–75° S for the Middle Triassic, based on reconstructions in Smith *et al.*, 1981; Kutzbach and Ziegler, 1993; Golonka *et al.*, 1994), it is clear that the Late Permian and Triassic plants of Antarctica were living under a strongly seasonal climate regime. This allows for more accurate interpretation of paleoclimate based on tree ring features, and also permits the use of other biological indicators in land plants (anatomy, morphology, etc.) that thrived in Antarctica during these time periods.

FORMATION OF TREE RINGS

In woody plants stem diameter increases as a result of new secondary xylem (= wood) cells being produced by a lateral meristem termed the vascular cambium. In seasonal climates large wood cells with thin walls are produced when light, temperature, and soil moisture are favorable, such as typically occurs at the beginning of the growing season. This component of wood growth is termed springwood or earlywood. As the season continues, growth begins to slow, with the resulting cells becoming smaller in size and forming thicker cell walls. This later growth is termed summerwood or latewood. In a seasonal climate, growth ceases in the fall, when either the leaves are abscised (if the plant is seasonally deciduous) or the temperature falls below 0 °C (if the plant is evergreen). When growth is initiated the next season, there is a pronounced boundary between the small, thick-walled cells of the latewood that marked the end of the previous growing season and the large, thin-walled cells of the earlywood that represent the new season's growth (Fig. 11.1). Larson (1964) and others have correlated the production of earlywood with growth in length at the shoot tips; the change to latewood occurs with the cessation of elongation growth. This repeated production of large and small cells that reflect the growth phases of a woody plant is termed a growth ring (Fig. 11.1). Growth rings can be analyzed based on the width of the rings during each growth cycle, numbers of cells produced in each phase, and any abnormal features that reflect an interruption in the normal growth cycle. Such features could be caused by drought, unseasonable frost, or defoliation as a result of herbivory, and can leave a recognizable signature in the growth ring.

The genetics of the plant play a critical role in the production of growth rings, since they control the cell cycle in the meristematic zone by means of the production and distribution of plant hormones. Differential growth and cell production can result in a variety of anatomical differences in cell number and size, wall thickness, cell type, etc. of woody species growing within the same community. Therefore only trees that vary ring production in relation to climate are suitable for dendroclimatic and dendrochronologic studies. Ring sensitivity was developed as a simple measure of the amount of variance in ring width from year to year (e.g., Fritts, 1976). Trees that vary little on an annual basis are said to be complacent and are poor choices for dendroclimatic or dendrochronologic studies. The best specimens for a dendroclimatic study are individuals that vary in growth from year to year (so-called 'sensitive' trees), and that are growing close to the limits of their physiologic tolerances. In these marginal environments (e.g., high altitude, high latitude, water-stressed sites), the plants, and thus the tree rings, track their environment most closely and leave behind the strongest climate signal. Especially significant is the degree of resolution possible in tree ring data: the cells within each ring can represent an almost daily record of the environment in which the plant grew. However, it is equally important to remember when working with fossil material that rings are produced only during the growing season so they cannot provide details about winter climate or environmental events that take place in the non-growth phases of a woody plant. For this reason, it is critical to place any fossil tree ring analysis within the context of sedimentology and of interpretations based on paleoecology and fossil biotas.

In addition to the endogenous (i.e., genetic) factors that affect growth ring production, it is clear from the influence of seasonal climate on the formation of growth rings that exogenous factors leave a recognizable signature in the rings. There are three primary environmental influences that affect the formation of cells and thus tree rings in woody plants. These include available soil moisture, light regime, and temperature. At marginal sites, tree growth is generally limited by one or more of these factors, often interacting in complex ways. For example, in high latitude trees today, such as those in the Canadian Arctic, Alaska, or Siberia, wood production is generally limited by a combination of water and temperature. Since there is widespread permafrost in high latitudes today, the availability of water for plant growth,

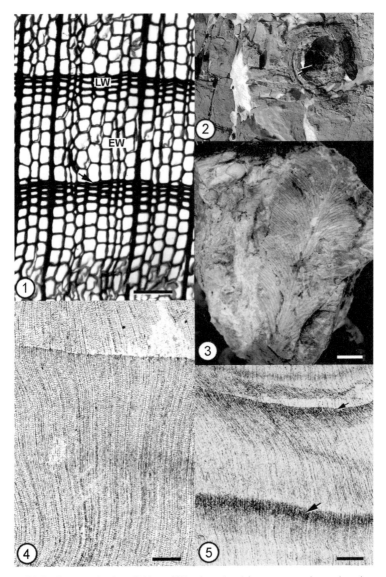

Figure 11.1. Extant *Larix sibirica* (Siberian larch) cross-section showing ring boundary (arrow), earlywood (EW), and latewood (LW). A single year's growth extends from the arrow to LW. Scale bar = 100 μm.

Figure 11.2. Permian fossil forest from Mount Achernar: Stump *in situ* (arrow). The ring around the stump is a depositional feature. Diameter of stump ~14 cm.

Figure 11.3. Surface view of permineralized *Glossopteris* foliage (Permian; Skaar Ridge, Beardmore Glacier area, Antarctica). Scale bar = 1.0 mm.

Figure 11.4. Permian fossil forest from Mount Achernar: Cross-section of wood through parts of two rings showing an exceptionally wide growth ring with ring boundary near the top of the photograph. Only at the boundary is latewood present; the remainder of the photograph is earlywood. Scale bar = 0.5 mm.

Figure 11.5. Permian fossil forest from Mount Achernar: Cross-section through two rings showing crushing near ring boundary (smaller arrow) and false ring (larger arrow) in the center of a single year's growth. Scale bar = 1.0 mm.

and thus the length of the growing season, is also strongly influenced by temperature. Although day length is a distinct factor in regulating many aspects of plant growth at all latitudes (e.g., seed germination and reproductive timing), it does not appear to be important in the onset or cessation of growth in woody plants in the Arctic. This conclusion is based on the fact that wood production apparently ceases in the fall long before the light levels become too low for photosynthesis to occur (Oswalt, 1960). By contrast, light *does* appear to be the determining factor in controlling the production of cell growth in the fossil wood that has been examined from high latitude sites thus far (Taylor, 1989; Taylor and Taylor, 1993). In addition to the environmental factors that affect wood growth, when interpreting fossil tree rings and paleoclimate it is important to consider whether the plants were deciduous or evergreen. All of these environmental parameters will be considered relative to the analysis of growth rings in Permian and Triassic wood from various sites in Antarctica.

ANTARCTIC FOSSIL WOOD

Fossil wood is relatively common in Permian and Triassic rocks of the Transantarctic Mountains, and occurs in a variety of depositional settings. This evidence of former forests on the Antarctic continent occurs most commonly as isolated fragments of wood or small trunks within fluvial sandstones and mudstones. Almost all of the wood is extensively decorticated, i.e., the outer tissues (bark) have been lost as a result of transport. While such wood can provide general climate information, it is possible to have trunks in a single deposit that have been transported from some distance (e.g., see 'Fremouw Peak' below; Van Stone, 1958; Francis, 1986). In marine deposits, wood can be transported hundreds or even thousands of miles (e.g., Barber *et al.*, 1959; Smith *et al.*, 1989; Dickson, 1992) and thus may not be a good indicator of local conditions.

In addition to isolated wood or trunk fragments, there are also several deposits of permineralized peat that contain woody plants. These are important in that they provide evidence for identifying the wood (e.g., leaves or reproductive organs), as well as evidence of the flora associated with the woody plants. Knowledge of the composition of the flora and the structure of the plants within it can provide additional paleoclimate information and can help to corroborate conclusions based on tree rings alone. The most important deposits of permineralized wood are those of *in situ* forests, i.e., trees that were permineralized at the site and in the position in which they grew. In addition to paleoclimate data from the tree rings themselves, these sites also provide data on the spacing of the trees, and thus the density and productivity of the forest. The designation '*in situ*' may have several meanings relative to fossil wood and trees. For most geoscientists, the term '*in situ*'' simply means that the wood is still contained within the sediment matrix rather than having been removed by erosion. The meaning given to '*in situ*' by biologists, however, refers to fossils that remain in their original position of growth. The fossil forests discussed here fit the biological definition, i.e., the stumps are rooted in the place in which they grew. While identifying fossil trees in growth position would appear to be a relatively simple task, the aftermath of the

1982 Mount St Helens eruption (Cascade Range, USA) demonstrates that it can often be difficult to determine whether or not log and stump deposits are *in situ*. Using the deposits generated by the eruption, for example, Fritz (1980) demonstrated that stumps with attached roots can be transported considerable distances and ultimately deposited in an upright position. To avoid any confusion, he suggests that multiple criteria should be used to determine whether a forest is *in situ* (Fritz and Harrison, 1985). These include the presence of intact large roots that extend into the sediment, a certain percentage (>15%) of trunks that are upright rather than horizontal, and finally a depositional environment that would indicate forest growth, such as mudstones or lacustrine rocks rather than channel sandstones.

PERMIAN *IN SITU* FORESTS

Mount Achernar

A fossil forest was measured and sampled during the 1991–1992 field season on a nearly horizontal bench near the peak of Mount Achernar (84° 22′ 23″ S, 164° 37′ 56″ E) (Buckley Island Quadrangle; Barrett and Elliot, 1973). The site is in the central Transantarctic Mountains, overlooking the Law Glacier (Taylor *et al.*, 1992) (Figs. 11.6 and 11.7). The forest occurs within the upper part of the Buckley Formation, and consists of 15 trunks preserved in growth position within a shaley floodplain deposit. Based on palynomorphs recovered from the Buckley Formation at nearby sites (Farabee *et al.*, 1991) and lithostratigraphy, the age of the forest is considered to be Late Permian. The trunks range from 9 to 18 cm in diameter and are preserved close to ground level; the tallest specimen is only 6–8 cm high (Fig. 11.2). The shale matrix contains roots of *Vertebraria* and abundant *Glossopteris* leaf impressions, suggesting that the trees bore this leaf type (Cúneo *et al.*, 1993) (Fig. 11.3). Statistical analysis of the spacing of the trunks indicates that the density of the forest was very high, with close to 2000 individuals per hectare. Mean growth ring width is 4.5 mm, with the largest ring extending to 11.38 mm wide. The rings characteristically exhibit a large amount of earlywood and very little latewood (only 2–3 cells per year); the maximum number of rings in a single specimen is seven. Combining these data with the large size of the individual rings (Fig. 11.4), the forest has been reconstructed as a stand of rapidly growing young trees (i.e., saplings).

There are a number of areas in the wood of these trees where the cell walls are thickened and thus resemble ring boundaries (Fig. 11.5). However, there is no true cessation of growth in these areas, indicating that these represent false rings, areas of latewood-like cells within a single ring. In modern trees, false rings are developed in response to an environmental stimulus, such as drought, low temperatures, or defoliation during the growing season (Norton and Ogden, 1987). They can also form as a result of periodic rapid growth of the apex and young leaves (Larson, 1969; Telewski and Lynch, 1991). These multiple growth flushes would involve elongation by the apex and production of another whorl of leaves. It is not possible to determine with certainty the cause of the false rings in these specimens. However, it is easy to speculate that since these trees were young, and growing so rapidly as to average 4.5 mm of wood production per year, they could easily have produced multiple flushes of growth. At the same time their water requirements would have been extensive and they may have experienced periodic water stress.

The wood from Mount Achernar also contains zones of crushed cells that at first glance resemble latewood bands (Fig. 11.5, smaller arrow). These zones may have developed in response to stress or torsion to the trunk of the tree, such as strong wind or some other form of mechanical pressure. However, it is more likely that they were formed after the depositional event that initially buried the trees and before fossilization, i.e., at a period when the trees were waterlogged and perhaps being compacted by the weight of the sediment surrounding them. In some areas, the radial lines of wood cells are compressed in a chevron (zigzag) pattern, which appears to be fairly common among permineralized *in situ* trunks (e.g., Francis, 1986; Basinger, 1991). The occurrence of this pattern in disparate fossil forests suggests that it is probably due to taphonomic events and not to the particular structure of the wood itself.

McIntyre Promontory forest

During the 1995–1996 field season, in association with a remote camp located adjacent to the Shackleton Glacier, standing trunks were discovered on a steep slope on the eastern face of McIntyre Promontory (84° 56′ 41″ S, 179° 45′ 58″ E). The trunks are *in situ* within a carbonaceous shale and the paleoenvironment is characterized as transitional from lacustrine to fluvial (J. L. Isbell, personal communication). The trunks occur within the upper part of the Lower Buckley Formation and therefore are older than the Mount Achernar forest. The trees, up to 0.5 m in diameter, occur at different levels, suggesting that they represent more than a single forest occurrence (Taylor *et al.*, 1997). Within the shales are numerous *Glossopteris* leaves. Owing to the inaccessibility of the site and the weight of the trunks, only a single specimen was collected. It represents the smallest of the trees at the site and measures 42 cm long by 31 cm wide through the bole-root transition zone. Rings are preserved in portions of the axis although overall preservation is poor. They average 2.8 mm, with the widest ring being 6 mm. Preliminary analysis of the silicified wood indicates that the growth rings are similar in structure to those that occur in the trees from the other Permian and Triassic sites in Antarctica. In the lacustrine deposits below the trunks at McIntyre Promontory are dropstones. Since evidence of continental glaciers is lacking from the sedimentary record, it is suggested that these have resulted from the seasonal break up of river ice (J. L. Isbell, field observations). The paleolatitude for the site is similar to that for Mount Achernar, approximately 80–85° S, but McIntyre Promontory was at least 160 km further south than Mount Achernar. At such a high latitude, even 160 km would make a difference in the growing season. The McIntyre Promontory forest is particularly significant, not only because it represents an *in situ* Permian forest from a more southerly site, but also because it is the first example of a mature forest from the Permian of Antarctica.

TRIASSIC WOOD AND *IN SITU* FORESTS

Allan Hills

A Triassic site with abundant silicified wood occurs on the eastern arm of the Allan Hills in southern Victoria Land (Gabites, 1985) (Fig. 11.6). The sediments are included within Member B of the Lashly Formation (Triassic) and contain point bar surfaces that are characteristic of meandering streams (Isbell *et al.*, 1990). Associated with the logs are hummocks of silicified peat in which the various plant parts are highly fragmented, suggesting appreciable transport; these peat deposits also show considerable degradation from microbes and fungi. While logs and peat from sites like the Allan Hills locality provide useful information about features of the wood and the nature of the growth rings, such specimens have a limited value in that it is impossible to determine much about the life association of the plants, because of transport prior to deposition.

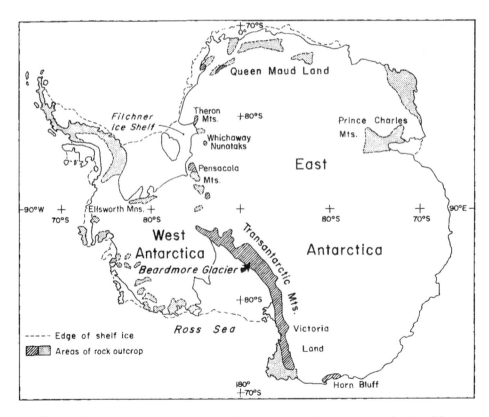

Figure 11.6. Map of Antarctica and Transantarctic Mountains showing localities mentioned in the text.

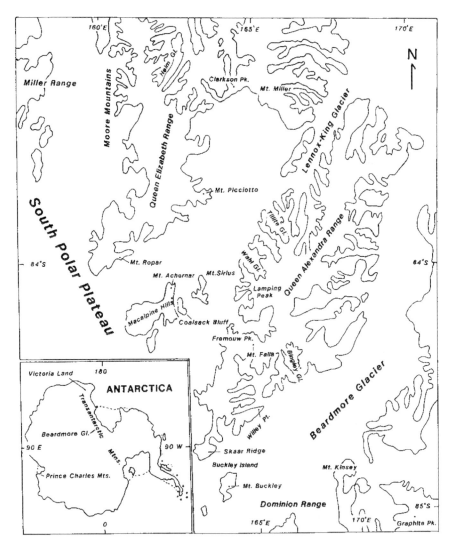

Figure 11.7. Map of Beardmore Glacier area, central Transantarctic Mountains.

Fremouw Peak

The Fremouw Peak site occurs in the Queen Alexandra Range of the central Transantarctic Mountains (Figs. 11.6 and 11.7) and has been extensively studied because of the presence of large, silicified peat rafts containing abundant, well-preserved plants and fungi (e.g., Taylor and Taylor, 1990, and references therein). In addition to the peat rafts, this early Middle Triassic locality contains logs of various dimensions (Fig. 11.10), stumps (Fig. 11.8), and large, woody roots (Fig. 11.9), all of which were deposited in a large stream bed. It is suggested that the peat rafts represent a portion of a forested island or river levee that was undercut during periods of flooding, moved some distance, and subsequently deposited on the chan-

Figures 11.8–11.13. Triassic permineralized wood from col north of Fremouw Peak, central Transantarctic Mountains.

Figure 11.8. Trunk in fossil stream bed. Although the center of the trunk is not preserved, tree rings are visible on the periphery (next to radio antenna), and large pieces of wood are visible in the foreground. The radio is ~20 cm in length without antenna.

Figure 11.9. Cross-section of fossil root wood showing varying width of tree rings; center of axis to bottom of photograph. Scale bar = 1.0 mm.

Figure 11.10. Large, shattered trunk in stream bed (~20 m long).

Figure 11.11. Cross-section of oldest stem known from this site (85 rings total). This trunk also contains frost rings. Scale bar = 2.0 mm.

nel floor (Collinson *in* E. Taylor *et al.*, 1989*a*). Present at this locality are woody twigs only a few millimeters in diameter (within the peat) and logs greater than 1.5 m in diameter (in the paleochannel). Attesting to the size of some of the trees that once inhabited this area is one specimen exposed in the paleochannel which measures more than 20.2 m long (Barrett, 1969) (Fig. 11.10). Additionally, the specimen shows very little taper along its entire preserved length, indicating that the tree probably exceeded 30 m in length. Unfortunately, this specimen is too fragmented by weathering processes to be useful for ring analysis. The approximate diameters of other, less complete stems may be determined by plotting the radius of curvature of the rings. It is important to point out that the woody stems that have been examined to date have not been taxonomically separated; rather ring characters and sizes have been averaged from a collection of specimens that may represent more than a single biological species. As additional studies define the taxonomic boundaries of these various specimens, there may be some slight differences in ring production among species, but this will not alter the basic paleoclimatic signals that can be determined from the wood.

All of the woody axes at the Fremouw site show well-developed growth rings that range from 0.1 mm to 7.2 mm in width (Fig. 11.11). An analysis of specimens both within the peat and in the sandstone channel indicates that the rings from these two sources are similar in structure and size, the only difference being the presence of larger, older trunks in the stream bed. This would be expected in that the stems within the peat are limited by the maximum size of the peat hummocks ($1 \times 2.5 \times 2.5$ m). The largest intact stem contains 85 rings that average 2.1 mm in width (Fig. 11.11); the largest root has 215 rings averaging 0.6 mm. The average ring width for stems from this site is 2.08 mm. These Triassic rings are similar to those of other previously described, high latitude fossil woods in that there is abundant, thin-walled earlywood (up to 90% of the ring width) and very little latewood present in each ring (Fig. 11.12) (e.g., Jefferson, 1982; Parrish and Spicer, 1988; Taylor, 1989; Spicer and Parrish, 1990; Taylor and Taylor, 1993). The cell walls in the latewood are not appreciably thickened as they are in temperate woods and, in the Antarctic material, there is very little change in cell size from earlywood to latewood.

In addition to numerous specimens of 'normal' wood from Fremouw Peak, there is a single woody axis collected from the stream channel which contains multiple examples of frost rings (Figs. 11.11 and 11.13). Frost rings are distorted or otherwise crushed cells that appear at the beginning or the end of a single year's growth increment. They form when there is an unseasonable frost, either late in the spring or early in the fall. The low temperatures disrupt and/or kill the actively dividing cells in portions of the vascular cambium. When growth eventually resumes, the first wood cells formed are misshapen and usually not aligned in well-defined radial rows. Evidence for the presence of frost rings is most commonly found as a narrow line of distorted or enlarged cells within the earlywood. Frost rings provide an additional climatic signature that may appear in fossil wood, but which to date has not been extensively analyzed.

Figure 11.12. Detail of wood cross-section in *Rhexoxylon*-like stem (Taylor, 1992), showing ring boundary (at LW). Center of stem to the bottom. Note large amounts of earlywood (above and below boundary) and small amount of latewood. Scale bar = 200 μm.

Figure 11.13. Cross-section from trunk in Fig. 11.11 showing frost damage (above). Arrow below marks normal ring boundary. Scale bar = 1.0 mm.

Figure 11.14. Standing trunk at Gordon Valley (Triassic) site. Trunk is permineralized in place and rooted in shales containing abundant *Dicroidium* foliage. Trunk height ~0.5 m.

Figure 11.15. Compressed foliage of *Dicroidium* (Triassic; Gordon Valley). Scale bar = 1.0 cm.

Figure 11.16. Cross-section of wood showing response to wounding (center). Normal growth at bottom; curving ring boundaries on right and left indicate growth over damaged area. Scale bar = 1.0 mm.

Why are frost rings absent in the other samples from this site, including those preserved in the peat clasts? One plausible explanation may be related to the altitude at which this particular tree was growing. Since only a slight elevational difference at such a high latitude (70–75° S paleolatitude) can have a marked influence on climate and thus on growth ring production, the possibility exists that the tree with frost rings grew at a higher elevation where intermittent frost occurred, whereas the remainder of the woody axes present at the Fremouw site grew in communities at lower altitudes which experienced no frost during the growing season.

Gordon Valley forest

This site is in the upper Gordon Valley, near Mount Falla (central Transantarctic Mountains) (84° 11′ 10″ S, 164° 54′ 28″ E) (Taylor *et al.*, 1991) (Fig. 11.7), and occurs within the upper part of the Fremouw Formation. In this area, this part of the Fremouw is interpreted as Middle Triassic in age (Barrett *et al.*, 1986; Farabee *et al.*, 1990). Occurring at this site are 98 standing trunks, ranging from 13 to 61 cm in diameter (Fig. 11.14), that probably grew along a river levee in a proximal floodplain environment (Cúneo *et al.*, 1999). Based on trunk diameters it is believed that some of the trees attained heights in excess of 20 m; the largest preserved height is 0.5 m. The Gordon Valley trees have growth rings with the same structure (large amount of earlywood, small amount of latewood) as those from other woods of both Permian and Triassic age in Antarctica. When compared with modern gymnospermous wood types, these fossil trees share the largest number of anatomical characters with wood of the Podocarpaceae (del Fueyo *et al.*, 1995). Based on the abundant impression-compression foliage of *Dicroidium* in the rocks at the base of the stumps (Fig. 11.15), it appears probable that the Gordon Valley trees bore leaves of this type. Because the leaves occur in extensive mats in the sediment, it has been suggested that the trees were seasonally deciduous (Taylor, 1996).

The large number of *in situ* stumps preserved at the Gordon Valley site also provides a unique opportunity to examine the density of a high latitude forest stand. Using the point-centered quarter method (Cottam and Curtis, 1956), the forest is estimated to have included approximately 257 trees per hectare (Cúneo *et al.*, 1999). It is interpreted as a mature forest, based on the size of the trunks, the density of the trunks, and the tree rings, which indicate that the oldest preserved tree had 86 rings. Ring widths average from 0.92 to 2.54 mm (overall range 0.23–7.1 mm). The upper end of this range is comparable to the largest rings in the Mount Achernar wood. However, unlike Mount Achernar, the Gordon Valley forest was a mature forest. Many trees produce large rings when they are young, especially if growing in favorable conditions, but gradually form narrower rings as they age. The presence of consistently large rings in the trunks from Gordon Valley reinforces the interpretation of the climate as very advantageous for tree growth. Clearly the largest ring width (7.11 mm) is not simply an anomaly or a mismeasurement. This stem (specimen 11,471) includes a total of 7 rings that are wider than 4.0 mm and 17 that are wider than 3.0 mm, representing a tremendous amount of growth at a paleolatitude of 70–75° S. As a comparison, modern trees in Alaska average a few tenths of a millimeter of wood growth each summer (0.25–0.9 mm (Giddings, 1943) and 0.31–1.69 mm in trees growing in a valley bottom with silty soil (Oswalt, 1958)).

INTERPRETATION OF FOSSIL TREE RING DATA

There are a number of factors that affect tree ring formation and plant growth that cannot be determined from ring analysis alone. Whether woody plants are seasonally deciduous or evergreen can have a tremendous effect on the environmental limits of the species. In addition, the size of the leaves in the canopy will affect not only the density of the forest but also the nature of the understorey. Although some tropical plants can drop their leaves throughout the growing season, most temperate plants are seasonally deciduous, dropping their leaves at the end of the fall or, more rarely, during a mid-season drought. There are basically three methods that can be used to determine whether or not a fossil plant from a temperate paleoclimate is seasonally deciduous: (1) comparison to nearest living relative (NLR), (2) anatomical evidence of an abscission zone in the leaf or stem, and (3) presence of leaves in varved deposits, which is relatively rare, or in mats, which is more common in the depositional record (Spicer, 1989). As in paleoclimate determination, the NLR method is more accurate with geologically younger fossils, which are more likely to have close living relatives, such as Tertiary angiosperms. In general, the older the fossil specimens, the more likely it is that the climatic tolerances of the fossil and the extant plants may have diverged. Spicer and Herman (1996) have recently drawn attention to this problem in the Mesozoic cycad *Nilssoniocladus*. Unlike extant cycads, which are evergreen, *Nilssoniocladus* is interpreted as being deciduous, and thus was able to grow at very high latitudes in a strongly seasonal climate regime. Modern cycads are tropical to subtropical in their distribution.

The major challenge facing the plants that lived in Antarctica during the Permian and Triassic was their need to balance photosynthesis and water loss (via transpiration) in a habitat that included months of winter darkness. Trees growing at high latitudes today face somewhat the same challenge (although at slightly lower latitudes), and have adapted in part by reducing leaf surface and therefore reducing respiration. The majority of high latitude modern trees are evergreen conifers that bear needle- or awl-shaped leaves. This allows them to reduce respiration, survive freezing temperatures, and photosynthesize within a limited growing season. However, both *Glossopteris*, the dominant leaf type in Gondwana during the Permian, and *Dicroidium*, which filled the same niche in the Triassic, were broad, expanded leaf types (Figs. 11.3 and 11.15). *Dicroidium* leaves are most similar to modern fern fronds, and the lanceolate-shaped *Glossopteris* leaves extended up to 30 cm in length. It is thus highly unlikely that either of these plants with such large leaf surfaces were evergreen, as transpiration rates would have been too high for them to survive not only winter but also the periods of low light each fall and spring. We have reconstructed these Antarctic forest plants as deciduous (Taylor, 1996). Both *Dicroidium* and *Glossopteris* leaves are found in dense mats, not only in Antarctica but also on other Gondwanan continents. *Glossopteris* leaves have also been noted in varved silts by Gunn and Walcott (1962) and in varved lake sediments in Australia (Retallack, 1980). Finally, permineralized stems of *Kykloxylon*, which are believed to have borne *Dicroidium* leaves, exhibit a periderm layer beneath the leaf bases. Since the growth of this layer would have severed all contact of the leaf with the living tissue of the stem, the presence of a periderm beneath the leaf bases indicates a regular and programmed abscission of the leaves (Meyer-Berthaud *et al.*, 1992, 1993). Spicer and Parrish (1986) and Spicer (1989) have noted that leaves which are dropped in response to cold tend to accumulate in mats, because of a lower rate of decay.

The presence of large-leaved, deciduous forests at high latitudes fits neither the predictions of modern biome-based plant distributions nor the vegetation types predicted by Wolfe's leaf physiognomy techniques (e.g., Wolfe, 1985, 1993 and references therein). It does fit the predictions of some earlier studies on fossil floras, however, which noted that predominantly deciduous floras occurred at high latitudes in times of polar warmth, while evergreens predominated in times with cold poles, such as the present (Wolfe, 1987; Spicer and Chapman, 1990). Wolfe (1985) noted a similar deviation from predictions at high latitudes in his review of Tertiary floras of the northern hemisphere. Where leaf physiognomy data predicted notophyllous (7.6–12.7 cm long) broad-leaved evergreen forest, Wolfe found large-leaved deciduous plants and suggested that perhaps low winter light levels prevented the development of evergreen forests. Creber and Chaloner (1984*a*) have pointed out, however, that the amount of light received near the poles is equivalent to that received at lower latitudes, but it is simply distributed differently throughout the growing season. We believe a more likely scenario is that, in the case of both Wolfe's Tertiary forests and the Permian/Triassic Antarctic forests, the trees were deciduous because at least some part of the dark winter, including early spring and late fall when light levels are very low, was above freezing. Within such a climatic regime, evergreens would continue to respire during periods of low light and exhaust their stored food reserves, whereas deciduous trees could become truly dormant.

It is also important when considering leaf physiognomy and climate at high latitudes to relate leaf morphology to the architecture of the trees and the density of the forest. As Creber and Chaloner (1984*b*) have discussed, at high latitudes the tall, conical architecture of individual trees and their spacing is clearly related to the low angle of solar radiation. Jefferson (1982), in his description of a fossil forest from the Lower Cretaceous of Alexander Island, Antarctica, also reconstructed the trees as conical. More recent simulation studies utilizing different crown architectures and angles of sunlight confirm these reconstructions (Kuuluvainen, 1992). Concerning leaf size, Wolfe (1985) suggests that large leaves in polar forests represent an adaptation to the low angle of light. He also notes that, because of the cooler temperatures near the poles, the leaves would not be subjected to overheating, as is common with large leaves in more temperate or tropical habitats. Certainly, the size of the Permian *Glossopteris* leaves and the Triassic *Dicroidium* leaves from Antarctica agrees with Wolfe's observations on northern hemisphere polar fossil floras.

All of the Antarctic woods discussed in this paper grew at high polar latitudes above 70° S. In addition to other Antarctic fossil woods (e.g., Jefferson, 1982; Jefferson *et al.*, 1983; Francis, 1986, 1991*a*; Francis *et al.*, 1993), Cretaceous wood from high paleolatitudes in the Arctic also shows well-defined growth rings (e.g.,

Parrish and Spicer, 1988; Spicer and Parrish, 1990). Eocene *in situ* forests have been found on Ellesmere and Axel Heiberg islands in the Canadian Arctic (Basinger, 1991; Francis, 1988, 1991*b*), at a paleolatitude of 74–80° N (Irving and Wynne, 1991). These trunks contain growth rings up to 1.0 cm in width, indicating that climatic conditions were favorable for periods of rapid growth. In overall structure, the rings from the Arctic polar forests, both Cretaceous and Tertiary, are identical to the Antarctic specimens, with a large amount of earlywood and small percentage of latewood in each ring. Clearly, this type of ring structure is common among the high latitude fossil floras; but is it present in modern forests?

It is common in paleontology when trying to decipher features in fossil organisms to look for similar patterns in modern analogs. This method has met with varying degrees of success, depending upon the age of the fossil and the questions being posed. In the case of the Permian and Triassic floras from Antarctica, neither of the dominant woody plant groups and very few of the other fossils in this flora have modern relatives. For the interpretation of paleoclimate based on tree ring analysis, a potentially more rewarding approach might be to examine more distantly related plants with a similar growth habit, i.e., woody, deciduous gymnosperms from seasonal, high latitude climate regimes. Using this approach, the closest modern analog would be *Larix sibirica*, the Siberian larch, one of only three genera of extant deciduous gymnosperms. *Larix sibirica* grows today between about 60 and 72° N latitude in Siberia. Two species of *Larix*, along with several species of *Salix*, the Arctic willow (a flowering plant), represent the highest latitude woody vegetation today, extending as far as 72° N. At this latitude there is a reduced period of available light, which is analogous to the photoperiod in which the trees grew during the Permian and Triassic in Antarctica. A transverse section of Siberian larch wood (Fig. 11.1) shows that there is a gradual thickening of cell walls in the transition from earlywood to latewood, with a corresponding diminution in cell radial diameter (Taylor and Putz, 1993). The fossil wood, however, from both Permian and Triassic deposits, shows a different pattern of growth, with a rather abrupt change from earlywood to latewood (Figs. 11.4 and 11.12) and little change in cell diameter (Fig. 11.17).

The pattern seen in living *L. sibirica* is typical of temperate wood today, even at the highest modern latitudes. As noted above, most modern high latitude trees are limited in their growth by availability of water and by temperature. Since the two factors interact so intimately, it is often difficult to determine which is more important. If water were increasingly scarce during the growing season in the Permian and Triassic woods, one would expect wood production similar to that found in extant temperate woods – gradual decrease in cell size and gradual increase in cell wall thickness. We have previously suggested (Taylor, 1989; Taylor and Taylor, 1993) that light was the limiting factor in the growth of the trees in Antarctica. As light levels decreased each fall, they reached a minimum level when photosynthesis was no longer energy-effective. At this point, the leaves abscised and the tree became dormant. With the low angle of the sun at 70–85° S paleolatitude, this change was no doubt very dramatic, and the tree rings reflect this sudden change. In addition, there are other aspects of the wood structure that suggest

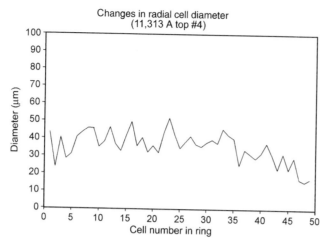

Figure 11.17. Changes in radial cell diameter across the width of a single ring, showing little change throughout the growing season.

water was not a limiting factor. When water is abundant, trees tend to produce large diameter, thin-walled cells, and these are seen throughout the Antarctic growth rings, rather than just in the earlywood, as is common in modern high latitude plants. Briand *et al.* (1993) note that in fast-growing trees, cell diameter is relatively constant for most of the ring width, with a sharp decline before the initiation of latewood. This type of anatomy can be correlated with trees that have abundant water for growth.

ADDITIONAL PALEOCLIMATE SIGNALS FROM WOOD

While the presence of growth rings in fossil wood has been the most widely used paleoclimatic signal, there are several other wood features that may yield information about the climate in which the plant lived. These include different types of scars in the wood which may be fairly common in many fossil woods. Each scar represents a cessation of vascular cambium activity and may be continuous around the circumference of the stem or occur only in a single sector. Recently, Putz and Taylor (1996) analyzed fossilized stems of Triassic age from Antarctica and identified four basic scar patterns that were the result of wound responses to both biotic and abiotic factors. The most common pattern consists of scars that are triangular in outline on the stem surface and in transverse section. This type of scar is formed when a strip of vascular cambium is destroyed and the subsequent ring growth develops over it from both sides (Fig. 11.16). The wound may or may not have been closed at the time of fossilization, but the curved pattern of subsequent annual rings is very characteristic.

Another pattern in the Antarctic woods consists of open wounds which never closed by the addition of subsequent growth. Such scars are generally elongate and remain open until the death of the tree, although partial closure as a result of cambial activity and wound periderm may be evident. There is a striking similarity between the triangular scars on the Antarctic trees and those present on modern trees that have undergone fire damage. Unfortunately ancillary evidence in the form of fusain in the sediments or in the surrounding wood could not be conclusively demonstrated, and thus it is impossible to determine whether the scars were the result of fires in the Antarctic forest or some type of mechanical damage. Such damage could include slope movements, ice or debris damage in floods, or destruction by animals (Putz and Taylor, 1996).

Wounds and other abnormalities in wood may also be the result of various fungi that gain entrance to the plant stem by some abiotic event, such as mechanical damage. In some cases the response of the host, such as the formation of barrier zones or subsequent activities of the vascular cambium, can produce features that document these interruptions in normal wood ring development (Blanchette, 1992). One of the most common indications of fungal interaction in Antarctic woods is the presence of spindle-shaped cavities up to 3.0 mm in diameter and 3.0 cm long. These structures are present in Permian and Triassic woody axes of both stems and roots and constitute evidence of the activity of particular basidiomycetous fungi (Stubblefield and Taylor, 1986). These fungi degrade areas of the wood, resulting in a condition termed white pocket rot. Present in the same Antarctic wood is evidence of other fungi, including those that selectively degrade lignin and cellulose. These fungi produce different 'symptoms' in the wood, such as the separation of the middle lamella from the primary cell wall or pale cell walls due to the degradation of lignin alone. These white rot fungi and the pocket rot fungi exist today in temperate forests and produce the same symptoms as those found in the Antarctic woods (Blanchette, 1991). Although there are few examples of the fungi themselves preserved in the woods from Antarctica, the identical symptoms clearly indicate the existence of these organisms during the Permian and Triassic of Antarctica, and thus offer another set of biological indicators that may prove useful in establishing paleoclimatic parameters (e.g., temperature, moisture content).

Other biotic factors, such as attack by herbivores, may cause disruptions in wood formation and thus offer information useful in paleoclimatic reconstructions. While some of the scars reported from the Antarctic woods could be caused by various arthropods and small animals, to date there are no reliable data that can be adequately documented (Putz and Taylor, 1996). Nevertheless, the presence of various types of frass (coprolites) associated with the plants in the Antarctic peat (Taylor *et al.*, 1998) indicates the presence of an arthropod fauna, some members of which may have interacted with woody plants as herbivores (Krause and Morin, 1995), or as burrowing organisms that caused abnormal cambial growth patterns (Raitio, 1992). While scars and abnormal growth characteristics in fossil wood may be difficult to interpret, they do need to be described and assembled with other lines of both biotic and abiotic evidence to more accurately characterize the paleoecosystem and paleoenvironment.

OTHER FLORAL EVIDENCE

The study of tree rings alone cannot provide a complete or perhaps even a totally accurate picture of paleoclimate at a particular point in geologic time.

Previous studies on plants of Permian and Triassic age from Antarctica have provided basic information about the biology and evolution of these floras. On a broader scale, however, the reconstruction of several of the woody plants in these floras, based on *in situ* forests (Taylor, 1996), material from permineralized peat (Table 11.1), and compression fossils, has made it possible to more accurately characterize the dominant plants in the landscape, and their associations with other organisms. As a result it is clear that the climate during both the Late Permian and the Middle Triassic was hospitable for growth of a number of different plant types.

Late Permian

The Permian floras from Antarctica, as elsewhere in Gondwana, are of relatively low diversity, especially when compared with those of similar age in the northern hemisphere. Some of the diversity is probably masked, however, by the dominance of glossopterids in both abundance and number of species. Few of the plants in this assemblage have been reconstructed in their entirety, and thus it is not currently possible to delimit the true diversity of the floras. Silicified peat from a Permian site on Skaar Ridge in the central Transantarctic Mountains (E. Taylor *et al.*, 1989*a*; Taylor and Taylor, 1990; Table 11.1) includes at least two species of *Glossopteris* leaves (Pigg and Taylor, 1993), as well as several distinctive types of glossopterid reproductive organs and seeds (Taylor and Taylor, 1992; Zhao *et al.*, 1995), indicating that the plants that produced the glossopterid leaf type were highly variable in habit and habitat. In addition to above-ground woody axes of the *Araucarioxylon* type from this site, there are numerous examples of *Vertebraria*, the roots of glossopterids. These woody subaerial axes also contain evidence of growth periodicity in the form of rings in their wood. In some cases the root wood exhibits distinct ring boundaries; in other cases the boundaries are less distinct, as is typical in many living roots. This anatomical disparity could possibly reflect a difference in habitat.

In addition to various plant organs assignable to the Glossopteridales, the Skaar Ridge peat contains several small filicalean ferns (Galtier and Taylor, 1994), the moss *Merceria* (Smoot and Taylor, 1986), and several types of fungi (Table 11.1). Although not well dated, the Skaar Ridge site is thought to be Late Permian, based on palynomorphs (Farabee *et al.*, 1991).

Permian age compression floras from the Beardmore Glacier area have been little studied to date (T. Taylor *et al.*, 1989). Rocks of comparable age in southern Victoria Land have yielded several species of *Glossopteris* leaves, *Plumsteadia*-type fructifications, and *Noeggerathiopsis* leaves (Kyle, 1974; E. Taylor *et al.*, 1989*b*; Cúneo *et al.*, 1993). Although the latter has not yet been discovered in the Skaar Ridge peat, it has been described from the Upper Permian Bainmedart Coal Measures from the Prince Charles Mountains in East Antarctica (McLoughlin and Drinnan, 1996).

Middle Triassic

By early Middle Triassic time Antarctica supported a more diverse vegetation than was present in the Permian, based on permineralized plants contained in peat from Fremouw Peak (Table 11.1) and compression floras from both the central Transantarctic Mountains and southern Victoria Land. Glossopterid seed ferns were replaced by corystospermalean seed ferns. In the permineralized flora were numerous gymnosperms, including several types of seed ferns (Taylor *et al.*, 1994), conifers (Meyer-Berthaud and Taylor, 1991; Yao *et al.*, 1997), and a cycad (Smoot *et al.*, 1985). The dominant corystosperms were represented by leaves of *Dicroidium*, three

Table 11.1. *Floras from permineralized peat in the central Transantarctic Mountains, Antarctica*

Skaar Ridge (Permian):
Glossopteridales (seed ferns):
    Glossopteris skaarensis (leaves, stems) (Pigg, 1990*a*, Pigg and Taylor, 1993)
    Glossopteris schopfii (leaves, stems) (Pigg, 1990*a*; Pigg and Taylor, 1993)
    Vertebraria (roots)
    Plectilospermum elliotii (seeds) (Taylor and Taylor, 1987)
    Araucarioxylon-type wood (Pigg and Taylor, 1993)
Bryophyta (mosses):
    Bryidae - *Merceria augustica* (leaves, axes, rhizoids) (Smoot and Taylor, 1986)
Filicales (ferns) - Skaaripteridaceae
    Skaaripteris minuta (stems, roots, sporangia) (Galtier and Taylor, 1994)
Fungi - Basidiomycetes:
    white rot and pocket rot in *Vertebraria* (Stubblefield and Taylor, 1986)

Fremouw Peak (Triassic):
Corystospermales (seed ferns):
    Dicroidium fremouwensis (fronds) (Pigg, 1990*b*)
    Kykloxylon fremouwensis (stems) (Meyer-Berthaud *et al.*, 1993)
    Pteruchus fremouwensis (pollen organs) Yao *et al.*, 1995)
    Jeffersonioxylon gordonense (woody stems) (del Fueyo *et al.*, 1995)
    Rhexoxylon-like axis (Taylor, 1992)
Petriellales (seed ferns):
    Petriellaea triangulata (ovulate cupules) (Taylor *et al.*, 1994)
Coniferales (conifers):
    Podocarpaceae - *Notophytum krauselii* (stems, roots) (Meyer-Berthaud and Taylor, 1991)
    Taxodiaceae - *Parasciatopitys aequata* (seed cones) (Yao *et al.*, 1997)
Cycadales (cycads):
    Stangeriaceae/Zamiaceae - *Antarcticycas schopfii* (stems, roots) (Smoot *et al.*, 1985)
Other seed plants:
    Ignotospermum monilii (isolated seeds) (Perovich and Taylor, 1989)
Filicales (true ferns):
    Gleicheniaceae? - *Antarctipteris sclericaulis* (rhizomes) (Millay and Taylor, 1990)
    Cyatheaceae?/Pteridaceae? - *Schopfiopteris repens* (rhizomes) (Millay and Taylor, 1990)
    Matoniaceae? - *Soloropteris rupex* (stems) (Millay and Taylor, 1990)
    Osmundaceae - *Osmundacaulis beardmorensis* (stems) (Schopf, 1978)
    Other ferns: *Fremouwa inaffecta* (rhizomes) (Millay and Taylor, 1990)
        Schleporia incarcerata (stems) (Millay and Taylor, 1990)
Marattiales - *Scolecopteris antarctica* (pinnules, sporangia) (Delevoryas *et al.*, 1992)

Table 11.1. (*cont.*)

Fremouw Peak (Triassic) (cont.):
Equisetales (articulates):
    Spaciinodum collinsonii (stems) (Osborn and Taylor, 1989)
    Spaciinodum collinsonii (cones) (Phipps *et al.*, 1995)
Fungi - Basidiomycetes:
    Palaeofibulus antarctica (Osborn *et al.*, 1989)
    wood rot in *Araucarioxylon* (Stubblefield and Taylor, 1986)
    Zygomycetes (Endogonales - endomycorrhizae):
        Sclerocystis-like (Stubblefield *et al.*, 1986)
        Mycocarpon asterineum (Taylor and White, 1989)
        Endogone-like zygospores (White and Taylor, 1989*a*)
        Glomus-like chlamydospores (White and Taylor, 1989*a*)
        trichomycete (White and Taylor, 1989*b*)
        Gigasporites myriamyces (Phipps and Taylor, 1996)
        Glomites cycestris (Phipps and Taylor, 1996)
    Ascomycetes? - *Endochaetophora antarctica* (White and Taylor, 1988)
    Chytridiomycetes - endobiotic resting spores (Taylor and Stubblefield, 1987)

types of stems, and *Pteruchus* pollen organs (Table 11.1). Also present were ferns (Schopf, 1978; Millay and Taylor, 1990), a sphenophyte (Osborn and Taylor, 1989), and a diverse assemblage of fungi (e.g., Osborn *et al.*, 1989; White and Taylor, 1989*a,b*; Table 11.1). Compression floras typically show somewhat more diversity than permineralized peat (e.g., Axsmith *et al.*, 1995; Boucher *et al.*, 1995). The diversity of plants in Antarctica during the Triassic, especially in a peat deposit (Table 11.1), is noteworthy considering the high paleolatitude of these deposits (70–75° S). However, a total of only 28–30 taxa (10 of which are fungi) in the Fremouw peat still represents a lower diversity than compression floras of comparable age at lower paleolatitudes, such as in Australia (e.g., Jones and de Jersey, 1947; Rigby, 1977). Unfortunately, there is no comparable Triassic peat deposit which could be used for a more meaningful comparison of floral diversity.

    The anatomical features of several of the permineralized plants can also be useful in providing information about paleoclimate. For example, the spongy, parenchymatous stem of *Antarcticycas* (Smoot *et al.*, 1985) is almost identical to that of some living cycads which are restricted to tropical and subtropical environments. The presence of tree rings from different sources (permineralized peat and fluvial deposits), all with the same anatomy, clearly indicates a strongly seasonal environment. The anatomy of individual rings suggests that light was the most important factor in leaf abscission and thus in seasonality, at least as far as the plant is concerned. Because of the extreme light regime, it is difficult to imagine that the region could have supported an aseasonal tropical or subtropical flora such as those associated with modern cycads. Woody plants in these climates today, with only a few exceptions (e.g., Buckley *et al.*, 1995), produce anatomically different types of growth rings, if they produce them at all (Tomlinson and Craighead, 1972; Eckstein *et al.*, 1995). So how could this spongy cycad stem have survived a winter at 70–75° S latitude? If winters were cold, the cycads would need to withstand frost.

If winters were relatively warm, light levels were still probably too low for photo-synthesis, so desiccation and respiration were major obstacles to overcome. The answer to this question is still very speculative, but perhaps the stem was subterra-nean with deciduous leaves, or perhaps *Antarcticycas*, like *Nilssoniocladus* (Spicer and Herman, 1996), had different environmental tolerances than its nearest extant relative. The large number of roots in the matrix around *Antarcticycas* stems and the absence of any evidence of attached leaves provides some support for a subterranean habit.

In roots associated with *Antarcticycas* is a clearly defined zone of thickened cells termed a *phi* layer (Millay *et al.*, 1987). In modern plants such layers function to regulate lateral water movement in the root. Since *phi* layers occur only in modern plants that are subject to fluctuating water levels, the presence of this layer in the fossil roots indicates that they periodically underwent some degree of water stress, and in fact may have existed for a portion of their life in a supersaturated soil. This anatomical evidence, together with the types of plants present at the Fremouw site (e.g., spongy cycads, sphenophytes), indicates that there was sufficient water and that temperatures were conducive for extensive plant growth.

While the number of species present during the Late Permian in the central Transantarctic Mountains is small by comparison with those in the Middle Triassic, the paleoclimate during both of these time periods was sufficiently hospitable to support a luxuriant flora. In the Permian, leaf and root fossils are found abundantly in different depositional environments, i.e., floodplains, abandoned channels, and lacustrine environments (Cúneo *et al.*, 1993). Despite the restrictions of the light regime during both periods, the wide growth rings in the silicified wood indicate exceptional growth during the summer season.

PALEOCLIMATE MODELS

The paleoclimate information from the Permian and Triassic floras of Antarctica is at odds with most paleoclimate models based on physical parameters (i.e., general circulation models, or GCMs). For example, early models proposed that winter temperatures in Antarctica averaged −30 to −40 °C and that summer temperatures never exceeded 0 °C (Kutzbach and Gallimore, 1989). More recent models of the Permian have incorporated paleotopography, and Yemane (1993) found that the addition of large lakes in South Africa greatly ameliorated the climate in this region of Gondwana. Unfortunately, the climate in higher latitudes, i.e., Antarctica, was not affected by these changes. Ziegler (1990) superimposed modern, climatically defined floral biomes on the Permian world as a means of visualizing plant distribution at that time. Later, Kutzbach and Ziegler (1993) coupled this biome model with an atmosphere–ocean climate model which included topography and Yemane's (1993) geographic modifications. Although seasonal extremes were lessened to some degree, these authors noted that the temperatures at very high latitudes (>60° S) still did not conform to the climate evidence from fossil floras. Kutzbach and Ziegler's model produced a polar flora of climate-biome 8 (cold temperate with boreal coniferous forests), while the fossils indicated a probable climate-biome 6 (cool temperate with nemoral broadleaf deciduous forests). As presented here, the Late Permian of Antarctica above 80° S was a time of abundant growth that included not only deciduous trees, but also ferns, mosses, and numerous fungi. Additional climatic evidence occurs in the presence of coal deposits, which are common in Permian rocks from Antarctica (as elsewhere in Gondwana), and indicate the development of considerable plant biomass.

One of the major problems in including fossil plants within a biome classification based on extant floras is that almost all fossil taxa represent isolated organs. The glossopterids are an excellent example of a fossil group that is generally considered to be uniform in habit and habitat because of the similarity of their leaves. When the reproductive organs of these plants are considered, however, it is apparent that the glossopterids were diverse, including at least two natural groups (Surange and Chandra, 1975; Taylor, 1996). They also occupied a large number of habitats, and must have possessed an extensive range of climatic tolerances. Since the glossopterids are still very poorly known as complete plants, relatively little is known about their biology, and even less about the extent of their climatic ranges.

Similar inconsistencies exist between models and paleobotanical evidence during the Triassic. For example, Wilson et al. (1994) suggest winter lows of $-50\,°C$ with summer highs up to $25\,°C$ in polar latitudes during the Triassic. Crowley (1994) draws attention to the very large annual temperature range in models for both the Permian and the Triassic, due to the low heat capacities of large land areas. He mentions that it is difficult to reconcile the models with geologic data, especially at high latitudes. As is the case for the Permian, these difficulties are also present in the Triassic of Antarctica. The paleobotanical evidence indicates that it is unlikely that the climatic parameters suggested by Wilson et al. (1994) existed. While it is beyond the scope of this paper to attempt to reconcile the differences between the GCMs and the paleontological data relative to paleoclimates, there are a few points that need to be considered. The first of these concerns the differences of scale that exist between the broadly defined GCMs and more highly resolved data sets assembled from paleontological and sedimentological sources. Only when hundreds of paleontological localities are used is it possible to accurately define the continental habitat in which some of the plants lived. It is still very difficult to define the microhabitats that included the plants. The recent study of post-glacial Permian geology and paleoclimate of central and southern Africa presents an example of more detailed resolution in a paleoclimatic reconstruction (Visser, 1995). Here, in a 3000-km^2 region of southwestern Gondwana, six different climatic zones (semiarid to cold and wet) are recorded.

Secondly, other factors that are both physical and biological need to be incorporated into the models to better characterize the paleoclimate, as was done, for example, with the use of lakes and cloud cover in central Gondwana during the Permian (Yemane, 1993), or Parrish's use of climatically sensitive sediments to delimit paleoclimate (e.g., Parrish et al., 1982, 1996). Major differences can be anticipated when a single new parameter such as paleotopography is added to a model (e.g., Moore et al., 1992). In a similar context the existing vegetation of a region can have a strong feedback effect on the climate at high latitudes, not only through albedo changes (Otto-Bliesner and Upchurch, 1997) but also through the effects of

344 EDITH L. TAYLOR ET AL.

increased (or decreased) transpiration. Thirdly, there needs to be increased dialogue among the various disciplines with data sets useful for paleoclimate reconstruction. Whether physical models, paleontological data, or geological and sedimentological information, all are necessary and must be incorporated in deciphering past climates. This is the challenge we all face and which will ultimately provide the increased resolution necessary to understand the complex interactions that existed between the organisms and the world in which they lived.

ACKNOWLEDGEMENTS

We would like to acknowledge the following for assistance during several different field seasons: Ana Archangelsky, Brigitte Meyer-Berthaud, Lisa M. Boucher, David Buchanan, Tim Culley, Georgina del Fueyo, John L. Isbell, Hans Kerp, and Jeffrey M. Osborn; also the employees of Helicopters New Zealand, and the members of VXE-6 Squadron, US Navy. We would especially like to acknowledge the work of former students Elizabeth McAllister, Michael Benedict and Michelle Putz, and H.C. Fritts for the specimen of *Larix sibirica* wood. This work was partially supported by the National Science Foundation (OPP-9315353).

REFERENCES

Axsmith, B. J., Taylor, T. N., Taylor, E. L., Boucher, L. D., Rothwell, G. W. and Cúneo, N. R. (1995). Triassic conifer seed cones from the Lashly Formation, southern Victoria Land. *Antarctic Journal of the U.S.*, **30**, 44–6.

Barber, H. N., Dadswell, H. E. and Ingle, H. D. (1959). Transport of driftwood from South America to Tasmania and Macquarie Island. *Nature*, **184**, 203–4.

Barrett, P. J. (1969). Stratigraphy and petrology of the mainly fluviatile Permian and Triassic Beacon rocks, Beardmore Glacier area, Antarctica. *Ohio State University, Institute of Polar Studies, Report*, **34**, 1–132.

Barrett, P. J. and Elliot, D. H. (1973). *Reconnaissance Geologic Map of the Buckley Island Quadrangle, Transantarctic Mountains, Antarctica.* US Geological Survey Antarctic Geologic Map A-3.

Barrett, P. J., Elliot, D. H. and Lindsay, J. F. (1986). The Beacon Supergroup (Devonian–Triassic) and Ferrar Group (Jurassic) in the Beardmore Glacier Area, Antarctica. In *Geology of the Central Transantarctic Mountains*, Antarctic Research Series, vol. 36, no. 14, pp. 339–428. Washington, DC: American Geophysical Union.

Basinger, J. F. (1991) The fossil forests of the Buchanan Lake Formation (early Tertiary), Axel Heiberg Island, Canadian Arctic Archipelago: preliminary floristics and paleoclimate. *Geological Survey of Canada Bulletin*, **403**, 39–65.

Blanchette, R. A. (1991). Delignification by wood-decay fungi. *Annual Review of Phytopathology*, **29**, 381–98.

Blanchette, R. A. (1992). Anatomical responses of xylem to injury and invasion by fungi. In *Defense Mechanisms of Woody Plants against Fungi*, eds. R. A. Blanchette and A. R. Biggs, pp. 76–95. Berlin: Springer-Verlag.

Boucher, L. D., Taylor, E. L., Taylor, T. N., Cúneo, N. R. and Osborn, J. M. (1995). *Dicroidium* compression floras from southern Victoria Land. *Antarctic Journal of the U.S.*, **30**, 40–1.

Briand, C. H., Posluszny, U. and Larson, D. W. (1993). Influence of age and growth rate on radial anatomy of annual rings of *Thuja occidentalis* L. (eastern white cedar). *International Journal of Plant Sciences*, **154**, 406–11.

Buckley, B. M., Barbetti, M., Watanasak, M., D'Arrigo, R., Boonchirdchoo, S. and Sarutanon, S. (1995). Dendrochronological investigations in Thailand. *IAWA Journal*, **16**, 393–409.

Cottam, G. and Curtis, J. T. (1956). The use of distance measures in phytosociological sampling. *Ecology*, **37**, 451–60.

Creber, G. T. and Chaloner, W. G. (1984*a*). Climatic indications from growth rings in fossil woods. In *Fossils and Climate*, ed. P. Brenchley, pp. 49–74. London: John Wiley & Sons.

Creber, G. T. and Chaloner, W. G. (1984*b*). Influence of environmental factors on the wood structure of living and fossil trees. *Botanical Review*, **50**, 357–448.

Crowley, T. J. (1994). Pangean climates. In *Pangea: Paleoclimate, Tectonics and Sedimentation during Accretion, Zenith, and Breakup of a Supercontinent*. Geological Society of America Special Paper No. 288, ed. G. D. Klein, pp. 25–39. Boulder, CO: Geological Society of America.

Cúneo, N. R., Isbell, J. L., Taylor, E. L. and Taylor, T. N. (1993). The *Glossopteris* flora from Antarctica. *Taphonomy and paleoecology: Proceedings of the IX International Congress on Carboniferous/Permian Stratigraphy and Geology (Buenos Aires, September 1991)*, **2**, 13–40.

Cúneo, N. R., Taylor, E. L. and Taylor, T. N. (1999). *In situ* fossil forest from the Middle Triassic Fremouw Formation, Antarctica: paleoenvironmental setting and paleoclimate analysis. *Palaeogeography, Palaeoclimatology, Palaeoecology* (submitted).

del Fueyo, G., Taylor, E. L., Taylor, T. N. and Cúneo. N. R. (1995). Triassic wood from the Gordon Valley, central Transantarctic Mountains, Antarctica. *IAWA Journal*, **16**, 111–26.

Delevoryas, T., Taylor, T. N. and Taylor, E. L. (1992). A marattialean fern from the Triassic of Antarctica. *Review of Palaeobotany and Palynology*, **74**, 101–7.

Dickson, J. H. (1992). North American driftwood, especially *Picea* (spruce), from archaeological sites in the Hebrides and Northern Isles of Scotland. *Review of Palaeobotany and Palynology*, **73**, 49–56.

Eckstein, D., Sass, U. and Baas, P. (eds.) (1995). Growth periodicity in tropical trees. *IAWA Journal*, **16**, 323–442.

Farabee, M. J., Taylor, E. L. and Taylor, T. N. (1990). Correlation of Permian and Triassic palynomorphs from the central Transantarctic Mountains. *Review of Palaeobotany and Palynology*, **65**, 257–65.

Farabee, M. J., Taylor, E. L. and Taylor, T. N. (1991). Late Permian palynomorphs from the Buckley Formation in the central Transantarctic Mountains, Antarctica. *Review of Palaeobotany and Palynology*, **69**, 353–68.

Francis, J. E. (1986). Growth rings in Cretaceous and Tertiary wood from Antarctica and their palaeoclimatic implications. *Palaeontology*, **29**, 665–84.

Francis, J. E. (1988). A 50-million-year-old fossil forest from Strathcona Fiord, Ellesmere Island, Arctic Canada: evidence for a warm polar climate. *Arctic*, **41**, 314–18.

Francis, J. E. (1991*a*). Palaeoclimatic significance of Cretaceous–early Tertiary fossil forests of the Antarctic Peninsula. In *Geological Evolution of Antarctica (Proceedings of the Fifth International Symposium on Antarctic Earth Sciences, 23–28 August 1987)*, eds. M. R. A. Thomson, J. A. Crane and J. W. Thomson, pp. 623–7. Cambridge: Cambridge University Press.

Francis, J. E. (1991*b*). The dynamics of polar fossil forests: Tertiary fossil forests of Axel Heiberg Island, Canadian Arctic Archipelago. *Geological Survey of Canada Bulletin*, **403**, 29–38.

Francis, J. E., Woolfe, K. J., Arnot, M. J. and Barrett, P. J. (1993). Permian forests of Allan Hills, Antarctica. The palaeoclimate of Gondwanan high latitudes. *Special Papers in Palaeontology*, **49**, 75–83.

Fritts, H. C. (1976). *Tree Rings and Climate*. New York: Academic Press, Inc.

Fritz, W. J. (1980). Stumps transported and deposited upright by Mount St. Helens mud flows. *Geology*, **8**, 586–8.

Fritz, W. J. and Harrison, S. (1985). Transported trees from the 1982 Mount St. Helens sediment flows: Their use as paleocurrent indicators. *Sedimentary Geology*, **42**, 49–64.

Gabites, H. I. (1985). *Triassic Paleoecology of the Lashly Formation, Transantarctic Mountains, Antarctica*. Unpublished MS Thesis, Victoria University of Wellington, New Zealand.

Galtier, J. and Taylor, T. N. (1994). The first record of ferns from the Permian of Antarctica. *Review of Palaeobotany and Palynology*, **83**, 227–39.

Giddings, J. L. (1943). Some climatic aspects of tree growth in Alaska. *Tree-Ring Bulletin*, **9**, 26–32.

Golonka, J., Ross, M. I. and Scotese, C.R. (1994). Phanerozoic paleogeographic and paleoclimatic modeling maps. In *Pangea: Global Environments and Resources*. Canadian Society of Petroleum Geologists Memoir No. 17, eds. A. F. Embry, B. Beauchamp and D. J. Glass, pp. 1–47. Calgary, AB: Canadian Society of Petroleum Geologists.

Gunn, B. M. and Walcott, R. I. (1962). The geology of the Mt. Markham region, Ross Dependency, Antarctica. *New Zealand Journal of Geology and Geophysics*, **5**, 407–26.

Irving, E. and Wynne, P. J. (1991). The paleolatitude of the Eocene fossil forests of Arctic Canada. *Geological Survey of Canada Bulletin*, **403**, 209–11.

Isbell, J. L., Taylor, T. N., Taylor, E. L., Cúneo, N.R. and Meyer-Berthaud, B. (1990). Depositional setting of Permian and Triassic fossil plants in the Allan Hills, southern Victoria Land. *Antarctic Journal of the U.S.*, **25**, 22.

Jefferson, T. H. (1982). Fossil forests from the Lower Cretaceous of Alexander Island, Antarctica. *Palaeontology*, **25**, 681–708.

Jefferson, T. H., Siders, M.A. and Haban, M.A. (1983). Jurassic trees engulfed by lavas of the Kirkpatrick Basalt Group, northern Victoria Land. *Antarctic Journal of the U.S.*, **18**, 14–16.

Jones, O. A. and de Jersey, N. J. (1947). The flora of the Ipswich Coal Measures – morphology and floral succession. *University of Queensland Department of Geology, Papers, N.S.*, **3/4**, 1–88.

Krause, C. and Morin, H. (1995). Impact of spruce budworm defoliation on the number of latewood tracheids in balsam fir and black spruce. *Canadian Journal of Forest Research*, **25**, 2029–34.

Kutzbach, J. E. and Gallimore, R. G. (1989). Pangaean climates: Megamonsoons of the megacontinent. *Journal of Geophysical Research*, **D3**, no. 94, 3341–57.

Kutzbach, J. E. and Ziegler, A. M. (1993). Simulation of Late Permian climate and biomes with an atmosphere–ocean model: comparisons with observations. *Philosophical Transactions of the Royal Society of London*, **341B**, 327–40.

Kuuluvainen, T. (1992). Tree architectures adapted to efficient light utilization: is there a basis for latitudinal gradients? *Oikos*, **65**, 275–84.

Kyle, R. A. (1974). *Plumsteadia ovata* n. sp., a glossopterid fructification from South Victoria Land, Antarctica (note). *New Zealand Journal of Geology and Geophysics*, **17**, 719–21.

Larson, P. R. (1964). Some indirect effects of environment on wood formation. In *The Formation of Wood in Forest Trees*, ed. M. H. Zimmermann, pp. 345–65. New York: Academic Press.

Larson, P. R. (1969). Wood formation and the concept of wood quality. *Yale University School of Forestry and Environmental Studies, Bulletin*, **74**, 1–54.

McLoughlin, S. and Drinnan, A. N. (1996). Anatomically preserved Permian *Noeggerathiopsis* Feistmantel leaves from East Antarctica. *Review of Palaeobotany and Palynology*, **92**, 207–27.

Meyer-Berthaud, B., Taylor, E. L. and Taylor, T. N. (1992). Reconstructing the Gondwana seed fern *Dicroidium*: Evidence from the Triassic of Antarctica. *Géobios*, **25**, 341–4.

Meyer-Berthaud, B. and Taylor, T. N. (1991). A probable conifer with podocarpacean affinities from the Triassic of Antarctica. *Review of Palaeobotany and Palynology*, **67**, 179–98.

Meyer-Berthaud, B., Taylor, T. N. and Taylor, E. L. (1993). Petrified stems bearing *Dicroidium* leaves from the Triassic of Antarctica. *Palaeontology*, **36**, 337–56.

Millay, M. A. and Taylor, T. N. (1990). Fern stems from the Triassic of Antarctica. *Review of Palaeobotany and Palynology*, **62**, 41–64.

Millay, M. A., Taylor, T. N. and Taylor, E. L. (1987). *Phi* thickenings in fossil seed plants from Antarctica. *IAWA Bulletin*, **8**, 191–201.

Moore, G. T., Sloan, L. C., Hayashida, D. N. and Umrigar, N. P. (1992). Paleoclimate of the Kimmeridgian/Tithonian (Late Jurassic) world: II. Sensitivity tests comparing three different paleotopographic settings. *Palaeogeography, Palaeoclimatology, Palaeoecology*, **95**, 229–52.

Norton, D. A. and Ogden, J. (1987). Dendrochronology: A review with emphasis on New Zealand applications. *New Zealand Journal of Ecology*, **10**, 77–95.

Osborn, J. M. and Taylor, T. N. (1989). Structurally preserved sphenophytes from the Triassic of Antarctica: Vegetative remains of *Spaciinodum* gen. nov. *American Journal of Botany*, **76**, 1594–1601.

Osborn, J. M., Taylor, T. N. and White, J. F. (1989). *Palaeofibulus* gen. nov., a clamp-bearing fungus from the Triassic of Antarctica. *Mycologia*, **81**, 622–6.

Oswalt, W. H. (1958). Tree-ring chronologies in south-central Alaska. *Tree-Ring Bulletin*, **22**, 16–22.

Oswalt, W. H. (1960). The growing season of Alaskan spruce. *Tree-Ring Bulletin*, **23**, 3–9.

Otto-Bliesner, B. L. and Upchurch, G. R. (1997). Vegetation-induced warming of high-latitude regions during the late Cretaceous period. *Nature*, **385**, 804–7.

Parrish, J. T., Bradshaw, M. T., Brakel, A. T., Mulholland, S. M., Totterdell, J. M. and Yeates, A. N. (1996). Palaeoclimatology of Australia during the Pangean interval. *Palaeoclimates*, **1**, 241–81.

Parrish, J. T. and Spicer, R. A. (1988). Middle Cretaceous wood from the Nanushuk Group, central North Slope, Alaska, *Palaeontology*, **31**, 19–34.

Parrish, J. T., Ziegler, A. M. and Scotese, C. R. (1982). Rainfall patterns and the distribution of coals and evaporites in the Mesozoic and Cenozoic. *Palaeogeography, Palaeoclimatology, Palaeoecology*, **40**, 67–101.

Perovich, N. E. and Taylor, E. L. (1989). Structurally preserved fossil plants from Antarctica. IV. Triassic ovules. *American Journal of Botany*, **76**, 992–9.

Phipps, C. J., Osborn, J. M. and Taylor, T. N. (1995). Structurally preserved sphenophytes from the Triassic of Antarctica: Reproductive remains of *Spaciinodum* (abs.). *American Journal of Botany*, **82** (No. 6, Supplement), 89.

Phipps, C. J. and Taylor, T. N. (1996). Mixed arbuscular mycorrhizae from the Triassic of Antarctica. *Mycologia*, **88**, 707–14.

Pigg, K. B. (1990*a*). Anatomically preserved *Glossopteris* foliage from the central Transantarctic Mountains. *Review of Palaeobotany and Palynology*, **66**, 105–27.

Pigg, K. B. (1990*b*). Anatomically preserved *Dicroidium* foliage from the central Transantarctic Mountains. *Review of Palaeobotany and Palynology*, **66**, 129–45.

Pigg, K. B. and Taylor, T. N. (1993). Anatomically preserved *Glossopteris* stems with attached leaves from the central Transantarctic Mountains, Antarctica. *American Journal of Botany*, **80**, 500–16.

Putz, M. K. and Taylor, E. L. (1996). Wound response in fossil trees from Antarctica and its potential as a paleoenvironmental indicator. *IAWA Journal*, **17**, 77–88.

Raitio, H. (1992). Anatomical symptoms in the wood of Scots pine damaged by frost and pine bark bugs. *Flora*, **186**, 187–93.

Retallack, G. J. (1980). Late Carboniferous to Middle Triassic megafossil floras from the Sydney Basin. In *A Guide to the Sydney Basin*. Geological Survey of New South Wales Bulletin, No. 26, eds. C. Herbert and R. J. Helby, pp. 384–430. NSW: Geological Survey of New South Wales.

Rigby, J. F. (1977). New collections of Triassic plants from the Esk Formation, southeast Queensland. *Queensland Government Mining Journal*, **78**, 320–5.

Schopf, J. M. (1978). An unusual osmundaceous specimen from Antarctica. *Canadian Journal of Botany*, **56**, 3083–95.

Smith, A. G., Hurley, A. M. and Briden, J. C. (1981). *Phanerozoic Paleocontinental World Maps*. London: Cambridge University Press.

Smith, J. M. B., Rudall, P., and Keage, P. L. (1989). Driftwood on Heard Island. *Polar Record*, **25**, 223–8.

Smoot, E. L. and Taylor, T. N. (1986). Structurally preserved fossil plants from Antarctica. II. A Permian moss from the Transantarctic Mountains. *American Journal of Botany*, **73**, 1683–91.

Smoot, E. L., Taylor, T. N. and Delevoryas, T. (1985). Structurally preserved fossil plants from Antarctica. I. *Antarcticycas*, gen. n., a Triassic cycad stem from the Beardmore Glacier area. *American Journal of Botany*, **71**, 410–23.

Spicer, R. A. (1989). The formation and interpretation of plant fossil assemblages. *Advances in Botanical Research*, **16**, 96–191.

Spicer, R. A. and Chapman, J. L. (1990). Climate change and the evolution of high latitude terrestrial vegetation and floras. *Trends in Ecology and Evolution*, **5**, 279–84.

Spicer, R. A. and Herman, A. B. (1996). *Nilssoniocladus* in the Cretaceous Arctic: new species and biological insights. *Review of Palaeobotany and Palynology*, **92**, 229–43.

Spicer, R. A. and Parrish, J. T. (1986). Paleobotanical evidence for cool North Polar climates in middle Cretaceous (Albian–Cenomanian) time. *Geology*, **14**, 703–6.

Spicer, R. A. and Parrish, J. T. (1990). Latest Cretaceous woods of the central North Slope, Alaska. *Palaeontology*, **33**, 225–42.

Stubblefield, S. P. and Taylor, T. N. (1986). Wood decay in silicified gymnosperms from Antarctica. *Botanical Gazette*, **147**, 116–25.

Stubblefield, S. P., Taylor, T. N. and Seymour, R. L. (1986). A possible endogonaceous fungus from the Triassic of Antarctica. *Mycologia*, **79**, 905–6.

Surange, K. R. and Chandra, S. (1975). Morphology of the gymnospermous fructifications of the *Glossopteris* flora and their relationships. *Palaeontographica*, **149B**, 153–80.

Taylor, E. L. (1989). Tree-ring structure in woody axes from the central Transantarctic Mountains, Antarctica. In *Proceedings of the International Symposium on Antarctic Research (Hangzhou, P.R. China, May 1989)*, pp. 109–13. Tianjin: China Ocean Press.

Taylor, E. L. (1992). The occurrence of a *Rhexoxylon*-like stem in Antarctica. *Courier Forschungsinstitut Senckenberg*, **147**, 183–9.

Taylor, E. L. (1996). Enigmatic gymnosperms? Structurally preserved Permian and Triassic seed ferns from Antarctica. *Review of Palaeobotany and Palynology*, **90**, 303–18.

Taylor, E. L. Cúneo, R. and Taylor, T. N. (1991). Permian and Triassic fossil forests from the central Transantarctic Mountains. *Antarctic Journal of the U.S.*, **26**, 23–4 (published December 1992).

Taylor, E. L., Harter, C. M. and Taylor, T. N. (1998). Plant–animal interactions in the Triassic of Antarctica (abs.). *American Journal of Botany*, **85** (No. 6, Supplement), p. 82.

Taylor, E. L. and Putz, M. K. (1993). Wood anatomy and tree ring structure in *Larix sibirica*: an analogue for high latitude fossil forest growth (abs.). *American Journal of Botany*, **80** (No. 6, Supplement), 94.

Taylor, E. L. and Taylor, T. N. (1990). Structurally preserved Permian and Triassic floras from Antarctica. In *Antarctic Paleobiology and its Role in the Reconstruction of Gondwana*, eds. T. N. Taylor and E. L. Taylor, pp. 149–63: New York: Springer-Verlag.

Taylor, E. L. and Taylor, T. N. (1992). Reproductive biology of the Permian Glossopteridales and their suggested relationship to flowering plants. *Proceeding of the National Academy of Sciences*, **86**, 11495–7.

Taylor, E. L. and Taylor T. N. (1993). Fossil tree rings and paleoclimate from the Triassic of Antarctica. In *The Nonmarine Triassic*. New Mexico Museum of Natural History and Science Bulletin No. 3, eds. S. G. Lucas and M. Morales, pp. 453–5. Albuquerque, NM: New Mexico Museum of Natural History.

Taylor, E. L., Taylor, T. N. and Collinson, J. W. (1989*a*). Depositional setting and paleobotany of Permian and Triassic permineralized peat from the central Transantarctic Mountains, Antarctica. *International Journal of Coal Geology*, **12**, 657–79.

Taylor, E. L., Taylor, T. N. and Cúneo, N. R. (1992). The present is not the key to the past: A polar forest from the Permian of Antarctica. *Science*, **257**, 1675–7.

Taylor, E. L., Taylor, T. N. and Isbell, J. L. (1997). Evidence of Permian forests from the Shackleton Glacier area, Antarctica (abs.). *American Journal of Botany*, **84** (No. 6, Supplement), 143.

Taylor, E. L., Taylor, T. N., Isbell, J. L. and Cúneo, N. R. (1989*b*). Fossil floras of southern Victoria Land – 2. Kennar Valley. *Antarctic Journal of the U.S.*, **24**, 26–8.

Taylor, T. N., del Fueyo, G. M., and Taylor, E. L. (1994). Permineralized seed fern cupules from the Triassic of Antarctica: Implications for cupule and carpel evolution: *American Journal of Botany*, **81**, 666–77.

Taylor, T. N. and Stubblefield, S. P. (1987). A fossil mycoflora from Antarctica. *VII Simposio Argentino de Paleobotánica y Palinologia, Actas*, 187–90.

Taylor, T. N. and Taylor, E. L. (1987). Structurally preserved fossil plants from Antarctica. III. Permian seeds. *American Journal of Botany*, **74**, 904–13.

Taylor, T. N., Taylor, E. L. and Isbell, J. L. (1989). Glossopterid reproductive organs from Mt. Achernar, Antarctica. *Antarctic Journal of the U.S.*, **24**, 28–30.

Taylor, T. N. and White, J. F., Jr (1989). Fossil fungi (Endogonaceae) from the Triassic of Antarctica. *American Journal of Botany*, **76**, 389–96.

Telewski, F. W. and Lynch, A. M. (1991). Measuring growth and development of stems. In *Techniques and Approaches in Forest Tree Ecophysiology*, eds. J. P. Lassoie and T. M. Hinckley, pp. 503–55. Boca Raton, FL: CRC Press.

Tomlinson, P. B. and Craighead, F. C., Sr. (1972). Growth-ring studies on the native trees of sub-tropical Florida. In *Research Trends in Plant Anatomy*, eds. A. K. M. Ghouse and M. Yunus, pp. 39–51. Bombay: Tata McGraw-Hill Publishing Co. Ltd.

Van Stone, J. W. (1958). The origin of driftwood on Nunivak Island, Alaska. *Tree-Ring Bulletin*, **22**, 12–15.

Visser, J. N. J (1995). Post-glacial Permian stratigraphy and geography of southern and central Africa: Boundary conditions for climate modeling. *Palaeogeography, Palaeoclimatology and Palaeoecology*, **118**, 213–43.

White, J. F., Jr and Taylor, T. N. (1988). Triassic fungus from Antarctica with possible ascomycetous affinities. *American Journal of Botany*, **75**, 1495–500.

White, J. F. and Taylor, T. N. (1989a). Triassic fungi with suggested affinities to the Endogonales (Zygomycotina). *Review of Palaeobotany and Palynology*, **61**, 53–61.

White, J. F. and Taylor, T. N. (1989b). A trichomycete-like fossil from the Triassic of Antarctica. *Mycologia*, **81**, 643–6.

Wilson, K. M., Pollard, D., Hay, W. W., Thompson, S. L. and Wold, C.N. (1994). General circulation model simulations of Triassic climates: Preliminary results. In *Pangea: Paleoclimate, Tectonics and Sedimentation during Accretion, Zenith, and Breakup of a Supercontinent*. Geological Society of America Special Paper No. 288, ed. G. Klein, pp. 91–116. Boulder, CO: Geological Society of America.

Wolfe, J. A. (1985). Distribution of major vegetational types during the Tertiary. In *The Carbon Cycle and Atmospheric CO_2: Natural Variations Archean to Present*. Geophysical Monograph, vol. 32, pp. 357–75. Washington, DC: American Geophysical Union.

Wolfe, J. A. (1987). Late Cretaceous–Cenozoic history of deciduousness and the terminal Cretaceous event. *Paleobiology*, **13**, 215–26.

Wolfe, J. A. (1993). A method of obtaining climatic parameters from leaf assemblages. *U.S. Geological Survey Bulletin*, **2040**, 1–71.

Yao, X., Taylor, T. N. and Taylor, E. L. (1995). The corystosperm pollen organ *Pteruchus* from the Triassic of Antarctica. *American Journal of Botany*, **82**, 535–46.

Yao, X., Taylor, T. N. and Taylor, E. L. (1997). A taxodiaceous seed cone from the Triassic of Antarctica. *American Journal of Botany*, **84**, 343–54.

Yemane, K. (1993). Contribution of Late Permian palaeogeography in maintaining a temperate climate in Gondwana. *Nature*, **361**, 51–4.

Zhao, L., Taylor, T. N. and Taylor, E. L. (1995). Cupulate glossopterid seeds from the Permian Buckley Formation, central Transantarctic Mountains. *Antarctic Journal of the U.S.*, **30**, 54–5.

Ziegler, A. M. (1990). Phytogeographic patterns and continental configurations during the Permian Period. In *Palaeozoic Palaeogeography and Biogeography*. Geological Society of America Memoir No. 12, eds. W. S. McKerrow and C. R. Scotese, pp. 363–79. Boulder, CO: Geological Society of America.

12

Organic carbon burial and faunal dynamics in the Appalachian Basin during the Devonian (Givetian–Famennian) greenhouse: an integrated paleoecological and biogeochemical approach

ADAM E. MURPHY, BRADLEY B. SAGEMAN, CHARLES A. VER STRAETEN, AND DAVID J. HOLLANDER

ABSTRACT

The Middle to Late Devonian was characterized by widespread organic carbon burial and a major biotic crisis lasting as long as 7 million years. Oceanographic and climatic mechanisms for these two phenomena have been advanced and inconclusively debated. Although both imply significant changes in marine ecosystems, there has been relatively little biogeochemical analysis of the interval, and little integration of geochemical and paleoecological data. In this study, a multi-proxy methodology for assessing changes in carbon cycling, depositional and early burial redox conditions, and nutrient dynamics through stratigraphic intervals which record changes in organic carbon burial and faunal change is developed and applied to Givetian–Frasnian strata of western New York. Analyses of core samples for amount and isotopic composition of organic materials, degree of pyritization, and organic and inorganic phosphorus content suggest that enhanced organic carbon burial was the result of efficient recycling of the limiting nutrient element, P, within the shallow marine ecosystem of the Appalachian Basin. This interpretation does not support water column stratification, but does provide a mechanism to explain changes in brachiopod and goniatite diversity which occur across the study interval. Ultimately this methodology will provide a means of investigating biogeochemical dynamics in the context of both short-term, regional events and long-term global phenomena.

INTRODUCTION

Over the past two decades there has been a steady increase in studies of warm climates in the geologic past, stimulated in part by concern over rising levels of greenhouse gases in the atmosphere since the Industrial Revolution (e.g., Houghton *et al.*, 1990). Associated with this, a revised view of the earth as an integrated physical, chemical, and biological system has emerged (i.e., earth system science). Major climatic changes associated with ancient greenhouse events are now viewed in the context of global biogeochemical cycles. In particular, changes in the storage and flux of carbon among large reservoirs in the lithosphere, atmosphere, hydrosphere, and biosphere, and a variety of feedback mechanisms that appear to regulate fluxes, are thought to have played critical roles in the earth's history. Figure 12.1 (modified

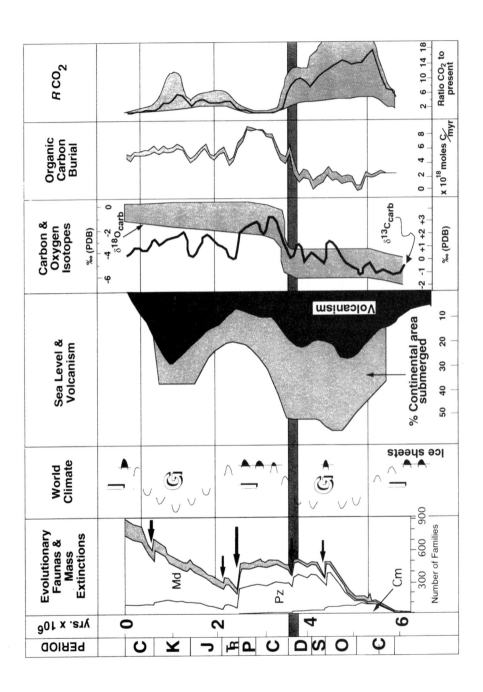

from Fischer, 1981, 1982) summarizes this view of ancient greenhouse climates. They were characterized by elevated volcanism and CO_2 output, increased poleward latent heat transport and enhanced global warmth, lack of continental ice sheets, high global sea level, widespread flooding of continental interiors, and large-scale deposition of marine organic carbon (OC) thought to be associated with benthic oxygen deficiency. Despite increasing consensus on the general history of greenhouse events, however, there has continued to be much debate concerning specific issues, such as the mechanisms of OC accumulation. Although many authors have speculated on the relationship between oxygenation dynamics driven by the carbon cycle and faunal events such as mass extinctions, some of which characterize greenhouse times (e.g., Elder, 1989; McGhee, 1996), relatively little direct research has addressed this problem. The purpose of our study was to investigate the linkage between carbon burial processes and community dynamics during the late phase of the Paleozoic greenhouse episode.

Driven by uniformitarian constraints and the need for well-preserved materials for geochemical analyses, many studies have been focused on the younger of the two greenhouse intervals shown in Fig. 12.1. However, extensive stratigraphic and paleontologic analyses of stratal successions recording the Paleozoic greenhouse provide an excellent foundation upon which to expand the investigation of global warmth into the geologic past. For example, the Devonian rocks from many localities include OC-rich marine strata recording deposition from Givetian through Famennian time. This interval represents the waning of the Paleozoic greenhouse (Fischer, 1981, 1982), and has been interpreted as having high but decreasing atmospheric pCO_2 levels (Berner, 1990), high rates of OC burial (Berner and Raiswell, 1983), and a protracted faunal crisis lasting approximately 4–7 million years and culminating in the Frasnian–Famennian (F/F) extinction event (McGhee, 1982, 1989).

Among the most stratigraphically complete, well-exposed, richly fossiliferous, and least tectonically disturbed sections of the Givetian–Famennian interval in North America are those of the northern Appalachian Basin in central and western New York (Fig. 12.2). Consequently, these are among the most intensively studied and documented rocks of this age (Woodrow *et al.*, 1988; Roen, 1993 and historical references therein), and among the best constrained with respect to depositional environments (Woodrow, 1985; Brett and Baird, 1994; Kirchgasser *et al.*, 1994),

Figure 12.1. Compilation of Phanerozoic paleoenvironmental data showing long-term climate states (G, greenhouse; I, icehouse), sea level, and volcanism (based on Fischer, 1981, 1982). Sepkoski's (1984) evolutionary faunas represent Phanerozoic biotic history (shaded area represents addition of known diversity of soft-bodied, poorly preserved taxa to establish maximum diversity estimate; arrows highlight five largest mass extinctions). Geochemical data relevant to carbon cycling include brachiopod carbon and oxygen isotopes (Veizer *et al.*, 1986), estimate of organic carbon burial through time (Berner and Raiswell, 1983; Arthur *et al.*, 1987) and atmospheric CO_2 concentration as ratio to present (Berner, 1990). The Middle to Late Devonian study interval is highlighted. Md, Modern fauna; Pz, Paleozoic fauna; Cm, Cambrian fauna.

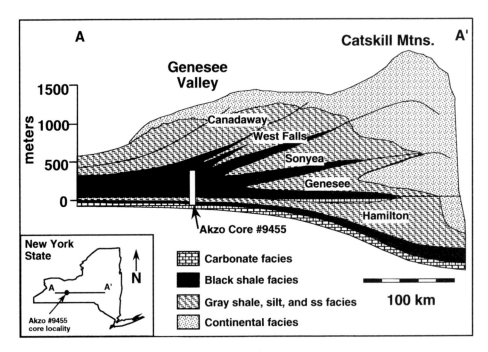

Figure 12.2. Schematic cross-section (A-A′) of the Catskill Delta Complex of New York, showing the overall stratal architecture and relative location of Akzo core #9455 (modified from House and Kirchgasser, 1993). Inset shows map of New York State, with approximate line of section A-A′ and core location indicated. ss, sandstone.

relative sea level (House and Kirchgasser, 1993; Brett and Baird, 1994; Kirchgasser *et al*., 1994), and tectonic history (Ettensohn, 1985a,b; Faill, 1985). Despite this volume of literature, however, several outstanding controversies remain that are particularly relevant to the question of ancient warm climates. These include the cause and effect of OC burial in the Appalachian Basin, and the cause of Middle to Late Devonian community overturn events.

Leading hypotheses for both of these phenomena invoke significant changes in paleoenvironments and biogeochemical dynamics of the marine ecosystem. For example, although the cause of oxygen deficiency in the Appalachian Basin is commonly attributed to the standard water column stratification model (e.g., Ettensohn *et al*., 1988; Kepferle, 1993), high primary productivity due to increased nutrient supply (e.g., Pederson and Calvert, 1990) has recently been suggested (Algeo *et al*., 1995). To account for the Late Devonian biotic crises, causes as disparate as global warming (Thompson and Newton, 1988), global cooling (McGhee, 1982; McLaren, 1982; Copper, 1986), oceanic anoxia (Buggisch, 1991; Joachimski and Buggisch, 1993), and extraterrestrial impact (McLaren, 1970; McGhee, 1982) have been suggested. In each case, distinctive changes in marine ecosystems and biogeochemistry would have occurred. Although numerous workers have concentrated on the measurement of geochemical proxies (e.g., carbon isotopic compositions of organic materials and carbonates) immediately adjacent to the F/F boundary (Wang *et al*.,

1991; Joachimski and Buggisch, 1993; Yan *et al.*, 1993), there has been relatively little detailed geochemical analysis of the interval preceding the boundary. Thus 'boundary event' interpretations often lack historical context, without which it is difficult to assess the significance of rapid environmental changes relative to long-term, perhaps incremental, changes.

The strongest approach to the analysis of environmental dynamics during ancient greenhouse periods, where modern uniformitarian analogs become difficult to apply, is the integration of multiple proxies including both geochemical and paleontological data, within a high resolution stratigraphic framework (e.g., Arthur and Dean, 1991; Sageman and Hollander, 1999). The Givetian–Famennian interval in the Appalachian Basin provides an ideal target for an integrated study of this kind. The stratigraphic section, representing the distal portion of the Catskill Deltaic Complex (Fig. 12.2), includes small- and large-scale alternations between black, dark gray, and gray shale facies (e.g., House and Kirchgasser, 1993), as well as major overturns in benthic and other faunal assemblages (Sutton and McGhee, 1985; Brett and Baird, 1996*b*; House, 1996).

In this paper we provide a framework for application of the multi-proxy approach to the Givetian–Famennian interval by reviewing salient features of the Late Devonian Appalachian Basin, and then presenting new observations to test the linkages between biogeochemical dynamics and the history of paleocommunities. These results focus on a major faunal boundary and OC burial event preceding the F/F extinction. They include geochemical analyses of a core (Akzo #9455) drilled in the Genesee River Valley of western New York (Fig. 12.2), and a compilation of paleontologic data from the literature. The results suggest a revision of previous interpretations of the depositional environment of the Late Givetian Geneseo Shale. Significant changes in biogeochemical cycles within the basin accompanied enhanced burial of OC and were coincident with declines in benthic brachiopod diversity and contemporaneous diversification of nektonic (or nektobenthonic?) goniatites. Based on these observations, we suggest a model, drawing significantly on the work of Ingall *et al.* (1993), for Late Devonian environmental dynamics in the Appalachian Basin.

GEOLOGICAL FRAMEWORK

In the following section we provide an overview of background information on the Devonian Appalachian Basin (DAB) which is essential to understanding the questions of OC burial and faunal change. This background includes paleogeography (Fig. 12.3), paleoclimate, tectonics, relative sea level, sedimentation history, and faunal characteristics of the DAB. The succession of stratigraphic units is illustrated in Fig. 12.2 (group level) and Fig. 12.4 (formation level).

Paleogeography

The recognition that paleomagnetic data from the New York region probably represent a late Paleozoic remagnetization (Miller and Kent, 1986; Witzke and Heckel, 1988) stimulated workers to re-evaluate these data, and better integrate them with paleoenvironmental indicators from the stratigraphic record. Recent reconstructions have consequently converged on a position of approximately 30–35° S, with a latitudinally subparallel orientation (Scotese and McKerrow, 1990; Witzke, 1990; see Fig. 12.3). This conclusion has significant implications for paleoenvironmental interpretations. For example, Parrish (1982) suggested that an equatorially positioned DAB was characterized by upwelling, and Ormiston and Oglesby (1995) interpreted water column stagnation in a more subtropically located DAB.

Paleoclimate

The most effective means of reconstructing Paleozoic climates has been through the use of climatically sensitive lithologies (e.g., coals, evaporites) and of the distribution of narrowly tolerant marine faunas (e.g., corals). Paleozoic climate modeling relies heavily on geologic evidence for the establishment of boundary conditions such that the results cannot be viewed as entirely independent evidence. Despite this necessary dependence on the rock record, however, models may provide otherwise unattainable constraints on the distributions of temperature and precipitation as well as wind patterns (e.g., Ormiston and Oglesby, 1995; see below). The interpretation of climate changes in the Middle to Late Devonian depends on constraining climatic conditions at as many specific times during this interval as is possible. This has proven to be very difficult for a time in the distant past when so little detailed, unequivocal climatic information is available. Thus controversy about Late Devonian climate change has remained. Generalizations about the regional climatic conditions can be viewed in terms of two groups, with one favoring a warming trend and the other favoring a cooling trend (Fig. 12.4).

GIVETIAN/FRASNIAN

DEEP ⟶ SHALLOW NONMARINE

Figure 12.3. Paleogeography of the Middle to Late Devonian Appalachian Basin (modified from Ettensohn and Barron (1981), Witzke and Heckel (1988), and Leventhal (1987)). Location of Akzo core #9455 is shown by the dot at 30° S.

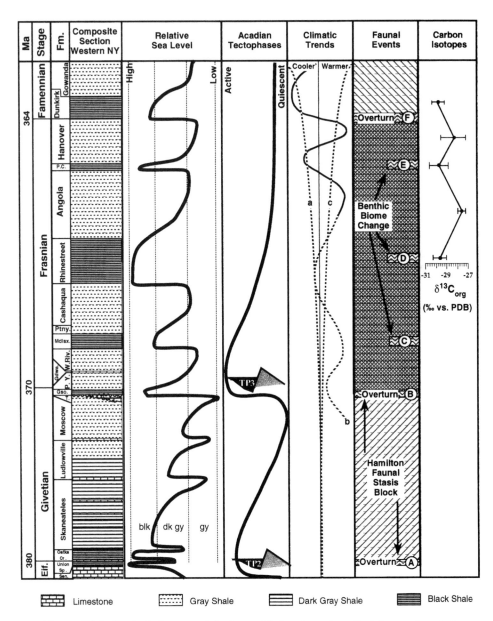

Figure 12.4. Geologic framework for latest Eifelian through earliest Famennian strata of western New York, including geochronology (Gradstein and Ogg, 1996), lithostratigraphy and lithofacies (Kirchgasser *et al.*, 1994; this study), interpreted relative sea level changes (House and Kirchgasser, 1993; Brett and Baird, 1996*a*), tectonic events (tectophases of Ettensohn *et al.*, 1988), climatic trends (see text for sources), major faunal events (see text for sources), and carbon isotope data (Maynard, 1981).

Based on the presence of carbonate paleosols, evaporites, thick carbonates, and trace fossils of dipnoan fish in the proximal Catskill Delta, Woodrow *et al.* (1973) interpreted temperatures in the Givetian–Famennian Appalachian Basin to have been warm to hot. Witzke and Heckel (1988) interpreted that the basin lay on the northern border of a humid, warm temperate belt characterized by strong seasonality. This is consistent with evidence for wet–dry climate alternations noted by Woodrow (1985) and the abundance of storm deposits cited by McCollum (1988). Scotese and McKerrow (1990) suggest Late Devonian cooling as temperatures changed from hot in the Givetian–Frasnian to warm in the Famennian (Fig. 12.4, curve a), and this is supported by faunal data (Copper, 1986), as warm-water faunas suffered much greater reductions in diversity than their higher-latitude counterparts during this time. Although Buggisch (1991) interpreted significant fluctuations approaching the F/F boundary (Fig. 12.4, curve b), the general trend of his curve is consistent with Late Devonian cooling. Work by Isaacson *et al.* (1997) has suggested that recently discovered tillites in South America may be as old as early Frasnian. This represents compelling evidence for Late Devonian cooling and may indicate that permanent snow cover existed in Gondwanaland as early as the late Middle Devonian.

In contrast to many of these interpretations, the results from a Late Devonian general circulation model (GCM) climate simulation (Ormiston and Oglesby, 1995) show little seasonal temperature variation at the general latitude of the DAB, with summer values around 25 °C and winter values around 20 °C. This model also shows little seasonal variation in precipitation and predicts that the Acadian Mountains would have produced a strong orographic effect, receiving far more precipitation than the adjacent Appalachian Basin. Strong, year-round southwesterlies and northeasterlies converged near the coast of North America, and were slightly more vigorous in summer. Interestingly, the model produced no summer snow cover in either hemisphere, thus failing to produce Gondwanan glaciation in the Late Devonian as suggested by Caputo (1985) and by Isaacson *et al.* (1997). In addition, Popp *et al.* (1986), Veizer *et al.* (1986), and Brand (1989) have interpreted oxygen isotope data from brachiopods to suggest relative warming in the Late Devonian (Fig. 12.4, curve c). The paleoecological work of Thompson and Newton (1988) on Late Devonian tropical species extinction supports this warming model because if water temperatures (interpreted from the aforementioned brachiopod studies) were truly as high as 37 °C, this would have been sufficiently hot at least to prevent reproduction in tropical marine organisms if not kill them outright.

Despite specific differences, interpretations of Middle to Late Devonian climate from both geological evidence and modeling results suggest that subtropical conditions prevailed. If the Ormiston and Oglesby (1995) model reflects true Frasnian–Famennian conditions, then the Late Devonian world was ice-free and high-frequency glacio-eustatic control of sea level fluctuations is unlikely. Although this may argue for a tectono-eustatic mechanism for sea level changes, such as that proposed by Johnson *et al.* (1985), or a more locally controlled, tectonic driver, the question remains unresolved, particularly in the light of emerging evidence for Gondwanan glaciation in the Late Devonian (Isaacson *et al.*, 1997).

Tectonics

Devonian tectonism and the uplift of a major orogenic belt along the eastern margin of North America exerted the primary influence on the subsidence of the Appalachian foreland basin as well as the supply of clastic sediments to the basin. The onset of the Acadian Orogeny resulted from the oblique collision of North America with one or more microcontinents, sometimes referred to as the Avalon terrane (Ettensohn, 1985a, 1987; Faill, 1985; Woodrow, 1985; Rast and Skehan, 1993). Orogenic uplift and thrust-loading to the east and southeast of New York produced cratonic downwarping across eastern North America (Quinlan and Beaumont, 1984; Beaumont et al., 1988) and subsidence of the Appalachian foreland basin. The highlands in the east provided a source of siliciclastics which were shed into the subsiding foreland, resulting in the eastward-thickening clastic wedge known as the Catskill Deltaic Complex.

The orogeny appears to have been a stepwise process, consisting of several episodes of active tectonism separated by relatively quiescent times that Ettensohn (1985a) termed tectophases (Fig. 12.4). This produced episodic foreland infilling and the individual progradational packages which comprise the major stratigraphic units of the Catskill Delta shown in Fig. 12.2 (Rickard, 1975; Brett and Baird, 1994; Ver Straeten and Brett, 1995; Ver Straeten, 1996).

Sea level and stratigraphy

Middle to Upper Devonian strata of the Catskill Delta contain a hierarchy of transgressive and regressive (T–R) packages (Johnson et al., 1985; Brett and Baird, 1994, 1996a; Filer, 1994). Two large-scale T–R cycles comprising the Late Eifelian to Givetian Hamilton Group and the overlying Late Givetian to Famennian strata (Tully Formation to Conewango Group) are separated by a major lowstand in sea level (Taghanic Unconformity) (Fig. 12.4). Maximum deepenings are thought to have occurred in the Late Eifelian and Late Givetian, during deposition of the lower parts of the Hamilton Group (Union Springs Formation) and the Genesee Group (Geneseo Formation), respectively (Brett and Baird, 1994, 1996a). Another notable deepening event is associated with the Rhinestreet Shale, which is one of the most extensive of the Late Devonian black shales of the DAB (Pepper and de Witt, 1951; House and Kirchgasser, 1993). Several scales of higher-frequency sea level fluctuations are superposed over these two cycles, but not shown in Fig. 12.4.

The overall stratigraphic architecture of the Catskill Delta is an eastward-thickening, dominantly clastic wedge, representing the punctuated progradational infilling of the Acadian foreland basin (Fig. 12.2). Late Eifelian/Givetian through early Famennian strata of the DAB in western New York are assigned to, in ascending order, the Hamilton, Genesee, Sonyea, West Falls, Java, and Canadaway groups.

Each of the major stratigraphic units of the Catskill Deltaic Complex in distal portions of the basin contains a basal black shale, which is overlain by gray shales and mudstones, often capped by small amounts of siltstone (Figs. 12.2 and 12.4). This overall pattern has been interpreted to represent initial sediment starvation associated with a sea level rise, producing the basal black shales which often have disconformable lower contacts (Baird et al., 1988). These are later buried by more abundant and coarser clastics during subsequent delta progradation (Brett and Baird, 1994; Kirchgasser *et al.*, 1994) as sediment supply increased during a later highstand to fall in relative sea level.

Marine faunas

The benthic faunas of the DAB in central and western New York developed during the acme of Sepkoski's (1984) Paleozoic evolutionary fauna (Fig. 12.1) and are dominated by representatives of it (e.g., brachiopods) with sparse representation of the modern fauna (e.g., bivalves). They have been extensively studied for spatial trends and temporal changes in community structure at stratigraphic member (if not finer) resolution through most of the Givetian (e.g., Brett *et al.*, 1986, 1991; papers in Brett, 1986 and Landing and Brett, 1991; Brower *et al.*, 1988) and at the formation level in the Late Devonian (Sutton *et al.*, 1970; Bowen *et al.*, 1974; Thayer, 1974; McGhee and Sutton, 1981, 1983, 1985; Sutton and McGhee, 1985). All of these studies revealed relatively similar onshore to offshore trends in community structure and diversity, with delta platform to delta front facies (siltstone to silty mudstone) containing the most trophically complex, diverse assemblages (up to ≥ 60 species) and basinal facies (dark gray to black mudstones) the least (1–15 species). These trends have been interpreted to represent the ecological consequences of lateral physical and/or chemical stress gradients, including changes in substrate, turbidity, and dissolved oxygen concentration. In addition, Brett and Baird (1996*c*) have shown lateral migration of these faunal assemblages in association with interpreted relative sea level changes, referring to this pattern as 'faunal tracking'.

Some useful generalizations can be made about benthic fossil assemblages in Givetian through Frasnian basinal black and gray shales. Black shales, when fossiliferous, contain low diversity assemblages of dominantly epifaunal to semi-infaunal brachiopods. Rare infaunal elements are almost exclusively nuculoid bivalves. There is little evidence of soft-bodied burrowers in the black shale facies. These features suggest that the sediment column was dominantly anoxic to sulfidic although bottom-water turbidity may also have played a role. Gray mudstones are typically associated with somewhat higher diversity levels, except in the Hamilton Group where they are often much more diverse. Faunas in the gray mudstones are more evenly distributed between epifaunal and infaunal life habits and, although bivalves are more common, brachiopods remain dominant. Homogeneous textures suggest that many of these gray intervals may have been completely bioturbated. This suggests that sediments were well-aerated during the deposition of gray mudstones.

Planktonic communities in the DAB are relatively poorly known as their fossil record is sparse and often taxonomically equivocal. The most abundantly preserved phytoplankton include acritarchs (Wicander, 1975) and cysts of the green algae *Tasmanites* and *Leiosphaeridia* (e.g., Obermajer *et al.*, 1997), both of which are ascribed to the class Prasinophyceae. Acritarchs have been interpreted to represent algal cysts of unknown affinity (Loeblich, 1970). Well-preserved zooplankton include styliolinids, which have small, calcareous, conical shells and may have been protists (Yochelson and Lindemann, 1986), as well as rare radiolarians (Martin, 1995).

The present work is principally concerned with the record of paleoenvironmental changes in offshore facies (i.e., mudstones and shales). The specific goal is to assess biogeochemical records across horizons of documented turnover in benthic faunas. Brett and Baird (1996b) recognized a Givetian faunal block separated from underlying Eifelian and overlying Frasnian communities by major overturn events (Fig. 12.4, A and B in 'faunal events' column). In addition, Sutton and McGhee (1985) and House (1996) identified a series of smaller-scale faunal changes within the Frasnian (Fig. 12.4, C–E), culminating in the F/F extinction (Fig. 12.4, F).

Evolving terrestrial ecosystems

Although plants may have moved onto the land as early as the Middle Ordovician (Gray, 1985), fossil evidence suggests that they did not vascularize until the latest Silurian (Stewart, 1983; Gensel and Andrews, 1984; DiMichele *et al.*, 1992) and remained very ecologically restricted until the Middle to Late Devonian (Chaloner and Lacey, 1973; Chaloner and Sheerin, 1979; Gensel and Andrews, 1984). It is during this latest time interval that land plants are thought to have achieved arboreal stature, as well as ecological tolerances sufficiently broad and modes of reproduction adequately efficient to permit their widespread geographic distribution.

These early plants are thought to have exerted a strong stabilizing effect on terrestrial environments, providing ground cover and soil retention, despite the likelihood that they also enhanced rates of chemical weathering (Beerbower, 1985). Middle to Late Devonian soils appear to have supported arthropod communities very similar to modern soil and litter communities, and included arachnids, centipeds, arthropleurids, and possibly insects (Shear *et al.*, 1996). The activities of these detritivores, and possibly herbivores, would have increased rates of decay of plant material, speeding soil development and possibly enhancing the flux of nutrients and refractory terrigenous organic matter to shallow marine ecosystems.

MARINE FAUNAL CHANGE

Temporal diversity trends

Many studies have focused on long-term changes in taxonomic diversity from the Givetian through the Famennian. For example, work by Boucot (1975), Bayer and McGhee (1986), and Linsley (1994) has shown that significant reductions in faunal diversity occurred from the Givetian to the Famennian. In the case of the brachiopods, data compiled by Dutro (1981) and Linsley (1994) have shown that the Late Givetian diversity reduction was much greater than that at the F/F boundary in the DAB. Most recently, Brett and Baird (1996b) have documented a series of faunal stasis blocks, separated by relatively rapid and comprehensive overturn events near the base and top of the Givetian in western New York (overturn A and B, Fig. 12.4). One of these overturns (Late Givetian, Geneseo Shale) corresponds to the time of greatest brachiopod diversity loss in the New York DAB (Fig. 12.5). Data compiled by Linsley (1994) show that this interval is also coeval with the time of greatest ammonoid diversification in the Middle to Late DAB (Fig. 12.5).

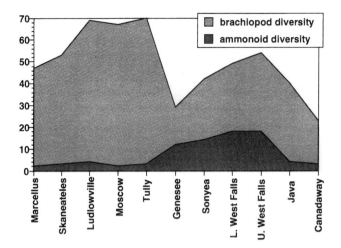

Figure 12.5. Changes in species diversity for New York brachiopods and goniatites through Middle to Late Devonian strata. The disparity in the stratigraphic resolution of data between the Hamilton Group and all overlying groups reflects the lack of formation-level distinction in Upper Devonian studies of western New York. Data are from Linsley (1994).

In the interval overlying the Geneseo Shale a number of changes in the composition of benthic communities have been documented in association with black shale deposition (Fig. 12.4). It must be noted that paleoecological work in the Upper Devonian is less detailed than that in the Middle Devonian, and that much of what has been done is not recent work. Thus many new concepts have not been applied to these benthic community types, and it is possible that Upper Devonian faunal changes associated with black shale deposition represent only responses to environmental changes (i.e., faunal tracking), rather than ecological and/or evolutionary changes.

Much effort has been focused on the F/F boundary (e.g., Copper, 1986; Goodfellow *et al.*, 1988; McGhee, 1988, 1991, 1992, 1996; Wang *et al.*, 1991; Joachimski and Buggisch, 1993). This biotic crisis had a survival rate of 30% for Appalachian Basin species (McGhee, 1982) and is unusual in that the highly diverse, largely endemic faunas of the Middle Devonian had been supplanted by the end of the Givetian (Boucot, 1988) such that the taxa which vanished in the Late Devonian were largely cosmopolitan and low in diversity, although high in abundance (Boucot, 1975). Organisms most affected in this extinction are confined to tropical and subtropical zones (McLaren, 1982). Most notably, reef-building organisms such as the stromatoporoids (McLaren, 1982) and corals (Pedder, 1982) virtually disappeared. Only 10 of 71 Frasnian brachiopod genera survived to the Famennian (Johnson and Boucot, 1973). It is significant that this brachiopod diversity loss at the F/F boundary is similar in magnitude to that noted by Brett and Baird (1996*b*) at the end of the Hamilton faunal stasis block discussed below ('Organic carbon burial, Importance').

Conceptual models

It is clear that there is substantial evidence for complex faunal changes in the Givetian–Famennian section of the DAB, some of which predate the major extinction boundary by as much as 6–12 million years. To what extent do models proposed to explain the extinction apply to the preceding interval? A great diversity of alternate hypotheses have been proposed to account for Late Devonian extinctions. There is general consensus that the animals most affected by the extinction were shallow-water filter feeders with planktonic larval stages (McLaren, 1982). Their collective demise could have been the result of a sea level change restricting habitat (Johnson, 1974; Schlager, 1981), an impact event that caused rapid climatic and/or oceanographic changes (McLaren, 1970; McGhee, 1982), oceanic anoxia (Heckel and Witzke, 1979; Joachimski and Buggisch, 1993), or climate change unrelated to impact, including warming (Thompson and Newton, 1988) or cooling (Copper, 1986). Each of these mechanisms has been advocated for the F/F, but the supporting evidence remains inconclusive. One of the aims of the present investigation is to integrate our understanding of excess organic carbon burial and major faunal changes preceding the F/F boundary in western New York. By establishing a historical context for Late Devonian extinctions, the significance and causes of the 'boundary event' may become clearer.

ORGANIC CARBON BURIAL

Importance

In terms of the carbon cycle, the Givetian–Famennian black shales of western and central New York are significant in several respects: (1) they were deposited during a time when atmospheric CO_2 content is thought to have achieved its maximum rate of decrease, and fallen from $10\times$ to $6\times$ that at present (Berner, 1990); (2) they reflect the burial of significant amounts of organic carbon in basinal sediments (Robl and Barron, 1988; Ettensohn, 1992), and (3) they are associated with faunal changes which are sometimes profound (i.e., Union Springs and Geneseo formations, Brett and Baird, 1996*b*) and sometimes relatively minor (i.e., Middlesex, Rhinestreet, Pipe Creek, and Dunkirk formations; Sutton and McGhee, 1985), suggesting shifts in trophic and community structure. The degree to which these events reflect causes or consequences of change in carbon utilization and carbon transport between the atmosphere, hydrosphere, biosphere, and lithosphere has not been investigated.

Studies of Quaternary (Barnola *et al.*, 1987) and Mesozoic (Freeman and Hayes, 1992) environments have clearly established a relationship between changes in atmospheric pCO$_2$ and changes in global climate at a variety of time scales. Long-term trends, such as greenhouse and icehouse phases, correlate well with intervals thought to be characterized by high and low atmospheric pCO$_2$, respectively (Berner, 1990; Arthur *et al.*, 1987) (Fig. 12.1). Transitions between these states of the earth system are the consequences of fundamental shifts in storage and flux of carbon between organic and inorganic reservoirs. The largest carbon reservoir which may rapidly (on the order of 10^5 years) interact with and influence climate is that of marine sedimentary organic carbon (Sundquist, 1993). Widespread burial of organic carbon (OC) in marine muds and its subsequent preservation in black shales during the Middle to Late Devonian (Fischer, 1981; Ettensohn, 1992) may therefore have played a significant role in the inception of the late Paleozoic icehouse.

Whatever its role in changing global climate, carbon burial was certainly associated with major regional faunal events. Each black shale in the Givetian–Famennian of western New York marks a regional faunal boundary across which the composition of benthic communities changes to a varying extent (Fig. 12.4). For example, the Union Springs Formation black shale at the base of the Hamilton Group is associated with a faunal turnover event and marks the inception of the Hamilton faunal stasis block (Brett and Baird, 1996*b*). The black shale of the Geneseo Formation marks the end of the Hamilton faunal stasis block defined by Brett and Baird (1996*b*), and is characterized by a major diversity change documented in Linsley's (1994) compilation of brachiopod and goniatite ranges.

Maynard (1981), with only a limited number of samples, documented negative shifts in the isotopic composition of OC in several of these Frasnian black shales (Fig. 12.4), but interpreted the trends to be a consequence of changes in relative proportions of terrestrial vs. marine OC. Without the benefit of modern analytical techniques (e.g., Isotope Ratio Monitoring–Gas Chromatography Mass Spectrometry, or IRM-GCMS) and our much-improved understanding of C isotopes in organic materials, Maynard (1981) could not adequately consider factors other than organic matter sources. The carbon isotopic record of DAB black shales needs to be re-evaluated in a more integrated, paleoenvironmental framework, using improved techniques and understanding.

Conceptual models

The mechanisms by which organic carbon is buried in marine sediments have been debated vigorously in recent years. The dominant opposing hypotheses include the 'pycnocline' (or preservation) model (e.g., Demaison and Moore, 1980) and the 'productivity' model (e.g., Pederson and Calvert, 1990; Fig. 12.6). Recently, a 'feedback' model was proposed by Ingall *et al.* (1993), which integrates bottom-water anoxia with enhanced productivity via nutrient (P) cycling.

The pycnocline model explains high rates of OC burial resulting in black shale formation as the result of dysoxic to anoxic bottom waters, isolated from mixing with well-oxygenated surface waters by density stratification of the water column (Fig. 12.6a). This condition is typically thought to be established during marine transgressions and sea level highstands. The pycnocline may be maintained by any combination of temperature and salinity differences between surface- and bottom-water masses. Anoxia precludes the aerobic benthos which would normally consume most deposited organic materials and destroy the sediment's laminated fabric. Anaerobic bacteria, particularly sulfate-reducers, can become the dominant consumers of organic matter (OM) in the benthic environment, producing H_2S as a metabolic byproduct, ensuring the demise of benthic aerobic life (H_2S destroys the oxygen carrying capacity of hemoglobin, resulting in molecular suffocation of aerobic organisms). The deep-water conditions thought to be asso-

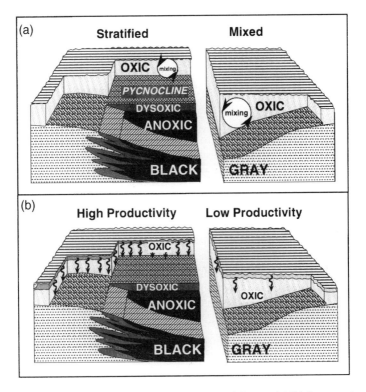

Figure 12.6. Stratified Basin model (a) and productivity model (b) for organic matter accumulation and development of oxygen deficiency in shallow epicontinental basins (modified from Brooks *et al.*, 1987).

ciated with pycnocline establishment also produce siliciclastic starvation in the deep basin. Thus OM may become relatively concentrated, adding to the OC richness of deep-water black shales. This model is based largely on modern environments, such as the California Borderland basins and the type euxinic basin, the Black Sea, in which laminated, OC-rich muds are accumulating beneath a stably stratified water column.

The productivity model (Fig. 12.6b) invokes enhanced surface-water productivity to explain elevated OC burial. Higher productivity increases the flux of organic materials to the depositional interface. This creates a greater demand for oxygen by metazoans and aerobic bacteria in the benthic environment. If high productivity is maintained, as in upwelling zones, this demand eventually exceeds the advective resupply from surface waters, particularly at depths relatively isolated from surface mixing effects (below about 200 m). Although oxygen continues to be circulated into the environment, it is outpaced by the supply of OM. The oxic–anoxic interface, or the redox boundary (defined as the boundary between free oxygen and free hydrogen sulfide, which are effectively mutually exclusive), then migrates upward, creating toxic, sulfidic conditions in the sediment column. If bottom currents are sufficiently weak, this boundary may rise into the overlying water column, as has happened to an extreme in the modern Black Sea, affecting nektonic and planktonic life.

The feedback model of Ingall *et al.* (1993) suggests a mechanism by which black shale deposition, once begun, may be perpetuated. Anoxic bottom-water conditions appear to facilitate the preferential release of P (relative to C) from sedimentary OM during microbial decomposition. The release of P into the water column and its transport back into the photic zone provides what is typically a limiting nutrient in marine ecosystems, and allows elevated productivity. The consequent high flux of OM to the benthic environment maintains bottom-water anoxia. This efficient recycling of P enables a strong biological pumping mechanism for exporting OC to sediments and preserving it.

The black shales of the DAB have dominantly been interpreted to be the result of water column stratification (Ettensohn, 1985*b*, 1992; Woodrow *et al.*, 1988; Kepferle, 1993) and consequent bottom-water anoxia. It has recently been suggested by Algeo *et al.* (1995) that high productivity may have resulted from an increased flux of land-derived nutrients to the DAB from developing soils produced by the rapidly diversifying terrestrial plants of the Middle to Late Devonian. Neither of these models has been explicitly tested using biogeochemical techniques in the black shales of the DAB.

BIOGEOCHEMICAL PROXIES

To test integrated models for OC burial and local to regional faunal change, we have investigated carbon cycling, depositional and early burial redox conditions, and nutrient dynamics as the three most significant biogeochemical parameters. Our proxy materials for tracking changes in the fundamental processes affecting these parameters are, respectively, (1) the amount and carbon isotopic composition of organic materials, (2) degree of pyritization, and (3) the amounts of organic and inorganic phosphorus.

Organic materials: carbon cycling

The amount and carbon isotopic composition of organic materials provide information about carbon sources, sinks, and cycling in ecosystems. Ideally, these isotopic compositions are interpreted relative to that of the dissolved inorganic carbon (DIC) reservoir, which is determined by analysis of biogenic, planktic carbonates thought to precipitate largely in equilibrium with surface waters, although there are complicating, species-dependent vital effects. This is very difficult in the Paleozoic because of the paucity of calcareous plankton and the effects of diagenetic recrystallization of shell carbonates on their carbon isotopic compositions. Our attempts to characterize changes in the $\delta^{13}C$ of surface-water DIC, using whole-rock carbonates as well as samples drilled from atrypid and chonetid brachiopods (although benthic, these are the only arguably 'primary' carbonates present), have yielded data with no systematic variation with respect to lithology or the $\delta^{13}C$ record from OM. We conclude that these carbonates are probably of mixed biogenic and authigenic origin and are of dubious utility in the characterization of any single

DIC reservoir. We therefore present and interpret only the relative changes in the carbon isotopic compositions of the organic materials.

Temporal changes in the ^{13}C content of organic materials from marine primary producers depend principally on changes in the carbon isotopic composition of the photosynthetic reservoir, the size of that reservoir, and the growth rate of photosynthetic organisms (Hayes, 1993; Goericke and Fry, 1994; Laws et al., 1995). Phytoplankton preferentially incorporate ^{12}C into their biomass as biochemical kinetics favor the lighter isotope. The extent of this discrimination, however, is influenced by extremes in supply of and/or demand for photic zone CO_2, which change as a function of overall productivity and phytoplanktonic growth rates. If the size of this CO_2 reservoir becomes smaller or the growth rate greater, fractionation decreases and the $\delta^{13}C$ of primary biomass will increase. Conversely, increased supply and/or decreased growth rate will lead to carbon isotopic depletion of primary biomass. If the effect of carbon isotopic changes in the DIC reservoir is greater than the effects of changes in $[CO_2]_{aq}$ and algal growth rate, the $\delta^{13}C$ of organic materials may be used to qualitatively assess changes in the surface-water DIC reservoir.

In earlier studies, the ^{13}C content of total OC (TOC) has been used to assess changes in OM sources (i.e., marine vs. terrestrial) (Maynard, 1981) and in primary production (Caplan et al., 1996) in samples of Devonian age. The $\delta^{13}C$ of TOC, however, is a composite signal containing variable contributions from different primary producers, an array of heterotrophs and diverse bacteria. The isotopic composition of individual, stable organic compounds representing specific groups of organisms (biomarkers) can provide unique constraints to the interpretation of changes in $\delta^{13}C$ of TOC, including the relative contributions of primary producers to preserved organic materials and the significance of heterotrophic and/or bacterial reworking (Hayes et al., 1990).

Two widely used, saturated hydrocarbon biomarker compounds – the 19-carbon normal alkane ($n\text{-}C_{19}$, a simple chain of singly bonded carbons with all remaining bonding sites occupied by hydrogens) (Gelpie et al., 1970; Collister et al., 1994) and phytane (Hayes et al., 1990), a 20-carbon isoprenoid – are here used to interpret changes in primary biomass. Both of these compounds are thought to be dominantly produced by the diagenetic alteration of algally produced compounds. Carbon isotopic compositions are conserved through these alteration processes. Odd carbon-numbered n-alkanes are derived principally from the decarboxylation of even carbon-numbered fatty acids. Fatty acids with chain lengths of 20 or fewer carbons are typically produced by marine phytoplankton. Therefore n-alkanes with 15, 17, or 19 carbons can be interpreted as representing primary, marine biomass (Gelpie et al., 1970; Collister et al., 1994). Because $n\text{-}C_{19}$ is the dominant n-alkane in most samples from our study interval and has very similar carbon isotopic values to $n\text{-}C_{15}$ and $n\text{-}C_{17}$ we have chosen it as representative of the 'planktonically sourced' n-alkanes. Phytane is produced largely from the diagenetic cleaving and reduction of the phytol chains present in many chlorophylls, but may also derive from lipids present in anoxygenic photosynthetic, methanogenic, and halophilic bacteria (i.e., Archaea). The use of these two compounds together limits multiple-sourcing difficulties and provides a constraint on changes in photic zone CO_2 sources and photosynthetic utilization.

Degree of pyritization: depositional and early burial redox conditions

One of the most significant factors affecting OC preservation in marine sediments is thought to be bottom-water oxygen content (Demaison and Moore, 1980). Based on modern studies of the Black Sea and the California Borderland Basins, well-oxygenated bottom waters have been associated with diverse benthic communities, bioturbated sediments, and low OC preservation. The other endmember in the benthic oxygenation continuum is characterized by depauperate or absent benthic communities, laminated sediments, and high preservation of OC (Rhoads and Morse, 1971; Arthur and Sageman, 1994; Wignall, 1994). Benthic biodiversity and life habits, as well as OC content, can be the consequence of many factors, such as turbidity and sedimentation rate, from which a redox-specific signal is difficult to extract. Many studies have also revealed puzzling complications to the biofacies approach to bottom-water oxygenation, such as the persistence of calcified epibenthos in the absence of a soft-bodied infauna (e.g., Sageman, 1989; Wignall, 1990). For these reasons, the best approach to assessing benthic redox conditions is the utilization of multiple lines of evidence, including geochemistry. Because bottom-water oxygenation plays an important role in the fates of both sedimentary OM and benthic communities, an assessment of redox conditions through the study interval is essential. In a recent review of eight common geochemical paleoredox proxies for ancient mudstones, Jones and Manning (1994) concluded that degree of pyritization (DOP) was among the three more reliable parameters assessed and, in fact, used it as the standard against which the others were measured.

The degree of pyritization is defined as (Berner, 1970; Raiswell and Berner, 1985):

$$DOP = \text{sulfide Fe}/(\text{sulfide Fe} + \text{reactive Fe}).$$

It represents the extent to which the iron available for early diagenetic sulfide mineral formation (in sedimentary rocks, this is synonymous with pyrite as intermediate, metastable forms such as greigite and mackinawite will not persist on geologic time scales in the presence of sulfide) has, in fact, formed pyrite. DOP values of 0 to 0.45 are typical of normal, well-oxygenated, marine environments, while values from 0.45 to 0.75 represent marine environments with restricted bottom-water oxygenation and values from 0.75 to 1.00 typify deposition under euxinic conditions (Raiswell *et al.*, 1988; Canfield *et al.*, 1996). Thus lower bottom-water oxygen concentration leads to more sulfidic conditions and more complete pyritization, while well-oxygenated bottom waters produce less sulfidic conditions and leave more iron unpyritized.

Organic and inorganic phosphorus: nutrient supply and recycling

In recent years, the importance of changes in primary productivity in controlling the OC content of ancient mudrocks has been argued with increasing frequency and rigor (Pederson and Calvert, 1990, and references therein). Rates of primary productivity in modern marine environments are limited by the availability of the nutrient elements N and P (Holland, 1978; Broecker and Peng, 1982). Although these nutrients appear to be depleted simultaneously (Redfield, 1958), P is thought to have exerted stronger limitations on productivity over geologic time (Broecker and Peng, 1982). The reason for this is that while certain bacteria (including cyanobacteria) are capable of fixing N from an essentially inexhaustible gaseous reservoir for biological use, there is no such mechanism for introducing additional P into the system (Jahnke, 1992). Examining the relative extent to which P has been sequestered in sediments in organic matter and in inorganic form provides insight into the release of P back into the biologically active water column reservoir. The record of P burial in organic and mineral forms is thus a powerful tool in the reconstruction of the supply and biogeochemical cycling of this limiting nutrient, which provides constraints on changes in P bioavailability and, therefore, productivity (Ingall and Van Capellan, 1990; Ingall *et al.*, 1993; Van Capellan and Ingall, 1996). Because phosphorus arrives at the sediment–water interface principally in the form of OM, and OC is a more conservative measure of original sedimentary OM concentration than organic P, organic P results have been normalized to organic carbon (Ingall and Van Capellan, 1990; Ingall and Jahnke, 1994; Ingall *et al.*, 1993).

STUDY INTERVAL

A 330-m-long core of Middle and Upper Devonian strata from the Genesee Valley in western New York was obtained from the Akzo Nobel Salt Company. The strata examined in this study comprise the uppermost Windom Shale Member of the Moscow Formation (Hamilton Group) and the overlying Geneseo and lowermost Penn Yan Formations (Genesee Group; Fig. 12.7). The unconformable Windom–Geneseo contact marks the boundary between the Hamilton and Genesee faunal stasis blocks defined by Brett and Baird (1996*b*). The conformable Geneseo–Penn Yan contact corresponds to the Givetian–Frasnian (Middle–Upper Devonian) boundary (Kirchgasser *et al.*, 1988; Kirchgasser and Oliver, 1993) in western New York. Characteristics of the Geneseo Shale were described earlier. To test explicitly the models for black shale deposition in the context of documented faunal changes, high resolution samples were collected through the study interval spanning lithologic transitions from gray mudstone to black shale and back to gray mudstone.

MATERIALS AND METHODS

The study interval was physically described at millimeter to centimeter scale for lithology, color, degree of bioturbation, ichnofauna and macrofauna present, and other physical characteristics. Samples were then cut from a 15-m section spanning the study interval, rinsed with acetone and ground and homogenized in a Spex Mixer-Mill. Organic carbon content was determined by standard coulometric techniques (Huffman, 1977) with an average error of less than 0.01 wt%. Samples for organic carbon isotopic analysis were treated overnight with excess 1 N hydrochloric acid to ensure removal of carbonate minerals. Isotopic analyses were then performed in triplicate on a VG Optima mass spectrometer following combustion at 1020 °C in a Fisons ISOCHROM-EA elemental analyzer. Average reproducibility was within

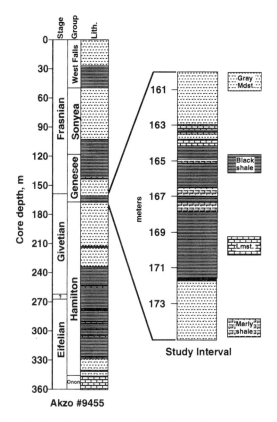

Figure 12.7. Graphic log of Akzo core #9455, from Genesee River Valley, western New York, with expanded view of Windom-Geneseo-Penn Yan study interval.

0.2‰ vs. Pee Dee Belemnite (PDB). Lipids were extracted, and the hydrocarbon fraction isolated using soxhlet and long-column chromatographic methods, respectively, as described by Wakeham and Pease (unpublished). Two hydrocarbon extracts, representing lithologic endmembers, were molecularly characterized using GCMS facilities at Indiana University. Compound identification in all other samples was done by retention time matching. Carbon isotopic compositions of individual hydrocarbon biomarkers were determined by duplicate analyses on a Fisons ISOCHROM-GC gas chromatograph and VG Optima mass spectrometer. Precision for these analyses also averaged 0.2‰ vs. PDB. The data needed for making degree of pyritization calculations were obtained using the chromium reduction technique of Canfield *et al.* (1986) for determining total inorganic sulfur and the hot acid extraction method described by Raiswell *et al.* (1988) for extracting reactive iron. Iron concentrations were determined using photometric techniques (Stookey, 1970) on a Shimadzu spectrophotometer. Based on duplicate analyses, the average reproducibilities were within 3% and 6%, respectively. Phosphorus extractions were performed using procedures described by Aspila *et al.* (1976). Total phosphorus was extracted using 1 N HCl after ashing for 2 h at 550 °C. Inorganic phosphorus was

extracted using 1 N HCl only. Concentrations were determined using spectrophoto-
metry (Murphy and Riley, 1967) which carries an inherent 8% analytical uncertainty
for P. Organic phosphorus content was determined by difference.

RESULTS

Carbon isotopic analyses

Carbon isotopic changes in this interval broadly concur with lithologic
changes and can be divided into samples which are (1) ^{13}C-depleted and rich in
organic carbon and (2) ^{13}C-enriched and poor in OC, with some 'transitional' samples (Fig. 12.8). The trends in the δ^{13}C values of TOC through the study interval can
be summarized in five stratigraphically ascending segments. The lowermost segment
is characterized by fairly consistent values averaging −28‰. This is followed by a
brief, but large, negative shift to values of about −31‰. Above this, values of
−29.5‰ to −30.5‰ are seen throughout Geneseo deposition and into the basal
Penn Yan. The fourth segment is characterized by increasing δ^{13}C back to pre-
Geneseo values. Above this, consistent values of about −28‰ are re-established.

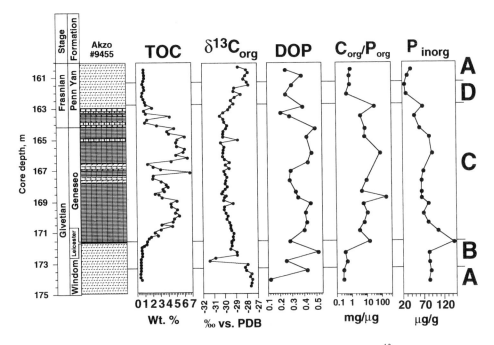

Figure 12.8. Stratigraphic plot of total organic carbon (TOC), δ^{13}C of organic
carbon, degree of pyritization (DOP), organic phosphorus to organic carbon ratio
(note that scale is logarithmic), and inorganic phosphorus data for the Windom-
Geneseo-Penn Yan study interval. Subintervals A, B, C, and D correspond to
interpreted biogeochemical conditions that are illustrated in Fig. 12.10. Lithologic
symbols as in Fig. 12.7.

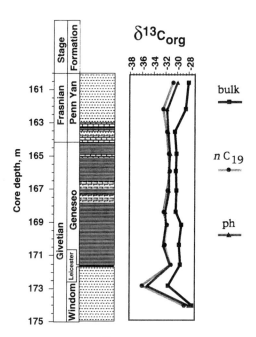

Figure 12.9. Stratigraphic plot of $\delta^{13}C$ data of total organic carbon (bulk) compared with $\delta^{13}C$ for two hydrocarbon biomarkers (C_{19} n-alkane and phytane) for the Windom-Geneseo-Penn Yan study interval. Results are reported in ‰ vs. PDB. Lithologic symbols as in Fig. 12.7.

The $\delta^{13}C$ values of n-C_{19} and phytane show similar values to each other (Fig. 12.9). This suggests a common, phytoplanktonic source for both compounds. These hydrocarbons also show trends which are qualitatively similar to the trends in TOC isotopic values. This suggests that the carbon isotopic trends observed in TOC derive primarily from changes in surface-water conditions which influenced primary production. The variable differences between the algally sourced compounds and the bulk signal may be the result of changes in heterotrophic or bacterial reworking.

The carbon isotopically depleted values associated with OC enrichment in this interval may be the consequence of three phenomena: (1) isotopic depletion in the photosynthetic C reservoir, (2) increased size of the photic zone CO_2 reservoir and/or decreased algal growth rate, or (3) enhanced preservation of isotopically light organic constituents (i.e., lipids) under anoxic conditions. Based on the results of compound-specific carbon isotope analyses through this interval, we conclude that it is not the amount of lipidic material preserved in a sample but the isotopic composition of that material that confers changes in the carbon isotopic composition of bulk OC, and we discount enhanced preservation as a mechanism for TOC carbon isotopic depletion. Although we cannot dismiss the possibility of lower growth rates for photosynthetic organisms, based on other results discussed below, our favored hypothesis is that carbon isotopic depletion associated with OC enrichment is probably the result of increased size and/or isotopic depletion of the photic zone CO_2 reservoir.

Degree of pyritization analyses

DOP trends show a general correspondence to TOC content and can be categorized by lithology (Fig. 12.8). Windom Shale values are highly variable, but average around 0.3. The Geneseo Shale contains much less DOP variability, and averages a somewhat higher 0.4. Penn Yan values are more variable than those of the Geneseo, but less so than those of the Windom, and average about 0.25.

Although DOP values are typically accepted as being indicative of paleo-redox conditions, there exist at least two additional influences that may affect them. It is possible that the amount of metabolizable organic matter reaching the bacterial sulfate-reduction zone may have been limited during Geneseo deposition. This would have placed limitations on the bacterial production of sulfide, potentially making OM rather than reactive Fe the limiting reactant in pyrite formation. A second phenomenon with the same possible consequences for DOP values is the incorporation of sulfur into OM during kerogen formation. If kerogen becomes a significant sink for S during diagenesis, pyritization could become sulfide limited.

If DOP values throughout the study interval are taken to reflect depositional and early burial redox conditions, the values are quite low for a black shale commonly interpreted to represent basinal anoxia. This suggests that bottom water during Geneseo deposition was dominantly oxygenated, normal marine water.

Organic and inorganic phosphorus analyses

Windom Shale samples containing less than about 1 wt% organic carbon have organic C/P values on the order of 0.1 mg μg^{-1} (Fig. 12.8). Samples from the Windom, Geneseo, and Penn Yan shales containing at least 1 wt% organic carbon have organic C/P values two to three orders of magnitude greater. The end of the zone of OC enrichment, near the base of the Penn Yan Shale, marks the return of low organic C/P values averaging about 0.5 mg μg^{-1}.

Possible causes for this pattern include (1) a primary difference in C/P composition of organic materials preserved during black shale deposition, and (2) a secondary difference in C/P composition of organic materials due to different rates of P release from organic matter relative to C during decomposition under different depositional and early burial conditions (e.g., oxic vs. anoxic conditions in the sediment column). Because there is no primary OM with such high C/P ratios (Lerman, personal communication, 1997), we believe that increased organic C/P represents the preferential release of P from organic matter in the sediment column. This phenomenon has been documented by Aller (1994) in laboratory experiments and by Ingall *et al.* (1993) in a survey of ancient mudstones. The fate of this released P was either immobilization in an authigenic mineral phase or release to the water column, depending on the position of the redox boundary relative to the sediment–water interface.

Inorganic P concentrations in the Windom Shale are consistently around 80 μg P g^{-1} sample, rising to 145 μg P g^{-1} sample in the uppermost Windom sample, probably in association with the Windom–Geneseo unconformity (Fig. 12.8). The interval of OC enrichment, including the basal Penn Yan, shows an overall decrease in inorganic P. The two main sources of inorganic P in sediments are (1) mineralization of P released from organic matter, and (2) direct delivery to sediments as phosphate adsorbed on iron oxy-hydroxides, which is released into solution if the Fe is reduced (Jahnke, 1992). The trend of decreasing inorganic P may thus represent the failure to retain P released during OM decomposition (or the failure to release it in the first place) in authigenic mineral form. Because higher organic C/P results from the preferential release of P from OM, we conclude that the trend of decreasing inorganic P is the result of P loss from OM as well as from adsorbed phosphate.

INTERPRETATION

Based on the relationships among the parameters measured, the study interval is divided stratigraphically into four ascending units, A, B, C, and D (Fig. 12.8), from which are interpreted four biogeochemical states (Fig. 12.10). Units A and C represent the 'gray shale' and 'black shale' endmembers, respectively, while B and D represent transitional states 'from gray to black' and 'from black to gray', respectively.

Unit A is characterized by poorly laminated sediment fabric, low TOC content, isotopically enriched TOC and phytoplanktonic biomarkers, low DOP, low organic C/P and moderate inorganic P content. From these observations, we

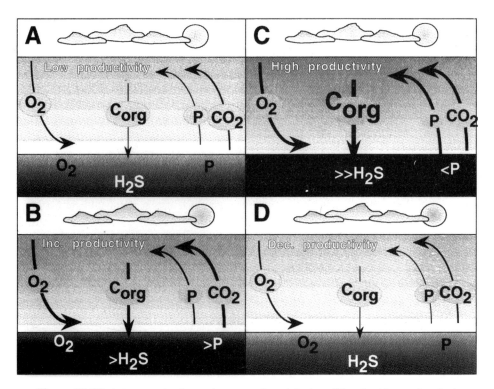

Figure 12.10. Interpreted paleoenvironmental models describing the biogeochemical processes that characterize units A, B, C, and D (from Fig. 12.8) for the Windom-Geneseo-Penn Yan study interval. See text for explanation.

interpret biogeochemical conditions as follows. Low DOP suggests deposition under a well-mixed, fully oxygenated water column. This mixing delivered oxygen to the benthic environment and returned nutrients, such as P, and CO_2 from aerobic respiration to surface waters. Modest productivity resulted in a relatively small flux of OM to the sediment–water interface (SWI). The rate of oxygen flux was easily able to keep pace with the metabolic needs of benthic heterotrophs which consumed the already small amount of OM being sequestered in the mud. This kept surface sediments aerated and P was retained there in both organic and inorganic forms. The zone of bacterial sulfate reduction was kept well below the SWI as sulfate reducers are obligate anaerobes (Fig. 12.10).

Unit B is characterized by increasing TOC content, carbon isotopic compositions of TOC and phytoplanktonic markers which become very depleted and then stabilize at an intermediate value, variable, but still low, DOP, increasing organic C/P, and inorganic P content whose high value just below the Windom–Geneseo contact may represent anomalous P enrichment at the base of the pyritic lag which marks this unconformity. DOP values for Unit B still indicate an oxygenated water column, but may suggest an expanding or unstable sulfide zone within the sediment column. The dramatic carbon isotopic depletion in organic materials suggests an influx of ^{12}C-rich CO_2 into the photic zone. Because respired CO_2 is isotopically depleted and abundant beneath the photic zone, it is the most plausible source in the absence of other evidence. This enhancement of bottom-water CO_2 supply to the photic zone implicitly suggests that water column mixing became more vigorous, which also increased the rate at which nutrients were cycled back into surface waters. Greater nutrient renewal rate produced somewhat higher productivity, which ultimately drove carbon isotopic compositions back toward more enriched values, as demand for the isotopically depleted photic zone CO_2 became higher and increased the flux of OM to the SWI. The effect of changing the carbon isotopic composition of surface-water DIC must have been greater than the kinetic isotope effect of higher phytoplanktonic growth rates. This larger flux produced an expanded zone of sulfide production which was still sufficiently below the SWI to retain P in the sediment and resulted in higher rates of OM burial. Upward migration of the redox boundary brought the horizon at which dissolved P was mineralized closer to the SWI, resulting in increasing amounts of P being trapped with increasing proximity to the water column (Fig. 12.10).

Unit C is characterized by well-laminated sediment fabric, high TOC content, isotopically depleted TOC and phytoplanktonic biomarkers, less variable and slightly higher DOP, high organic C/P, and a trend of decreasing inorganic P content. DOP values in Unit C suggest that deposition was still occurring under an oxygenated water column. There is no evidence for the establishment of anoxic conditions under a pycnocline during Geneseo deposition. Instead, Unit C appears to represent a positive feedback loop, wherein increasing OM flux to the sediments caused the sulfidic zone to expand to the SWI. This resulted in the preferential release of P from an increasing pool of OM as well as the release of adsorbed phosphate, which had previously been trapped in the sediment column. The transfer of P into the water column stimulated high surface-water productivity, which ensured a high flux of OM to the bottom. The amount of OM was such that the entire sediment column remained anoxic, precluding benthic organisms, maintaining a short surface-sediment residence time for P and minimizing its burial in organic and inorganic forms. Throughout the deposition of Unit C, the amount of isotopically depleted CO_2 cycled back to surface waters must have been sufficiently greater than P to allow higher phytoplankton productivity and growth rates to result in isotopically depleted OM. The redox boundary was largely maintained at the SWI (prevented from migrating often or far up into the water column) by the frequent introduction of oxygenated water to the benthic environment (Fig. 12.10).

Unit D is characterized by decreasing TOC content, increasing $\delta^{13}C$ of TOC and phytoplanktonic biomarkers, variable, but lower, DOP, decreasing organic C/P, and inorganic P concentrations which fall to their lowest value in the study interval. It is marked by a reversal of the apparently stable conditions which prevailed throughout the deposition of Unit C and establishment of the positive feedback which ultimately results in the return to conditions similar to those present during the deposition of Unit A. Increasing $\delta^{13}C$ values result from the reduction in the delivery of isotopically depleted CO_2 to the photic zone. This reduction may signify a relative decrease in the vigor of water column mixing, which, in turn, lowered the rate of nutrient recycling. This produced a decrease in productivity and a subsequent decrease in the delivery of OM to the SWI. A degree of oxygenation returned to the near-surface sediments. The reduction of the sulfide zone produced lower DOP values and P began to be more strongly retained in OM. This reduced the availability of P in the biologically active nutrient reservoir and productivity fell back to pre-Geneseo levels. Conditions that characterized Unit A were re-established and mineral P began to accumulate once again, at the expense of the already-depleted bioavailable P reservoir.

The more pervasively sulfidic conditions in the sediment column during Geneseo deposition, coupled with high surface-water productivity, may also explain the coincident decline in brachiopod diversity and proliferation of goniatite species. Although the brachiopods were largely epifaunal or, in a minority of cases, semi-infaunal benthos, their proximity to an unstable toxic chemocline made substrate colonization only episodically viable for the most tolerant taxa, which appear to have been the rhynchonellid brachiopods of the family Leiorhynchidae. By contrast, the nektonic life habit of the goniatites enabled them to escape redox boundary migrations while high productivity ensured abundant nourishment at and near the hub of the marine food web.

CONCLUSIONS AND ONGOING WORK

Preliminary results from the application of our methodology to the Givetian–Frasnian overturn event indicate the following: (1) new data from the Geneseo black shale are difficult to explain using the traditional pycnocline model; (2) organic carbon burial through the Geneseo Shale may have been a function of surface-water productivity driven by the efficient recycling of the limiting nutrient element, P; (3) the intrabasinal cycling of nutrients represents a fundamentally different model from that suggested by Algeo *et al.* (1995), which relied on a terrestrial nutrient source; and (4) patterns of change in benthonic and nektonic faunas are consistent with our model for biogeochemical changes through the study interval. Sulfide may have been a critical limiting factor for biotas, but only at or near the sediment–water interface. By contrast, elevated productivity may have enhanced ecological opportunities for some species that occupied niches higher in the water column.

The degree to which this model has explanatory power for other Late Devonian black shales in the Devonian Appalachian Basin remains to be seen. Ultimately, collection of data for the entire section shown in Fig. 12.4, and evaluation of these data in the context of subsidence, relative sea level, and climate history for the basin, will allow a test of its applicability. Doubtless corrections and modifications will be made. Nevertheless, it is our hope that this model will serve as an effective working hypothesis for studying the dynamics of the Paleozoic greenhouse earth.

ACKNOWLEDGEMENTS

The authors wish to thank Dr Timothy W. Lyons at the University of Missouri for generously providing analytical facilities and instruction for the DOP and P work presented here. This paper was greatly improved by the critical reviews of Drs. Carleton E. Brett and Katherine H. Freeman.

REFERENCES

Algeo, T. J., Berner, R. A., Maynard, J. B. and Scheckler, S. E. (1995). Late Devonian oceanic anoxic events and biotic crises: 'Rooted' in the evolution of vascular land plants? *GSA Today*, **5**, 64–6.

Aller, R. C. (1994). Bioturbation and remineralization of sedimentary organic matter: effects of redox oscillation. *Chemical Geology*, **114**, 331–45.

Arthur, M. A. and Dean, W. E. (1991). An holistic geochemical approach to cyclomania: examples from Cretaceous pelagic limestone sequences. In *Cycles and Events in Stratigraphy*, eds. G. Einsele, W. Ricken and A. Seilacher, pp. 126–66. Berlin: Springer-Verlag.

Arthur, M. A. and Sageman, B. B. (1994). Marine black shales: a review of depositional mechanisms and significance of ancient deposits. *Annual Review of Earth Planetary Science*, **22**, 499–551.

Arthur, M. A., Schlanger, S. O. and Jenkyns, H. C. (1987). The Cenomanian–Turonian Oceanic Anoxic Event II: Paleoceanographic controls on organic matter production and preservation. In *Marine Petroleum Source Rocks*, eds. J. Brooks and A. Fleet, pp. 401–20. London: Geological Society of London.

Aspila, K. I., Agemian, H. and Chau, A. S. Y. (1976). A semi-automated method for the determination of inorganic, organic and total phosphate in sediments. *Analyst*, **101**, 187–97.

Baird, G. C., Brett, C. E. and Kirchgasser, W. T. (1988). Genesis of black shale-roofed discontinuities in the Devonian Genesee Formation, western New York. In *Devonian of the World, II*, Canadian Society of Petroleum Geologists Memoir No. 14, eds. N. J. McMillan, A. F. Embry and D. J. Glass, pp. 357–75. Calgary, Alberta: Canadian Society of Petroleum Geologists.

Barnola, J. M., Raynaud, D., Korotkevich, Y. S. and Lorius, C. (1987). Vostok ice core provides 160,000 year record of atmospheric CO_2. *Nature*, **329**, 408–14.

Bayer, U. and McGhee, G. R. (1986). Cyclic patterns in the Paleozoic and Mesozoic: Implications for time scale calibrations. *Paleoceanography*, **1**, 383–402.

Beaumont, C., Quinlan, B. and Hamilton, J. (1988). Orogeny and stratigraphy: Numerical models of the Paleozoic in the eastern interior of North America. *Tectonics*, **7**, 389–416.

Beerbower, R. (1985). Devonian continental ecosystems and habitats. In *The Catskill Delta*, Geological Society of America Special Paper No. 201, eds. D. L. Woodrow and W. D. Sevon, p. 123. Boulder, CO: Geological Society of America.

Berner, R. A. (1970). Sedimentary pyrite formation. *American Journal of Science*, **268**, 1–23.

Berner, R. A. (1990). Atmospheric carbon dioxide levels over Phanerozoic time. *Science*, **249**, 1382–6.

Berner, R. A. and Raiswell, R. (1983). Burial of organic carbon and pyrite sulfur over Phanerozoic time: a new theory. *Geochimica et Cosmochimica Acta*, **47**, 855–62.

Boucot, A. J. (1975). *Evolution and Extinction Rate Controls*. Amsterdam: Elsevier.

Boucot, A. J. (1988). Devonian biogeography: an update. In *Devonian of the World, III*, Canadian Society of Petroleum Geologists Memoir No. 14, eds. N. J. McMillan, A. F. Embry and D. J. Glass, pp. 211–28. Calgary, Alberta: Canadian Society of Petroleum Geologists.

Bowen, Z. P., Rhoads, D. C. and McAlester, A. L. (1974). Marine benthic communities in the Upper Devonian of New York. *Lethaia*, **7**, 93–120.

Brand, U. (1989). Global climatic changes during the Devonian–Mississippian: Stable isotope biogeochemistry of brachiopods. *Palaeogeography, Palaeoclimatology, Palaeoecology*, **75**, 311–29.

Brett, C. E. (ed.) (1986). *Dynamic Stratigraphy and Depositional Environments of the Hamilton Group (Middle Devonian) in New York State, Part 1*. New York State Museum Bulletin No. 457. Albany, NY: New York State Museum.

Brett, C. E. and Baird, G. C. (1994). Depositional sequences, cycles and foreland basin dynamics in the late Middle Devonian (Givetian) of the Genesee Valley and Western Finger Lakes Region. In *New York State Geological Association, 66th Annual Meeting, Field Trip Guidebook*, eds. C. E. Brett and J. Scatterday, pp. 505–85. New York: New York State Geological Association.

Brett, C. E. and Baird, G. C. (1996a). Middle Devonian sedimentary cycles and sequences in the northern Appalachian basin. In *Paleozoic Sequence Stratigraphy*, Geological Society of America Special Paper No. 306, eds. B. J. Witzke, G. Ludvigson and J. E. Day, pp. 213–42. Boulder, CO: Geological Society of America.

Brett, C. E. and Baird, G. C. (1996b). Coordinated stasis and evolutionary ecology of Silurian–Devonian faunas in the Appalachian Basin. In *New Approaches to Speciation in the Fossil Record*, eds. D. H. Erwin and R. L. Anstey, pp. 285–315. New York: Columbia University Press.

Brett, C. E. and Baird, G. C. (1996c). Epiboles, outages, and ecological evolutionary bioevents: Taphonomic, ecological, and biogeographic factors. In *Paleontological Events: Stratigraphic, Ecological and Evolutionary Implications*, eds. C. E. Brett and G. C. Baird, pp. 249–84. New York: Columbia University Press.

Brett, C. E., Baird, G. C. and Miller, K. B. (1986). Sedimentary cycles and lateral facies gradients across a Middle Devonian shelf-to-basin ramp: Ludlowville Formation, Cayuga Basin. *New York State Geological Association, 58th Annual Meeting, Guidebook*, pp. 81–127. Albany, NY: New York State Geological Association.

Brett, C. E., Dick, V. B. and Baird, G. C. (1991). Comparative taphonomy and paleoecology of Middle Devonian dark gray and black shales from Western New York. In *Dynamic Stratigraphy and Deposiitonal Environments of the Hamilton Group*

(Middle Devonian) of New York State, Part 2, eds. E. Landing and C. E. Brett, pp. 5–36. Albany, NY: New York State Museum.

Broecker, W. S. and Peng, T. H. (1982). *Tracers in the Sea*. New York: Eldigio Press.

Brooks, J., Cornford, C. and Archer, R. (1987). The role of hydrocarbon source rocks in petroleum exploration. In *Marine Petroleum Source Rocks*, Geological Society Special Publication No. 26, eds. J. Brooks and A. J. Fleet, pp. 17–46. London: Geological Society.

Brower, J. C., Thomson, J. A. and Kile, K. M. (1988). The paleoecology of a Middle Devonian regression. In *Devonian of the World, III*, Canadian Society of Petroleum Geologists Memoir No. 14, eds. N. J. McMillan, A. F. Embry and D. J. Glass, pp. 243–56. Calgary, Alberta: Canadian Society of Petroleum Geologists.

Buggisch, W. (1991). The global Frasnian–Famennian "Kellwasser Event". *Geologische Rundschau*, **80**, 49–72.

Canfield, D. E., Lyons, T. W. and Raiswell, R. (1996). A model for iron deposition to euxinic black sea sediments. *American Journal of Science*, **296**, 818–34.

Canfield, D. E., Raiswell, R., Westrich, J. T., Reaves, C. M. and Berner, R. A. (1986). The use of chromium reduction in the analysis of reduced sulfur in sediments and shales. *Chemical Geology*, **54**, 149–55.

Caplan, M. L., Bustin, R. M. and Grimm, K. A. (1996). Demise of a Devonian–Carboniferous carbonate ramp by eutrophication. *Geology*, **24**, 715–18.

Caputo, M. V. (1985). Late Devonian glaciation in South America. *Palaeogeography, Palaeoclimatology, Palaeoecology*, **51**, 291–317.

Chaloner, W. G. and Lacey, W. S. (1973). The distribution of Late Paleozoic floras. In *Organisms and Continents through Time*, Special Papers in Palaeontology No. 12, ed. N. F. Hughes, pp. 271–89. London.

Chaloner, W. G. and Sheerin, A. (1979). *Devonian Macrofloras*, Special Papers in Palaeontology No. 23, pp. 145–61. London.

Collister, J. W., Rieley, G., Stern, B., Eglington, G. and Fry, B. (1994). Compound-specific $\delta^{13}C$ analysis of leaf lipids from plants with different carbon dioxide metabolisms. *Organic Geochemistry*, **21**, 619–27.

Copper, P. (1986). Frasnian–Famennian mass extinction and cold-water oceans. *Geology*, **14**, 835–9.

Demaison, G. J. and Moore, G.T. (1980). Anoxic environments and oil source bed genesis. *AAPG Bulletin*, **64**, 1179–209.

DiMichele, W. A. and Hook, R. W. (1992). Paleozoic terrestrial ecosystems. In *Terrestrial Ecosystems through Time*, eds. A. K. Behrensmeyer, J. D. Damuth and W. A. DiMichele, pp. 205–92. Chicago: University of Chicago Press.

Dutro, J. T. (1981). Devonian brachiopod biostratigraphy. In *Devonian Biostratigraphy of New York, Part 1*, IUGS Subcommission on Devonian Stratigraphy, eds. W. A. Oliver, Jr. and G. Klapper, pp. 67–82. Washington, DC.

Elder, W. P. (1989). Molluscan extinction patterns across the Cenomanian–Turonian stage boundary in the Western Interior of the United States. *Paleobiology*, **15**, 299–320.

Ettensohn, F. R. (1985a). The Catskill Delta complex and the Acadian Orogeny: A model. In *The Catskill Delta*. Geological Society of America, Special Paper No. 201, eds. D. L. Woodrow and W. D. Sevon, pp. 39–49. Boulder, CO: Geological Society of America.

Ettensohn, F. R. (1985b). Controls on development of Catskill Delta complex basin-facies. In *The Catskill Delta*. Geological Society of America, Special Paper No. 201, eds. D. L. Woodrow and W. D. Sevon, pp. 65–77. Boulder, CO: Geological Society of America.

Ettensohn, F. R. (1987). Rates of relative plate motion during the Acadian orogeny based on the spatial distribution of black shales. *Journal of Geology*, **95**, 572–82.

Ettensohn, F. R. (1992). Controls on the origin of the Devonian–Mississippian oil and gas shales, east-central United States. *Fuel*, **71**, 1487–92.

Ettensohn, F. R. and Barron, L. S. (1981). Depositional model for the Devonian–Mississippian black shales of North America: a paleoclimatic-paleogeographic approach. In *GSA Fieldtrip Guidebook*, ed. T. J. Roberts, pp. 344–61. Cincinatti, OH: American Geological Institute.

Ettensohn, F. R., Miller, M. L., Dillman, S. B. *et al.* (1988). Characterization and implications of the Devonian–Mississippian black shale sequence, eastern and central Kentucky, U.S.A.: Pycnoclines, transgression, and tectonism. In *Devonian of the World, II* Canadian Society of Petroleum Geologists, Memoir No. 14, eds. N. J. McMillan, A. F. Embry and D. J. Glass, pp. 323–45. Calgary, Alberta: Canadian Society of Petroleum Geologists.

Faill, R. T. (1985). The Acadian orogeny and the Catskill Delta. In *The Catskill Delta*, Geological Society of America Special Paper No. 201, eds. D. L. Woodrow and W. D. Sevon, pp. 15–38. Boulder, CO: Geological Society of America.

Filer, J. K. (1994). High frequency eustatic and siliciclastic sedimentation cycles in a foreland basin, Upper Devonian, Appalachian Basin. In *Tectonic and Eustatic Controls on Sedimentary Cycles*, SEPM, Concepts in Sedimentology and Paleontology, vol. 4, eds. J. M. Dennison and F. R. Ettensohn, pp. 133–45. Tulsa, OK: SEPM.

Fischer, A. G. (1981). Climatic oscillations in the biosphere. In *Biotic Crises in Ecological and Evolutionary Time*, ed. M. Nitecki, pp. 103–31. New York: Academic Press.

Fischer, A. G. (1982). Long-term climatic oscillations recorded in stratigraphy. In *Climate in Earth History*, eds. W. Berger and H. Crowell, pp. 97–104, Washington, DC: National Academy Press.

Freeman, K. H. and Hayes, J. M. (1992). Fractionation of carbon isotopes by phytoplankton and estimates of ancient CO_2 levels. *Global Biogeochemical Cycles*, **6**, 185–98.

Gelpie, E., Schneider, H., Mann, J. and Oro, J. (1970). Hydrocarbons of geochemical significance in microscopic algae. *Phytochemistry*, **9**, 603–12.

Gensel, P. G. and Andrews, H. S. (1984). *Plant Life in the Devonian*. New York: Praeger Scientific.

Goericke, R. and Fry, B. (1994). Variations of marine plankton $\delta^{13}C$ with latitude, temperature, and dissolved CO_2 in the world ocean. *Global Biogeochemical Cycles*, **8**, 85–90.

Goodfellow, W. D., Geldsetzer, H. H. J, McLaren, D. J., Orchard, M. J. and Klapper, G. (1988). The Frasnian–Famennian extinction: Current results and possible causes. In *Devonian of the World, III*, Canadian Society of Petroleum Geologists, Memoir No. 14, eds. N. J. McMillan, A. F. Embry and D. J. Glass, pp. 9–21. Calgary, Alberta: Canadian Society of Petroleum Geologists.

Gradstein, F. M. and Ogg, J. (1996). A Phanerozoic time scale. *Episodes*, **19**, 3–5.

Gray, J. (1985). The microfossil record of early land plants: Advances in understanding early terrestrialization. *Philosophical Transactions of the Royal Society of London*, **B309**, 167–95.

Hayes, J. M. (1993). Factors controlling ^{13}C contents of sedimentary organic compounds: Principles and evidence. *Marine Geology*, **113**, 111–25.

Hayes, J. M., Freeman, K. H., Popp, B. N. and Hoham, C. (1990). Compound-specific isotopic analyses: a novel tool for the reconstruction of ancient biogeochemical processes. In Advances in Organic Geochemistry 1989, eds. B. Durand and F. Behar, *Organic Geochemistry*, **16**, 1115–28.

Heckel, P. H. and Witzke, B. J. (1979). Devonian world paleogeography determined from distribution of carbonates and related lithic paleoclimate indicators. In *The Devonian System*, Special Papers in Palaeontology No. 23, eds. M. R. House, C. T. Scrutton and M. G. Bassett, pp. 99–123. London.

Holland, H. D. (1978) *The Chemistry of the Atmosphere and the Oceans*. New York: Wiley.

Houghton, J. T., Jenkins, G. J. and Ephamaums, J. J. (eds.) (1990). *Climate Change: The IPCC Scientific Assessment*. Cambridge: Cambridge University Press.

House, M. R. (1996). Juvenile goniatite survival strategies following Devonian extinction events. In *Biotic Recovery from Mass Extinction Events*, Geological Society Special Publication No. 102, ed. M. B. Hart, pp. 163–85. Boulder, CO: Geological Society of America.

House, M. R. and Kirchgasser, W. T. (1993). Devonian goniatite biostratigraphy and timing of facies movements in the Frasnian of eastern North America. In *High Resolution Stratigraphy*, Geological Society Special Publication No. 70, eds. E. A. Hailwood and R. B. Kidd, pp. 267–92. Boulder, CO: Geological Society of America.

Huffman, E. W. D., Jr. (1977). Performance of a new carbon dioxide coulometer, *Microchemistry Journal*, **22**, 567–73.

Ingall, E. D., Bustin, R. M. and Van Capellan, P. (1993). Influence of water column anoxia on the burial and preservation of carbon and phosphorus in marine shales. *Geochimica et Cosmochimica Acta*, **57**, 303–16.

Ingall, E. D. and Jahnke, R. A. (1994). Evidence for enhanced phosphorus regeneration from marine sediments overlain by oxygen-depleted waters. *Geochimica et Cosmochimica. Acta*, **58**, 2571–5.

Ingall, E. D. and Van Capellan, P. (1990). Relation between sedimentation rate and burial of organic phosphorus and organic carbon in marine sediments. *Geochimica et Cosmochimica Acta*, **54**, 373–86.

Isaacson, P. E., Grader, G. W. and Diaz-Martinez, E. (1997). Late Devonian (Famennian) glaciation in Gondwana and forced marine regression in North America. *Geological Society of America Abstracts with Programs*, **29**(6), 117.

Jahnke, R. A. (1992). The phosphorus cycle. In *Global Biogeochemical Cycles*, eds. S. S. Butcher, R. J. Charlson, G. H. Orians and G. V. Wolfe, pp. 301–13. New York: Academic Press.

Joachimski, M. M. and Buggisch, W. (1993). Anoxic events in the late Frasnian – causes of the Frasnian–Famennian faunal crisis? *Geology*, **21**, 675–8.

Johnson, J. G. (1974). Extinction of perched faunas. *Geology*, **2**, 479–82.

Johnson, J. G. and Boucot, A. J. (1973). Devonian brachiopods. In *Atlas of Paleobiogeography*, ed. A. Hallam, pp. 89–96. Amsterdam: Elsevier.

Johnson, J. G., Klapper, G. and Sandberg, C. A. (1985). Devonian eustatic fluctuations in Euroamerica. *Geological Society of America Bulletin*, **96**, 567–87.

Jones, B. and Manning, D. A. C. (1994). Comparison of geochemical indices used for the interpretations of paleoredox conditions in ancient mudstones. *Chemical Geology*, **111**, 111–29.

Kepferle, R. C. (1993). A depositional model and basin analysis for the gas-bearing black shale (Devonian and Mississippian) in the Appalachian Basin. In *Petroleum Geology of the Devonian and Mississippian Black Shale of Eastern North America*, US Geological Survey Bulletin No. 1909, eds. J. B. Roen and R. C. Kepferle, pp. F1–F23. Reston, VA: US Geological Survey.

Kirchgasser, W. T., Baird, G. C. and Brett, C. E. (1988). Regional placement of the Middle/Upper Devonian (Givetian–Frasnian) boundary in western New York State. In *Devonian of the World, III*, Canadian Society of Petroleum Geologists, Memoir No. 14, eds. N. J. McMillan, A. F. Embry and D. J. Glass, pp. 113–18. Calgary, Alberta: Canadian Society of Petroleum Geologists.

Kirchgasser, W. T. and Oliver, W. A., Jr. (1993). Correlation of stage boundaries in the Appalachian Devonian, eastern United States. *Subcommission on Devonian Stratigraphy, Newsletter*, **10**, 5–8.

Kirchgasser, W. T., Over, D. J. and Woodrow, D. L. (1994). Frasnian (Upper Devonian) strata of the Genesee River Valley, Western New York. In *New York State Geological Association, 66th Annual Meeting, Field Trip Guidebook*, eds. C. E. Brett and J. Scatterday, pp 325–58. New York: New York State Geological Association.

Landing, E. and Brett, C. E. (eds.) (1991). *Dynamic Stratigraphy and Depositional Environments of the Hamilton Group (Middle Devonian) in New York State, Part 2*. New York State Museum Bulletin No. 469. Albany, NY: New York State Museum.

Laws, E. A., Popp, B. N., Bidigare, R. R., Kennicutt, M. L. and Macko, S. A. (1995). Dependence of phytoplankton carbon isotope composition on growth rate and $[CO_2(aq)]$: Theoretical considerations and experimental results. *Geochimica et Cosmochimica Acta*, **59**, 1131–8.

Leventhal, J. S. (1987). Carbon and sulfur relationships in Devonian shales from the Appalachian Basin as an indicator of environment of deposition. *American Journal of Science*, **287**, 33–49.

Linsley, D.M. (1994). *Devonian Paleontology of New York*. Paleontological Research Institution Special Publication No. 21. Ithaca, NY: Paleontological Research Institution.

Loeblich, A. R., Jr. (1970). Morphology, ultrastructure and distribution of Paleozoic acritarchs. *Proceedings of the North American Paleontological Convention, 1969*, **Part G**, 705–88.

Martin, R. E. (1995). Cyclic and secular variation in microfossil biomineralization: Clues to the biogeochemical evolution of Phanerozoic oceans. *Global and Planetary Change*, **11**, 1–23.

Maynard, J. B. (1981). Carbon isotopes as indicators of dispersal patterns in Devonian–Mississippian shales of the Appalachian Basin. *Geology*, **9**, 262–5.

McCollum, L. B. (1988). A shallow epeiric sea interpretation for an offshore middle Devonian black shale facies in eastern North America. In *Devonian of the World, II*, Canadian Society of Petroleum Geologists Memoir No. 14, eds. N. J. McMillan, A. F. Embry and D. J. Glass, pp. 347–55. Calgary, Alberta: Canadian Society of Petroleum Geologists.

McGhee, G. R. (1982). The Frasnian–Famennian extinction event: a preliminary analysis of Appalachian marine ecosystems. In *Geological Implications of Impacts of Large Asteroids and Comets on the Earth*, Geological Society of America Special Paper No. 190, pp. 491–500. Boulder, CO: Geological Society of America.

McGhee, G. R. (1988). Evolutionary dynamics of the Frasnian–Famennian extinction event. In *Devonian of the World, III*, Canadian Society of Petroleum Geologists Memoir No. 14, eds. N. J. McMillan, A. F. Embry and D. J. Glass, pp. 23–8. Calgary, Alberta: Canadian Society of Petroleum Geologists.

McGhee, G. R. (1989). The Frasnian–Famennian extinction event. In *Mass Extinction: Processes and Evidence*, ed. S. K. Donovan, pp. 133–51. New York: Columbia University Press.

McGhee, G. R. (1991). Extinction and diversification in the Devonian brachiopoda of New York State: No correlation with sea level? *Historical Biology*, **5**, 215–27.

McGhee, G. R. (1992). Evolutionary biology of the Devonian brachiopoda of New York State: No correlation with rate of change of sea level? *Lethaia*, **25**, 165–72.

McGhee, G. R. (1996). *The Late Devonian Mass Extinction*. New York: Columbia University Press.

McGhee, G. R. and Sutton, R. G. (1981). Late Devonian marine ecology and zoogeography of the central Appalachians and New York. *Lethaia*, **14**, 27–43.

McGhee, G. R. and Sutton, R. G. (1983). Evolution of Frasnian (Late Devonian) marine environments in New York and the central Appalachians, *Alcheringa*, **7**, 9–21.

McGhee, G.R. and Sutton, R.G. (1985). Late Devonian marine ecosystems of the lower West Falls Group in New York. In *The Catskill Delta*, Geological Society of America Special Paper No. 201, eds. D. L. Woodrow and W. D. Sevon, pp. 199–210. Boulder, CO: Geological Society of America.

McLaren, D. J. (1970). Time, life and boundaries. *Journal of Paleontology*, **44**, 801–15.

McLaren, D. J. (1982). Frasnian–Famennian extinctions. In *Geological Implications of Impacts of Large Asteroids and Comets on the Earth*, Geological Society of America Special Paper No. 190, pp. 447–84. Boulder, CO: Geological Society of America.

Miller, J. D. and Kent, D. V. (1986). Synfolding and prefolding magnetizations in the Upper Devonian Catskill Formation of eastern Pennsylvania: implications for the tectonic history of Acadia. *Journal of Geophysical Research*, **91**, 791–812.

Murphy, J. and Riley, J. P. (1967). A modified single solution method for the determination of phosphate in natural waters. *Analytica Chimica Acta*, **27**, 31–6.

Obermajer, M., Fowler, M. G., Goodarzi, F. and Snowdon, L. R. (1997). Organic petrology and organic geochemistry of Devonian black shales in southwestern Ontario, Canada, *Organic Geochemistry*, **26**, 229–46.

Ormiston, A. R. and Oglesby, R. J. (1995). Effect of Late Devonian paleoclimate on source rock quality and location. In *Paleogeography, Paleoclimate and Source Rocks*, AAPG Studies in Geology No. 40, ed. A. Y. Huc, pp. 105–32. Tulsa, OK: AAPG.

Parrish, J.T. (1982). Upwelling and petroleum source beds, with reference to the Paleozoic. *AAPG Bulletin*, **66**, 750–74.

Pedder, A. E. H. (1982). The rugose coral record across the Frasnian–Famennian boundary. In *Geological Implications of Impacts of Large Asteroids and Comets on the Earth*; Geological Society of America Special Paper No. 190, pp. 485–9. Boulder, CO: Geological Society of America.

Pederson, T. F. and Calvert, S. E. (1990). Anoxia vs. productivity: What controls the formation of organic-carbon-rich sediments and sedimentary rocks? *AAPG Bulletin*, **74**, 454–66.

Pepper, J. F. and de Witt, W., Jr. (1951). *The Stratigraphy of the Perrysburg Formation of Late Devonian Age in Western and West-central New York*. US Geological Society Oil and Gas Investigations Chart OC-45, 2 sheets. Reston, VA.

Popp, B. N., Anderson, T. F. and Sandberg, P. A. (1986). Brachiopods as indicators of original isotopic compositions in some Paleozoic limestones. *Geological Society of America Bulletin*, **97**, 1262–19.

Quinlan, G. M. and Beaumont, C. (1984). Appalachian thrusting and the Paleozoic stratigraphy of the eastern interior of North America. *Canadian Journal of Earth Science*, **21**, 973–94.

Raiswell, R. and Berner, R. A. (1985). Pyrite formation in euxinic and semi-euxinic sediments. *American Journal of Science*, **285**, 710–24.

Raiswell, R., Buckley, F., Berner, R. A. and Anderson, T. F. (1988). Degree of pyritization of iron as a paleoenvironmental indicator of bottom-water oxidation. *Journal of Sedimentary Petrology*, **58**, 812–19.

Rast, N. and Skehan, J. W. (1993). Mid-Paleozoic orogenesis in the North Atlantic. In *The Acadian Orogeny: Recent Studies in New England, Maritime Canada and the Autochthonous Foreland*, Geological Society of America Special Paper No. 275, eds. D. C. Roy and J. W. Skehan, pp. 153–64. Boulder, CO: Geological Society of America.

Redfield, A. C. (1958). The biological control of chemical factors in the environment. *American Journal of Science*, **46**, 205–21.

Rhoads, D. C. and Morse, J. M. (1971). Evolutionary and ecologic significance of oxygen-deficient marine basins. *Lethaia*, **4**, 413–28.

Rickard, L. V. (1975). *Correlation of the Devonian Rocks in New York State*: New York Museum and Science Service, Map and Chart Series No. 24, Reston, VA.

Robl, T. L. and Barron, L. S. (1988). The geochemistry of Devonian Black Shales in central Kentucky and its relationship to inter-basinal correlation and depositional environment. In *Devonian of the World, II*, Canadian Society of Petroleum Geologists Memoir No. 14, eds. N. J. McMillan, A. F. Embry and D. J. Glass, pp. 377–92. Calgary, Alberta: Canadian Society of Petroleum Geologists.

Roen, J. B. (1993). Introductory review – Devonian and Mississippian black shale, eastern North America in the Appalachian Basin. In *Petroleum Geology of the Devonian and Mississippian Black Shale of Eastern North America*, US Geological Survey Bulletin No. 1909, eds. J. B. Roen and R. C. Kepferle, pp. A3–A8. Reston, VA: US Geological Survey.

Sageman, B. B. (1989). The benthic boundary biofacies model: Hartland Shale Member, Greenhorn Formation (Cenomanian), Western Interior, North America. *Palaeogeography, Palaeoclimatology, Palaeoecology*, **74**, 87–110.

Sageman, B. B. and Hollander, D. J. (1999). Integration of paleoecological and geochemical proxies: a holistic approach to the study of past global change. In *The Evolution of Cretaceous Ocean/Climate Systems*, Geological Society of America Special Paper, eds. C. J. Johnson and E. Barrera. Boulder, CO: Geological Society of America (in press).

Schlager, W. (1981). The paradox of drowned reefs and carbonate platforms. *Geological Society of America Bulletin*, **92**, 197–211.

Scotese, C. R. and McKerrow, W. S. (1990). Revised world maps and introduction. In *Paleozoic Paleogeography and Biogeography*, Geological Society Memoir No. 12, eds. W. S. McKerrow and C. R. Scotese, pp. 1–21. London: Geological Society.

Sepkoski, J. J., Jr. (1984). A kinetic model of Phanerozoic taxonomic diversity III: Post-Paleozoic families and mass extinctions. *Paleobiology*, **10**, 246–67.

Shear, W. A., Gensel, P. G. and Jeram, A. J. (1996). Fossils of large terrestrial arthropods from the Lower Devonian of Canada. *Nature*, **384**, 555–7.

Stewart, W. N., (1983). *Paleobotany and the Evolution of Plants*. New York: Cambridge University Press.

Stookey, L. L. (1970). Ferrozine – a new spectrophotometric reagent for iron. *Analytical Chemistry*, **42**, 779–81.

Sundquist, E. T. (1993). The global CO_2 budget. *Science*, **259**, 934–41.

Sutton, R. G., Bowen, Z. P. and McAlester, A. L. (1970). Marine shelf environments of the Upper Devonian Sonyea Group of New York. *Geological Society of America Bulletin*, **81**, 2975–92.

Sutton, R. G. and McGhee, G. R. (1985). The evolution of Frasnian marine "community-types" of south-central New York. In *The Catskill Delta*, Geological Society of America Special Paper No. 201, eds. D. L. Woodrow and W. D. Sevon, pp. 211–24. Boulder, CO: Geological Society of America.

Thayer, C. W. (1974). Marine paleoecology in the Upper Devonian of New York. *Lethaia*, **7**, 121–55.

Thompson, J. and Newton, C. (1988). Late Devonian mass extinction: episodic climatic cooling or warming? In *Devonian of the World, III*, Canadian Society of Petroleum Geologists Memoir No. 14, eds. N. J. McMillan, A. F. Embry and D. J. Glass, pp. 29–34. Calgary, Alberta: Canadian Society of Petroleum Geologists.

Van Capellan, P. and Ingall, E. D. (1996). Redox stabilization of the atmosphere and oceans by phosphorus-limited marine productivity. *Science*, **271**, 493–6.

Veizer, J., Fritz, P. and Jones, B. (1986). Geochemistry of brachiopods: Oxygen and carbon isotopic records of Paleozoic oceans. *Geochimica et Cosmochimica Acta*, **50**, 1679–96.

Ver Straeten, C.A. (1996). *Stratigraphic Synthesis and Tectonic and Sequence Stratigraphic Framework, upper Lower and Middle Devonian, Northern and Central Appalachian Basin*. Unpublished PhD Dissertation, University of Rochester, Rochester.

Ver Straeten, C. A. and Brett, C. E. (1995). Lower and Middle Devonian foreland basin fill in the Catskill Front: Stratigraphic synthesis, sequence stratigraphy, and the Acadian Orogeny. In *New York State Geological Association, 67th Annual Meeting, Field Trip Guidebook*, eds. J. I. Garver and J. A. Smith, pp. 313–56. New York: New York State Geological Association.

Wang, K., Orth, C. J., Attrep, M., Chatterton, B. D. E., Hou, H. and Geldsetzer, H. H. J. (1991). Geochemical event for a catastrophic biotic event at the Frasnian–Famennian boundary in south China. *Geology*, **19**, 776–9.

Wicander, E. R. (1975). Fluctuations in a Late Devonian–Early Mississippian phytoplankton flora of Ohio, USA. *Palaeogeography, Palaeoclimatology, Palaeoecology*, **17**, 89–108.

Wignall, P. B. (1990). *Benthic Paleoecology of the Late Jurassic Kimmeridge Clay in England*. Palaeontological Association Special Paper No. 43. London.

Wignall, P. B. (1994). *Black Shales*. Oxford: Clarendon Press.

Witzke, B. J. (1990). Paleoclimatic constraints for Paleozoic paleolatitudes of Laurentia and Euramerica. In *Paleozoic Paleogeography and Biogeography*, Geological Society Memoir No. 12, eds. W. S. McKerrow and C. R. Scotese, pp. 57–74. London: Geological Society.

Witzke, B. J. and Heckel, P. H. (1988). Paleoclimatic indicators and inferred Devonian paleo-latitudes of Euramerica. In *Devonian of the World, I*, Canadian Society of Petroleum Geologists Memoir No. 14, eds. N. J. McMillan, A. F. Embry and D. J. Glass, pp. 49–63. Calgary, Alberta: Canadian Society of Petroleum Geologists.

Woodrow, D. L. (1985). Paleogeography, paleoclimate, and sedimentary processes of the Late Devonian Catskill Delta. In *The Catskill Delta*, Geological Society of America Special Paper No. 201, eds. D. L. Woodrow and W. D. Sevon, pp. 51–63. Boulder, CO: Geological Society of America.

Woodrow, D. L., Dennison, J. M., Ettensohn, F. R., Sevon, W. T. and Kirchgasser, W. T. (1988). Middle and Upper Devonian stratigraphy and paleogeography of the central and southern Appalachians and eastern Midcontinent, U.S.A. In *Devonian of the World, I*, Canadian Society of Petroleum Geologists Memoir No. 14, eds. N. J. McMillan, A. F. Embry and D. J. Glass, pp. 277–301. Calgary, Alberta: Canadian Society of Petroleum Geologists.

Woodrow, D. L., Fletcher, F. W. and Ahrnsbrak, W. F. (1973). Paleogeography and paleo-climate at the depositional sites of the Devonian Catskill and Old Red Facies. *Geological Society of America Bulletin*, **84**, 3051–64.

Yan, Z., Hou, H. and Ye, L. (1993). Carbon and oxygen isotope event markers near the Frasnian–Famennian boundary, Luoxiu section, China. *Palaeogeography, Palaeoclimatology, Palaeoecology*, **104**, 97–104.

Yochelson, E. L. and Lindemann, R. H. (1986). Considerations on the systematic placement of the Styliolines (*incertae sedis*: Devonian). In *Problematic Fossil Taxa*, Oxford Monographs in Geology and Geophysics No. 5, eds. A. Hoffman and M. H. Nitecki, pp. 45–58. Oxford: Oxford University Press.

13

Glaciation in the early Paleozoic 'greenhouse': the roles of paleogeography and atmospheric CO$_2$

MARK T. GIBBS, KAREN L. BICE, ERIC J. BARRON, AND LEE R. KUMP

ABSTRACT

The Late Ordovician (Hirnantian) glaciation was the only major glaciation during the generally ice-free early Paleozoic. It had a profound effect on the earth system and is closely associated with a major Phanerozoic mass extinction. This glaciation has attracted attention in recent years because geochemical models and proxy data indicate ~14–16 times the present atmospheric level (PAL) of atmospheric CO$_2$, apparently coincident with the glaciation. Previous climate modeling studies postulated that glaciation was possible at high CO$_2$, given the paleogeographic configuration of the Late Ordovician. With the coast of Gondwana positioned close to the South Pole, the ocean's thermal inertia would have prevented summer temperatures from rising above freezing, thus allowing snow to survive and glacial inception. However, recent stratigraphic and stable isotope studies indicate that major glaciation lasted perhaps less than 1 million years, meaning that mechanisms operating on tectonic timescales are unlikely to be the primary cause of the glaciation. A change in oceanographic circulation which led to an increase in organic carbon burial has recently been proposed as the cause of a brief drawdown of atmospheric CO$_2$, which resulted in the Late Ordovician glaciation.

We present new general circulation model results for both Late Ordovician and Early Silurian paleogeographies, using identical solar luminosities (a 4.5% reduction from modern) and atmospheric CO$_2$ levels. These experiments allow us to address the relative importance of paleogeography and CO$_2$ in forcing the development of a positive snow mass balance on Gondwana. Both paleogeographies are ice free at 18× PAL and glaciation is indicated for both at 10× PAL. Thus, if CO$_2$ is high/low enough, ice-free/glacial conditions result; the paleogeographic configuration is irrelevant at these levels. To explain predominantly ice-free conditions for the early Paleozoic, atmospheric CO$_2$ must have been continuously high. However, glaciation is indicated by the model at 14× PAL for the Late Ordovician but not for the Early Silurian. As has been postulated, this is a consequence of Gondwana's position over the South Pole. Our results indicate that the Late Ordovician configuration may have acted to precondition the earth system for glaciation, making it particularly sensitive to a drop in atmospheric CO$_2$.

INTRODUCTION

Widespread glaciogenic deposits are an unequivocal indicator of paleoclimate, representing the presence of large quantities of permanently frozen water on land. They are therefore a key indication of the earth's climatic state, representing the most fundamental difference between a cold climate ('icehouse' world) and a warm climate ('hothouse' or 'greenhouse' world). Here our goal is to explore the relative importance of differences in atmospheric $p\mathrm{CO_2}$ and paleogeography (two important controls of paleoclimate) between the Late Ordovician, a time of major glaciation, and the succeeding interval, the ice-free Early Silurian. Thus, unlike other chapters in this volume which examine the 'second order' patterns and aspects of warm climates in earth history, this chapter is concerned with explaining the 'first order' difference between glacial and warm, ice-free climates.

The Late Ordovician glaciation is one of three major glaciations in the Phanerozoic (Hambrey and Harland, 1981). This glaciation has left an excellent record on Gondwana and nearby areas and has also been closely tied to one of the largest mass extinctions in the Phanerozoic (see review by Brenchley, 1989). There are few other good paleoclimate indicators available for the early Paleozoic (i.e., Cambrian–Silurian) compared with more recent intervals. Vascular plants were not abundant, and high resolution studies of sea surface temperatures based on foraminiferal stable isotope analyses are not possible. As a result of this coarse record and lack of typical indicators we must ask broader questions. What caused the Late Ordovician glaciation, and does this fit with our present understanding of major controls on climate throughout earth history? This glaciation has recently been shown to be of very short duration; the rest of the early Paleozoic was essentially ice free. Another question must therefore be asked: Why didn't extensive glaciation occur in the rest of the early Paleozoic? In other words, we must explain the broad climatic evolution of the early Paleozoic and why the climate system only once passed briefly through an important threshold that allowed the build-up of substantial ice sheets, otherwise remaining ice free.

The first part of this paper reviews recent observational and modeling studies of the Late Ordovician glaciation. Geochemical estimates (Yapp and Poths, 1992; Berner, 1994) indicate that levels of atmospheric $\mathrm{CO_2}$ were apparently very high during the Late Ordovician ($\sim$14–16 times present atmospheric level (PAL) defined here as the pre-industrial level of 280 ppm). Solar physics predicts an increase in solar luminosity through earth history (e.g., Newman and Rood, 1977; Endal and Sofia, 1981; Bahcall and Ulrich, 1988), a consequence of stellar evolution along the main sequence, as hydrogen is converted into helium by fusion. Therefore because solar luminosity was lower in the early Paleozoic compared with today, a relatively high level of atmospheric $\mathrm{CO_2}$ (i.e., at least several times PAL) would be required to prevent a runaway icehouse from occurring (Gibbs et al., 1997). Previous climate modeling studies (Crowley and Baum, 1991, 1995) indicate that, with solar luminosity reduced by 4.5%, glaciation could have occurred at $14\times$ PAL $\mathrm{CO_2}$ in the Late Ordovician when the South Pole lay close to the coast of the Gondwanan supercontinent. This result follows the hypothesis of Crowley et al. (1987), that with such a paleogeographic configuration, the ocean's high thermal inertia could prevent summertime temperatures at the pole from rising above freezing. Snow could then survive year-round and build up, so permitting glacial inception.

Recently published stratigraphic and stable isotope evidence (e.g., Brenchley *et al.*, 1994; Paris *et al.*, 1995) suggests major glaciation occurred at the very end of the Ordovician, lasting approximately 1 Myr. Such a short duration would mean that the geochemical estimates of high atmospheric CO_2 levels (calculated over broader time intervals) were not necessarily coeval with the glaciation; indeed, atmospheric pCO_2 could have been relatively low at the time of glaciation. Furthermore, mechanisms acting on tectonic timescales (>1 Myr), such as changes in paleogeography, appear to be precluded as the primary cause of the glaciation. The new carbon isotope data indicate a major perturbation to the carbon cycle (such as an oceanographic event that increased rates of organic carbon burial) and indicate that a brief drawdown of atmospheric CO_2 could have caused the glaciation (e.g., Brenchley *et al.*, 1994; Kump *et al.*, 1995). This could explain the abrupt onset, short duration, and demise of the glaciation, and would be consistent with the paradigm that the level of atmospheric CO_2 is a major control of paleoclimate.

Given, then, that atmospheric pCO_2 may not have been 14× PAL at the time of the glaciation, Gibbs *et al.*'s (1997) aim in a climate model study of the Late Ordovician was to investigate the potential effects of a range of different atmospheric CO_2 levels. They found that the range of pCO_2 values for ice-free (18×) to runaway icehouse (8×) climates lies within the range of uncertainty (8× to >20× PAL) in the geochemical estimates of pCO_2. These results also indicate that glacial inception in the Late Ordovician could have been very sensitive to modest changes in the forcing from atmospheric pCO_2. The question remains as to whether this sensitivity was uniquely related to the Late Ordovician paleogeography. Thus the second part of this paper reports results from new general circulation model experiments for the Early Silurian, an interval for which there is no record of major continental glaciation (Hambrey and Harland, 1981). As in the work of Gibbs *et al.* (1997), atmospheric pCO_2 is varied between experiments. Because all other factors are held constant, by comparing results from these experiments and those of Gibbs *et al.* (1997) we can evaluate the roles of paleogeography and pCO_2. The relative importance of these two factors has been a key issue arising from previous climate modeling studies (e.g., see Crowley, 1993; Crowley *et al.*, 1993*a*). We ask (1) whether a similar critical pCO_2 threshold for glaciation existed for the Early Silurian, and (2) how important for explaining glaciation are the differences in the Late Ordovician and Early Silurian paleogeographies. Following the pole-positioning hypothesis of Crowley *et al.* (1987), is the paleogeographic configuration of a pole-edged Gondwana a factor that allows glaciation only in the Late Ordovician, or does atmospheric pCO_2 have to remain high to ensure ice-free conditions in the Early Silurian?

THE GEOLOGIC RECORD OF THE LATE ORDOVICIAN GLACIATION AND RELATED EVENTS

Compared with the Permo-Carboniferous glaciation, the Late Ordovician glaciation has been recognized relatively recently, but it has attracted much interest.

A glacial event was first postulated by Spjeldnaes (1961) based on a contraction of faunal belts, which he attributed to a global cooling. No direct evidence of glaciation was known at the time. Detailed studies of actual glacial deposits in the Sahara arose from French petroleum exploration efforts in Algeria in the 1960s (richly documented by Beuf *et al.*, 1971). Practically all types of evidence seen today in glacial environments are recognized, including pingos, drumlins, and U-shaped valleys, and the event is often invoked as the archetypal ancient glaciation. More recent investigations have confirmed and extended the known limits of the glaciation to Arabia (Vaslet, 1990; Abed *et al.*, 1993; Powell *et al.*, 1994) and to the Argentinian Precordillera (Sánchez *et al.*, 1991; Buggisch and Astini, 1993). Crowley and Baum (1991) derived an area of $\sim 12 \times 10^6 \, \text{km}^2$ for the Late Ordovician ice sheet by considering the locations of most of the major global deposits and allowing for some erosion. By comparison, the Pleistocene Laurentide covered $11.6 \times 10^6 \, \text{km}^2$ (Denton *et al.*, 1971) and the present-day East Antarctic ice sheet covers $10.5 \times 10^6 \, \text{km}^2$ (Paterson, 1972).

A Late Ordovician to Early Silurian age was initially established for the glaciation (Beuf *et al.*, 1971; Berry and Boucot, 1973), suggesting a duration of 35 Myr. This duration, along with paleomagnetic results indicating very high latitudes for the glacial deposits (see review by Van der Voo, 1993), led many authors to propose that the glaciation was a result of the polar positioning of Gondwana (e.g., Frakes, 1979; Hambrey and Harland, 1981; Crowell, 1983; Caputo and Crowell, 1985; Hambrey, 1985; Frakes *et al.*, 1993). More recent stratigraphic studies (e.g., Destombes, 1981; McClure, 1988; Hiller, 1992; Buggisch and Astini, 1993; Legrand, 1995; Paris *et al.*, 1995) support the idea of major glaciation lasting for just part of the Hirnantian, the latest stage of the Ashgill, which is in turn the latest series of the Ordovician. The Ashgill has a duration of 4 Myr (Harland *et al.*, 1990; Tucker *et al.*, 1990), and a duration of 1 Myr is typically given for the Hirnantian, based on the fact that the Ashgill contains four stages. More precise ages are clearly needed: the Hirnantian could easily have lasted 500 kyr or 3.5 Myr. However, also clearly, the glaciation was short, certainly lasting less than 4 Myr.

As regards deposits older than Hirnantian, only the glacial and/or periglacial Melez–Chograne Formation in Libya (Havlicek and Massa, 1973) is still regarded as Caradocian (early Late Ordovician). But it should be noted that Paris *et al.* (1995), in an extensive study of chitinozoans from Late Ordovician glacial deposits in North Africa, were unable to recover suitable samples to verify this age. Semtner and Klitzsch (1994) have argued for a second ice advance in the Early Silurian in Egypt and Sudan. However, their dating is based on trace fossils and these deposits might be of Late Carboniferous age (Legrand, 1995). Only in the Amazonas and Parnaíba basins in Brazil are there glacial deposits of proven Early Silurian age (Llandovery and possibly earliest Wenlock; Grahn and Caputo, 1992). There is evidence for an Early Cambrian glaciation in West Africa (Bertrand-Sarfati *et al.*, 1995), but otherwise no other early Paleozoic deposits have been unambiguously identified (Hambrey and Harland, 1981; Frakes *et al.*, 1993).

Recent studies indicate major positive excursions in marine oxygen (δ^{18}O) of $\sim +4\text{‰}$ and inorganic carbon ($\delta^{13}\text{C}_{\text{carbonate}}$) of $\sim +5$ to $+7\text{‰}$ (relative to Pee Dee

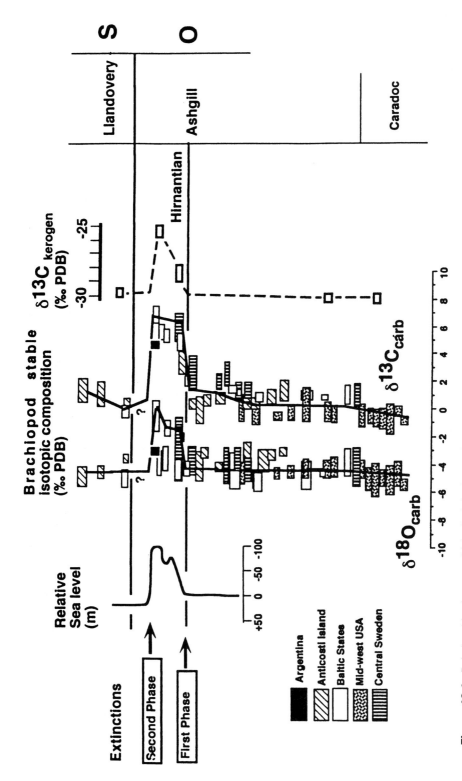

Figure 13.1. Relationships between biological, bathymetric, and stable isotope changes in the Late Ordovician. Stable isotope curves are interpreted composites based on data from a range of sites. From Marshall *et al.* (1997).

Belemnite (PDB)) that are confined to the Hirnantian (Brenchley *et al.*, 1994, 1995; Kump *et al.*, 1995; Marshall *et al.*, 1997; Fig. 13.1). These excursions are super-imposed on a longer-term Ordovician baseline of $\sim-5‰$ for $\delta^{18}O$ and $\sim 0‰$ for $\delta^{13}C_{carbonate}$ (Qing and Veizer, 1994). That the same signals are seen from a world-wide range of sites and from whole-rock samples, as well as individual brachiopod shells, argues for these data representing a global, rather than regional, event (Kump *et al.*, 1995; Marshall *et al.*, 1997). Additionally, Wang *et al.* (1993, 1997) have found a positive excursion of similar magnitude in organic carbon ($\delta^{13}C_{org}$) from deep-water black shales from China. Marshall *et al.* (1997) examined thermally immature organic matter from Estonia and found a shift in $\delta^{13}C_{org}$, parallel to the shift in $\delta^{13}C_{carbonate}$ observed earlier in brachiopod shells from the same basin (Brenchley *et al.*, 1994). Underwood *et al.* (1997) have found that the $\delta^{13}C_{org}$ excursion lies mainly within the *extraordinarius* graptolite biozone (i.e., the lower part of the Hirnantian) at Dob's Linn in Scotland, site of the Ordovician/Silurian boundary stratotype (Cocks, 1985), which substantially assists global correlations.

In summary, the vast majority of early Paleozoic glacial deposits are con-fined to the Hirnantian, suggesting that although glaciation was intensive it was short, lasting perhaps less than 1 Myr. If a glaciation is defined as the presence of significant terrestrial ice somewhere on the globe, then this short duration is in remarkable contrast to the other main Phanerozoic glaciations. The Permo-Carboniferous glaciation lasted for 60 Myr (Veevers and Powell, 1987). Cenozoic glaciation began on Antarctica in the early Oligocene around 34 Ma (Miller *et al.*, 1987, 1991; Zachos *et al.*, 1992) or possibly earlier, in the Late Eocene (Miller *et al.*, 1991; Zachos *et al.*, 1994). Other major environmental events also occurred at the end of the Ordovician. A major global regression and transgression of sea level occurred in the Hirnantian which is closely tied to the stable isotope excursions and is very convincingly glacio-eustatic in origin (Brenchley *et al.*, 1994; Fig. 13.1). The size of the oxygen isotope excursion requires both extensive tropical cooling as well as the removal of isotopically depleted sea water to form continental ice sheets that would account for the observed sea level fall of 50–100 m (Brenchley *et al.*, 1995). The carbon isotope data point to an association between the glaciation and a major (but short) perturbation of the global carbon cycle. Also, two phases of mass extinction are observed in the Hirnantian, the first related to cooling and sea level fall at the start of the glaciation, the second during the transgression at the end of the glaciation (Brenchley, 1989; Fig. 13.1). Oceanographic circulation and pro-ductivity changes have also been proposed as causes of these extinctions (Brenchley *et al.*, 1995). This range of evidence strongly indicates that the Hirnantian climate was unique compared with the rest of the early Paleozoic.

LATE ORDOVICIAN GLACIATION AT HIGH ATMOSPHERIC pCO$_2$?

Increased levels of atmospheric CO_2 and the resulting enhanced greenhouse effect have often been invoked to explain warm, ice-free climates of intervals such as the Mesozoic, while, conversely, times of extensive continental glaciation are ascribed to relatively low levels of atmospheric CO_2 and a reduced greenhouse effect (e.g., Chamberlin, 1899; Roberts, 1971; Berner *et al.*, 1983; Fischer, 1984; Barron and Washington, 1985; Veevers, 1990; Raymo, 1991; Freeman and Hayes, 1992; Frakes *et al.*, 1993; Barron *et al.*, 1995; DeConto *et al.*, Chapter 9, this volume). Net consumption of atmospheric CO_2 occurs during the chemical weathering of silicate minerals on land and precipitation of carbonates in the ocean (Urey, 1952). Essential to all models of the carbonate–silicate cycle is that on long (>1 Myr) timescales the rate of volcanic outgassing of CO_2 must be balanced by the rate of removal of atmospheric CO_2 by silicate weathering on land followed by carbonate precipitation in the ocean (e.g., Walker *et al.*, 1981). In Berner *et al.*'s (1983) BLAG model the level of atmospheric CO_2 which led to this balance was primarily dictated by the land area available for weathering. Berner's (1994) GEOCARB II model incorporates the effects of increased solar luminosity through time, different rates of continental uplift, vascular plant evolution, and partitioning of organic and inorganic carbon burial based on the $\delta^{13}C$ record. Its predictions of variations in atmospheric pCO_2 are consistent with the general evolution of climate during the Phanerozoic; for example, low pCO_2 is predicted for the intervals of Permo-Carboniferous and late Cenozoic glaciation. There is agreement with proxy indicators (e.g., Berger and Spitzy, 1988; Cerling, 1991; Freeman and Hayes, 1992) and general circulation model (GCM) modeling of the warm, ice-free climates of the Mesozoic (Barron *et al.*, 1995, and references therein). For the Late Ordovician, GEOCARB II yields a midpoint estimate of $\sim 14\times$ PAL CO_2.

Fossil soil $\delta^{13}C_{carbonate}$ isotope profiles and the concentration and carbon isotopic composition of carbonate substituted in goethites in ancient soils have been proposed as atmospheric pCO_2 barometers (e.g., Cerling, 1991; Mora *et al.*, 1991, 1996; Yapp and Poths, 1992, 1996). Mora *et al.*'s (1996) results show a decline of pCO_2 from $\sim 14\times$ PAL in the Late Silurian to $\sim 1\times$ PAL in the Carboniferous, in good agreement with GEOCARB II (Berner, 1997). But very few paleosols have been identified from the Late Ordovician. Yapp and Poths (1992) studied the goethite in a weathering profile that sits on a Late Ordovician unconformity atop the Neda Formation in Wisconsin (Paull, 1977). They assumed that variations in the $\delta^{13}C$ isotopic composition and mole fraction of the $Fe(CO_3)OH$ component in the goethite represent variations in ambient CO_2, and that the goethite was formed in a paleosol. Adapting Cerling's (1991) model they obtained a value of $16\times$ PAL for atmospheric pCO_2 in the Late Ordovician.

These geochemical estimates of high atmospheric pCO_2 at the time of the Late Ordovician glaciation imply an intriguing paradox. High atmospheric pCO_2 is often given as a major explanation for warm, ice-free climates (see above); high pCO_2 and glaciation would thus seemingly be incompatible. However, two important factors could be expected, at least partially, to explain this paradox. The first is a decrease of 3.5–5.0% in solar luminosity for the Late Ordovician/Early Silurian compared with today (Crowley and Baum, 1991). The second is the general logarithmic relationship between atmospheric pCO_2 and radiative forcing (Berner and Barron, 1984; Kiehl and Dickinson, 1987): that is, large changes in atmospheric pCO_2 at high levels result in relatively small changes in forcing, compared with the same-sized changes in atmospheric pCO_2 at lower levels.

But even allowing for these factors such high levels of atmospheric CO_2 would apparently not be consistent with a major glaciation. By making a global average energy balance calculation Crowley and Baum (1991) found that the combined effect of $14\times$ PAL CO_2 and a 4.5% lower solar luminosity relative to the present day would be comparable to the combined effect for the Cretaceous with $4\times$ PAL CO_2 and a 1% reduction in solar luminosity. As Crowley and Baum (1991) noted, the climate of the ice-free Cretaceous was very different from the glaciated Late Ordovician. Thus the focus of modeling efforts of Late Ordovician climates in the early 1990s was how to explain glaciation at high atmospheric CO_2.

Crowley *et al.* (1986, 1987) and Hyde *et al.* (1990) have emphasized the importance of seasonality as modified by continent size. In particular, the location of a supercontinent relative to the pole will affect polar continentality, and whether ice sheets can develop. Crowley *et al.* (1987) developed the hypothesis that glaciation could occur with high atmospheric pCO_2 if the pole were located close to the edge of a supercontinent. In such a case the high thermal inertia of the ocean (compared with land) prevents summer temperatures near the coast from rising much above freezing, if at all. This allows snow to remain year-round (a major prerequisite for glaciation), rather than melting under high summer temperatures that would occur when the pole is in the center of the continent. To investigate the climate of the Late Ordovician, Crowley and Baum (1991) used an energy balance model (EBM) developed by Hyde *et al.* (1990), which incorporates the well-known ice–albedo feedback. They concluded that the Late Ordovician glaciation occurred as a result primarily of geographic positioning, as predicted by Crowley *et al.* (1987).

Energy balance models can only predict the spatial distribution of surface temperature and cannot resolve the hydrologic cycle and therefore snow balance (Barron, 1987). Recognizing this, Crowley and Baum (1995) readdressed the problem using the GENESIS GCM (described in Thompson and Pollard, 1995*a*) to investigate the Late Ordovician glaciation. They found that their original hypothesis still held, albeit with a more restricted range of boundary conditions. They demonstrated that glaciation could indeed occur at $14\times$ PAL CO_2, the value predicted by Berner's (1994) geochemical cycling model, although an elevated topography of 500 m for most land grid points (vs. none at all) was required (Fig. 13.2).

Barron (1987) has expressed concern about the assumption that continental positioning is a sufficient condition for glaciation. For example, he noted that the South Pole moved over Antarctica and associated parts of Gondwana during the Jurassic and Cretaceous, times that apparently had warm, ice-free climates. Crowley and Baum (1991) acknowledged Barron's criticism and postulated potential errors in Cretaceous paleomagnetic pole positions as a possible explanation for this observation. But such errors are greatly amplified for the Ordovician. Crowley and Baum (1995) also argued that since they found that glaciation in the Late Ordovician was only likely in a limited area of parameter space, slight differences in key forcing factors may have been important. For instance, the net radiative forcing from atmospheric CO_2 and solar luminosity combined may have been slightly higher in the Cretaceous. Also, topography, which Crowley and Baum (1995) found to be critical for allowing snow cover to survive, may have been lower in the Cretaceous. Another

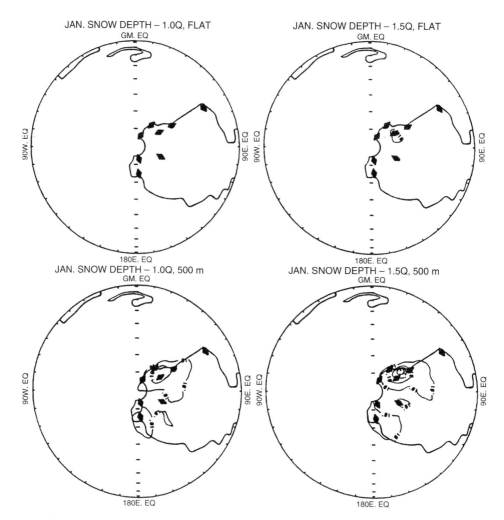

Figure 13.2. January snow thickness in meters for the Late Ordovician predicted by the GENESIS climate model (south polar projection) with a 4.5% reduction in solar luminosity relative to the present day and 14× PAL CO_2. Q represents the poleward ocean heat transport term (1.0× or 1.5× present-day control values); flat and 500 m mean no elevation or a 500 m elevation for land grid points respectively. Black diamonds represent glacial deposits. From Crowley and Baum (1995).

point that Crowley and Baum (1995) drew attention to is the presence of extensive vegetation in the Cretaceous, presumably a major difference with the Late Ordovician. Vegetation at high latitudes can significantly affect climate. It can prevent permanent snow cover building up by lowering albedo, both directly compared with a bare surface, and indirectly through taller vegetation (trees) protruding through the snow pack (e.g., Lovelock, 1988; Bonan *et al*., 1992; Foley *et al*., 1994; Dutton and Barron, 1996; Otto-Bliesner and Upchurch, 1997).

As a demonstration of how pole-positioning could explain the Late Ordovician glaciation, Crowley (1993) described the results of a series of EBM sensitivity experiments in which Gondwana was moved over the South Pole at 5.5 cm year^{-1}. Maximum permanent snow cover on land was greatest when the pole was closest to the edge of the supercontinent, and the variation in permanent snow cover matched the frequency of occurrences of glacial deposits through the Late Ordovician to Early Silurian summarized in Frakes *et al.* (1993). But as discussed above, a large number of these deposits have since been assigned strictly Hirnantian ages. Extremely rapid plate movement (>20 cm year^{-1}) would be required for the edge of Gondwana to pass on and off the South Pole and through the critical area where glaciation would occur. But plate speeds have rarely exceeded 10 cm year^{-1} in the Mesozoic and Cenozoic (Gibbs *et al.*, 1995). More reasonable (i.e., lower) speeds would mean that glaciation would be expected to last longer than the observed 1 Myr. Thus the very short duration of the glaciation means that the Crowley *et al.* (1987) hypothesis of a purely paleogeographic explanation for the glaciation is unlikely.

LATE ORDOVICIAN GLACIATION CAUSED BY A BRIEF DRAWDOWN OF ATMOSPHERIC CO₂

Crowley and Baum's (1995) demonstration that glaciation at high atmospheric pCO$_2$ is feasible is a significant contribution. But in light of recent stratigraphic and stable isotope studies (discussed above) that suggest major glaciation was confined to the Hirnantian and lasted for less than 1 Myr, atmospheric pCO$_2$ may not have been as high during the glaciation as the geochemical estimates indicate. Berner's (1994) GEOCARB II model is executed at 10 Myr timesteps, and would be unable to resolve changes in atmospheric CO$_2$ occurring on the 1 Myr timescale of the Hirnantian glaciation, unless the forcing functions had this level of resolution. Furthermore, Berner's (1994) own estimates of uncertainty allow for an atmospheric CO$_2$ level of 5× PAL, which with reduced solar luminosity compared with the present day would result in a net radiative forcing close to that of today's glacial climate (Kasting, 1992). And although Yapp and Poths' (1992) paleosol-derived estimate of atmospheric pCO$_2$ seems to confirm Berner's (1994) predictions, it should be noted that the range of possible values associated with their technique allows a lower bounding estimate for pCO$_2$ of ~8× PAL. Also, Yapp and Poths (1992) assume that δ^{13}C$_{air}$ reflects the prevailing δ^{13}C of surface ocean waters. If indeed the paleosol atop the Neda Formation is directly contemporaneous with the short-lived Hirnantian glaciation (the age of the paleosol can only be constrained as Late Ordovician to Early Silurian; Paull, 1977), their estimate of pCO$_2$ would be reduced if the δ^{13}C excursion recently found by several groups (see above) is considered.

Thus, even if both techniques for deriving pCO$_2$ have sufficient temporal resolution to resolve changes during the Hirnantian, their ranges of uncertainty could allow values of pCO$_2$ that are consistent with glaciation, under the condition of reduced solar luminosity. Indeed, given the seeming robustness of the CO$_2$–climate paradigm in explaining many key features of the earth's climate history, a short-lived drawdown of atmospheric CO$_2$ is a reasonable explanation for the initiation, short duration, and rapid demise of the Late Ordovician glaciation. A timescale of less than 1 Myr does not point towards a plate tectonic mechanism. This includes mechanisms that would affect climate directly (e.g., continental position such as presence of land in polar regions, development of high topography elevating land above the snow line). Also, assuming that the carbonate–silicate cycle is primarily controlled by tectonic mechanisms (e.g., rate of volcanic outgassing, amount of land area exposed, rate of continental uplift) then it is unlikely to have a significant effect on atmospheric pCO_2 on timescales of less than 1 Myr. The most likely explanation appears to be a perturbation to the organic carbon cycle, which would be consistent with the new carbon isotopic evidence discussed above. Such a perturbation could be induced by changes in ocean circulation that led to a major imbalance in the inputs and ouputs of carbon or a critical nutrient such as phosphorus. The timescale of this perturbation would be dependent upon the residence time (τ), defined as the reservoir size divided by the input or output rate (at steady state). In today's ocean, τ is 100 ka for the total carbon in the system, as well as for phosphorus (Broecker and Peng, 1982). However, oceanic reservoir sizes (as well as input and output rates), and therefore τ, may have been different in the early Paleozoic from today. For instance, because of carbonate equilibria, an atmospheric CO_2 level of $10\times$ PAL would result in a $\sim$2–3 times increase in total oceanic carbon compared with today. Nevertheless, an 'internal' oceanographic mechanism could have a sufficiently short timescale ($\sim$100 kyr), so that atmospheric pCO_2 could be affected on the required timescale ($<$1 Myr). Plate tectonic factors could still be involved, in that conditions may have evolved to bring the climate system close to a threshold where a change in oceanographic processes could have effected glaciation.

A reasonable hypothesis (Brenchley *et al.*, 1994, 1995; Qing and Veizer, 1994; Kump *et al.*, 1995; Wang *et al.*, 1997) is that the Late Ordovician glaciation occurred as the result of a combination of favorable positioning of Gondwana, and a sequence of events affecting the carbon cycle which led to a short-lived drawdown of atmospheric CO_2 and ultimately extensive glaciation. The oceanography of the early Paleozoic, and in particular the early Late Ordovician, was significantly different from today: oxygen isotopic data suggest that evaporative, low latitude seas generated warm, saline waters which filled the deep basins of Iapetus, Proto-Tethys, and perhaps the Boreal Ocean (Railsback *et al.*, 1990). Permanent snow cover may have begun to develop in the southern polar regions as Gondwana moved toward the South Pole in the Late Ordovician (Crowley and Baum, 1995; see also discussion section below). Kump *et al.* (1995) have argued that there may also have been a long-term decline of atmospheric pCO_2 from the Middle to Late Ordovician due to decreasing global volcanism (Stillman, 1984; Huff *et al.*, 1992) and the increased weatherability (see Kump and Arthur, 1997) resulting from the Taconic orogeny (Richter *et al.*, 1992; Khain and Seslavinskii, 1994; van Staal, 1994). All of these mechanisms would have led to high latitude cooling and could have allowed sea ice to form. This would have promoted deep convection in high latitudes, eventually causing a switch to a thermohaline circulation (Brenchley *et al.*, 1994, 1995). Another mechanism for an oceanographic change could have been the opening or closing of a critical oceanic gateway.

Although the actual trigger is poorly known, enhanced rates of oceanic circulation in the Hirnantian have often been inferred (e.g., Sheehan, 1979, 1988; Wilde *et al.*, 1990; Wilde, 1991). This switch in circulation mode could explain the positive carbon isotope excursion, argued Brenchley *et al.* (1994), because the new mode would promote greater upwelling of nutrient-rich bottom water and higher productivity, and thus, presumably, higher rates of organic carbon burial. The rapid burial of large quantities of organic carbon (representing a sink for atmospheric CO_2) over ~1 Myr during the Hirnantian could have had a significant effect on atmospheric pCO_2, and allowed glaciation to ensue.

The duration of the enhanced organic carbon burial, atmospheric CO_2 drawdown, and thus glaciation would ultimately have been limited by the availability of phosphate (e.g., Broecker and Peng, 1982; Kump, 1989). The earlier, more 'sluggish' halothermal ocean would have developed phosphate-rich deep waters (Sarmiento *et al.*, 1988). The new, more vigorously circulating thermohaline ocean would tend to have lower phosphate contents at steady state. The Hirnantian glaciation may represent the transition between these two states. The enhanced organic carbon burial during the transition would cause a progressive reduction in the ocean's phosphate content, reducing the productivity of the surface waters. Presuming no change in burial efficiency, the burial rate of organic carbon (and phosphorus) would diminish until the output of phosphate with burial once again balanced the riverine input.

It would be pleasing to find extensive black shale deposits associated with this proposed drawdown of atmospheric CO_2. It is intriguing that such deposits are a key global feature of the very latest Hirnantian and Early Silurian, that is, associated with, or after, the deglacial transgression (e.g., Leggett, 1980, Klitzsch, 1981; Ulmishek and Klemme, 1990; Mahmoud *et al.*, 1992). Two somewhat unsatisfactory solutions can be provided. Firstly, effectively no Ordovician ocean floor has been preserved. Secondly, though black shales have high weight percentages of total organic carbon, they are not necessarily the largest repository of organic carbon and do not necessarily imply high accumulation rates of organic carbon (Berner, 1982). To establish the total amount of carbon buried, rates of total sediment accumulation need to be determined. Such a task would require a global survey, and may be simply unresolvable for such a short-lived event.

EXPERIMENTAL DESIGN

The goal of the GCM experiments presented here is to investigate the sensitivity of the early Paleozoic climate to changes in atmospheric pCO_2 and paleogeography. What reasonable bounds can be placed on these factors that are consistent with the geologic evidence for a major glaciation in the Hirnantian at the end of the Ordovician, but otherwise also with the rest of the early Paleozoic that was essentially ice free? Crowley and Baum (1995) have shown that glaciation is possible at $14\times$ PAL CO_2 for the Late Ordovician paleogeography with a 4.5% reduction in solar luminosity. This study expands upon the Late Ordovician experiments described in Gibbs *et al.* (1997) who found that a lower value of $10\times$ PAL better matched the observed distribution of glacial deposits and that a value of $18\times$ PAL led to an ice-free climate. This range of values is within the uncertainty of the geochemical estimates for $p\mathrm{CO_2}$ at this time, demonstrating the utility of GCMs in constraining a paleoclimate problem. Here we evaluate the effects of the differences in paleogeography between the Late Ordovician and Early Silurian (in particular, the movement of Gondwana over the South Pole). Does the same range of atmospheric $\mathrm{CO_2}$ levels for the Early Silurian paleogeography result in the same glacial/ice-free states as with the Late Ordovician paleogeography? The Crowley *et al.* (1987) pole-positioning hypothesis would lead us to expect that, for the same $\mathrm{CO_2}$ level, there is a difference in climate due to paleogeography, and that the Late Ordovician was more sensitive to glaciation at high $\mathrm{CO_2}$ levels than the Early Silurian. In total, six experiments are discussed here, three Late Ordovician and three Early Silurian experiments, with the same range of atmospheric $\mathrm{CO_2}$ levels (10×, 14×, and 18× PAL). The experiments are referred to here as ORD-10×, ORD-14×, ORD-18×, SIL-10×, SIL-14×, and SIL-18×.

Computing restrictions mean that GCMs cannot be directly coupled with glaciological and lithosphere models to fully model the growth of an ice sheet. Most workers therefore assume, as we do here, that the presence of permanent snow cover or an increase in the net snow balance (i.e., snow survives year-round, and is able to build up) over land in a GCM experiment indicates a critical condition for ice-sheet initiation (Crowley *et al.*, 1993a,b, 1994; Dong and Valdes, 1995; Gallimore and Kutzbach, 1996; Otto-Bliesner, 1996; Bice, 1997; Gibbs *et al.*, 1997).

Model

All experiments in this study were conducted with the GENESIS (Global Environmental and Ecological Simulation of Interactive Systems) global climate model, v. 1.02A. GENESIS, an atmospheric GCM (AGCM) coupled to a sophisticated land surface model ('LSX'), is described in Pollard and Thompson (1995) and Thompson and Pollard (1995a,b). The atmospheric model is coupled to a 50 m mixed-layer ('slab') ocean to represent the thermal inertia of the oceans. (The 50 m thickness provides the best match with the present-day observed climate.) Ocean heat transport is incorporated following Covey and Thompson (1989). Predicting rather than prescribing sea surface temperatures (SSTs) is an advantage given our very sparse knowledge of the Late Ordovician. The presence of sea ice is predicted thermodynamically, after Semtner (1976). Sea ice advection is not modeled in these experiments.

All experiments were started from initial conditions of a motionless atmosphere, present-day annual-average zonal temperatures, no snow cover on land, and 2 m thick sea ice at all ocean gridpoints above 60° latitude. This method is computationally more expensive than other methods such as using an EBM to prescribe initial SSTs (e.g., Crowley *et al.*, 1994) or starting from an earlier experiment with almost the same boundary conditions that has come to equilibrium (e.g., Crowley and Baum, 1995). However, it ensures consistency and eliminates concern about sensitivity to initial conditions. Both the AGCM and LSX components of GENESIS were run at a resolution of ~7.5° longitude × 4.5° latitude. All experiments were run for 30 model years, ensuring that a steady-state condition was reached for annual surface temperature at all latitudes. The last five model years were then used for analysis.

Boundary conditions

Our Late Ordovician boundary conditions, apart from atmospheric CO_2 level and a slight difference in poleward ocean heat transport (see discussion), are the same as Crowley and Baum's (1995). Early Silurian boundary conditions are from Moore et al. (1994) but adapted where necessary to be consistent with the Late Ordovician experiments. The key boundary conditions are shown in Table 13.1 and justifications for their selection are discussed below, listing those that remain constant between experiments first. Table 13.2 shows the combinations of paleogeography and atmospheric pCO_2 for each experiment.

Solar luminosity

Estimates for solar luminosity in the Late Ordovician range from 3.5% to 5.0% lower than today (Crowley and Baum, 1991), depending on assumptions of the initial composition of the sun. Here we use a 4.5% reduction (0.955×1370 W m^{-2})

Table 13.1. *Boundary conditions that must be specified for a GENESIS climate model experiment and how these are varied for the experiments reported in this study*

Boundary condition	How varied (see text for explanation)
Held constant for all experiments:	
Solid luminosity	4.5% reduction (0.955×1370 W m^{-2})
Orbital parameters	'Cold summer orbit' for southern hemisphere:
	eccentricity: 0.06
	obliquity: 22°
	perihelion: southern hemisphere winter solstice
Ozone mixing ratios	Held constant at present-day values
Soil texture	Intermediate value
Soil color	Intermediate value
Vegetation type	None (bare soil)
Oceanic poleward heat flux	Present-day control values, i.e., 0.15× Carissimo et al. (1985) estimates, symmetrically distributed about the equator
Rotation rate	24 h
Varied between experiments:	
Paleogeography	Late Ordovician:
	Scotese and Golonka (1992) for 458 Ma, adapted by Crowley and Baum (1995) for 440 Ma
	Early Silurian:
	Cocks and Scotese (1991) adapted by Moore et al. (1994)
Atmospheric CO_2	18×
(× 280 ppm)	14×
	10×

Table 13.2. *Experiments reported in this study*

Experiment	Atmospheric CO_2 level ($\times$ PAL)	Global mean annual surface temperature (°C)	Comments
Late Ordovician (440 Ma)			
ORD-18×	18×	18.9	ice free
ORD-14×	14×	17.6	glaciates
ORD-10×	10×	13.8	glaciates
Early Silurian (425 Ma)			
SIL-18×	18×	19.1	ice free
SIL-14×	14×	18.2	ice free
SIL-10×	10×	15.1	glaciates

following Crowley and Baum (1995) for both the Late Ordovician and the Early Silurian.

Orbital parameters

Orbital configuration is known from Quaternary studies (e.g., Imbrie *et al.*, 1992) and Carboniferous studies (Crowley *et al.*, 1993a,b) to be an important factor for glacial inception. The direct record of Late Ordovician glacial deposits is too coarse to permit direct identification of Milankovitch cycles. Nevertheless, within some of the African strata multi-tillite successions are seen, suggesting at least several individual glacial advances and retreats (Hambrey, 1985; Hiller, 1992). Williams (1991) has reported geochemical variations in bedded halite deposits from the Canning Basin in Australia (paleolatitude of approximately 15°) on potential Milankovitch timescales for this period. Armstrong (1995) has found five discrete conodont extinctions within the early Hirnantian. Timescales in both studies match the different periods of orbital variations that would be expected for the early Paleozoic (Berger, 1989; Berger *et al.*, 1989). Glacial cycles induced by Milankovitch orbital variations are thus plausible for the Late Ordovician; certainly, there is no reason why such orbital variations did not occur at this time. However, the amplitudes of these orbital variations in the early Paleozoic are uncertain. Following Crowley and Baum (1991, 1995), maximum Pleistocene values for a cold summer orbit (CSO) for the southern hemisphere (Berger, 1978) are used here. A glaciation is most likely to start during a cold summer orbit, since this orbital configuration optimizes summertime snow survival.

Land surface

With little or no information on early Paleozoic soil types, the same median soil parameters are prescribed for every land grid point. These are from Dickinson *et al.* (1986): soil texture of class 6 and soil color of class 4. The timing and extent of colonization of the land surface by plants in the early Paleozoic is unclear (see discussion). Also, the question of how widespread lichen and fungi coverage would have been is not easy to answer (e.g., Schwartzman and Volk, 1989). In this study a bare soil is prescribed globally for all experiments to allow intercomparison of results for the two intervals.

Poleward oceanic heat transport

The model-simulated meridional sea surface temperature gradient (and therefore polar temperatures) is largely controlled by the specified magnitude of the oceanic heat transport. The oceanic heat transport therefore represents a tunable model parameter which has not been varied from the modern 'control' values in any simulations described here. The true magnitude of ocean heat transport is unconstrained for past time periods, especially for those such as the Late Ordovician lacking quantitative marine temperatures inferred from depth-stratified foraminiferal isotopic analyses. Present-day GENESIS control values for poleward ocean heat transport, symmetrically distributed about the equator, were used here to minimize the boundary conditions being changed. These values are 0.15 times the estimates of Carissimo *et al.* (1985) and were chosen to achieve the best fit between the model and the present-day observed climate (Thompson and Pollard, 1995*a*).

Rotation rate

The earth's rotation rate has been steadily dampened by tidal resistance. Although a daylength of 21 h is likely for the Ordovician (Scrutton, 1978), here a rotation rate of 24 h was used, again to minimize the boundary conditions being changed.

Paleogeography

Our Late Ordovician paleogeography is the same as used by Crowley and Baum (1995). Because no detailed paleogeography is available for the Ashgill (440 Ma), Crowley and Baum (1995) took Scotese and Golonka's (1992) paleogeography for 458 Ma, rotated Gondwana further north, and moved Baltica closer to Laurentia to represent the closure of the Iapetus ocean (Fig. 13.3a). Our Early Silurian paleogeography is the same as Moore *et al.*'s (1994) who adapted Cocks and Scotese's (1991) paleogeography for the Wenlock (425 Ma). A major difference between this geography (Fig. 13.3b) and the Late Ordovician geography is that Gondwana is now located over the South Pole, with all area above 70° S being land. In addition, the closure of Iapetus was well advanced by the Early Silurian. Terranes such as North China which had rifted away from Gondwana by this time are now included. Undoubtedly, errors exist in both these paleogeographies. For instance, the representation of Kazakhstan in the Early Silurian is larger than in more recent maps (e.g., Scotese and Golonka, 1992). However, our goal here is the comparison of two reasonable early Paleozoic geographies, where at least the major continents differ in a reasonable and coherent manner.

A uniform elevation of 500 m was specified for land grid points (except for coastal areas where 250 m was specified) for both paleogeographies. In the Early Silurian paleogeography higher elevations (≥ 2000 m) were specified for a few areas such as the Taconic orogenic belt where Laurentia and Baltica were colliding. The

(a) Late Ordovician

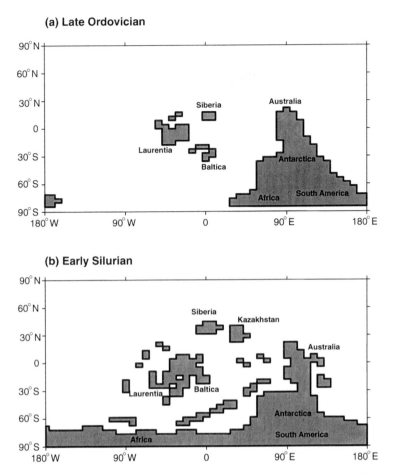

(b) Early Silurian

Figure 13.3. Land-sea distributions used in this study. (a) Late Ordovician (after Crowley and Baum, 1995); (b) Early Silurian (after Moore *et al.*, 1994).

initiation of Pleistocene ice sheets is believed to be sensitive to topography (e.g., Budd and Smith, 1987; Dong and Valdes, 1995) and the distribution of Early Permian ice sheets appears to be related to paleotopography (Ziegler *et al.*, 1997). However, there is no evidence of an area of high topography in the vicinity of Gondwana that experienced glaciation in the Late Ordovician (A. M. Ziegler, personal communication, 1996). Furthermore, the goal of these experiments is to address the gross sensitivity to the potential for glacial inception.

Atmospheric pCO₂

Evidence for early Paleozoic atmospheric CO_2 levels was discussed above. In this study values of 18×, 14× (same as Crowley and Baum, 1995), and 10× PAL were used. PAL is defined here as 280 ppm (pre-industrial concentration).

RESULTS

The key results of this study are shown in Fig. 13.4, the annual cycle of average monthly snow depths over land in high southern latitudes. Glaciation can clearly be inferred for experiments where snow cover survives through the summer months and its depth increases through time (i.e., the $10\times$ CO_2 experiments for both paleogeographies). The SIL-$10\times$ experiment has almost double the rate of accumulation of the ORD-$10\times$ experiment. Ice-free conditions (i.e., glaciation would not be expected to occur) are inferred for experiments where there is complete meltback in the austral summer and the average snow depth is effectively zero everywhere. This is the case for the SIL-$14\times$ experiment (Fig. 13.4b), as well as both $18\times$ CO_2 experiments (not shown). The most interesting aspect of these results are those for the ORD-$14\times$ experiment where the average monthly snow depth above $60°$ S does not reach zero (Fig. 13.4a). This implies that there must be some amount of snow cover somewhere that is at least surviving (if not actually accumulating) through the austral summer.

Average February snow cover (i.e., for the end of the austral summer), which we assume to reflect permanent snow cover, is shown in Fig. 13.5. Only one grid cell in the ORD-$18\times$ experiment has an average snow cover that is just over 1 cm (Fig. 13.5a); effectively it is ice free. In the ORD-$14\times$ experiment a small area in southwestern Gondwana (eastern North Africa today) has a snow cover greater than 1 m. Both $10\times$ experiments (Figs. 13.5e and f) have large areas with in excess of 1 m snow depth. Both the SIL-$14\times$ and SIL-$18\times$ experiments have no February snow cover over 1 cm, and would be regarded as completely ice free.

Although the presence of a permanent summer snow cover implies possible glaciation, study of net snow accumulation rates allows more certainty in predicting glaciation. Average yearly net accumulation rates (liquid water equivalent) are shown in Fig. 13.6. As would be inferred from Fig. 13.4, the SIL-$10\times$ experiment has the highest accumulation rates, reaching 18 cm year^{-1} in places (Fig. 13.6f). Rates in the ORD-$10\times$ experiment are not as high, but still reach 10 cm year^{-1} (Fig. 13.6e). These accumulation rates compare better with observational and modeling studies of 5–20 cm year^{-1} for the present-day East Antarctica ice sheet, vs. 10–100 cm year^{-1} for the present-day Greenland ice sheet (Thompson and Pollard, 1997, and references therein). Both experiments display a 'hole' of low accumulation in the middle of Gondwana. The ORD-$14\times$ experiment has a small area in southwestern Gondwana with net snow accumulation greater than 1 cm year^{-1} (Fig. 13.6c), which matches the area of thickest February snow depths (Fig. 13.5c). There are no areas with significant accumulation rates in the other experiments.

Maps of seasonal averages for June–July–August (JJA) and December–January–February (DJF) of surface temperature (Fig. 13.7, see color plate) show expected seasonal variations for each experiment. For instance, all experiments show a seasonal cycle of $\sim 40\,°C$ in the high-latitude core of Gondwana. The strong meridional surface temperature gradient in the austral winter (JJA) in all experiments leads to strong zonal winds (westerlies) over Gondwana (not shown). Most of the northern hemisphere is ocean in both paleogeographies; consequently there is a virtual absence of any temperature seasonality in the northern hemisphere ocean in

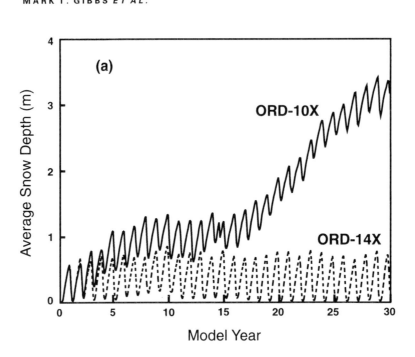

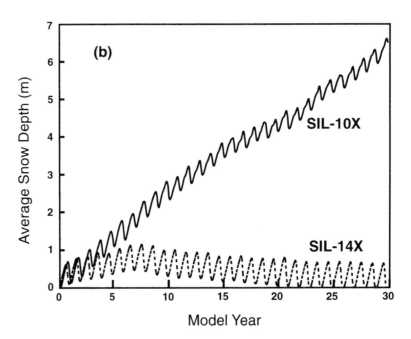

Figure 13.4. Average monthly snow depth over land between 90° S and 60° S for the (a) ORD-10× and ORD-14×, and (b) SIL-10× and SIL-14× experiments. Note the difference in scales for the *y*-axes.

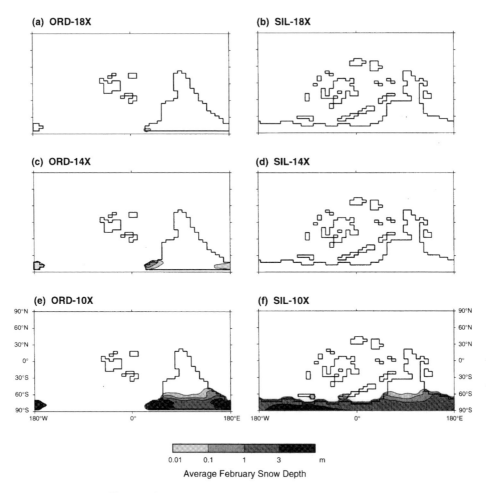

Figure 13.5. Average February snow depth over land.

the 18× and 14× experiments. Along with the associated lack of change in winds, this means that it is highly unlikely that there would have been a seasonal change in ocean circulation in the northern hemisphere, except in the 10× experiments where northern polar temperatures were low enough for the formation of considerable sea ice in winter. In addition, variations between experiments, due to geography and atmospheric pCO_2 level, can also be seen in Fig. 13.7. Here, given the focus of this paper on snow survival and glacial initiation, we examine the differences and similarities in seasonal surface temperatures in high southern latitudes.

A strong land/sea contrast in austral winter (JJA) temperatures at high southern latitudes can be seen in the two 18× CO_2 experiments (Figs. 13.7a and b). Both 18× CO_2 experiments reach temperatures lower than −25 °C over land. However, over ocean the 0 °C isotherm penetrates to a higher latitude and temperatures are not as intensely cold in ORD-18× compared with the equivalent areas

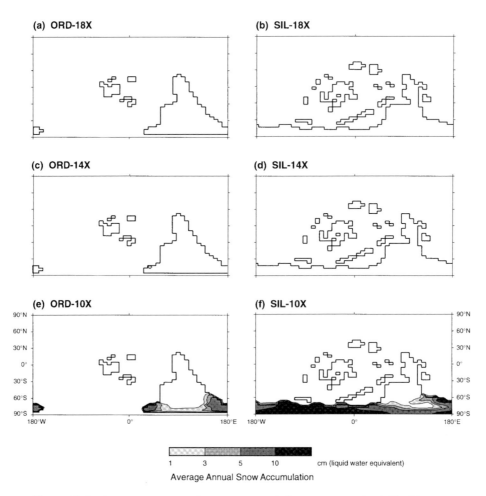

Figure 13.6. Average net annual snow accumulation rates over land (liquid water equivalent).

which are land in SIL-18×. In these two experiments, the DJF temperatures are particularly different between experiments in the high southern latitudes (Figs. 13.7e and f). Whereas all of Gondwana is above 5 °C in SIL-18×, in ORD-18× the 10 °C isotherm lies further equatorward and parts of southern Gondwana, though not freezing, are cool with temperatures below 5 °C. A small area of sea ice survives in ORD-18×, as shown by the region with subzero temperatures off the southeastern tip of Gondwana.

In the 14× CO_2 experiments little difference is seen in JJA temperatures in high southern latitudes (Figs. 13.7c and d); sea ice and land experience the same temperatures. As with the 18× experiments a contrast in DJF temperatures is seen between ORD-14× and SIL-14× (Figs. 13.7i and j) which matches the differences in the land/sea distribution. In ORD-14×, all ocean areas above ~70° S are below 0 °C and are covered by sea ice (Fig. 13.7i), and a small area of land on the southwestern tip of Gondwana is below freezing. This area matches that with the thickest snow cover (Fig. 13.5c) and is an area of positive snow accumulation (Fig. 13.6c). In SIL-14×, though large parts of high latitude Gondwana experience cool summers ($<5\,°C$), no area lies below freezing.

Austral winters in the 10× CO_2 experiments are both very cold (Figs. 13.7e and f): JJA temperatures are less than $-45\,°C$ in the high latitude core of Gondwana. Austral summers in these experiments are also similar (Figs. 13.7k and l); the freeze line lies at about 60° S in both experiments (allowing snow cover to build up rapidly; Figs. 13.6b and c), and there is little difference in temperature due to surface type.

DISCUSSION

These experiments demonstrate a large sensitivity to reductions in atmospheric CO_2 for both the Late Ordovician and Early Silurian paleogeographies. Along with dramatic differences in permanent snow cover and accumulation (i.e., ice-free vs. glaciation), a reduction from 18× to 10× CO_2 leads to 5 °C and 4 °C reductions in global annual-average surface temperature for the Late Ordovician and Early Silurian experiments respectively (Table 13.2). In these experiments the reduction in the actual forcing from CO_2 is in fact modest for two reasons. Firstly, the temperature response to increases in atmospheric pCO_2 (i.e., the greenhouse effect) follows a general logarithmic relationship as the 'strong' 15 μm band saturates (Berner and Barron, 1984; Kiehl and Dickinson, 1987). Thus, according to the CO_2 concentration/radiative forcing relationship of Kiehl and Dickinson (1987), a change in atmospheric pCO_2 from 10× to 18× represents a similar change in forcing to that resulting from a doubling from 1× to 2× (4.9 W m^{-2} vs. 4.8 W m^{-2} respectively). Secondly, GENESIS uses the same parameterization of absorption of radiation by atmospheric CO_2 used in the earlier Community Climate Model (CCM1; Kiehl et al., 1987). This parameterization does not explicitly account for the weaker absorption bands of CO_2, which do not saturate at higher levels of CO_2 and continue to provide a linear greenhouse response to increasing atmospheric CO_2 levels (Kiehl et al., 1987). Consequently, CCM1 and GENESIS would be expected to underestimate the greenhouse effect of CO_2 for the high concentrations considered here.

A similar sharp response in permanent snow cover due to modest changes in forcing has been seen in other GENESIS GCM experiments for the Westphalian (Late Carboniferous). Crowley et al. (1994) found a discontinuous increase in snow cover in response to incremental reductions in solar luminosity for the Westphalian paleogeography. With a cold summer orbit for the southern hemisphere, the area of permanent snow cover increased from near zero to $>30 \times 10^6$ km^2 when solar luminosity was reduced from 98.5% to 98% of the present-day value. Other Westphalian experiments with a 3% solar luminosity reduction but with a range of different orbital configurations also show large differences in permanent snow cover. A cold summer orbit for the southern hemisphere and the present-day orbit both yield extensive permanent snow cover, in areas that are in good agreement with the geological record of glacial deposits (Crowley et al., 1993b; Otto-Bliesner, 1996). A warm summer orbit for the southern hemisphere is ice free (Crowley et al., 1993a,b; Valdes and Crowley, 1998), which has interesting implications for explaining glacial/interglacial fluctuations. Crowley *et al.* (1994), while noting that a GCM cannot be used to fully explore parameter space, drew parallels between this behavior and the 'small ice cap instability' (SICI) that is a feature of many EBM studies (e.g., North, 1990). The key premise of SICI is that a modest reduction in insolation, atmospheric CO_2, or even stochastic variations in atmospheric circulation brings the climate system below a critical threshold that permits the development of permanent snow cover, initially in a small region. This permanent snow cover raises the local surface albedo and leads to a cooling over an area larger than the original region. The ice–albedo feedback thus strongly amplifies the response to the original modest change in forcing and rapidly takes the climate into a new, colder state. This behavior could provide an explanation for times of rapid climate transitions (Crowley and North, 1988; Baum and Crowley, 1991), such as the initiation of the Late Ordovician glaciation (Gibbs *et al.*, 1997).

As discussed in a previous section, the short duration of the Late Ordovician glaciation means that the Crowley *et al.* (1987) pole-positioning hypothesis (see also Crowley and Baum, 1991, 1995; Crowley, 1993; Crowley *et al.*, 1993*a*) cannot wholly explain this glaciation. The GCM experiments reported here indicate that, with solar luminosity reduced by several percent relative to today, glacial or fully ice-free conditions will result, depending on the atmospheric CO_2 concentration specified. Our results lend support to the idea that a brief drawdown of atmospheric CO_2 is most likely to be the primary cause of the glaciation (Brenchley *et al.*, 1994, 1995; Kump *et al.*, 1995). However, these experiments also indicate that the Late Ordovician paleogeography may have acted to precondition the earth system for glaciation (see Crowley and Baum, 1995). For the same atmospheric CO_2 level ($14\times$ PAL) glaciation is predicted for the Late Ordovician (albeit not as dramatically as for lower CO_2 levels), but not for the Early Silurian. Thus the ice-free/glaciation threshold occurs at a higher atmospheric CO_2 level for the Late Ordovician (Table 13.2), which must result from differences in paleogeography. In the Late Ordovician, the ocean's thermal inertia permits colder summertime (DJF) temperatures near the South Pole along the coast of Gondwana (Fig. 13.7i). When the pole is located nearer the center of a supercontinent as in the Early Silurian, continentality causes higher summertime temperatures (Fig. 13.7j) and snow is unable to survive. The largest differences in DJF temperatures at high southern latitudes between ORD-$14\times$ and SIL-$14\times$ are in the center of Gondwana, i.e., between $\sim 90°$ E and $180°$ E. As predicted by Hyde *et al.* (1990), the other parts of high latitude Gondwana are sufficiently close enough (< 2000 km) to the ocean that they can still 'feel' some effect of the ocean's higher thermal inertia. Though neither $18\times$ CO_2 experiments indicate glaciation, a similar difference in summer temperatures due to continentality can be seen between those experiments. DJF temperatures at high southern latitudes in ORD-$18\times$ (Fig. 13.7g) are colder than in SIL-$18\times$ (Fig. 13.7h). The $14\times$ and $18\times$ CO_2 experiments clearly support the Crowley *et al.* (1987) hypothesis that a supercontinent with its edge located close to the pole can allow glaciation to occur at a higher atmospheric CO_2 level.

By contrast, little difference in DJF temperatures due to paleogeography is seen between the two $10\times$ CO_2 experiments (Figs. 13.7k and l). In ORD-$10\times$

temperatures are cold enough for the ice–albedo feedback to lead to permanent sea ice cover above 60° S. Consequently, both experiments have a similar area that remains below freezing year-round over which snow can (rapidly) accumulate. (Indeed, SIL-10× 'glaciates' at roughly twice the rate of ORD-10×.) The Crowley *et al.* (1987) pole-positioning hypothesis apparently does not apply at this level of atmospheric CO_2; both experiments effectively have the same paleogeography at high southern latitudes. Almost all the area above 60° S is covered by the same cold, high-albedo surface. It does not matter for the climate that sea ice or land lie beneath the accumulating snow.

For a given CO_2 level, a supercontinent with its edge located close to the pole can make glacial initiation more likely compared with a situation with a pole-centered supercontinent. However, with a sufficiently high CO_2 level both types of paleogeography will be ice free; likewise with a low enough CO_2 level both types of paleogeography will glaciate. In these experiments the paleogeographic configuration plays a lesser role in that though it can alter the location of the ice-free/glaciated threshold, it cannot prevent glaciation from occurring if pCO_2 is lowered sufficiently. Thus an important implication of this study is that atmospheric pCO_2 has to be high to ensure ice-free conditions in the early Paleozoic. This value could vary from $\sim14\times$ to $\sim18\times$ PAL depending on the paleogeography.

The relatively minor role of paleogeography compared with atmospheric pCO_2 in these experiments is a consequence of the relatively minor paleogeographic differences between the Late Ordovician and the Early Silurian; Gondwana can only move so far over the pole between these intervals. GCM experiments with other more 'extreme' pole-centered paleogeographic scenarios, such as a lower sea level (leading to a larger continental area) or a more truly pole-centered supercontinent, have resulted in ice-free conditions for much lower net forcing from CO_2 and solar luminosity than that considered here. In these cases, paleogeography does have a larger role. For instance, with a 3% reduction in solar luminosity the Westphalian (Late Carboniferous) is ice free, but only with a hot summer orbit for the southern hemisphere. This occurs with GENESIS (Crowley *et al.*, 1993*a,b*; see also Otto-Bliesner, 1996), as well as with the UK Universities Global Atmospheric Modelling Programme (UGAMP) GCM (Valdes and Crowley, 1998), indicating that this is not a model-dependent result. Another Westphalian experiment with present-day solar luminosity (but a cold summer orbit) also yields ice-free conditions (Crowley *et al.*, 1994). This is a significant result because this experiment has the same orbital configuration as those reported here. Yet, when considering the global average energy balance, the net radiative forcing for the Late Ordovician from $14\times$ CO_2 and a 4.5% reduction in solar luminosity is $\sim9\,\mathrm{W\,m^{-2}}$ *higher* than that for the present day (see fig. 1 in Crowley and Baum, 1995). This would be roughly equivalent to a $4\times$ CO_2 increase for the present day. Thus, as Crowley *et al.* (1993*a*) observed, for certain paleogeographic configurations the Crowley *et al.* (1987) pole-positioning hypothesis can be very effective in 'over-riding' low net forcing levels and maintaining ice-free conditions, whereas a pole-edged location is definitely required if glaciation is to occur at higher net forcing levels.

Additional climate-forcing factors that could affect the actual position of the ice-free/glaciation threshold include solar luminosity, the earth's rotation rate, and topography. For instance, estimates of the reduction in solar luminosity for the Late Ordovician compared with the present day range from 3.5% to 5% (Crowley and Baum, 1991). A smaller reduction in solar luminosity than that used here (e.g., 3.5%) would require an atmospheric CO_2 level below $10\times$ PAL in order for glaciation to occur. Jenkins *et al.* (1993) investigated the effects of a Precambrian 14 h rotation rate with a GCM and found a cooling effect in high latitudes. They found that the steeper equator-to-pole temperature gradient relative to the present-day control could be attributed to reduced meridional heat transport. A rotation rate of 21 h (Scrutton, 1978) would mean similar, albeit smaller, cooling effects could be expected for the early Paleozoic. The exact elevation of polar Gondwana will also be important to some extent, allowing snow to survive with lapse-rate cooling. In Crowley and Baum's (1995) GENESIS study of the Late Ordovician, at $14\times$ CO_2 snow survived the summer with a 500 m elevation, but not with 0 m elevation, for Gondwana. Though all these factors would be expected to change the ice-free/glaciation threshold, they apply to both the Late Ordovician and the Early Silurian. (For instance, the main North African and South American components of Gondwana were tectonically inactive and remained in high latitudes throughout this time.) Little change in these particular factors would be expected between these two intervals. Their incorporation in a modeling study would shift the glaciation threshold for *both* intervals.

Following Crowley *et al.*'s (1993*a,b*) GCM studies of the Late Carboniferous, a different orbital configuration could be expected to change the threshold for glaciation. For instance, a hot summer orbit (HSO) for the southern hemisphere, as opposed to the CSO used here, might be ice free at $10\times$ PAL CO_2. This is an important next step for climate modeling, though it may be difficult to identify Milankovitch advances and retreats within the record of the short Hirnantian glaciation. Here, we held the orbital configuration constant, to evaluate the roles of paleogeography and atmospheric pCO_2 between the Late Ordovician and the Early Silurian. It may be possible that the Early Silurian never actually experienced an extreme CSO, whereas the Late Ordovician orbital elements may have fortuitously interfered at 440 Ma to create an extreme CSO. This would have further added to the unique conditions (e.g., paleogeographic configuration and atmospheric CO_2 drawdown) that caused the intensive Hirnantian glaciation within the otherwise generally ice-free early Paleozoic.

Other contributions to an explanation for the Late Ordovician glaciation could be differences in oceanic circulation and heat transport, as well as differences in vegetation cover between the Late Ordovician and other intervals such as the Early Silurian. Poleward oceanic heat transport has been proposed as an important climate-forcing factor (e.g., Barron, 1983; Covey, 1991; Chandler *et al.*, 1992; Barron *et al.*, 1993, 1995; Sloan *et al.*, 1995; MacLeod *et al.*, Chapter 8, this volume; DeConto *et al.*, Chapter 9, this volume). A clue to the importance of this factor can be seen by comparing the Ordovician experiments presented here with those of Gibbs *et al.* (1997). These experiments are completely similar in every respect except for a slightly different means of distributing the present-day control ocean heat flux for the Late Ordovician paleogeography. This minor difference results in the experiments reported here having slightly warmer mean annual global temperatures ($0.5\,^{\circ}C$ in each case), and accordingly reduced snow cover and sea ice extent in the austral summer. Current GCMs generally prescribe poleward oceanic heat transport as a latitudinal average. A better tool would ultimately be a fully coupled atmosphere–ocean model, capable of resolving ocean circulation patterns and their effects on climate (DeConto et al., Chapter 2, this volume). It is already well known that ocean circulation during the Hirnantian was very different from the rest of the early Paleozoic, based on faunal and sedimentological arguments (e.g., see Sheehan, 1979, 1988; Wilde et al., 1990; Wilde, 1991). Was there a warm polar current that kept the rest of the early Paleozoic ice free?

The colonization of the land surface by vascular plants took place at some point in the early Paleozoic. This time of colonization has now been pushed back to sometime in the Middle or early Late Ordovician (Gray et al., 1982; Gray, 1985). In the latest Ordovician 'polsterlands,' that is, plant formations dominated by multicellular plants lacking roots or rhizomes, may have been widespread (Retallack, 1992). By the Early Silurian (Wenlock) multicellular plants were present at many sites on Laurentia, Baltica, and Gondwana (fig. 3. in Moore et al., 1994). The development of a significant plant cover could have affected climate in two important ways. The first way is via altering the carbonate–silicate cycle and atmospheric pCO_2. Plants increase weathering rates by elevating soil pCO_2 and making organic acids available. But plants may not have had an important effect on the carbonate–silicate cycle until the development of major root systems (trees) in the Devonian (Algeo et al., 1995; Berner et al., 1996; Berner, 1997). The second way is via their direct effects on climate, i.e., albedo, evapotranspiration, roughness length, etc. But the precise physical characteristics of this primitive vegetation clearly belong in the realm of the unknown. Nonetheless, a suitable suite of sensitivity experiments that spanned an extreme range of potential characteristics, designed in a similar manner to this study, could be very interesting.

To fully explain glaciation it is critical to evaluate the 'net' combination of the plausible range of these factors, rather than advocating the primacy of one particular factor above all others. A matrix of experiments for different scenarios, that can be constrained by critical geological evidence (e.g., glaciation as considered here, or warm polar floras in the middle Cretaceous; Barron et al., 1995), can guide our understanding of the *relative* importance of different factors for controlling paleoclimate. Another advantage is the potential for tighter estimates of CO_2 than current geochemical cycling models such as Berner's (1994) GEOCARB II.

Did glaciation occur in the Early Silurian? Llandovery and earliest Wenlock glacial deposits in South America (Grahn and Caputo, 1992) have been ascribed to mountain or localized glaciation (Hambrey, 1985; Brenchley et al., 1995), since their distribution is much restricted compared with the Hirnantian deposits and South America was in high latitudes throughout the Late Ordovician and Early Silurian. Markwick and Rowley (1998) have pointed out that in the Mesozoic and Cenozoic, an ice sheet on Antarctica cannot be ruled out, but it must have been sufficiently small not to have reached the coast and left a record. Likewise, no unambiguous Silurian glacial deposits are known from outside South America, but a *small* ice sheet in central Africa cannot be ruled out. It is clear that these deposits do not represent as profound events as the Hirnantian glaciation, but perhaps they cannot be dismissed too easily. They are interbedded with marine units: therefore, the glaciers were of sufficient size to reach close to sea level. Johnson (1996) has tied these glaciations to short-lived (<1 Myr) eustatic drawdowns in sea level seen in the Llandovery from many locations globally.

Our results do not rule out glaciation for any interval in the early Paleozoic if atmospheric pCO_2 is lowered sufficiently. Were the events of the Hirnantian unique, or a more general feature of the early Paleozoic? Intriguingly, Wenzel and Joachimski (1996) have reported parallel positive excursions in $\delta^{18}O$ and $\delta^{13}C$ from Silurian (latest Llandovery, Wenlock, and Ludlow) brachiopods from Gotland, Sweden. These excursions occurred at the same time as sea level lowstands. The first pair of excursions could be correlated with the youngest glacial advance recorded from South America (Grahn and Caputo, 1992). The largest excursions, which are of similar magnitude to those known from the Hirnantian, are in the late Ludlow (Late Silurian), but this is an interval from which no glacial record is known at all.

Mora *et al.*'s (1996) range of values (11–17× PAL) for Late Silurian pCO_2 based on paleosol isotope proxies would be consistent with an ice-free climate. Their results are from the Bloomsberg Formation (Pennsylvania) which is late Ludlovian to early Pridolian in age (Driese *et al.*, 1992). It is therefore possible that this pCO_2 estimate is contemporaneous with the large, positive excursions in $\delta^{18}O$ and $\delta^{13}C$ (and glaciation?) in the late Ludlow (Wenzel and Joachimski, 1996). But as discussed above with respect to Yapp and Poth's (1992) results for the Late Ordovician, this estimate could be lowered if account was made of the $\delta^{13}C_{carbonate}$ excursion reported by Wenzel and Joachimski (1996). More stable isotope data spanning all of the Silurian, and particularly the Llandovery, are urgently needed from other sites: do variations exist that match the timings of all the glacial advances seen by Grahn and Caputo (1992)? Are there similar events to that in the Hirnantian, albeit probably of a reduced magnitude, in the Llandovery?

CONCLUSIONS

New stratigraphic and carbon isotope evidence indicates that although the Late Ordovician glaciation (~440 Ma) left an excellent record, particularly on Gondwana, it was very short lived. The main part of the glaciation was confined to the Hirnantian (latest stage of the Ordovician), lasting less than 1 Myr. Geochemical indicators imply high atmospheric CO_2 levels at apparently the same time as a major glaciation, and, given a paleogeographic configuration in which the edge of Gondwana is close to the South Pole, previous climate model studies have shown that glaciation is just possible at 14× PAL. However, its short duration means that the glaciation may not have occurred under high atmospheric CO_2 levels (~14× PAL) as has been previously suggested (although a lower solar luminosity for that time meant that the minimum prevailing level of atmospheric CO_2 was still substantially higher than the present day). Furthermore, mechanisms acting on tectonic timescales (> 1 Myr), such as changes in paleogeography, would be unable to explain the sudden initiation, short duration, and rapid termination of the Late Ordovician glaciation. Nevertheless, such mechanisms might have created the necessary preconditioning that moved the earth system close to a critical glaciation threshold. For example, declining volcanism and increasing orogeny may have set the stage by leading to a long-term decline in atmospheric $p\mathrm{CO_2}$ throughout the Late Ordovician.

New carbon isotope studies point to a major perturbation of the carbon cycle during the Late Ordovician glaciation, including suggestion of a drawdown of atmospheric CO_2. An oceanographic event that promoted increased oceanic productivity, atmospheric CO_2 consumption, and organic carbon burial that was then terminated by exhaustion of nutrients would be consistent with the timing and record of the Late Ordovician glaciation. This explanation would be consistent with the CO_2–climate paradigm that can explain much of the earth's climate evolution.

GCM experiments were conducted for different atmospheric CO_2 levels with both Late Ordovician and Early Silurian paleogeographies. With a 4.5% reduction in solar luminosity relative to today, experiments with both geographies are effectively ice free at $18\times$ PAL CO_2, and clearly indicate that glacial inception (defined as an increase in the depth of permanent snow cover) would occur at $10\times$. Our results indicate a very strong sensitivity to changes in atmospheric CO_2 for both paleogeographies. Ice–albedo feedback is the main amplifier of the modest reductions in direct radiative forcing from atmospheric CO_2. This strong sensitivity is suggestive of SICI behavior found in EBMs, which has been proposed as an explanation for sudden climate change.

With a low enough net forcing from solar luminosity and atmospheric CO_2, glaciation will occur, regardless of the paleogeographic configuration (given, of course, the presence of some land in high latitudes where ice can accumulate). Likewise with a high enough net forcing, the particular position of the supercontinent relative to the pole is unable to prevent ice-free conditions from occurring. However, in the experiments reported here the Early Silurian paleogeography is ice free at $14\times$ PAL CO_2, whereas the Late Ordovician glaciates at this CO_2 level (some snow cover survives, and depths increase in a small area). This difference can be attributed to the differences in paleogeography. As predicted by Crowley *et al.* (1987), land areas in high latitudes in the Silurian are warmer in summer. This is because of increased continentality, resulting from the South Pole being located nearer the center of Gondwana, rather than just off the edge of the supercontinent. A difference in summer temperatures in high latitudes due to continentality also exists for the $18\times$ PAL experiments. Although the Ordovician experiment is cooler, both experiments have no areas below freezing in summer.

Since solar luminosity was unlikely to have increased significantly between the Late Ordovician and the Early Silurian, changes in atmospheric $p\mathrm{CO_2}$ are likely to be the main cause of the rapid movement in and out of glacial conditions. Clearly, though, at certain $p\mathrm{CO_2}$ levels paleogeographic differences are important; the Late

Ordovician paleogeography would have been more sensitive to a decline in atmospheric pCO_2 since the ice-free/glaciation threshold occurs at a higher atmospheric CO_2 level than for the experiments with an Early Silurian geography. To explain the ice-free conditions that apparently existed for most of the early Paleozoic, atmospheric CO_2 levels must therefore have been continuously high.

The causes of glaciation during a 'greenhouse' climate is an appropriate problem to address with GCMs. As computing costs continue to decrease, more GCM experiments can be performed. The present study illustrates an approach to the complex problem of multiple unknowns that utilizes a 'matrix' of sensitivity experiments which consider different factors simultaneously. In such a manner, the major uncertainties in boundary conditions for the Paleozoic can be surmounted, and a plausible range of parameter space can be delineated. For example, here we test a range of atmospheric CO_2 levels and paleogeographies to evaluate the relative importance of each factor for explaining early Paleozoic glacial and ice-free climates. In addition, we can place narrower bounds on atmospheric CO_2 levels from climate model studies than those from geochemical studies. Atmospheric GCM experiments such as these can complement and serve as a prelude to thorough explorations with carbon cycle box models and ocean GCM experiments. Ultimately, to fully explain the Hirnantian glaciation a fully coupled ocean–atmosphere model that includes an interactive biosphere and carbon cycle (primitive as it may have been) and a dynamic ice-sheet component will be required.

ACKNOWLEDGEMENTS

This work was partly supported by National Science Foundation grant EAR 92-20008. The GENESIS climate model was developed at the National Center for Atmospheric Research by Starley Thompson and Dave Pollard. We thank Tom Crowley, Steve Baum, and George Moore for providing boundary condition files. Computational and graphical assistance was provided by Bill Peterson and Jim Sloan respectively. Helpful reviews by Pat Brenchley and Tom Crowley improved the manuscript.

REFERENCES

Abed, A. M., Makhlouf, I. M., Amireh, B. S. and Khalil, B. (1993). Upper Ordovician glacial deposits in southern Jordan. *Episodes*, **16**, 316–28.

Algeo, T. J., Berner, R. A., Maynard, J. B. and Scheckler, S. E. (1995). Late Devonian oceanic anoxic events and biotic crises: "Rooted" in the evolution of vascular land plants? *GSA Today*, **5**, 64–6.

Armstrong, H. A. (1995). High-resolution biostratigraphy (conodonts and graptolites) of the Upper Ordovician and Lower Silurian – evaluation of the Late Ordovician mass extinction. *Modern Geology*, **20**, 41–68.

Bahcall, J. N. and Ulrich, R. K. (1988). Solar models, neutrino experiments, and helioseismology. *Reviews of Modern Physics*, **60**, 297–372.

Barron, E. J. (1983). A warm equable Cretaceous: the nature of the problem. *Earth Science Reviews*, **19**, 305–38.

Barron, E. J. (1987). Explaining glacial periods. *Nature*, **329**, 764–5.

Barron, E. J., Fawcett, P. J., Peterson, W. H., Pollard, D. and Thompson, S. L. (1995). A "simulation" of mid-Cretaceous climate. *Paleoceanography*, **10**, 953–62.

Barron, E. J., Fawcett, P. J., Pollard, D. and Thompson, S. L. (1993). Past climate and the role of ocean heat transport: Model simulations for the Cretaceous. *Paleoceanography*, **8**, 785–98.

Barron, E. J. and Washington, W. M. (1985). Warm Cretaceous climates: High atmospheric CO_2 as a plausible mechanism. In *The Carbon Cycle and CO_2: Natural Variations Archean to Present*, Geophysical Monograph No. 32, eds. E. T. Sundquist and W. S. Broecker, pp. 546–53. Washington, DC: American Geophysical Union.

Baum, S. K. and Crowley, T. J. (1991). Seasonal snowline instability in a climate model with realistic geography: Application to Carboniferous (~ 300 Ma) glaciation. *Geophysical Research Letters*, **18**, 1719–22.

Berger, A. L. (1978). Long-term variations of caloric insolation resulting from the earth's orbital elements. *Quaternary Research*, **9**, 139–67.

Berger, A. L. (1989). The spectral characteristics of pre-Quaternary climatic records, an example of the relationship between the astronomical theory and geosciences. In *Climate and Geo-Sciences*, eds. A. Berger, S. Schneider and J. C. Duplessy, pp. 47–76. Amsterdam: Kluwer.

Berger, A. L., Loutre, M. F. and Dehant, V. (1989). Influence of the changing lunar orbit on the astronomical frequencies of pre-Quaternary insolation patterns. *Paleoceanography*, **4**, 555–64.

Berger, W. H. and Spitzy, A. (1988). History of atmospheric CO_2: Constraints from the deep-sea record. *Paleoceanography*, **3**, 401–11.

Berner, R. A. (1982). Burial of organic carbon and pyrite sulfur in the modern ocean: Its geochemical and environmental significance. *American Journal of Science*, **282**, 451–73.

Berner, R. A. (1994). GEOCARB II, A revised model of atmospheric CO_2 over Phanerozoic time. *American Journal of Science*, **294**, 56–91.

Berner, R. A. (1997). The rise of plants and their effect on weathering and atmospheric CO_2. *Science*, **276**, 544–6.

Berner, R. A. and Barron, E. J. (1984). Comments on the BLAG model: Factors affecting atmospheric CO_2 and temperature over the past 100 million years. *American Journal of Science*, **284**, 1183–92.

Berner, R. A., Cochran, M. F., Moulton, K. and Rao, J.-L. (1996). The quantitative role of plants in silicate weathering. In *Fourth International Symposium on the Geochemistry of the Earth's Surface*, ed. S. H. Bottrell, pp. 513–16. Ilkley, UK: University of Leeds.

Berner, R. A., Lasaga, A. C. and Garrels, R. M. (1983). The carbonate–silicate geochemical cycle and its effect on atmospheric carbon dioxide over the past 100 million years. *American Journal of Science*, **283**, 641–83.

Berry, W. B. N. and Boucot, A. J. (1973). Glacio-eustatic control of Late Ordovician–Early Silurian platform sedimentation and faunal changes. *Geological Society of America Bulletin*, **84**, 275–84.

Bertrand-Sarfati, J., Moussine-Pouchkine, A., Amard, B. and Aït Kaci Ahmed, A. (1995) First Ediacaran fauna found in western Africa and evidence for an Early Cambrian glaciation. *Geology*, **23**, 133–6.

Beuf, S., Biju-Duval, B., de Charpal, O., Rognon, P., Gariel, O. and Bennacef, A. (1971). *Les Grès du Paléozoïque inférieur au Sahara*. Paris: Editions Technip.

Bice, K. L. (1997). *An Investigation of Early Eocene Deep Water Warmth using Uncoupled Atmosphere and Ocean General Circulation Models: Model Sensitivity to Geography, Initial Temperatures, Atmospheric Forcing and Continental Runoff*. PhD thesis, Pennsylvania State University, University Park, PA.

Bonan, G. B., Pollard, D. and Thompson, S. L. (1992). Effects of boreal forest vegetation on global climate. *Nature*, **359**, 716–18.

Brenchley, P. J. (1989). The Late Ordovician extinction. In *Mass Extinctions: Proceses and Evidence*, ed. S. K. Donovan, pp. 104–32. London: Belhaven Press.

Brenchley, P. J., Carden, G. A. F. and Marshall, J. D. (1995). Environmental changes associated with the "first strike" of the Late Ordovician mass extinction. *Modern Geology*, **20**, 69–82.

Brenchley, P. J., Marshall, J. D., Carden, G. A. F. *et al.* (1994). Bathymetric and isotopic evidence for a short-lived Late Ordovician glaciation in a greenhouse period. *Geology*, **22**, 295–8.

Broecker, W. S. and Peng, T.-H. (1982). *Tracers in the Sea*. Palisades, NY: Lamont-Doherty Geological Observatory, Columbia University.

Budd, W. F. and Smith, I. N. (1987). Conditions for growth and retreat of the Laurentide ice sheet. *Géographie Physique et Quaternaire*, **41**, 279–90.

Buggisch, W. and Astini, R. (1993). The Late Ordovician ice age: New evidence from the Argentine Precordillera. In *Gondwana Eight*, eds. R. H. Findlay, R. Unrug, M. R. Banks and J. J. Veevers, pp. 439–47. Rotterdam: A. A. Balkema.

Caputo, M. V. and Crowell, J. C. (1985). Migration of glacial centres across Gondwana during the Paleozoic era. *Geological Society of America Bulletin*, **97**, 1026–36.

Carissimo, B. C., Oort, A. H. and Vonder Haar, T. H. (1985). Estimating the meridional energy transports in the atmosphere and ocean. *Journal of Physical Oceanography* **15**, 82–91.

Cerling, T. (1991). Carbon dioxide in the atmosphere: evidence from Cenozoic and Mesozoic paleosols. *American Journal of Science*, **291**, 377–400.

Chamberlin, T. C. (1899). An attempt to frame a working hypothesis of the cause of glacial periods on an atmospheric basis. *Journal of Geology*, **7**, 545–84.

Chandler, M. A., Rind, D. and Ruedy, R. (1992). Pangaean climate during the Early Jurassic: GCM simulations and the sedimentary record of paleoclimate. *Geological Society of America Bulletin*, **104**, 543–59.

Cocks, L. R. M. (1985) The Ordovician–Silurian boundary. *Episodes*, **8**, 98–100.

Cocks, L. R. M. and Scotese, C. R. (1991). The global biogeography of the Silurian Period. *Special Papers in Palaeontology*, **44**, 109–22.

Covey, C. (1991). Climate change: Credit the oceans? *Nature*, **352**, 196–7.

Covey, C. and Thompson, S. L. (1989). Testing the effects of ocean heat transport on climate. *Palaeogeography, Palaeoclimatology, Palaeoecology*, **75**, 331–41.

Crowell, J. C. (1983). Ice ages recorded on Gondwanan continents. *Transactions of the Geological Society of South Africa*, **86**, 237–62.

Crowley, T. J. (1993). Climate change on tectonic time scales. *Tectonophysics*, **222**, 277–94.

Crowley, T. J. and Baum, S. K. (1991). Towards reconciliation of Late Ordovician ($\sim$ 440 Ma) glaciation with very high CO_2 levels. *Journal of Geophysical Research*, **96**, 22597–610.

Crowley, T. J. and Baum, S. K. (1995). Reconciling Late Ordovician (440 Ma) glaciation with very high ($14\times$) CO_2 levels. *Journal of Geophysical Research*, **100**, 1093–101.

Crowley, T. J., Baum, S. K. and Kim, K.-Y. (1993*a*). General circulation model sensitivity experiments with pole-centered supercontinents. *Journal of Geophysical Research*, **98**, 8793–800.

Crowley, T. J., Mengel, J. G. and Short, D. A. (1987). Gondwanaland's seasonal cycle. *Nature*, **329**, 803–7.

Crowley, T. J. and North, G. R. (1988). Abrupt climate change and extinction events in earth history. *Science*, **240**, 996–1002.

Crowley, T. J., Short, D. A., Mengel, J. G. and North, G. R. (1986). Role of seasonality in the evolution of climate during the last 100 million years. *Science*, **231**, 579–84.

Crowley, T. J., Yip, K.-J. J. and Baum, S. K. (1993*b*). Milankovitch cycles and Carboniferous climate. *Geophysical Research Letters*, **20**, 1175–8.

Crowley, T. J., Yip, K.-J. J. and Baum, S. K. (1994). Snowline instability in a general circulation model: application to Carboniferous glaciation. *Climate Dynamics*, **10**, 363–76.

Denton, G. H., Armstrong, R. L. and Stuiver, M. (1971). The late Cenozoic glacial history of Antarctica. In *The Late Cenozoic Glacial Ages*, ed. K. K. Turekian, pp. 267–306. New Haven: Yale University Press.

Destombes, J. (1981). Hirnantian (Upper Ordovician) tillites on the north flank of the Tindouf Basin, Anti-Atlas, Morocco. In *Earth's Pre-Pleistocene Glacial Record*, eds. M. J. Hambrey and W. B. Harland, pp. 84–8. Cambridge: Cambridge University Press.

Dickinson, R. E., Henderson-Sellers, A., Kennedy, P. J. and Wilson, M. F. (1986). *Biosphere Atmosphere Transfer Scheme (BATS) for the NCAR Community Climate Model*. NCAR Technical Note NCAR/TN-275 + STR. Boulder, CO: National Center for Atmospheric Research.

Dong, B. and Valdes, P. J. (1995). Sensitivity studies of Northern Hemisphere glaciation using an atmospheric general circulation model. *Journal of Climate*, **8**, 2471–96.

Driese, S. G., Mora, C. I., Cotter, E. and Foreman, J. L. (1992). Paleopedology and stable isotope chemistry of Late Silurian vertic paleosols, Bloomsburg Formation, Central Pennsylvania. *Journal of Sedimentary Petrology*, **62**, 825–41.

Dutton, J. F. and Barron, E. J. (1996). GENESIS sensitivity to changes in past vegetation. *Palaeoclimates*, **1**, 325–54.

Endal, A. S. and Sofia, S. (1981). Rotation in solar-type stars. I. Evolutionary models for the spin down of sun. *Astrophysical Journal*, **243**, 625–40.

Fischer, A. G. (1984). The two Phanerozoic supercycles. In *Catastrophes in Earth History: The New Uniformitarianism*, eds. W. A. Berggren and J. A. Van Couvering, pp. 129–50. Princeton: Princeton University Press.

Foley, J. A., Kutzbach, J. E., Coe, M. T. and Levis, S. (1994). Feedbacks between climate and boreal forests during the Holocene epoch. *Nature*, **371**, 52–4.

Frakes, L. A. (1979). *Climate Throughout Geologic Time*. Amsterdam: Elsevier.

Frakes, L. A., Francis, J. E. and Sytkus, J. I. (1993). *Climate Modes of the Phanerozoic*. Cambridge: Cambridge University Press.

Freeman, K. H. and Hayes, J. M. (1992). Fractionation of carbon isotopes by phytoplankton and estimates of ancient CO_2 levels. *Global Biogeochemical Cycles*, **6**, 185–98.

Gallimore, R. G. and Kutzbach, J. E. (1996). Role of orbitally induced changes in tundra area in the onset of glaciation. *Nature*, **381**, 503–5.

Gibbs, M. T., Barron, E. J. and Kump, L. R. (1997). An atmospheric pCO_2 threshold for glaciation in the Late Ordovician. *Geology*, **25**, 447–50.

Gibbs, M. T., Rowley, D. B. and Ziegler, A. M. (1995). Constraining the position of Gondwana during the Paleozoic using lithologic paleoclimate indicators. *Geological Society of America Abstracts with Program*, **27**(6), A-205.

Grahn, Y. and Caputo, M. V. (1992). Early Silurian glaciations in Brazil. *Palaeogeography, Palaeoclimatology, Palaeoecology*, **99**, 9–15.

Gray, J. (1985). The microfossil record of early land plants: advances in understanding of terrestrialization, 1970–1984. *Philosophical Transactions of the Royal Society of London, Series B*, **309**, 167–95.

Gray, J., Massa, D. and Boucot, A. J. (1982). Caradocian land plant microfossils from Libya. *Geology*, **10**, 197–201.

Hambrey, M. J. (1985). The Late Ordovician–Early Silurian glacial period. *Palaeogeography, Palaeoclimatology, Palaeoecology*, **51**, 273–89.

Hambrey, M. J. and Harland, W. B. (1981). *Earth's Pre-Pleistocene Glacial Record*. Cambridge: Cambridge University Press.

Harland, W. B., Armstrong, R. L., Cox, A. V., Craig, L. E., Smith, A. G. and Smith, D. G. (1990). *A Geologic Time Scale 1989*. Cambridge: Cambridge University Press.

Havlicek, V. and Massa, D. (1973). Brachiopodes de l'Ordovicien supérieur de Libye occidentale: Implications stratigraphiques régionales. *Géobios*, **6**, 267–90.

Hiller, N. (1992). The Ordovician system in South Africa: a review. In *Global Perspectives on Ordovician Geology*, eds. B. D. Webby and J. R. Laurie, pp. 473–85. Rotterdam: A.A. Balkema.

Huff, W. D., Bergstrom, S. M. and Kolata, D. R. (1992). Gigantic Ordovician volcanic ash fall in North America and Europe: Biological, tectonomagmatic, and event-stratigraphic significance. *Geology*, **20**, 875–8.

Hyde, W. T., Kim, K.-Y., Crowley, T. J. and North, G. R. (1990). On the relation between polar continentality and climate: Studies with a nonlinear energy balance model. *Journal of Geophysical Research*, **95**, 18653–68.

Imbrie, J., Boyle, E. A., Clemens, S. C. *et al.* (1992). On the structure and origin of major glaciation cycles: 1. The 100,000-year cycle. *Paleoceanography*, **7**, 701–38.

Jenkins, G. S., Marshall, H. G. and Kuhn, W. R. (1993). Precambrian climate: The effects of land area and Earth's rotation rate. *Journal of Geophysical Research*, **98**, 8785–91.

Johnson, M. E. (1996). Stable cratonic sequences and a standard for Silurian eustasy. In *Paleozoic Sequence Stratigraphy: Views from the North American Craton*, Geological Society of America Special Paper No. 306, eds. B. J. Witzke, G. A. Ludvigson and J. Day, pp. 203–11. Boulder, CO: Geological Society of America.

Kasting, J. F. (1992). Paradox lost and Paradox found. *Nature*, **355**, 676–7.

Khain, V. E. and Seslavinskii, K. B. (1994). Global rhythms of the Phanerozoic endogenic activity of the Earth. *Stratigraphy and Geological Correlation*, **2**, 520–41.

Kiehl, J. T. and Dickinson, R. E. (1987). A study of the radiative effects of enhanced atmospheric CO_2 and CH_4 on early earth surface temperatures. *Journal of Geophysical Research*, **92**, 2991–8.

Kiehl, J. T. Wolski, R. J., Briegleb, B. P. and Ramanathan, V. (1987). *Documentation of Radiation and Cloud Routines in the NCAR Community Climate Model (CCM1)*. NCAR Technical Note NCAR/TN-288 + IA. Boulder, CO: National Center for Atmospheric Research.

Klitzch, E. (1981). Lower Paleozoic rocks of Libya, Egypt, and Sudan. In *Lower Paleozoic of the Middle East, Eastern and Southern Africa, and Antarctica*, ed. C. H. Holland, pp. 131–63. New York: John Wiley & Sons.

Kump. L. R. (1989). Chemical stability of the atmosphere and ocean. *Global and Planetary Change*, **1**, 123–36.

Kump, L. R. and Arthur, M. A. (1997). Global chemical erosion during the Cenozoic: Weatherability balances the budgets. In *Tectonic Uplift and Climate Change*, ed. W. F. Ruddiman, pp. 399–426. New York: Plenum Press.

Kump, L. R., Gibbs, M. T., Arthur, M. A., Patzkowsky, M. E. and Sheehan, P. M. (1995). Hirnantian glaciation and the carbon cycle. In *Ordovician Odyssey; Short Papers for the Seventh International Symposium on the Ordovician System*. eds. J. D. Cooper, M. L. Droser and S. L. Finney, pp. 299–302. Fullerton, CA: Pacific Section, Society for Sedimentary Geology.

Leggett, J. K. (1980). British Lower Paleozoic black shales and their paleo-oceanographic significance. *Journal of the Geological Society*, **137**, 139–56.

Legrand, P. (1995). Evidence and concerns with regard to the Late Ordovician glaciation in North Africa. In *Ordovician Odyssey: Short Papers for the Seventh International Symposium on the Ordovician System*, eds. J. D. Cooper, M. L. Droser and S. L. Finney, pp. 165–9. Fullerton, CA: Pacific Section, Society for Sedimentary Geology.

Lovelock, J. E. (1988). *The Ages of Gaia.* New York: W.W. Norton.

Mahmoud, M. D., Vaslet, D. and Husseini, M. I. (1992). The Lower Silurian Qalibah Formation of Saudi Arabia: An important hydrocarbon source rock. *American Association of Petroleum Geologists Bulletin*, **76**, 1491–506.

Markwick, P. J. and Rowley, D. B. (1998). The geological evidence for Triassic to Pleistocene glaciations: Implications for eustacy. In *Paleogeographic Evolution and Non-Glacial Eustacy: Northern South America*, SEPM Special Publication No. 58, eds. J. L. Pindell and C. Drake, pp. 17–43. Tulsa, OK: SEPM.

Marshall, J. D., Brenchley, P. J., Mason, P. *et al.* (1997). Global carbon isotopic events associated with mass extinction and glaciation in the Late Ordovician. *Palaeogeography, Palaeoclimatology, Palaeoecology*, **132**, 195–210.

McClure, H. A. (1988). The Ordovician–Silurian boundary in Saudi Arabia. *Bulletin of the British Museum (Natural History), Geology Series*, **43**, 155–63.

Miller, K. G., Fairbanks, R. G. and Mountain, G. S. (1987). Tertiary oxygen isotope synthesis, sea-level history, and continental margin erosion. *Paleoceanography*, **2**, 1–19.

Miller, K. G., Wright, J. D. and Fairbanks, R. G. (1991). Unlocking the icehouse: Oligocene–Miocene oxygen isotopes, eustacy, and margin erosion. *Journal of Geophysical Research*, **96**, 6829–48.

Moore, G. T., Jacobson, S. R., Ross, C. A., and Hayashida, D. N. (1994). A paleoclimate simulation of the Wenlockian (late Early Silurian) world using a general circulation model with implications for early land plant paleoecology. *Palaeogeography, Palaeoclimatology, Palaeoecology*, **110**, 115–44.

Mora, C. I., Driese, S. G. and Colarusso, L. A. (1996). Middle to Late Paleozoic atmospheric CO_2 levels from soil carbonate and organic matter. *Science*, **271**, 1105–7.

Mora, C. I., Driese, S. G. and Seager, P. G. (1991). Carbon dioxide in the Paleozoic atmosphere: Evidence from C-isotopic compositions of pedogenic carbonate. *Geology*, **19**, 1017–20.

Newman, M. J. and Rood, R. T. (1977). Implications of solar evolution for the earth's early atmosphere. *Science*, **194**, 1413–14.

North, G. R. (1990). Multiple solutions in energy balance climate models. *Palaeogeography, Palaeoclimatology, Palaeoecology*, **82**, 225–35.

Otto-Bliesner, B. L. (1996). Initiation of a continental ice sheet in a global climate model (GENESIS). *Journal of Geophysical Research*, **101**, 16909–20.

Otto-Bliesner, B. L. and Upchurch, G. R. Jr (1997). Vegetation-induced warming of high-latitude regions during the Late Cretaceous period. *Nature*, **385**, 804–7.

Paris, F., Elaouad-Debbaj, Z., Jaglin, J. C., Massa, D. and Oulebsir, L. (1995). Chitinozoans and Late Ordovician glacial events on Gondwana. In *Ordovician Odyssey: Short Papers for the Seventh International Symposium on the Ordovician System*, eds. J. D. Cooper, M. L. Droser and S. L. Finney, pp. 171–6. Fullerton, CA: Pacific Section, Society for Sedimentary Geology.

Paterson, W. S. B. (1972). Laurentide ice sheets: Estimated volumes during late Wisconsin. *Reviews of Geophysics and Space Physics*, **10**, 885–917.

Paull, R. A. (1977). The Upper Ordovician Neda Formation of eastern Wisconsin. In *Geology of Southeastern Wisconsin, A Guidebook for the 41st Annual Tri-State Field Conference*, ed. K. G. Nelson, pp. C-1–C-18. Madison, WI: Geological and Natural History Survey, The University of Wisconsin.

Pollard, D. and Thompson, S. L. (1995). Use of a land-surface-transfer scheme (LSX) in a global climate model (GENESIS): The response to doubling stomatal resistance. *Global and Planetary Change*, **10**, 129–61.

Powell, J. H., Khalil Moh'd, B. and Masri, A. (1994). Late Ordovician–Early Silurian glacio-fluvial deposits preserved in paleovalleys in Southern Jordan. *Sedimentary Geology*, **89**, 303–14.

Qing, H. and Veizer, J. (1994). Oxygen and carbon isotopic composition of Ordovician brachiopods: Implications for coeval seawater. *Geochimica et Cosmochimica Acta*, **58**, 4429–42.

Railsback, L. B., Ackerly, S. C., Anderson, T. F. and Cisne, J. L. (1990). Palaeontological and isotope evidence for warm saline deep waters in Ordovician oceans. *Nature*, **343**, 156–9.

Raymo, M. E. (1991). Geochemical evidence supporting T.C. Chamberlin's theory of glaciation. *Geology*, **19**, 344–8.

Retallack, G. J. (1992). What to call early plant formations on land. *Palaios*, **7**, 508–20.

Richter, F. M., Rowley, D. B. and DePaolo, D. J. (1992). Sr isotope evolution of seawater: The role of tectonics. *Earth and Planetary Science Letters*, **109**, 11–23.

Roberts, J. D. (1971). Later Precambrian glaciation: an anti-greenhouse effect? *Nature*, **234**, 216–17.

Sánchez, M. T., Benedetto, J. L. and Brussa, E. (1991). Late Ordovician stratigraphy, palaeoecology and sea level changes in the Argentine Precordillera. In *Advances in Ordovician Geology*, Geological Survey of Canada Paper No. 90-9, eds. C. R. Barnes and S. H. Williams, pp. 245–58. Ottawa: Geological Survey of Canada.

Sarmiento, J. L., Herbert, T. and Toggweiler, J. R. (1988). Causes of anoxia in the world ocean. *Global Biogeochemical Cycles*, **2**, 115–28.

Schwartzman, D. W. and Volk, T. (1989). Biotic enhancement of weathering and the habitability of Earth. *Nature*, **340**, 457–60.

Scotese, C. R. and Golonka, J. (1992). *Paleogeographic Atlas*. Arlington, TX: Paleomap Project, University of Texas.

Scrutton, C. T. (1978). Periodic growth features in fossil organisms and the length of the day and month. In *Tidal Friction and the Earth's Rotation*, eds. P. Brosche and J. Sundermann, pp. 154–96. New York: Springer-Verlag.

Semtner, A. J. (1976). A model for the thermodynamic growth of sea ice in numerical investigations of climate. *Journal of Physical Oceanography*, **6**, 379–89.

Semtner, A. K. and Klitzsch, E. (1994). Early Paleozoic paleogeography of the northern Gondwana margin: New evidence for Ordovician–Silurian glaciation. *Geologische Rundschau*, **83**, 743–51.

Sheehan, P. M. (1979). Swedish Late Ordovician marine benthic assemblages and their bearing on brachiopod zoogeography. In *Historical Biogeography, Plate Tectonics, and the Changing Environment*, eds. J. Gray and A. J. Boucot, pp. 61–73. Corvallis, OR: Oregon State University Press.

Sheehan, P. M. (1988). Late Ordovician events and the terminal Ordovician extinction. In *Contributions of Paleozoic Paleontology and Stratigraphy in Honor of Rousseau H. Flower*, New Mexico Bureau of Mines and Mineral Resources Memoir No. 44, ed. D. L. Woberg, pp. 405–15. Socorro, NM: New Mexico Bureau of Mines and Mineral Resources.

Sloan, L. C., Walker, J. C. G. and Moore, T. C., Jr. (1995). Possible role of oceanic heat transport in early Eocene climate. *Paleoceanography*, **10**, 347–56.

Spjeldnaes, N. (1961). Ordovician climatic zones. *Norsk Geologisk Tidsskrift*, **21**, 45–77.

Stillman, C. J. (1984). Ordovician volcanicity. In *Aspects of the Ordovician System*, Palaeontological Contributions from the University of Oslo No. 295, ed. D. L. Bruton, pp. 183–94. Oslo: Universitetsforlaget.

Thompson, S. L. and Pollard, D. (1995a). A global climate model (GENESIS) with a land-surface transfer scheme (LSX). Part I: Present-day climate: *Journal of Climate*, **8**, 732–61.

Thompson, S. L. and Pollard, D. (1995b). A global climate model (GENESIS) with a land-surface transfer scheme (LSX). Part II: CO_2 sensitivity. *Journal of Climate*, **8**, 1104–21.

Thompson, S .L, and Pollard, D. (1997). Greenland and Antarctic mass balances for present and doubled atmospheric CO_2 from the GENESIS version-2 global climate model. *Journal of Climate*, **10**, 871–900.

Tucker, R. D., Krogh, T. E., Ross, R. J., Jr. and Williams, S. H. (1990). Time-scale calibration by high-precision U-Pb zircon dating of interstratified volcanic ashes in the Ordovician and Lower Silurian stratotypes of Britain. *Earth and Planetary Science Letters*, **100**, 51–8.

Ulmishek, G. F. and Klemme, H. D. (1990). *Depositional Controls, Distribution, and Effectiveness of World's Petroleum Source Rocks*. US Geological Survey Bulletin No. 1931. Denver, CO: US Geological Survey.

Underwood, C. J., Crowley, S. F., Marshall, J. D. and Brenchley, P. J. (1997). High resolution carbon isotope stratigraphy of the basal Silurian stratotype (Dob's Linn, Scotland) and its global correlation. *Geological Society of London Journal*, **154**, 709–18.

Urey, H. C. (1952). *The Planets, Their Origin and Development*. New Haven: Yale University Press.

Valdes, P. J. and Crowley, T. J. (1998). A climate model intercomparison for the Carboniferous. *Palaeoclimates*, **2**, 219–38.

Van der Voo, R. (1993). *Paleomagnetism of the Atlantic, Tethys and Iapetus Oceans*. Cambridge: Cambridge University Press.

van Staal, C. R. (1994). Brunswick subduction complex in the Canadian Appalachians: Record of the Late Ordovician to Late Silurian collision between Laurentia and the Gander margin of Avalon. *Tectonics*, **13**, 946–62.

Vaslet, D. (1990). Upper Ordovician glacial deposits in Saudi Arabia. *Episodes*, **13**, 147–61.

Veevers, J. J. (1990). Tectono-climate supercycle in the billion year plate tectonic eon: Permian Pangean icehouse alternates with Cretaceous dispersed continents greenhouse. *Sedimentary Geology*, **68**, 1–16.

Veevers, J. J. and Powell, C. McA. (1987). Late Paleozoic glacial episodes in Gondwanaland reflected in transgressive–regressive depositional sequences in Euramerica. *Geological Society of America Bulletin*, **98**, 475–87.

Walker, J. C. G., Hays, P. B. and Kasting, J. F. (1981). A negative feedback mechanism for the long-term stabilization of Earth's surface temperature. *Journal of Geophysical Research*, **86**, 9776–82.

Wang, K., Chatterton, B. D. E. and Wang, Y. (1997). An organic carbon isotope record of Late Ordovician to Early Silurian marine sedimentary rocks, Yangtze Sea, South China: Implications for CO_2 changes during the Hirnantian glaciation. *Palaeogeography, Palaeoclimatology, Palaeoecology*, **132**, 147–58.

Wang, K., Orth, C. J., Attrep, M., Jr., Chatterton, B. D. E., Wang, X. and Li, J. (1993). The great latest Ordovician extinction on the South China Plate: Chemostratigraphic studies of the Ordovician–Silurian boundary interval on the Yangtze Platform. *Palaeogeography, Palaeoclimatology, Palaeoecology*, **104**, 61–79.

Wenzel, B. and Joachimski, M. M. (1996). Carbon and oxygen isotopic composition of Silurian brachiopods (Gotland/Sweden): paleoceanographic implications. *Palaeogeography, Palaeoclimatology, Palaeoecology*, **122**, 143–66.

Wilde, P. (1991). Oceanography in the Ordovician. In *Advances in Ordovician Geology*, Geological Survey of Canada Paper No. 90-9, eds. C. R. Barnes and S. H. Williams, pp. 283–98. Ottawa: Geological Survey of Canda.

Wilde, P., Quinby-Hunt, M. S. and Berry, W. B. N. (1990). Vertical advection from oxic or anoxic water from the main pycnocline as a cause of rapid extinctions. In *Extinction Events in Earth History*, eds. E. G. Kaufmann and O. H. Walliser, pp. 85–98. Berlin: Springer-Verlag.

Williams, G. E. (1991). Milankovitch-band cyclicity in bedded halite deposits contemporaneous with Late Ordovician–Early Silurian glaciation, Canning Basin, Western Australia. *Earth and Planetary Science Letters*, **103**, 143–55.

Yapp, C. J. and Poths, H. (1992). Ancient atmospheric CO_2 pressures inferred from natural goethites. *Nature*, **355**, 342–4.

Yapp, C. J. and Poths, H. (1996). Carbon isotopes in continental weathering environments and variations in ancient atmospheric CO_2 pressure. *Earth and Planetary Science Letters*, **137**, 71–82.

Zachos, J. C., Breza, J. and Wise, S. W. (1992). Early Oligocene ice-sheet expansion on Antarctica: Stable isotopic and sedimentological evidence from Kerguelen Plateau, southern Indian Ocean. *Geology*, **20**, 569–73.

Zachos, J. C., Stott, L. D. and Lohmann, K. C. (1994). Evolution of early Cenozoic marine temperatures. *Paleoceanography*, **9**, 353–87.

Ziegler, A. M., Hulver, M. L. and Rowley, D. B. (1997). Permian world topography and climate. In *Late Glacial and Postglacial Environment Changes: Pleistocene, Carboniferous–Permian, and Proterozoic*, ed. I. P. Martini, pp. 111–46. Oxford: Oxford University Press.

V

Overview: climate across tectonic timescales

14

Carbon dioxide and Phanerozoic climate

THOMAS J. CROWLEY

ABSTRACT

Much has been written about the importance of past carbon dioxide fluctuations as a cause of Phanerozoic climate change. In this update I summarize the present level of data/CO_2 geochemical model agreement for the Phanerozoic. Although there is a striking first-order agreement between simulated changes in CO_2, proxy estimates of CO_2, and evidence for global climate change, I argue that additional factors need to be considered in order to derive a general theory for Phanerozoic climates that is in quantitative agreement with climate models, and that these factors fall into two levels of importance. The most important first-order considerations for global temperature involve land–sea distribution and solar luminosity changes. Changes in ocean heat transport, orography, and vegetation may be necessary for obtaining better agreement between spatial patterns of climate change and models, but they may have a second-order effect on global temperature except in some cases of high latitude vegetation cover and the effect of orography on CO_2 levels. A more complete explanation for climate on tectonic time scales has to incorporate all these factors in a parsimonious manner in order to avoid a proliferation of *ad hoc* explanations for past climate change. Enhanced statistical comparisons between models and data are essential for more stringent testing of model predictions.

INTRODUCTION

Since the groundbreaking work of Eric Barron, Robert Berner and colleagues in the early 1980s much has been learned about causes of climate change on tectonic time scales. The hypothesis that has attracted the most attention involves the role of carbon dioxide changes. A series of geochemical models predicts large changes in CO_2 throughout the Phanerozoic (Berner *et al.*, 1983; Berner, 1991, 1994). These geochemical models receive support from climate model studies, which suggest that increases in CO_2 are required to explain climates of very warm time periods such as the Cretaceous and Eocene (Barron and Washington, 1985; Barron *et al.*, 1995; Sloan and Rea, 1996). Although this volume focuses on warm time periods in earth history, it is also necessary to examine the 'flip side' of the problem (glaciations) in order to obtain a more comprehensive understanding of

Phanerozoic climate change. In this paper I will briefly summarize the present state of agreement between the CO_2 hypothesis and Phanerozoic evidence for climate change, and include discussions of both warm and cold time periods. Though an impressive first-order agreement is evident between CO_2 and climate, I will argue that additional factors need to be considered in order to obtain a quantitatively consistent agreement between data and models. Economic integration and testing of 'synthetic' models represents a challenge to observationalists and modelers.

PRESENT STATUS OF CO$_2$ PARADIGM

The most recent version of the geochemical model (Berner, 1994) predicts substantial changes in atmospheric CO_2 levels through the Phanerozoic. The modeled CO_2 changes result from changes in carbon burial on land (due to the evolution of land plants) and weathering changes caused by organic acids in soils. A long-term increase in solar luminosity (see below) also strongly influences climate through its impact on global average temperature, which affects weathering rates (Berner, 1994). Although a number of uncertainties apply to all these calculations, comparison of geochemical model predictions with proxy CO_2 estimates indicates a good first-order agreement (Berner, 1997). The latter estimates are derived mainly from various types of $\delta^{13}C$ analyses of soils, minerals, and oceanic carbon species whose ratio correlates with atmospheric CO_2 levels (e.g., Cerling, 1991; Freeman and Hayes, 1992; Yapp and Poths, 1992).

The CO_2 hypothesis also requires testing against climate data and models. One test involves comparison of model estimates with the Phanerozoic record of continental glaciation (Figs. 14.1a and d). A recent compilation (Crowley, 1998) of Phanerozoic evidence for continental glaciation represents the summation of a substantial level of effort by the entire community (e.g., Hambrey and Harland, 1985; Frakes and Francis, 1988; Frakes et al., 1992). Although open to challenge in detail, and subject to modification by other lines of paleoclimate evidence, the compilation nevertheless represents a useful starting point for future more detailed comparisons of CO_2 predictions with the paleoclimate record.

Although there is a first-order agreement between the geochemical model and glaciation (Figs. 14.1a and d), a more quantitative evaluation requires assessment of the CO_2 radiative forcing effect and also the role of solar luminosity changes. The CO_2 estimates from Berner (1994) can be converted into radiative forcing (Fig. 14.1b) using the logarithmic relationship discussed in Kiehl and Dickinson (1987). Such an adjustment substantially alters the relative magnitude of the mid-Mesozoic and early Paleozoic CO_2 curves. Comparison of estimated CO_2 radiative forcing variations with the glacial record still suggests good first-order agreement (Fig. 14.1). Both the late Cenozoic and Permo-Carboniferous glaciations occur during times of low CO_2. If rifting and weathering levels were as low for the proposed late Precambrian supercontinent (e.g., Bond et al., 1984; Dalziel, 1991) as they were for the late Paleozoic Pangea, then CO_2 levels may also have been low in the Neoproterozoic (Kaufman et al., 1993; cf. Knoll et al., 1986). This first-order 'supercycle' agreement (Fischer, 1982) supports the hypothesis that CO_2 fluctuations have played an important role in the long-term evolution of Phanerozoic

CO$_2$-GLACIATION COMPARISON

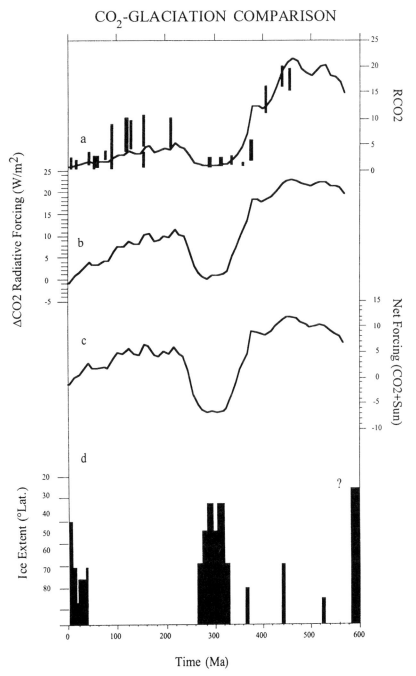

Figure 14.1. (a) Comparison of CO$_2$ concentrations from a geochemical model (Berner, 1994) with a compilation (Berner, 1997) of proxy CO$_2$ estimates (vertical bars). CO$_2$ (b) and solar radiative (c) forcing effects (W m^{-2}) as discussed in the text. Glaciological evidence for continental-scale glaciation (d) from Crowley (1998), which in turn was modified from a compilation from many sources. See text for further discussion.

SUMMER ΔT THROUGH ICE SHEET CENTER

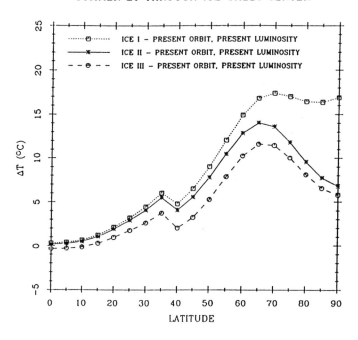

SUMMER ΔT THROUGH ICE SHEET CENTER

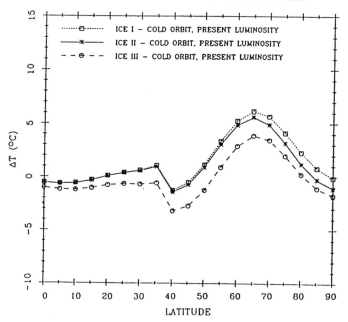

climate. Shorter-term variations in another trace gas, methane, may also be important in explaining episodic warm excursions such as that at the Paleocene–Eocene boundary (55 Ma) (Kennett and Stott, 1991; Dickens et al., 1997; Zachos et al., 1997).

INFLUENCE OF GEOGRAPHY AND SOLAR LUMINOSITY

Despite the agreement between CO_2 prediction and glaciation, a case can be made that additional factors are responsible for regulating first-order transitions into and out of glacial periods. Perhaps the most striking exception to CO_2/ice comparison is the Late Ordovician glaciation (440 Ma; Brenchley et al., 1994). Two other exceptions may involve the Late Devonian glaciation on South America (Caputo and Crowell, 1985) and the less well-documented Early Cambrian (530 Ma) glaciation on Africa (Briden et al., 1993; Bertrand-Sarfati et al., 1995). A somewhat more subtle exception involves the Permo-Carboniferous glaciation. Despite the fact that this glaciation occurs at a time of low modeled CO_2, climate model experiments (Baum and Crowley, 1991; Crowley et al., 1994; cf. Hyde et al., 1990) suggest that the very large land area on Gondwanaland, and the consequent strong summer warming, would have been sufficient to melt all winter snow cover during the Carboniferous at the present CO_2 level. Modeling studies indicate that for identical orbital configurations summer temperatures on Gondwanaland may have been 5–15 °C warmer than Pleistocene values (Fig. 14.2); such warming may have been sufficient to trigger melting of the Pleistocene ice sheets (cf. Tarasov and Peltier, 1997) or prevent initiation of Gondwanan glaciation (Crowley et al., 1991, 1994; Hyde et al., 1999).

To evaluate these complications it is necessary to consider additional factors contributing to long-term climate change. Because we are not at the stage where all of the different factors contributing to this problem have been sorted out quantitatively, the remaining part of the discussion represents my effort to try to impose some order on the problem without having definitive justification for all the points I make; it is in a sense a hypothesis awaiting further testing. I will first consider the roles of geography and solar luminosity, then pause to determine how much these

Figure 14.2. Evidence from climate model simulations that geography and solar luminosity may play a key role in explaining Permo-Carboniferous glaciation. This figure compares differences between two climate model simulations (Pleistocene and Carboniferous). The simulations use two orbital insolation configurations: the present orbit and a configuration yielding minimum summer insolation receipt (termed cold summer orbit) in the appropriate hemisphere. The figure illustrates the differences (Carboniferous-Pleistocene) in model-predicted summer temperatures on an equator-to-pole transect along the longitude of maximum equatorward ice extent for each glaciation. Because of uncertainties in Carboniferous ice extent, three different ice sheet reconstructions were utilized. The results indicate that, due to the very large land area of Pangea, summer warming over the Carboniferous ice sheet may have been ∼5–15 °C greater than over the Pleistocene ice sheet. This level of warming may have been sufficient to prevent ice growth in the Carboniferous (see Hyde et al., 1999). In order to explain Carboniferous ice extent, it is therefore necessary to invoke decreased solar luminosity. (From Crowley et al., 1991.)

additional mechanisms can explain. After that I will more briefly list factors whose contribution has not been completely assessed.

Geography

Changes in the land–sea distribution, including location and size of the continents, have long been thought to be important for the evolution of climate. Yet some climate modeling experiments suggest an ambiguous role for land–sea distribution. Fawcett and Barron (1998) illustrate little effect of land–sea distribution on global temperatures. In fact, simulated temperatures of land areas sometimes show trends opposite to that estimated from the geologic record; for example modeled Permian temperatures are higher than the Cretaceous. However, geography can significantly influence amounts and distribution of continental precipitation and runoff (Fawcett and Barron, 1998). Other experiments examining the seasonal cycle of temperatures suggest that in some cases (e.g., Greenland in the late Cenozoic) changes in land–sea distribution may have been important for summer cooling essential for glacial inception (Crowley et al., 1986; Hyde et al., 1990). As discussed above, the large seasonal cycle on Pangea may also have prevented ice inception on a polar Gondwana even with low CO_2 levels (Hyde et al., 1990; Crowley et al., 1994).

Solar luminosity

Another important factor to consider involves the long-term increase in solar luminosity. Almost all solar models predict a 25–40% increase in solar luminosity since the formation of the sun. The weakly exponential increase in solar luminosity (Fig. 14.3) is considered relatively robust because it is strongly dependent on the mean molecular mass of the sun's core (Endal and Sofia, 1981), which must increase with time as a result of fusion. The uncertainties in solar luminosity are mainly due to imprecise knowledge of the initial solar H/He value, which determines the early temperature of the sun's core. The predicted 5–6% increase in luminosity for the Phanerozoic is comparable in terms of radiative forcing changes (12–14 W m^{-2} net radiative forcing change, after accounting for the effects of planetary albedo) to the oscillatory changes in estimated CO_2 levels (Fig. 14.1b).

To account for the increase in solar luminosity through time, the CO_2 forcing curve (Fig. 14.1b) was adjusted for changes in solar luminosity. The latter term was estimated by taking the mean value of two end-member estimates of luminosity determined by Crowley et al. (1991), and applying a uniform 30% planetary albedo for all time periods. Although there are obvious uncertainties in the albedo correction, more sophisticated corrections do not seem warranted because of the uncertainties in the actual forcing terms themselves. The shortwave solar effect was then added to the CO_2 longwave effect to obtain a 'net CO_2' forcing. This approach is justifiable because climate model simulations indicate a similar system response to comparable solar and CO_2 radiative perturbations (Manabe and Wetherald, 1980; Hansen et al., 1984).

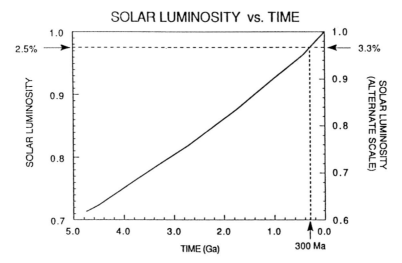

Figure 14.3. Evolution of solar luminosity vs. time (percent differences from the present) as computed from a model of solar evolution. Two different luminosity scales (y-axes) are used, representing uncertainties as to whether the initial luminosity may have been 25% (Endal and Sofia, 1981) or 40% (e.g., Bahcall and Ulrich, 1988) less than the present (see text). Dashed lines illustrate, for example, estimated luminosity (with uncertainties) for the Carboniferous glaciation. (Solar calculation from Endal and Sofia, 1981; alternate scaling from Crowley et al., 1991; figure from Crowley et al., 1991.)

Comparison of the 'net CO_2' forcing (Fig. 14.1c) model prediction with the geologic record indicates a general agreement in trend with the 'CO_2-only' term (Fig. 14.1b), except that the magnitude of the early Paleozoic forcing is substantially reduced. This reduction in forcing may be critical for explaining the Late Ordovician glaciation (Crowley and Baum, 1991, 1995; Gibbs et al., Chapter 13, this volume), and, by extension, the Late Devonian and Early Cambrian glaciations that have not yet been modeled. The very low 'net CO_2' forcing for the Permo-Carboniferous now suppresses summer warming on the Gondwanan land mass sufficiently to allow development of ice sheets (Crowley et al., 1994; Hyde et al., 1999). A coupled climate–ice sheet model study for the end of the Permo-Carboniferous glaciation suggests that a 2–3× CO_2 increase would be sufficient to melt the large Gondwanan ice sheets (Hyde et al., 1999). 'Net CO_2' levels are also relatively low between about 330 and 350 Ma, before the time of clear glaciologic evidence for significant ice volume. A modeling study (Crowley and Baum, 1992) suggests that such values would be conducive to glaciation at this time. Some new $\delta^{18}O$ evidence supports this prediction (Mii et al., 1999), but more data are needed to evaluate the problem. When coupled to a luminosity value 6% below present (Crowley and Baum, 1993), such forcing may also explain the very extensive late Precambrian glaciations, in which ice apparently extended into low latitudes (e.g., Hambrey and Harland, 1985).

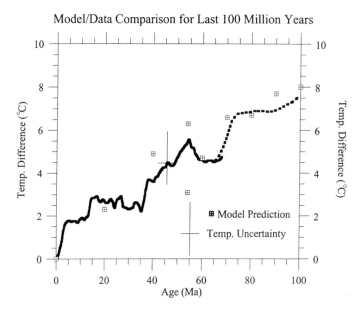

Figure 14.4. Comparison of model-predicted global average temperature differences from the present (squares) with temperature differences estimated by scaling the benthic $\delta^{18}O$ record (Douglas and Woodruff, 1981; Miller *et al.*, 1987) to global temperatures. The latter estimates were determined for the early Eocene (55 Ma) and Cenomanian (90 Ma) based on zonal $\delta^{18}O$ profiles (Crowley and Kim, 1995; Crowley and Zachos, Chapter 3, this volume). Uncertainties in 'observed' global temperature reflect the analysis of Crowley and Zachos (Chapter 3, this volume) but modified to account for the likelihood that the greatest uncertainty in sea surface temperature reconstruction involves the tropics (see text).

Comparison with data

It is now necessary to pause and evaluate how far the additional contributions of geography and luminosity increase our understanding of past climate change. As discussed above, the Paleozoic (and perhaps late Precambrian) glaciations may now be understandable as a result of the combined interactions of CO_2, luminosity, and geography. How well do these mechanisms explain Mesozoic–Cenozoic climate change? One way to address this problem is to compare CO_2-based model predictions with observed estimates of global temperature change for the last 100 million years (Fig. 14.4). The data are based on, but modified from, a reconstruction (Crowley and Kim, 1995) of global average temperatures and utilize the deep-sea benthic $\delta^{18}O$ record (Douglas and Woodruff, 1981; Miller *et al.*, 1987) for estimating the general trend in global temperatures. The modification used the more detailed $\delta^{18}O$ record for 0–70 Ma and spliced in the more uncertain Douglas and Woodruff (1981) Cretaceous data. The general trends agree with those inferred from northern hemisphere plants and high latitude southern hemisphere $\delta^{18}O$ records (Shackleton and Kennett, 1975a; Wolfe, 1978; Parrish and Spicer, 1988; Chase *et al.*, 1998).

A key adjustment of the benthic $\delta^{18}O$ record involves translating it into global temperature change. This is an uncertain exercise, with results subject to modification as we learn more about the assumptions. The basic idea is that the $\delta^{18}O$ record can be scaled to global temperature by using time slices from key horizons that have sufficient information on zonal temperature gradients to allow for estimation of global temperature. Such an approach can provide reasonable estimates of global temperature; for example an energy balance model fit for the Present yields 14.9 °C, only 0.2 °C greater than observed (W. Hyde, personal communication). Utilizing the same technique, Crowley and Kim (1995) calculated that the Early Eocene (54 Ma), a time of considerable interest in model studies (e.g., Sloan and Rea, 1996), was about 5.5 °C warmer than at present and the Cenomanian (90 Ma) about 7.5–8.5 °C warmer than at present. The latter value is comparable to the mid-Cretaceous temperature estimated by Barron (1983). The uncertainty in these estimates is very unclear; in this paper the value from Crowley and Zachos (Chapter 3, this volume) for individual sites is modified to take into account the fact that the greatest uncertainties in sea surface temperature (SST) estimates may apply to the tropics. The uncertainty level for global temperature is therefore about half the uncertainty for tropical SSTs. Applying these adjustments to the spliced $\delta^{18}O$ record, and giving more weight to the better-constrained Eocene record and Cenozoic benthic $\delta^{18}O$ data, leads to the global temperature (differences) plot in Fig. 14.4. This recalculation suggests that the mid-Cretaceous global temperature increase may have been slightly underestimated (~ 1 °C). This feature could reflect either an inadequate amount of $\delta^{18}O$ data for calibration or some breakdown in validity of the assumption of linearly scaling benthic $\delta^{18}O$ to global temperature.

The climate model results, which include the effects of geography, are taken primarily from Fawcett and Barron (1998), with two runs (Sloan and Rea, 1996) for the Early Eocene (54 Ma) added. All runs utilized the same GENESIS climate model (Thompson and Pollard, 1995), with present oceanic heat transport employed for all runs. This climate model has a sensitivity of ~ 2.0 °C for a doubling of CO_2 (Thompson and Pollard, 1995). The CO_2 forcing levels for the model runs (compared with the Present, which equals 340 ppm CO_2 in the control run) for all but the Early Eocene are from Berner (1991): 20 Ma (1×), 40 Ma (2×), 60 Ma (2×), 70 Ma (3×), 80 Ma (3.5×), 90 (3×), 100 (6×). The Early Eocene runs from Sloan and Rea (1996) employed values of 2× and 6× CO_2; subsequent work suggests the value may be closer to 3× CO_2 (J. Zachos et al., in preparation).

The agreement between the model runs and data for the last 100 million years is generally good, although it must be borne in mind that permanent ice on Greenland and Antarctica was stipulated for the Present. Allowing for all the uncertainties in these simulations (paleogeography, CO_2 levels, climate model sensitivity), the results suggest that, within our present level of knowledge, there is a first-order agreement between models and data for the last 100 million years with respect to global average temperature. The first-order agreement between models and data suggests that if the CO_2 estimates are at all reasonable (as suggested by the proxy CO_2 indices; Fig. 14.1), then a climate model climate sensitivity (°C temperature change per $W\,m^{-2}$ of forcing) of about 2.0 °C (for a doubling of CO_2) appears adequate to explain pre-Pleistocene climate change during the last 100 million years. This value is similar to that independently derived by Hoffert and Covey (1992). The uncertainties in CO_2 forcing and global temperature estimates place probably a 50% uncertainty on this mean value.

When these results are combined with the above-discussed progress in successfully simulating Paleozoic climate fluctuations, the following question inevitably arises: what else is needed to explain first-order trends in Phanerozoic climate? That is, if the combined effects of CO_2, geography, and luminosity appear to explain most of the observed changes in the mean climate state, why are other mechanisms needed? To address this question it is necessary to delve into yet another level of detail with respect to quantitative agreement between models and paleo-data.

ADDITIONAL PROCESSES OPERATING ON PHANEROZOIC TIME SCALES

Despite the good agreement between CO_2 and climate for global-scale temperature change, additional factors need to be explored to obtain a better understanding of regional climate trends inferred from paleoclimate records. In this section I will briefly discuss additional mechanisms that may be important. At this stage it is probably useful to point out that my views on this subject may differ significantly from other workers in the field. The reader should bear this in mind when reading this abbreviated discussion of additional processes.

Changes in orography

There has been a moderate amount of discussion with respect to the role of uplift and long-term climate change (e.g., Kutzbach *et al.*, 1989; Ruddiman and Kutzbach, 1991). Although uplift certainly affects regional climate it is less clear what the role is with respect to changes between ice-free and glaciated states. Local uplift may certainly be important with regards to glacial inception, but I am less convinced that far-field changes in geography can affect glaciation. For example, summer temperatures in eastern Canada, which are critical for glacial inception (Crowley *et al.*, 1986), are unlikely to be strongly affected by uplift of the western North America Cordillera. Inspection of the present seasonal cycle of temperature indicates that summer warming on continents is more driven by land–sea differences in heat capacity than by circulation changes. Orographically driven circulation changes can be important in winter, however. Topography may also play an important role in regulating CO_2 levels (see Raymo, 1991).

Ocean circulation

Changes in the circulation of the ocean have long been a topic of interest. Such changes could reflect the response to geography (e.g., Barron and Peterson, 1990) or opening and closing of ocean gateways such as the Central American isthmus, Drake Passage, and Indonesian straits (Maier-Reimer *et al.*, 1990; Hirst and Godfrey, 1993; Mikolajewicz *et al.*, 1993). For example, closure of the Central American seaway appears to have played a major role in development of the present thermohaline cell in the North Atlantic (Maier-Reimer *et al.*, 1990; Mikolajewicz and Crowley, 1997). A significant mid-Miocene increase in North Atlantic Deep

Water (NADW) production occurred 12–13 Ma (Wright and Miller, 1996), near the time of major increase in Antarctic ice volume (Shackleton and Kennett, 1975b). Conceivably the major increase in NADW production may have triggered mid-Miocene Antarctic ice advance; this idea obviously requires further testing. In general changes in ocean circulation do not have a large effect on global temperatures: heat gained in one region is lost from another. The principal exception to this pattern involves ice feedback in high latitudes, but even in this case the response is less than often conjectured. A collapse of the North Atlantic thermohaline cell results in only about a 0.2 °C change in global temperatures (R. Stouffer, personal communication, 1994, based on results in Manabe and Stouffer, 1988). This is because the collapse of NADW production (Fig. 14.5) eliminates the export of mass and more importantly heat from the southern hemisphere (Crowley, 1992), so that the northern hemisphere heat loss is therefore almost balanced by heat gain in the southern hemisphere.

Geologic data suggest that the ocean response to CO_2 increases appears to require some change in ocean heat transport in order to explain the large increases in high latitude temperatures (e.g., Barron, 1987a; Rind and Chandler, 1991; Barron et al., 1995; Sloan et al., 1995) and the stable tropical temperatures (Crowley and Zachos, 1998). One possible mechanism involves production of warm, saline bottom water in the tropics (Chamberlin, 1906; Brass et al., 1982; cf. MacLeod et al., Chapter 8, this volume). Although this is an intriguing idea, there are several substantial arguments that can be raised regarding the reality of this process (Crowley, 1999): for example that estimated surface water temperatures in the alleged subtropical source areas are about 10 °C warmer than observed bottom water temperatures

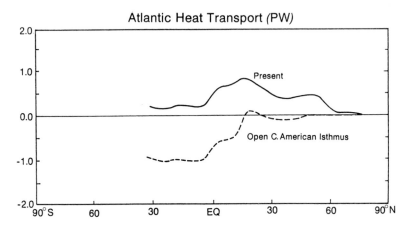

Figure 14.5. Comparison of poleward ocean heat transport (y-axis) in the Atlantic for an ocean general circulation model simulation with present boundary conditions and with an open Central American isthmus; positive values indicate northward-directed heat transport. Changes in heat transport reflect collapse of the North Atlantic thermohaline cell with an open Central American isthmus. Such changes may also have influenced climate variations on tectonic time scales. (From Crowley and North, 1991; based on results in Maier-Reimer et al., 1990).

(see Crowley and Zachos, Chapter 3, this volume) and that the ocean should have gone anoxic (Herbert and Sarmiento, 1991). The very fact that early Cenozoic deep waters were nearly the same temperature as high latitude surface waters (Shackleton and Kennett, 1975a) suggests that the most parsimonious explanation for warm deep waters involves the cooling of warm, high latitude surface waters (these would still be the densest water in the model ocean).

The precise mechanism by which ocean circulation is modified during warm time periods is still unclear. Many modelers essentially force the change in their models to agree better with observations without examining whether the forced changes are consistent with physical oceanographic theory and the overall weakened equator–pole temperature gradients. Some experiments (Baum and Crowley, 1997) suggest that weakened winds in the Eocene (Rea *et al.*, 1985) could have led to increased polar temperatures as a result of decreased evaporative heat loss from the ocean surface. Explicit examination of the polar heat transport problem represents an important aspect of understanding the climate of warm time periods and certainly requires more work.

An altered oceanic circulation may also eventually help explain a model–data discrepancy on land that is especially relevant to the climate of warm time periods. Geologic data often indicate warmer interior temperatures than predicted by models (e.g., Crowley *et al.*, 1989; Sloan and Barron, 1990; Wing and Greenwood, 1993; Greenwood and Wing, 1995; Sloan *et al.*, 1996). Although some of these discrepancies could reflect incorrect specification of local boundary conditions (e.g., lakes; Sloan, 1994), including a 'correct' equator–pole temperature gradient (i.e., very warm polar regions and stable tropics) may have a larger overall impact on temperatures in continental interiors. For example, Pliocene experiments by Sloan *et al.* (1996) indicate a substantial northward movement of the freezing line ($\sim 10°$ of latitude) for even 'modest' changes in SST as occurred in the Pliocene (Fig. 14.6). Proportionately larger changes should have occurred in the Eocene; this hypothesis is now being tested (L. Sloan, personal communication, 1998).

Vegetation

Owing to its effects on albedo and transpiration, changes in vegetation could significantly affect the surface energy balance. In particular, expansion of vegetation into high latitudes, such as occurred in the Mesozoic and early Cenozoic, may have led to a situation where more incoming solar insolation is absorbed, resulting in an increase of temperatures. Some initial experiments with a pole-centered Gondwanan supercontinent (Crowley and Baum, 1994) suggested the effect would be primarily in the summer; small changes occurred in the winter because sunlight levels are already low. However, sensitivity experiments with an interactive land–sea model (Otto-Bliesner and Upchurch, 1997; see DeConto *et al.*, Chapter 2, this volume) for a time period with smaller land masses in high latitudes (Late Cretaceous) suggest the vegetation feedback could sometimes translate into an increase in winter temperatures. For example, warmer temperatures in summer led to a decrease in sea ice area, which then led to more heat release from the ocean to the atmosphere in winter, warming the land. Similar changes may have occurred for the Early Miocene (20 Ma;

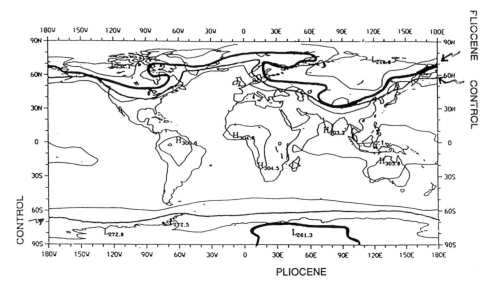

Figure 14.6. Example of how imposition of a 'realistic' paleo-sea surface temperature (SST) field can lead to significant changes in the freezing line in continental interiors. The figure shows changes for the mid-Pliocene (3 Ma) as calculated by Sloan et al. (1996), using the SST reconstruction of Dowsett et al. (1996). Larger changes in zonal temperature gradient (e.g., during the Eocene) may have an even larger effect on freezing line changes. (From Sloan et al., 1996.)

Dutton and Barron, 1997). This problem may also explain a conundrum in Paleozoic climate studies. Early work (Crowley et al., 1987) on modeling the early Paleozoic passage of Gondwana across the South Pole suggested that summer warming on land should be sufficient to prevent ice formation on a polar land mass. Subsequent experiments indicated that this response only applies to large land masses (Hyde et al., 1990); smaller ones do not heat up enough (Oglesby, 1991). The question therefore arises (Barron, 1987b) of why did the small land mass of Antarctica remain nearly ice free in the Cretaceous, even with higher CO_2 levels? Vegetation may be the answer; in effect it made Antarctica a larger continent (in terms of energy receipt).

Abrupt transitions

Finally, geological studies suggest that transitions between different climate states may be abrupt (e.g., North and Crowley, 1985; Crowley and North, 1988; Kennett and Stott, 1991). Such abrupt changes could be due to large and rapid changes in forcing but they could also reflect instabilities in the climate system. For example, the rapid temperature warming near the Paleocene–Eocene boundary (55 Ma) may reflect some global-scale excursion of the carbon reservoir associated with the $\sim 3‰$ $\delta^{13}C$ drop from the late Paleocene (Shackleton, 1985; Bralower et al., 1997). The Eocene–Oligocene (~ 32 Ma) expansion of the Antarctic ice sheet is another classic example (e.g., Zachos et al., 1992), as is the mid-Miocene expansion about 15 Ma (Shackleton and Kennett, 1975b). The demise of the Late Ordovician ice sheet may be another example of a rapid climate change (Brenchley et al., 1994).

Although much has been written on thermohaline instabilities (e.g., Broecker et al., 1985; Manabe and Stouffer, 1988) with respect to Pleistocene climate change, climatologists have known for more than two decades about instabilities related to snow albedo feedback in climate models (e.g., Budyko, 1969; Crowley et al., 1994; Otto-Bliesner, 1996). Small changes in boundary conditions can also cause an abrupt increase in snow cover or ice sheet volume (Hyde et al., 1999). The snow-line instability is related to high albedo over snow areas, which suppresses temperatures in nearby areas, allowing for snow to accumulate in those regions as well. Small changes in CO_2 or some other mechanism could trigger such changes; this may have been the case for the Late Ordovician glaciation (Gibbs et al., 1997). Kump et al. (1999) discuss how weathering changes induced by Ordovician glaciation could cause a negative feedback, raising CO_2 levels, and leading to the abrupt termination of the brief glacial event. These processes also need to be incorporated into any general theory for climate change.

CONCLUDING REMARKS

To summarize, an updated test of the climate/CO_2 hypothesis continues to provide support for the importance of this mechanism with respect to climate change on tectonic time scales. However, strong arguments can be made for the importance of changes in geography and solar luminosity for obtaining a better agreement between models and data for the Phanerozoic. Other mechanisms are also important: orography, ocean heat transport, vegetation, and instabilities in the climate system. Because much of the first-order effects of Phanerozoic climate change can be explained by geography–CO_2–luminosity, I conjecture that these additional processes either (a) are second-order feedbacks, (b) affect the timing of transitions, or (c) affect the spatial response of climate to large-scale forcing changes, with a secondary effect on global temperatures. Only when the climate is near a climate bifurcation point can these 'secondary' feedbacks drive the system into another climate state. In other words, changes in global temperature primarily require changes in global forcing; dynamical feedbacks involving the land–sea–air–ice system are needed to obtain a better spatial agreement between models and observations. Perhaps the primary exception to this conjecture involves the effects of orography on weathering and CO_2 levels.

With specific reference to warm time periods the most pressing climate problems involve understanding the factors responsible for changes in the equator–pole temperature gradient and winter warming in continental interiors. Sloan et al. (1996) suggest these problems may be linked, although the dynamics of the connection are not understood at the present time. It is my belief that deep-water changes are not primarily responsible for changes in the ocean temperature gradient and that unspecified surface changes are required to obtain an understanding of the important ocean processes occurring at this time. Some ocean model sensitivity experiments (Baum and Crowley, 1997) suggest that a weakening of winds associated with a decreased temperature gradient can yield increased polar temperatures as a result of decreased evaporative heat loss from the ocean and a spindown of the main subtropical gyres. More work is required in this area.

Although the whole range of 'secondary' mechanisms provide opportunities for stimulating inquiries into past climates, a problem arises due to a sheer abundance of explanations. In order to avoid a proliferation of *ad hoc* explanations for climate change, the different mechanisms must somehow be integrated in a systematic, parsimonious manner, with hypotheses presented in a testable, falsifiable manner. Only then can we be satisfied that a 'grand unification' has been achieved for climate change on tectonic time scales. This unified view can be accomplished via sensitivity experiments and more quantitative comparisons between models and data. In the latter regard, it is particularly important for the statistical significance of model results, and uncertainties in paleoclimate data estimates, to be included in making these comparisons.

ACKNOWLEDGEMENTS

This work was supported by NSF grant ATM-9529109. I thank B. Huber, W. Hyde, B. Otto-Bliesner, and C. Paulson for comments.

REFERENCES

Bahcall, J. N. and Ulrich, R. K. (1988). Solar models, neutrino experiments, and helioseismology. *Reviews of Modern Physics*, **60**, 297–372.

Barron, E. J. (1983). A warm, equable Cretaceous: The nature of the problem. *Earth Science Reviews*, **19**, 305–38.

Barron, E. J. (1987a). Eocene equator-to-pole surface ocean temperatures: A significant climate problem? *Paleoceanography*, **2**, 729–39.

Barron, E. J. (1987b). Explaining glacial periods. *Nature*, **329**, 764–865.

Barron, E. J., Fawcett, P. J., Peterson, W. H., Pollard, D. and Thompson, S. L. (1995). A "simulation" of Mid-Cretaceous climate. *Paleoceanography*, **10**, 953–62.

Barron, E. J. and Peterson, W. H. (1990). Mid-Cretaceous ocean circulation: Results from model sensitivity studies. *Paleoceanography*, **5**, 319–37.

Barron, E. J. and Washington, W. M. (1985). Warm Cretaceous climates: High atmospheric CO_2 as a plausible mechanism. In *The Carbon Cycle and Atmospheric CO_2: Natural Variations Archean to Present*, Geophysical Monograph No. 32, eds. E. T. Sundquist and W. S. Broecker, pp. 546–53. Washington, DC: American Geophysical Union.

Baum, S. K. and Crowley, T. J. (1991). Seasonal snow line instability in a climate model with realistic geography: Application to Carboniferous (~ 300 Ma) glaciation. *Geophysical Research Letters*, **18**, 1719–22.

Baum, S. K. and Crowley, T. J. (1997). Effect of weaker winds on meridional heat transport, with application to climatically warmer periods. *Eos (Transactions of the American Geophysical Union, Supplement to Fall Meeting)*, **78**, F358 (Abstract).

Berner, R. A. (1991). A model for atmospheric CO_2 over Phanerozoic time. *American Journal of Science*, **291**, 339–76.

Berner, R. A. (1994) GEOCARB II: A revised model of atmospheric CO_2 over Phanerozoic time. *American Journal of Science*, **294**, 56–91.

Berner, R. A. (1997). The rise of plants and their effect on weathering and atmospheric CO_2. *Science*, **276**, 544–6.

Berner, R. A., Lasaga, A. C. and Garrels, R. M. (1983). The carbonate–silicate geochemical cycle and its effect on atmospheric carbon dioxide over the last 100 million years. *American Journal of Science*, **283**, 641–83.

Bertrand-Sarfati, J., Moussine-Pouchkine, A., Amard, B. and Ahmed, A. A. K. (1995). First Ediacaran fauna found in western Africa and evidence for an Early Cambrian glaciation. *Geology*, **23**, 133–6.

Bond, G. C., Nickeson, P. A. and Kominz, M. A. (1984). Breakup of a supercontinent between 625 Ma and 555 Ma: New evidence and implications for continental histories. *Earth Planetary Science Letters*, **70**, 325–45.

Bralower, T. J., Thomas, D. J., Zachos, J. C. *et al.* (1997). High-resolution records of the late Paleocene thermal maximum and circum-Caribbean volcanism: Is there a causal link? *Geology*, **25**, 963–6.

Brass, G. W., Southam, J. R. and Peterson, W. H. (1982). Warm saline bottom waters in the ancient ocean. *Nature*, **296**, 620–3.

Brenchley, P. J., Marshall, J. D., Carden, G. A. F *et al.* (1994). Bathymetric and isotopic evidence for a short-lived Late Ordovician glaciation in a greenhouse period. *Geology*, **22**, 295–8.

Briden, J. C., McClelland, E. and Rex, D. C. (1993). Proving the age of a paleomagnetic pole: The case of the Ntonya Ring Structure, Malawi. *Journal of Geophysical Research*, **98**, 1473–9.

Broecker, W. S., Peteet, D. M. and Rind, D. (1985). Does the ocean–atmosphere system have more than one stable mode of operation? *Nature*, **315**, 21–6.

Budyko, M. I. (1969). The effect of solar radiation variations on the climate of the earth. *Tellus*, **21**, 611–19.

Caputo, M. V. and Crowell, J. C. (1985). Migration of glacial centers across Gondwana during the Paleozoic Era. *Geological Society of America Bulletin*, **96**, 1020–36.

Cerling, T. E. (1991). Carbon dioxide in the atmosphere: Evidence from Cenozoic and Mesozoic paleosols. *American Journal of Science*, **291**, 377–400.

Chamberlin, T. C. (1906). On a possible reversal of deep-sea circulation and its influence on geologic climates. *Journal of Geology*, **14**, 363–73.

Chase, C. G., Gregory, K. M., Parrish, J. T. and DeCelles, P. G. (1998). Topographic history of the Western Cordillera of North America and the controls on climate. In *Tectonic Boundary Conditions for Climate Reconstructions*, Oxford Monographs on Geology and Geophysics No. 39, eds. T. J. Crowley and K. Burke, pp. 73–99. New York: Oxford University Press.

Crowley, T. J. (1992). North Atlantic Deep Water cools the southern hemisphere. *Paleoceanography*, **7**, 489–97.

Crowley, T. J. (1998) Significance of tectonic boundary conditions for paleoclimate simulations. In *Tectonic Boundary Conditions for Climate Reconstructions*, eds. T. J. Crowley and K. Burke, pp. 3–17. New York: Oxford University Press.

Crowley, T. J. (1999). Paleomyths I have known. In *Modeling the Earth's Climate and Its Variability*, eds. W. R. Holland and S. Joussaume. North-Holland (in press).

Crowley, T. J. and Baum, S. K. (1991). Estimating Carboniferous sea level fluctuations from Gondwanan ice extent. *Geology*, **19**, 975–7.

Crowley, T. J. and Baum, S. K. (1992). Modeling late Paleozoic glaciation. *Geology*, **20**, 507–10.

Crowley, T. J. and Baum, S. K. (1993). Effect of decreased solar luminosity on late Precambrian ice extent. *Journal of Geophysical Research*, **98**, 16723–32.

Crowley, T. J. and Baum, S. K. (1994). General circulation model study of late Carboniferous interglacial climates. *Palaeoclimates: Data and Modelling*, **1**, 3–21.

Crowley, T. J. and Baum, S. K. (1995). Reconciling Late Ordovician (440 Ma) glaciation with very high (14×) CO_2 levels. *Journal of Geophysical Research*, **100**, 1093–101.

Crowley, T. J., Baum, S. K. and Hyde, W. T. (1991). Climate model comparison of Gondwanan and Laurentide glaciations. *Journal of Geophysical Research*, **96**, 9217–26.

Crowley, T. J., Hyde, W. T. and Short, D. A. (1989). Seasonal cycle variations on the supercontinent of Pangaea. *Geology*, **17**, 457–60.

Crowley, T. J. and Kim, K.-Y. (1995). Comparison of longterm greenhouse projections with the geologic record. *Geophysical Research Letters*, **22**, 933–6.

Crowley, T. J., Mengel, J. G. and Short, D. A. (1987). Gondwanaland's seasonal cycle. *Nature*, **329**, 803–7.

Crowley, T. J. and North, G. R. (1988). Abrupt climate change and extinction events in earth history. *Science*, **240**, 996–1002.

Crowley, T. J. and North, G. R. (1991). *Paleoclimatology*. New York: Oxford University Press.

Crowley, T. J., Short, D. A., Mengel, J. G. and North, G. R. (1986). Role of seasonality in the evolution of climate during the last 100 million years. *Science*, **231**, 579–84.

Crowley, T. J., Yip, K-Y. J. and Baum, S. K. (1994). Snowline instability in a general circulation model: Application to Carboniferous glaciation. *Climate Dynamics*, **10**, 363–76.

Dalziel, I. W. D. (1991). Pacific margins of Laurentia and East Antarctica–Australia as a conjugate rift pair: Evidence and implications for an Eocambrian supercontinent. *Geology*, **19**, 598–601.

Dickens, G. R., Castillo, M. M. and Walker, J. C. G. (1997). A blast of gas in the latest Paleocene: Simulating first-order effects of massive dissociation of oceanic methane hydrate. *Geology*, **25**, 259–62.

Douglas, R. G. and Woodruff, F. (1981). Deep sea benthic foraminifera. In *The Sea*, vol. 7, ed. C. Emiliani, pp. 1233–327. New York: Wiley-Interscience.

Dowsett, H., Barron, J. and Poore, R. (1996). Middle Pliocene sea surface temperatures: A global reconstruction. *Marine Micropaleontology*, **27**, 13–25.

Dutton, J. F. and Barron, E. J. (1997). Miocene to present vegetation changes: A possible piece of the Cenozoic cooling puzzle. *Geology*, **25**, 39–41.

Endal, A. S. and Sofia, S. (1981). Rotation in solar-type stars: I, Evolutionary models for the spin-down of the sun. *Astrophysics Journal*, **243**, 625–40.

Fawcett, P. J. and Barron, E. J. (1998). The role of geography and atmospheric CO_2 in longterm climate change: Results from model simulations for the Late Permian to the Present. In *Tectonic Boundary Conditions for Climate Reconstructions*, eds. T. J. Crowley and K. C. Burke, pp. 21–36. New York: Oxford University Press.

Fischer, A. G. (1982). Long-term climatic oscillations recorded in stratigraphy. In *Climate in Earth History*, eds. W. H. Berger and J. C. Crowell, pp. 97–104. Washington, DC: National Academy of Sciences.

Frakes, L. A. and Francis, J. E. (1988). A guide to Phanerozoic cold polar climates from high-latitude ice-rafting in the Cretaceous. *Nature*, **333**, 547–9.

Frakes, L. A., Francis, J. E. and Styktus, J. (1992). *Climate Modes of the Phanerozoic*. New York: Cambridge University Press.

Freeman, K. H. and Hayes, J. M. (1992). Fractionation of carbon isotopes by phytoplankton and estimates of ancient CO_2 levels. *Global Biogeochemical Cycles*, **6**, 185–98.

Gibbs, M. T., Barron, E. J. and Kump, L. R. (1997). An atmospheric pCO_2 threshold for glaciation in the Late Ordovician. *Geology*, **25**, 447–50.

Greenwood, D. R. and Wing, S. L. (1995). Eocene continental climates and latitudinal temperature gradients. *Geology*, **23**, 1044–8.

Hambrey, M. J. and Harland, W. B. (1985). The late Proterozoic glacial era. *Palaeogeography, Palaeoclimatology, Palaeoecology*, **51**, 255–72.

Hansen, J. E., Lacis, A., Rind, D. *et al.* (1984). Climate sensitivity: Analysis of feedback mechanisms. In *Climate Processes and Climate Sensitivity*, Geophysical Monograph No. 29, eds. J. E. Hansen and T. Takahashi, pp. 130–63. Washington, DC: American Geophysical Union.

Herbert, T. D. and Sarmiento, J. L. (1991). Ocean nutrient distribution and oxygenation: Limits of the formation of warm saline bottom water over the past 91 m.y. *Geology*, **19**, 702–5.

Hirst, A. C. and Godfrey, J. S. (1993). The role of Indonesian throughflow in a global ocean GCM. *Journal of Physical Oceanography*, **23**, 1057–86.

Hoffert, M. I. and Covey, C. (1992). Deriving global climate sensitivity from palaeoclimate reconstructions. *Nature*, **360**, 573–6.

Hyde, W. T., Crowley, T. J., Tarasov, L. and Peltier, W. R. (1999). Coupled climate ice sheet model simulations of the Pangean ice age. *Climate Dynamics* (in press).

Hyde, W. T., Kim, K.-Y., Crowley, T. J. and North, G. R. (1990). On the relation between polar continentality and climate: Studies with a nonlinear seasonal energy balance model. *Journal of Geophysical Research*, **95**, 18653–68.

Kaufman, A. J., Jacobsen, S. B. and Knoll, A. H. (1993). Vendian Sr and C isotopic variations in seawater: Implications for tectonics and paleoclimate. *Earth and Planetary Science Letters*, **20**, 409–30.

Kennett, J. P. and Stott, L. D. (1991). Abrupt deep-sea warming, palaeoceanographic changes and benthic extinctions at the end of the Palaeocene. *Nature*, **353**, 225–9.

Kiehl, J. T. and Dickinson, R. E. (1987). A study of the radiative effects of enhanced atmospheric CO_2 and CH_4 on early earth surface temperatures. *Journal of Geophysical Research*, **92**, 2991–8.

Knoll, A. H., Hayes, J. M., Kaufman, A. J., Swett, K. and Lambert, I. B. (1986). Secular variation in carbon isotope ratios from Upper Proterozoic successions of Svalbard and East Greenland. *Nature*, **321**, 832–8.

Kump, L. R., Arthur, M. A., Patzkowsky, M. E., Gibbs, M. T., Pinkus, D. S. and Sheehan, P. M. (1999). Glaciation at high atmospheric pCO_2 during the Late Ordovician (in preparation).

Kutzbach, J. E., Guetter, P. J., Ruddiman, W. F. and Prell, W. L. (1989). Sensitivity of climate to late Cenozoic uplift in southern Asia and the American west: Numerical experiments. *Journal of Geophysical Research*, **94**, 18393–7.

Maier-Reimer, E., Mikolajewicz, U. and Crowley, T. J. (1990). Ocean general circulation model sensitivity experiment with an open Central American isthmus. *Paleoceanography*, **5**, 349–66.

Manabe, S. and Stouffer, R. J. (1988). Two stable equilibria of a coupled ocean–atmosphere model. *Journal of Climate*, **1**, 841–66.

Manabe, S. and Wetherald, R. T. (1980). On the distribution of climate change resulting from an increase in CO_2 content of the atmosphere. *Journal of Atmospheric Science*, **37**, 99–118.

Mii, H.-S., Grossman, E. L. and Yancey, T. E. (1999). Carboniferous isotope stratigraphy of North America: Implications for Carboniferous paleoceanography and Mississippian glaciation. *Geological Society of America Bulletin* (in press).

Mikolajewicz, U. and Crowley, T. J. (1997). Response of a coupled ocean/energy balance model to restricted flow through the central American isthmus. *Paleoceanography*, **12**, 429–41.

Mikolajewicz, U., Maier-Reimer, E., Crowley, T. J. and Kim, K.-Y. (1993). Effect of Drake and Panamanian gateways on the circulation of an ocean model. *Paleoceanography*, **8**, 409–26.

Miller, K. G., Fairbanks, R. G. and Mountain, G. S. (1987). Tertiary oxygen isotope synthesis, sea level history, and continental margin erosion. *Paleoceanography*, **2**, 1–19.

North, G. R. and Crowley, T. J. (1985). Application of a seasonal climate model to Cenozoic glaciation. *Journal of the Geological Society (London)*, **142**, 475–82.

Oglesby, R. J. (1991). Joining Australia to Antarctica: GCM implications for the Cenozoic record of Antarctic glaciation. *Climate Dynamics*, **6**, 13–22.

Otto-Bliesner, B. L. (1996). Initiation of a continental ice sheet in a global climate model (GENESIS). *Journal of Geophysical Research*, **101**, 16909–20.

Otto-Bliesner, B. L. and Upchurch, G. R., Jr (1997). Vegetation-induced warming of high-latitude regions during the Late Cretaceous period. *Nature*, **385**, 804–7.

Parrish, J. T. and Spicer, R. A. (1988). Late Cretaceous terrestrial vegetation: A near-polar temperature curve. *Geology*, **16**, 22–5.

Raymo, M. E. (1991). Geochemical evidence supporting T. C. Chamberlin's theory of glaciation. *Geology*, **19**, 344–7.

Rea, D. K., Leinen, M. and Janecek, T. R. (1985). Geologic approach to the long-term history of atmospheric circulation. *Science*, **227**, 721–5.

Rind, D. and Chandler, M. (1991). Increased ocean heat transports and warmer climate. *Journal of Geophysical Research*, **96**, 7437–61.

Ruddiman, W. F. and Kutzbach, J. E. (1991). Plateau uplift and climatic change. *Scientific American*, **264**, 66–75.

Shackleton, N. J. (1985). Oceanic carbon isotope constraints on oxygen and carbon dioxide in the Cenozoic atmosphere. In *The Carbon Cycle and Atmospheric CO_2: Natural Variations Archean to Present*, Geophysical Monograph No 32, eds. E. T. Sundquist and W. S. Broecker, pp. 412–18. Washington, DC: American Geophysical Union.

Shackleton, N. J. and Kennett, J. P. (1975a). Paleotemperature history of the Cenozoic and the initiation of Antarctic glaciation: Oxygen and carbon isotope analysis in DSDP sites 277, 279, and 281. In *Initial Reports of the Deep-Sea Drilling Project*, No. 29, eds. J. P. Kennett *et al.*, pp. 743–55. Washington, DC: US Government Printing Office.

Shackleton, N. J. and Kennett, J. P. (1975b). Late Cenozoic oxygen and carbon isotope changes at DSDP site 284: Implications for glacial history of the northern hemisphere and Antarctica. In *Initial Reports of the Deep-Sea Drilling Project*, No. 29, eds. J. P. Kennett *et al.*, pp. 801–7. Washington, DC: U.S. Government Printing Office.

Sloan, L. C. (1994). Equable climates during the early Eocene: Significance of regional paleogeography for North American climate. *Geology*, **22**, 881–4.

Sloan, L. C. and Barron, E. J. (1990). "Equable" climates during earth history. *Geology*, **18**, 489–92.

Sloan, L. C., Crowley, T. J. and Pollard, D. (1996). Modeling the Middle Pliocene climate with the NCAR GENESIS general circulation model. *Marine Micropaleontology*, **27**, 51–61.

Sloan, L. C. and Rea, D. K. (1996). Atmospheric carbon dioxide and early Eocene climate: A general circulation modeling sensitivity study. *Palaeogeography, Palaeoclimatology, Palaeoecology*, **119**, 275–92.

Sloan, L. C., Walker, J. C. G. and Moore, T. C., Jr. (1995). Possible role of oceanic heat transport in early Eocene climate. *Paleoceanography*, **10**, 347–56.

Tarasov, L. and Peltier, W. R. (1997). Terminating the 100 kyr ice age cycle. *Journal of Geophysical Research*, **102**, 21665–93.

Thompson, S. L. and Pollard, D. (1995). A global climate model (GENESIS) with a land-surface-transfer scheme (LSX). 1, Present-day climate. *Journal of Climate*, **8**, 732–61.

Wing, S. L. and Greenwood, D. R. (1993). Fossils and fossil climate: The case for equable continental interiors in the Eocene. *Philosophical Transactions of the Royal Society of London Series B*, **341**, 243–52.

Wolfe, J. A. (1978). A paleobotanical interpretation of Tertiary climates in the Northern Hemisphere. *American Scientist*, **66**, 694–703.

Wright, J. D. and Miller, K. G. (1996). Control of North Atlantic Deep Water circulation by the Greenland–Scotland Ridge. *Paleoceanography*, **11**, 157–70.

Yapp, C. J. and Poths, H. (1992). Ancient atmospheric CO_2 pressures inferred from natural goethites. *Nature*, **355**, 342–4.

Zachos, J. C., Breza, J. R. and Wise, S. W. (1992). Early Oligocene ice-sheet expansion on Antarctica: Stable isotope and sedimentological evidence from Kerguelen Plateau, southern Indian Ocean. *Geology*, **20**, 569–73.

Zachos, J. C., Sloan, L. C., Crowley, T. J. and Hyde, W. T. (1997). The late Paleocene thermal maximum: An evaluation of the evidence for greenhouse gas forcing. *Eos (Transactions of the American Geophysical Union, Supplement to Fall Meeting)*, **78**, F363 (Abstract).

Index

AABW *see* Antarctic Bottom Water
Abathomphalus mayaroensis 248, 254–255,
    256, 258, 260, 267
abscission of leaves 334, 335, 341
abyssal 138, 142, 147, 245
Acadian Mountains 358
Acarinina 98, 107, 146
Acarinina mckannai 146, 149
Acarinina nitida 98
Acarinina primitiva 98
acritarchs 360
adiabatic transfer 287
aeolianites 38
Africa 13, 36, 86–87, 283, 285–286, 309, 342,
    343, 389, 400, 412
    African/South American continent 309,
        311, 312, 313, 314
    East 13, 86–87, 283, 286, 389, 400, 412,
        429
    Egypt 89, 389
    Goban Spur 111, 247
    South 87, 312, 342, 343
AGCM *see* models, atmospheric
Alaska 179, 198, 323, 333–334
albedo 6, 7, 162, 170, 283–284, 287
    ice 10, 12, 13, 85
    ice–albedo feedback 16, 31, 393,
        408–409, 413
    land surface 7, 32, 291, 408
    planetary 5, 8, 12, 16, 430
    snow 12, 13, 85, 170, 283, 287, 438
    vegetation 13–14, 32, 162, 170, 190,
        283–284, 343, 394, 411, 436
    high altitude 13, 34
    high latitude 16, 162, 190
Alexander Island, Antarctica 335

alkalinity 35, 50, 69, 263
Allan Hills, Antarctica 328
Alnus 226–227
Altiplano 163
altitude 34, 88, 163, 170, 175–182, 187, 189,
    323
    effect on growth rings 333
    of Laramide highlands and plains
        181–182
    relation of temperature to 312
Amazon Basin 169, 177, 389
ammonoid diversification 361
Andes 163, 169, 171, 173
angiosperms 283, 298, 314, 322, 334, 336
Angola Basin 289
anoxia
    benthic foraminifer response 138
    causes 364–366, 436
    Devonian oceans 354, 360, 363,
        372–376
    Eocene lakes 183
Antarctic Bottom Water 87, 139
Antarctic ice cap 9
Antarctic Peninsula 283, 285, 311
Antarctica 84, 87, 88, 102, 133, 143, 149, 169,
    267, 311–312, 393
    ice sheets 136, 141, 169, 279, 307, 310,
        389, 403, 411, 433, 435, 437
    tundra 283–285
    vegetation 287, 291, 306–307, 321–344
Antarcticycas 341–342
Antelope Creek 199, 202, 203, 207–210,
    217–219, 221, 224, 225, 228
AOGCM *see* models, atmosphere–ocean
Appalachian Basin 353–355, 356, 358, 362,
    377

arachnids 361
Araucaria 285
Araucarioxylon 339
Archaeoglobigerina spp. 258
Arctic 37, 54, 84, 122, 167, 323, 325, 335–337
Arctic Ocean 83, 84, 93, 122
Arctic willow *see Salix*
aridity *see* precipitation, aridity
arthropleurids 361
arthropod 338, 361
ash *see* volcanism, ash
Ashgill 389, 401
Asia 87, 167, 169, 172, 198, 212, 223,
    224–225, 283, 285–287, 309
asynchronous model *see* models,
    asynchronous
Atlantic Ocean 58–59, 86, 87, 99, 103, 105,
    116–119, 121–123, 139, 165, 249, 251,
    260, 262–263, 269, 435
        North Atlantic 16, 53–55, 57, 86–88,
            98, 105, 107, 110–111, 114, 123, 134,
            147, 165, 225, 242, 289, 434–435
        South Atlantic 86–87, 94, 107, 143, 243,
            264, 267, 289
Atmosphere–Ocean General Circulation
    Model *see* models, atmosphere–ocean
atmospheric circulation 7, 13, 14, 27, 38, 80,
    88, 162–164, 167, 168, 170, 171, 175, 190,
    277, 408
        Hadley and Ferrel cells 5
        mid-latitude westerlies 166–169, 403
        monsoonal circulation 167, 279
Atmospheric General Circulation Model *see*
    models, atmospheric
atmospheric humidity 133, 136, 170, 171,
    175, 180, 211
atmospheric model *see* models, atmospheric
austral 248, 261, 262, 403, 405, 407, 411
Australia 334, 341
Axel Heiberg Island, Canadian Arctic 336

bacterial loop 142
Bainmedart Coal Measures 339
Baja California *see* North America
Baltica 401, 411
baroclinicity 165
basalt 101, 147, 173, 176
basidiomycetous fungi 338
bathyal 138, 143, 148, 245, 248, 262, 264
        depths 139, 142, 147, 152, 260, 262, 269
        paleodepths 241, 244, 245, 269

bathymetry 25, 30, 37, 82, 88, 288, 289, 390
Bear Creek Mine 215–216
Beardmore Glacier 324, 329, 339
benthic 245, 364–366, 375
        faunas 355, 360–362, 364, 366, 368, 376
        oxygen 353, 368, 432
benthic foraminifera 39, 40, 56, 79, 94, 95, 98,
    99, 101, 106, 108, 111, 116, 135–153, 198,
    247, 253, 255–270
        Alabaminella weddellensis 138
        Bolivina 138
        Bolivinoides 139
        Brizalina 139
        Bulimina 138, 139, 145, 146
        Bulimina ovula 144, 146
        Bulimina semicostata 144
        Cassidulina 138
        Cibicidoides wuellerstorfi 138
        Coryphostoma 139
        Epistominella exigua 138
        extinction of 142–146, 149–152
        Gavelinella beccariiformis 144
        Lenticulina 136, 144, 146, 253
        Melonis 138
        Nuttallides truempyi 136, 139, 141,
            144–146, 150
        Nuttallides umbonifera 138, 139, 141
        O isotopic values 56, 102, 432, 433
        Pleurostomella 139
        Pullenia 138
        Stilostomella 139
        Tappanina 139
        uniserial lagenids 139
        Uvigerina 139
benthic–pelagic coupling 138
bicarbonate 97, 98
Bighorn Basin 188, 199–231
biogeochemical cycles 32, 38, 40, 351, 355,
    361, 366–369, 371, 374–377
biogeochemistry 21, 35–36, 354
biomes 33, 40, 279, 283, 285, 287, 298, 301,
    307–316, 335, 342–343
biosphere 24, 32, 277, 282, 351
bivalves *see also* invertebrates, fossil
        isotopic analysis 57, 188
black shales 137, 359–360, 363–366, 377, 391,
    397
Blake Nose 247
boreal forest 16, 285–287, 342–343
Boreal Ocean 396
boreal tundra 284

bottom water 64, 65, 81, 90, 99, 103, 105, 111, 114, 115, 122–123, 137, 139, 143, 243, 262, 263, 264, 375, 397, 435
anoxia 183, 364, 366, 368, 373
brachiopod 353, 355, 360–364, 366, 376, 391, 412
Brachyphyllum 302
Bridgerian 228
broadleaf evergreen trees 211, 226, 282, 286–287, 302, 308, 333–336, 342–343
bryophyte 284
Buckley Formation 326, 327
bulk carbonate 39, 65–68, 183
Bunophorus 204–206, 208, 220, 224

CA *see* Correspondence Analysis
calcite 35, 39, 57, 60, 64–67, 93–94, 111, 113, 184, 213–214
Calcite Compensation Depth 38
Cambrian 353, 389, 431
Campanian 244–246, 248–251, 255, 256, 258, 260–269, 276, 278–291
Canadaway Group 359
CANOCO *see* Canonical Community Ordination
Canonical Community Ordination 301
Caravaca 245, 246, 249–251, 254, 261–262
carbon-13 (δ^{13}C) *see* carbon isotopes
carbon cycle 281, 353, 363, 396, 414
global 36, 135
perturbations 388, 391, 396, 413
carbon dioxide 4, 34–36, 54, 55, 147, 242, 353
oceanic ridge sources 10
absorption during weathering 9, 35, 171, 392, 411, 426, 438
atmospheric 9–12, 34–36, 51–52, 54, 134, 162, 170, 171, 242, 245, 276–281, 283–286, 291, 363, 386–414
metabolic 60, 367–368, 372, 375–376
pCO$_2$ 63, 242, 245, 388, 392, 402–407
Phanerozoic changes in atmospheric concentration 70, 425–439
carbon isotopes 51, 61, 67, 97–98, 111, 134–136, 143–149, 151, 185–186, 245, 246–247, 251, 255, 262–263, 267, 281, 353, 357, 366–367, 371–372, 376, 389, 392, 395, 412, 426
carbonate 65, 80, 82, 93, 97, 105, 113, 175, 181, 184, 188, 222, 366, 388–392, 412, 413

excursions 67–68, 134–135, 144–146, 199–210, 215, 220, 223, 225, 228, 230, 251, 255, 262–264, 389, 395, 397, 412, 437
organic 97, 262, 371–372, 390–392
20-carbon isoprenoid 367
19-carbon normal alkane (*n*-C$_{19}$) 367
carbonate 38, 39, 54, 65–68, 80, 93–94, 98, 105, 113, 134, 135, 148, 175, 181, 186, 188, 222, 223, 281, 354, 358, 366, 369, 392
carbonate–silicate cycle 171, 392, 396, 411
Carboniferous 167, 389, 392, 400, 407–410, 429
Catskill Delta 354–359
CCD *see* Calcite Compensation Depth
Cenomanian 54, 56, 245, 251, 258, 262, 268, 432, 433
centipedes 361
Central Pangea Mountains *see* Pangea, Central Pangea Mountains
centroid, floral elements 302, 304
Cerling's method 361
CH$_4$ *see* methane
chaos theory 54, 56, 245, 251, 258, 262, 268, 432, 433
Chiloguembelina 361
China 54, 56, 245, 251, 258, 262, 268, 432, 433
chitinozoans 361
Cibicidoides 54, 56, 245, 251, 258, 262, 268, 432, 433
Cladophlebis 361
CLAMP *see* Climate Leaf Analysis Multivariate Program
Clarkforkian 200, 206, 209, 215, 217–218, 220, 222–230
Cf1 206, 215, 217
Cf3 206, 217, 226
Clarks Fork Basin 199–209, 215, 217–221
climate drift 30
Climate Leaf Analysis Multivariate Program 212, 298
climate model *see* models
climate proxies 172, 276, 299–300
geochemical 354–355, 366, 368
isotope 51, 213, 230, 283, 412
plants 179, 299–300, 322
climate regimes 3, 5, 6, 13, 300, 301, 310, 336
climate system 3–17, 23–28, 31, 32, 35, 37–40, 70, 80, 277, 387, 396, 408, 437–438

Climate System Model *see* models, Climate System
clouds 3, 5, 13, 34, 176–177, 187
    cover 5–6, 12–13, 16, 32, 34, 179–180, 190, 343
    models 13, 15, 32, 34, 51, 70, 291, 311
    noctilucent 34
    stratospheric 11, 16, 34–35
Cnemidaria 226-227
CO_2 *see* carbon dioxide
coal 38, 39, 200, 298, 299, 305–310, 339, 343, 356
cold summer orbit 399, 400, 407, 409, 410, 429
Colorado see North America
compression floras 228, 299, 333, 339–342
Conewango Group 359
Coniacian 57, 250, 260, 262
conifers 287, 302–308, 333, 334, 340
Coniopteris 303
conodonts 418
continental drift *see* paleogeography
continental runoff 30, 81, 83, 90, 93, 102, 107, 114, 122–123
Contusotruncana contusa 254
Contusotruncana spp. 258
convection 12, 34, 81, 83, 87, 90, 93, 132, 149, 289, 291, 311, 396 *see also* deep convection
cores *see* deep sea drilling cores; land-based cores
Coriolis force 163, 164
Cornell Ecology Program DECORANA 301
Cornus 226
correspondence analysis 226, 300–303, 306, 307
corrosivity 142, 148–149
corystosperms 340
Cretaceous 4, 8, 9, 10, 12, 17, 25, 28, 32, 39, 51, 54, 56, 63, 70, 137, 144, 150, 170, 217, 243–270, 275–292, 298, 309, 314–315, 335–336, 393–394, 411, 430, 432, 433, 436, 437
crocodiles 277, 285
crustal degassing 35
cryosphere 8, 22, 24, 40, 277
CSM *see* models, Climate System
CSO *see* cold summer orbit
Ctenis 302
currents 9, 13, 16, 27, 53, 57, 138–139, 143, 215, 288–289, 411
    Alaska Gyre 87
    Antarctic Circumpolar Current 86, 87, 289
    Brazil Current 86
    California Current 86, 87
    Canary Current 86
    Falkland Current 86
    Kuroshio Current 9, 86, 87
    Labrador Current 86, 87
    North Atlantic Drift 86
    North Pacific Current 86, 87
    Northern Equatorial Current 86, 87
    Oyashio Current 86, 87
    Peru Current 86, 87
    Southern Equatorial Current 87
    Tethys Current 88
cuticle 33, 299, 302, 322
cycadophytes 300, 302, 303, 306, 308
Cyperaceae 282
Czekanowskia 306
Czekanowskiales 298, 300

DAB *see* Devonian Appalachian Basin
deciduous 33, 83, 211, 226, 286, 287, 297, 302, 308, 322, 325, 333–334, 335–336, 342–343
deep convection 81, 90, 93, 123, 149, 275, 396
deep sea drilling cores
    Site 217 246
    Site 327 245, 249, 251, 254–255, 258, 260, 262–267
    Site 364 119, 123
    Site 390 245–249, 251, 254–255, 258, 260, 262–267, 269
    Site 462 246
    Site 511 245–246, 249, 251, 258, 261–262, 264–265
    Site 549 111, 113, 206
    Site 550 79, 98, 101, 105, 107, 110–111, 113, 119, 122–123, 201, 203, 206
    Site 690 96–97, 103, 106, 135, 140, 144–149, 151, 203, 248, 258, 260, 266–267, 269
    Site 738 103, 246
    Site 750 262–265
    Site 761 262–265
    Site 865 67–68, 103, 111, 113, 119, 123, 135, 140, 142–145, 148–149
defoliation 323, 326
degree of pyritization 351, 366, 368, 370–371, 373

dendroclimatology 323
Denver Basin 182
desert 278, 283, 285, 297, 309, 310, 313
    albedo 13
    of Africa 286
Desmiophyllum 302
detritivores 361
Devonian 32, 351, 353–356, 358–364, 367,
    369, 377, 411, 431 *see also* Devonian
    Appalachian Basin
Devonian Appalachian Basin 355–356,
    358–361, 363–364, 377
    black shales 366
diagenesis 50, 63–70, 111, 113, 181, 184, 247,
    266–267, 366–368, 373
DIC *see* dissolved inorganic carbon
Dicroidium 332, 333, 334, 335, 340
Dictyophyllum 298, 300
dimethyl sulfide 35
dissolved inorganic carbon 366–367, 375
diversity 57, 199, 225, 247–248, 260, 267, 321,
    351, 353, 355, 358, 360–364, 368, 376
        benthic foraminifera 142
        floras 215, 226, 228, 230, 297, 307, 310,
            321, 339–341
        Information Function 247
        inoceramid 245, 261
        species 135, 198
DMS *see* dimethyl sulfide
DOP *see* degree of pyritization
dropstones 327
drought 312, 322–323, 326, 334 *see also*
    precipitation, aridity
drumlins 389
Dunkirk Formation 363

E see equations, equitability
earlywood 321–323, 326, 331, 333, 336, 337
EBM *see* models, energy balance
eccentricity 8, 280, 399
Ectocion 208, 224
eddy 190, 290
        kinetic energy 165
        scale 27–28
Egypt 89, 389
Eifelian 357, 359, 361
Elatocladus 302
Elk Creek 199, 202, 203, 207–210, 217–219,
    221, 224, 225, 228
Ellesmere Island, Canadian Arctic 336
emissivity 5–7, 10–11, 16

energy balance models *see* models, energy
    balance
enthalpy 175–177, 182
Entrada Formation 309
Eocene 4, 8–9, 11–12, 25, 28, 32, 34, 50–51,
    54, 56–57, 65, 83, 94–96, 111, 114, 118,
    123, 132–133, 135, 150, 152
    deep-water formation 93
    faunas 139, 141
    foraminifera 93–94, 97–99, 107, 111,
        113, 132, 144, 149–151
    ocean circulation 79–80, 82–84, 86,
        89–90, 142
    Pacific Ocean 87
    paleoclimate 82, 84, 86, 132
    paleogeography 79, 81–82, 88
    paleotemperatures 102–103, 105, 107,
        111, 121, 123
    volcanism and seafloor spreading 102
Eocene Cordillera 181
eolian sediments 38
equations
    diffusion 246
    equitability 247
    flux correction 30
    Shannon diversity 247
Equilibrium Vegetation Ecology Model *see*
    models, Equilibrium Vegetation Ecology
Equis etites 303
equitability *see* equations, equitability
error estimate, SST 62
'*Eucommia*' serrata 226
Eurasia 9, 309–310, 312, 314
Europe 36–37, 122, 165, 172, 177, 198, 223,
    225, 309 *see also* Scotland; Spain
eustacy *see* sea level
eutrophic fauna 138, 144
evaporation 11, 39, 54, 188, 275, 289, 297,
    305, 312–313
        $^{16}O/^{18}O$ ratios 102, 116–117, 176, 181
        mountains 171
evaporites 10, 38–39, 54, 297–298, 305–306,
    310, 356, 358
EVE *see* models, Equilibrium Vegetation
    Ecology
evergreen 322, 325, 334–335
extinction
    conodont 400
    inoceramid 241, 245–246, 248–249, 251,
        255, 258, 260, 269

extinction (*continued*)
  Late Devonian *see* Famennian/
    Frasnian extinction
  mass 132, 353, 355, 386–387, 391
  plants 226
  benthic foraminifera 142–146, 149–152
extraterrestrial impact 354

F/F extinction *see* Frasnian/Famennian
    extinction
false rings 326
Famennian 357–359, 361–364
faunal belts, contraction 389
faunal stasis 361–362, 364, 369
faunal tracking 360, 362
feedback model 364, 366
ferns 339, 341, 343
'*Ficus*' *artocarpoides* 226
fire damage 338
first law of thermodynamics 6
flood basalts 147
floral gradients 300
floras, fossil 321, 325, 335, 338–343
Florissant lake beds 182, 191
flux correction 30
foliar physiognomy *see* leaf physiognomy
forb–fern, prairie 283, 285
forbs 282–283
forcing mechanisms 3–17, 284
forest
  boreal 337
  density 326–335
  fossil 325–327, 333–336, 339
  productivity 325
Fort Union Formation 200, 207, 214–215,
    217
Foster Gulch 199, 203–204, 207–209, 215,
    218, 224
Fourier 23
Frasnian 258, 260, 325, 361–362, 369, 376
Frasnian/Famennian extinction 353, 355,
    358, 361–362
freeze line 161, 407
Fremouw Formation 333
Fremouw Peak 329–333, 340–342
Frenelopsis 298
freshwater flux 114, 289
Front Range 180, 182
frost rings 331–333
fungal interaction 338
fungi 328, 329, 338–341, 343, 400
fusain 338

Gansserina gansseri Zone 248, 255
Gansserina weidenmayeri 258
gas hydrates 11 *see also* methane, clathrate
    release
Gaussian resolutions 24
GCM *see* models, General Circulation
Gebel Duwi 89
Genesee Group 359, 369, 371–372
Genesee River Valley 355, 369–370
Geneseo Formation 359, 363–364, 369
Geneseo Shale 355, 361–362, 369, 373, 376
GENESIS *see* models, GENESIS
Geochemical Ocean Sections Study 102–103,
    116, 118
Geomagnetic Polarity Time Scale 201, 203
GEOSECS *see* Geochemical Ocean Sections
    Study
ginkgophytes 297, 302–304, 306, 308
Givetian 351, 353, 355, 358–364, 369, 376
glacial inception 8, 386, 388, 400, 402, 413,
    430, 434
glacial sediments 38
glaciation 57, 358, 391, 425–427, 434
  Carboniferous 431
  Early Cambrian 389, 431
  late Cenozoic 392
  Permian 391, 429
  pre-Cambrian 431
  Ordovician 386–395, 397–398, 400,
    402–403, 407–414, 431, 438
glacio-eustatic 391
glacio-eustatic control 358
glendonites 38
Global Environmental Simulation Interactive
    System *see* models, GENESIS
Globigerina bulloides 62
Globigerinelloides 258
Globigerinelloides subcarinatus 267
Globigerinoides sacculifer 60, 104
Globotruncana falsostuarti Zone 248
Globotruncana spp. 258
Globotruncanita spp. 258
Glossopteridales 339
glossopterids 339, 341, 343
Glossopteris 324, 326, 327, 334, 335, 339
GMPTS *see* Geomagnetic Polarity Time
    Scale
Goban Spur 111, 247
goethite 214, 392
Gondwana 169, 311–314, 343, 406–407, 430,
    436

elevation 411
    floras 304, 306, 310, 334, 339, 342, 343, 411
    polar positioning 386, 389, 395–398, 413, 437
    glaciation 358, 386–389, 393, 401–403, 412, 429
Gondwanaland 358, 429, 431 *see also* Gondwana
goniatites 351, 355, 362, 364, 376
Gordon Valley, Antarctica 333
Graminaceae 282
graminoids 282
grassland 13, 170, 278, 282
gravity-wave parameterization 279
green algae 360
Green River Formation 161, 181–185, 188–189, 191
greenhouse gas 68, 85, 148, 162, 171, 242, 276, 351 *see also* carbon dioxide; methane
growth
    bivalves 188
    foraminifera 107, 141
    ice sheet 398, 429
    metabolic effect 60
    of Tibet 9
    photosynthetic organisms 367, 372, 375–376
    wood 183, 321–327, 331, 333–339, 342–343
Gulf Stream 9, 86
gymnosperms 333, 336, 341

Hadley and Ferrel cells 5
halothermal ocean *see* ocean circulation, halothermal
Hamilton Group 359–377
Haplomylus 208, 224
Haplomylus–Ectocion Zone 208
heat capacity
    land 4, 5
    ocean 12, 434
    sea-ice 12, 434
heat transport
    atmospheric 6, 25, 162
    latitudinal 5, 6, 27, 51, 53, 107, 134, 148, 190, 277, 283–286, 353, 436
    models 12, 14–16, 28, 30, 84, 278, 281–283, 288–291, 394, 398–401, 410–411, 433

oceanic 25, 80–82, 85, 115, 123, 242–243, 245, 281, 435, 438
hematite, isotopic composition 199, 214, 218–219, 221–223, 229
Heptodon 204, 205, 209, 224
Heterohelix globulosa 252, 255, 256, 258, 260, 264–265
H(S) *see* diversity, Information Function
H_2S *see* hydrogen sulfide
hydrological cycle 30, 39, 289
hyperthermal 150–152

Iapetus
    closure 401
    warm, saline seas 396
IBIS *see* Integrated Biosphere Simulator
ice sheets *see also* models, ice sheets
    albedo 10
    Cretaceous (lack of) 286
    Devonian (lack of) 353
    effect on productivity 141
    effect on zonal circulation 169
    Eocene 134
    Ordovician 387–412
in situ fossil forest *see* forest, fossil
Indian Ocean
    Cretaceous ocean circulation 289
    Eocene ocean circulation 87–88
Information Function 247
infrared emissivity *see* emissivity
inoceramids
    Cretaceous paleobiogeography 244–246, 248–251, 260–262
    extinction 251, 260
Inoceramus see inoceramids
insolation, variations of 8, 22, 53, 276–277, 287, 408, 429
Integrated Biosphere Simulator 33, 40
intermontane basins, estimating elevation of 174–191, 279
Intertropical Convergence Zone
    Eocene 85, 169
    Jurassic 309
    precipitation in 55
invertebrates, fossil
    brachiopods 353, 355, 360–364, 366, 376, 391, 412
    goniatites 351, 355, 362, 364, 376
    inoceramid bivalves 244–246, 248–251, 260–262
    Inoceramus 241, 245–246, 269

invertebrates, fossil (*continued*)
    nuculoid bivalves 360
    ostracodes 142
    rudist bivalves 57, 262, 266
    stromatoporoids 363
    styliolinids 360
isotopic composition *see also* carbon
    isotopes; oxygen isotopes
    glacial ice 39
    global carbon reservoir 134
    precipitation 39, 176, 189
    seawater 54, 93–94, 101–104, 113,
       116–124
    surface water 187, 213
ITCZ *see* Intertropical Convergence Zone

Java Group 359
jet stream, effects of mountain belt 163–165,
    170, 190–191
Jurassic climate and vegetation 169, 297–316

Kelvin–Helmholtz instabilities 7
Kerguelen Plateau 278
Kimmeridgian
    General Circulation Models 314
    paleogeography 311
Kykloxylon 334–335

Lake Flagstaff 184
Lake Gosiute 183, 189
Lambdotherium 204, 209, 224
Land Surface Transfer Scheme 32, 83,
    277–278, 398
land-based cores
    Energy Research and Development
      Administration (ERDA) White
      Mountain No. 1 183–186
    Union Pacific Railroad Company Blue
      Rim No. 44-19 183–184, 187
    United States Department of Energy
      (DOE) Currant Creek Ridge No. 1
      183–185
Laney Member 184–185
lapse rate 14–15, 174–181, 189–191
Laramide mountains 161–191
larch *see Larix sibirica*
Larix sibirica 336
Lashly Formation 328
Last Glacial Maximum 17, 37
Late Paleocene Thermal Maximum 70, 80,
    133, 437

latent heat 32, 33, 286–289, 291, 353
latewood 322–327, 331–333, 336–337
latitudinal gradients
    benthic 269
    sea surface temperatures *see*
      temperature, latitudinal gradients
Laurentia 401, 411
Law Glacier 326
leaf margin analysis 211–217, 220, 228–229
leaf physiognomy 266, 298, 314, 322, 335
Leiosphaeridia 360
LGM *see* Last Glacial Maximum
light regime 323, 341
    effect on tree ring growth 323, 341
LMA *see* leaf margin analysis
logs, fossil 326, 328–331
long-wave radiation 4, 6, 11, 13, 34, 280
Lostcabinian land mammal assemblage zone
    209, 220, 225, 228, 229
LPTM *see* Late Paleocene Thermal
    Maximum
LSX *see* Land Surface Transfer Scheme
lycophytes 302
Lygodium 226–227

Maastrichtian
    age of base 248
    inoceramid extinction 245, 249, 251,
      254, 261
    paleosols 281
    paleotemperature estimates of 56, 57,
      70, 243, 261, 265–269
    planktic foraminifer depth habitats 97
    planktic foraminifer population
      dynamics 255–260, 266–267
    stable isotopes 255, 263
    vertical isotopic gradients 265
Machaerium 228
macrophyllous 302–304, 306, 308
magnetochron
    C24n, correlation of 207
    C24n, faunas 224
    C24n, floras 225
    C24n, land-based sections 204, 208
    C24n, temperature estimates 220, 225,
      228
    C24r, correlation of 201–208
    C24r, land-based sections 210
    C24r, temperature estimates 220
    C25n, correlation of 201, 207
    C25n, faunas 223

C25n, land-based sections 206
C31n 248
C31r 248
C32n 248
MAT *see* mean annual temperature
McCullough Peaks 199, 202–204, 208–209, 224
McIntyre Promontory 328
mean annual temperature
　Campanian model simulations 285
　Eocene model simulations 84, 89–91, 107, 110, 115, 116
　Eocene paleobotanical estimates of 173–189, 198, 211–213, 215, 219–223, 228–229
　oxygen isotopic estimates of 69
Mediterranean 242, 243
Meeteetse Formation 217
Melez–Chograne Formation 389
Merceria 339
meridional thermal gradients *see* temperature, latitudinal gradient
meristem 322, 323
methane
　clathrate release 82, 135, 148, 149, 199
　estimates for variation of 11
　high latitude concentrations 34
　terrestrial sources 32
Micropaleontological Reference Center 245–246
microphyllous 302–304, 306, 308
mid-latitude westerlies 167
migration
　response to climate change 199, 223–228, 266–267, 276, 300, 309
　response to sea level change 360
Milankovitch cyclicity 134, 150, 201, 400, 410
Miocene
　Antarctic ice sheet growth 435, 437
　benthic foraminifer assemblages 139
　elevation estimates of lake sediments 175
　NADW increase 434–435
　temperature estimates 54–57, 70, 139
　vegetation changes 436
Mississippi River drainage 187
Mississippi, faunas from 223
mixed layer
　ocean models 25, 26, 53, 82–83, 87, 106–107

planktic foraminifer stable isotopes 54, 60, 61, 64–68, 103–124
planktic foraminifera 98–100
models
　asynchronous 28–29, 288
　atmosphere–ocean 28, 31
　atmospheric 22–23, 25–26, 28
　BLAG 392
　Climate System 31
　Community Climate 407
　deep ocean circulation 266
　energy balance 22, 28, 393, 395, 398, 408, 413
　Equilibrium Vegetation Ecology 33, 40, 278, 281–288
　General Circulation 6, 8, 11–16, 22–39, 191 *see also* AGCM, AOGCM, GENESIS
　GENESIS 23–26, 32, 39, 82–84, 107, 114, 123, 191, 242–243, 277–290, 393–394, 398–410, 433 *see also* models, Equilibrium Vegetation Ecology
　GEOCARB II 52, 281, 392, 395, 411
　hydrostatic primitive ocean 288
　ice-sheet 31–32, 431
　moisture balance 28
　Ocean General Circulation 22, 27–31, 36–40, 285–290
　Parallel Ocean Climate 82–84, 86–93, 117, 121–122
　periodically synchronous 28–29
　Proxy Formation 38–39
　slab ocean 25–28, 37, 83, 276–277, 281–282, 285, 398
　solar evolution 7, 280, 431
　solar physics 7, 387
　spectral General Circulation 23–24, 31
　steady-state 84, 396–397
　Universities Global Atmospheric Modelling Programme 409
Mongolia 167, 277, 285–286
monsoonal circulation 9, 163–164, 167–170, 231, 279, 286, 312–313 *see also* atmospheric circulation
Montana, fossil plants 189–190, 215
Morisson Formation 309
Morozovella aragonensis 98–99
Morozovella subbotinae 99
Morozovella velascoensis 98
morphocats 304, 306, 308

Moscow Formation 369
mosses 339, 340, 343
Mount Achernar, Antarctica 324, 326–327,
    333
Mount Falla, Antarctica 333
Mount St Helens 326
mountains, effects on climate 161–172,
    190–191, 358
Muricoglobigerina 98-99

n-C$_{19}$ *see* 19-carbon normal alkane
NADW *see* North Atlantic Deep Water
nahcolite 183
nannofossils 60
National Center for Atmospheric Research
    31, 82
Navajo Formation 390
NCAR *see* National Center for Atmospheric
    Research
nearest living relative 301, 322, 334
Neda Formation 392, 395
neritic 142, 245
New York 351, 353–357, 359–364, 369–370
Newton's laws of motion 6
Nilssonia 302
Nilssoniocladus 334, 342
nitrous oxide 35, 280
NLR *see* nearest living relative
N$_2$O *see* nitrous oxide
Noeggerathiopsis 339
Normalograptus extraordinarius Biozone 391
North America
    Acadian Orogeny 359
    annual temperature for snow stations
        179
    Givetian–Famennian boundary 353
    mammal immigration 199, 223–224
    paleoclimate 80, 161, 163–165, 169,
        181, 190–191, 198, 222, 229–230,
        277, 283, 285, 287, 297, 309, 312,
        358, 434
    paleoflora 172, 212, 225, 227
    red beds 208
    seasonal upwelling 89
    Baja California 223
    Colorado 180–182, 184, 191
    Mississippi 187, 223
    Montana 189, 215
    USA 36–37, 188, 228, 266, 326
    Wyoming 181–184, 189–191, 197–199,
        215, 228

North Atlantic
    Eocene model-predicted temperatures
        106–115
    Eocene ocean circulation 86–88
    Eocene oxygen isotope values 98, 105
    Eocene seawater isotopic composition
        121
    Eocene volcanism 134, 147
    latitudinal salinity pattern 55
    Mediterranean water mass 242
    sea surface temperatures 57
    thermohaline circulation 16, 53,
        434–435
North Atlantic Deep Water 55
North Dakota
    Paleocene–Eocene warming 198
    paleofloras 189, 228
northern hemisphere
    atmospheric circulation 163–165, 168
    Campanian paleogeography 276, 279
    Eocene vegetation 212, 225
    Jurassic floral gradient 305
    Jurassic vegetation localities 300–305,
        310, 314
    mean annual temperatures 212
    modeled ocean circulation 87
    modern vegetation 83
    mountain distribution in 168, 170
    Ordovician ocean circulation 404
    Ordovician paleogeography 403
    Ordovician seasonality 403
    seawater δ^{18}O 103, 110, 123
    Tertiary vegetation 335
NP9/NP10 boundary 201–203, 208
nutrients
    Devonian ocean 361, 366, 369, 375, 376
    Eocene ocean 136, 137, 142, 150
    modern ocean 36
    Ordovician ocean 413
Nuttalides see benthic foraminifera

O$_3$ *see* ozone
obliquity
    Cretaceous 280, 287
    Eocene 8
ocean circulation
    boundary currents during warm
        climates 53, 57, 289
    causes for changes 434–436
    deep-ocean reversal 152, 242, 269
    effect on climate 243

effects of paleogeography 9
Eocene 79–124
halothermal 242–243, 262, 264, 269, 397
inferred from biogeography 38
inferred from stable isotopes 264, 266
model *see* models, Ocean General Circulation
model predictions 25–31, 282–286, 290, 291, 405, 411
Ordovician 396
Pangean boundary currents 9
sluggish 137, 142
thermohaline 16, 53, 434–435
Ocean General Circulation Model *see* models, Ocean General Circulation
oceanic heat transport
Cretaceous 242, 282, 291
Eocene 81–82, 115, 123, 162
General Circulation Models 25, 242–243
OGCM *see* models, Ocean General Circulation
Oligocene
cooling 139, 391, 437
mean annual temperatures 191
sea surface temperatures 57
oligotrophic
foraminifera 141–144, 150, 248
oceans 138, 142
opportunistic foraminiferal taxa 138, 139, 143, 150, 151, 258
orbital effects on climate 7–8, 14, 22, 170
orbital parameters
Cretaceous 280
energy balance equation 5
GCM 24, 37, 278, 407, 409, 410, 429
Ordovician 399–400, 429
Ordovician
appearance of plants 361
Ashgill 389, 401
glaciation 386–414
δ^{18}O depth gradients 243
solar constant 5
organic carbon burial
Devonian Appalachian Basin 363–364
effects on carbon isotopes 97
organic matter in box models 36
orogeny
Acadian 359
Laramide 181–182
Taconic 396, 413

orography
effects on climate 5–7, 9–10, 161–191
in GCMs 14–15, 24
ostracodes, Paleocene–Eocene boundary 142
Otozamites 298, 300, 302
oxygen isotopes 135–136, 148, 258, 353, 388, 390–391
analytical error 58
and GCMs 39–40, 106–114
correction factors 94–95
Cretaceous 242, 247, 251–254, 259, 261–266
Cretaceous latitudinal temperature gradients 267–269
deep water 114–116, 136, 261, 264–266, 288
Devonian 353, 358
diagenetic effects 63–69
Eocene iron oxides 213–223, 228–230
Eocene marine carbonates 79–124, 198
Eocene paleosols 199, 213–214
Eocene sea surface temperature estimates 51–70, 79–131
foraminiferal depth stratification 96–101
fossil mammals 218–219
Holocene core tops 58–59
kinetic effects 63
meteoric water 213
Ordovician 391
paleoaltimetry estimates 175–191
salinity effects 63, 116–122
sea surface *see* sea surface temperature
seawater composition 101–104
temperature calibration 54–55, 93–101
temperature excursions 146, 150–152
transfer function 58
uncertainties 58–70, 104–106
vital effects 60–61
volcanic alteration effects 113
oxygen minimum zone 138
oxygenation
deep water 137, 141, 143, 149, 151, 152
Devonian Appalachian Basin 363–364
relation to carbon cycle 353
ozone
effect on climate 11, 280
GCM mixing ratio 399

Pachypteris 302
Pagiophyllum 302
paleoaltitude 161–191
paleobathymetry, Eocene 82
paleobiogeography of inoceramids 244–246,
    248–251, 260–262
Paleocene
    carbon isotopes 134
    mean annual temperature estimates 220
Paleocene, late
    intermontane lakes 181
    interspecies carbon isotope differences
        67
    paleogeography 88
    upwelling 89
    volcanism 102
    warming 198–199
Paleocene/Eocene boundary
    age of 113, 201–211
    benthic foraminifera 139–144
    Bighorn Basin 199
    event 67
    floras 225–228
    gas hydrate release 11, 82, 429 *see also*
        methane, clathrate release
    mammalian faunas 223–225
    sea level fluctuations 229
    stable isotopes 144–152, 218–219, 222,
        437
    unconformities 111, 113
paleoelevation *see* paleoaltitude
Paleogeographic Atlas Project 301, 305
paleogeography
    Campanian 244, 278–280
    Devonian 355–356
    effect on climate 9–10, 30, 31, 35–38,
        167–171, 277, 387–414
    Gondwanan 310
    Ordovician 387, 394, 397–398,
        401–402, 407–413
paleosalinity *see* oxygen isotopes, salinity
    effects
paleosol *see also* oxygen isotopes; hematite
    carbon dioxide indicator 10, 52
    climate indicator 38
paleotemperature, estimates of *see* oxygen
    isotopes; mean annual temperatures
paleotopography, effect on climate 161–191
palms, climate indicator 177, 189, 277
palynomorphs 326, 339
Panama

floras 198
isthmus closure 9
Pangea
    Central Pangea Mountains 167
    effect on summer warming 428–429
    ocean heat transport 9
    rifting and weathering levels 426
parameterization schemes 6, 14, 23–27, 32,
    36, 83, 311, 407
pCO_2 63, 242, 245, 388, 392, 402–407
peat 321, 325, 328–333, 334–335, 339–341
Penn Yan Formation 369–374
periodically synchronous model *see* models,
    periodically synchronous
Permian
    ice sheets 402
    modeled temperatures 342–344, 430
    paleoclimate 321–344
    vegetation 321–344
permineralized peat *see* peat
Persites argutus 226
phenology 278, 282
phi layer 342
Phoenicopsis 306
phosphate 374–375, 397
phosphorus
    inorganic 368–371, 373–374, 396–397
    organic 368–371, 373
photosynthesis 33, 211, 281, 325, 334, 336,
    342
physiognomy, foliar/leaf *see* leaf
    physiognomy
phytane 367, 372
phytochoria 300
phytodetritus, benthic foraminiferal species
    139, 142, 151
phytogeography, Jurassic 297–316
phytoplankton
    blooms 107
    Devonian 360, 367, 372–376
    producers of dimethyl sulfide gas 35
pingos, Ordovician 389
Pipe Creek Formation 363
Pityophyllum 302
planetary albedo 5, 8, 12–13, 16, 430
planktonic foraminifera
    Abathomphalus mayaroensis 248,
        254–255, 256, 258, 260, 267
    Acarinina 98, 107, 146
    Acarinina mckannai 146, 149
    Acarinina nitida 98

Acarinina primitiva 98
Archaeoglobigerina spp. 258
Chiloguembelina 361
Contusotruncana contusa 254
Contusotruncana spp. 258
Gansserina weidenmayeri 258
Globigerina bulloides 62
Globigerinelloides 258
Globigerinelloides subcarinatus 267
Globigerinoides sacculifer 60, 104
Globotruncana falsostuarti Zone 248
Globotruncana spp. 258
Globotruncanita spp. 258
Heterohelix globulosa 252, 255, 256,
    258, 260, 264–265
Morozovella aragonensis 98–99
Morozovella subbotinae 99
Morozovella velascoensis 98
Muricoglobigerina 98–99
Racemiguembelina fructicosa 257
Radotruncana calcarata 246, 256, 257
Rugoglobigerina 256–258
Subbotina 96–99, 106, 111
Subbotina eocaenica 99
Subbotina patagonica 98–99
Subbotina triangularis 98–99
Subbotina velascoensis 98–99
plant functional types 33
plants, fossil
    Alnus 226–227
    anatomy 211, 321–323, 333–334, 337,
        339, 341–342
    angiosperms 283, 298, 314, 322, 334,
        336
    Antarcticycas 341–342
    Araucaria 285
    Brachyphyllum 302
    Cladophlebis 361
    Cnemidaria 226–227
    Coniopteris 303
    Cornus 226
    corystospermalean seed ferns 340
    corystosperms 340
    Ctenis 302
    cycadophytes 300, 302, 303, 306, 308
    Czekanowskia 306
    Czekanowskiales 298, 300
    Desmiophyllum 302
    Dicroidium 332, 333, 334, 335, 340
    Dictyophyllum 298, 300
    Elatocladus 302

Equisetites 303
'*Eucommia*' *serrata* 226
ferns 339, 341, 343
'*Ficus*' *artocarpoides* 226
fossil forest 325–328, 333–336, 339
Frenelopsis 298
ginkgophytes 297, 302–304, 306, 308
glossopterid seed ferns 339–340, 341,
    343
Glossopteridales 339
Glossopteris 324, 326, 327, 334, 335,
    339
Jurassic 297–316
Kykloxylon 334–335
Leiosphaeridia 360
lycophytes 302
Lygodium 226-227
Merceria 339
mosses 339, 340, 343
Nilssonia 302
Nilssoniocladus 334, 342
Noeggerathiopsis 339
Otozamites 298, 300, 302
Pachypteris 302
Pagiophyllum 302
Paleocene–Eocene 197–230
palynomorphs 326, 339
peat 321, 325, 328–333, 334–335,
    339–341
Permian–Triassic 321–344
Persites argutus 226
Phoenicopsis 306
Pityophyllum 302
Platanus 215
Platycarya pollen 205, 208
Plumsteadia 339
Plumsteadia-type fructifications 339
Podocarpaceae 333, 340
Podocarpus 285
Podozamites 302, 306
Porosia verrucosa 226
Prasinophyceae 360
pteridosperms 297, 306
Pterophyllum 302
Pteruchus 340, 341
Ptilophyllum 300, 302
reproductive organs 325, 339, 343
roots 326, 329, 338–340, 342, 411
seed ferns 340
seeds 339
Sphenobaiera 302

plants, fossil (*continued*)
  sphenophytes 302–303, 306, 341–342
  Sphenopteris 299
  stumps 324–326, 329, 333
  Tasmanites 360
  Todites 303
  trunks 325–327, 331, 333, 336
  Vertebraria 326, 339
  Zamites 302
Platanus 215
Platycarya pollen 205, 208
Pleistocene
cool tropical SSTs 57
Laurentide ice sheet 389
Pliocene cool tropical SSTs 57
Plumsteadia 339
Plumsteadia-type fructifications 339
pocket rot fungi 338, 340
Podocarpaceae 333, 340
Podocarpus 285
Podozamites 302, 306
point-centered quarter method 333
polar regions 5, 6, 32, 169, 213, 242, 396, 436
Polecat Bench 199, 202–204, 206, 208, 209, 215, 217, 219
pollen
  Platycarya 205, 208
  Pteruchus 340, 341
pollen organs (*Pteruchus*) 340, 341
Porosia verrucosa 226
Prasinophyceae 360
precipitation 11, 13, 24, 30, 32, 33, 38, 39, 55, 64, 83–85, 90, 114, 147, 149, 161, 162, 165–167, 169–172, 179, 278, 279, 283, 286, 288, 289, 297, 298, 305, 307, 310–313, 315, 358, 430
  and leaf margins 211, 212
  aridity 13, 39, 134, 166, 167, 171, 181, 183, 188, 190, 191, 312, 322, 323, 326, 334, 343
  isotopic composition of 54, 102, 114, 116, 117, 175–178, 185, 187, 189, 213, 220, 222
precipitation rate 84, 85
preservation model 364
pressure systems 163, 167, 170, 171
primary productivity 132, 138, 143, 151, 246, 262, 354, 367–369, 372
Prince Charles Mountains, Antarctica 339

productivity 35, 36, 38, 132, 133, 135–138, 141–146, 149–152, 367, 375, 376, 377, 397
  continental 211, 297, 309, 325
  model 364–367
Proto-Tethys 396
psychrosphere 139, 141
pteridosperms 297, 306
Pterophyllum 302
Pteruchus (pollen organs) 340, 341
Ptilophyllum 300, 302
pycnocline model 364, 376
pyritization 368, 373 *see also* degree of pyritization

Q see oceanic heat transport
Queen Alexandra Range, Antarctica 329

Racemiguembelina fructicosa 257
Racemiguembelina fructicosa Zone 248, 254, 255
radiative forcing 34, 51, 281, 392, 393, 395, 407, 409, 413, 430
radiolarians 360
Radotruncana calcarata 246, 256–257
Radotruncana calcarata Zone 248, 256, 257
rain shadow 9, 165, 167, 170, 190, 286
rain-out effect 177
rainforest broadleaf trees 282
Rayleigh distillation 177, 213
redox boundary 365, 375, 376
redox conditions 351, 366, 368, 373
reefs 57, 262
reproductive organs, plants 325, 339, 343
respiration 287, 375
respiration, plants 334, 342
reworking 62, 246, 367, 372
Rhinestreet Formation 363
Rhinestreet Shale 359
rings *see* tree rings
Rocky Mountains 161, 163–165, 169, 179, 182, 189–191, 224, 228
roots, fossil 326, 329, 338–340, 342, 411
rotation rate 399, 401, 410
rudist bivalves 57, 262, 266
Rugoglobigerina 256–258
Rugoglobigerina spp. 252, 255, 256, 267
runoff 30, 39, 81, 83–85, 90, 93, 102, 107, 114, 122, 143, 147, 149, 152, 177, 179, 181, 188, 189, 199, 430

saline water masses 87, 90, 241–243, 262, 264, 269, 289, 396
salinity 27, 29–31, 50, 54, 55, 63–65, 79, 81, 83, 90, 92, 93, 97, 103, 105, 113, 118, 119, 122, 145, 152, 182, 264–268, 288, 289, 364
salinity effect 39, 57, 63, 69, 79, 102, 113, 116–124, 188, 243, 247, 264, 266, 269
Salix 336
Santonian 57, 245, 250, 260, 262, 268
scars, in wood 337, 338
Scotland, Dob's Lin 391
sea ice 4, 30–32, 84–85, 134, 396, 405–406, 409, 411, 436
    albedo feedback 12
sea level, Eocene 111
sea surface temperature
    in models 25, 34, 81, 288–289, 398, 433, 436–437
    isotopic estimates of 54–70
seasonal upwelling 61, 89
seasonality 322, 334, 341
seawater, isotopic composition of *see* oxygen isotopes, seawater composition
second law of thermodynamics 6
seed ferns
    corystospermalean 340
    glossopterid 339–341, 343
seeds, fossil 339
sensible heat 163
Shackleton Glacier 327
Shannon Index 247
short term climate variability 60, 62
short-wave radiation 34
shrubs 83, 228, 283
Siberia 323, 336
Siberian larch *see Larix sibirica*
Silurian 36, 387–389, 391–392, 395, 397–402, 407–414
Skaar Ridge, Antarctica 339, 340
small ice cap instability 408, 413
SMOW *see* standard mean ocean water
snow
    accumulation of 8, 9, 32, 84, 176, 177, 179, 189–190, 358, 388, 393–396, 398, 400, 403–409, 411, 413
    δ^{18}O 77, 184–185, 187–189, 213
    effects on albedo 12–13, 85, 170–171, 283, 287, 289, 438

    in models 38, 83, 85, 277–279, 283–284, 286–287, 289, 358, 393–395, 398, 400, 403–409, 411, 413, 429, 438
    line 176–190, 279, 396, 438
    melt 177–179, 181, 184–185, 187–190, 286–289
    snowline/MAT technique 180, 189, 190
soil
    as climate proxies 39, 133, 200–201
    ecosystems and weathering 361, 366, 411, 426
    moisture 162, 170–172, 312–313, 315, 322–323, 342
    nutrients 212
    properties in models 37, 38, 83, 277–278, 280, 283–284, 286, 399, 400–401
    stable isotopic composition 213–214, 222–223, 281, 392, 426
solar constant (S_0) 5, 7, 14, 37, 278, 280
    evolution of 7, 24, 280, 387, 392, 399, 407–408, 426, 430–431
solar radiation 4–7, 10, 12, 13, 162, 179, 280, 283, 287, 387, 392–395, 397, 407–410, 412, 413, 426–429, 436, 438
    variability and distribution 7, 22, 287, 335
solar system 7
Sonyea Group 359
South Atlantic 86–87, 94, 107, 243, 249, 264, 267, 289
Southeast Asia 286
Southern Ocean 80, 87–88, 90, 115, 123, 139, 262, 268
Spain 245, 249, 254
species richness (*S*) 223, 227, 247, 248, 258–259, 260, 266–267 *see also* diversity, species
Sphenobaiera 302
sphenophytes 302–303, 306, 341–342
Sphenopteris 299
SST *see* sea surface temperature
stable isotopes
    carbon *see* carbon isotopes
    osmium 136
    oxygen *see* oxygen isotopes
standard mean ocean water 93, 99, 101–104, 113, 115–116, 119, 175, 177–178, 185, 187, 189, 213
stomatal function 33
storm deposits 38, 358
storm track 165, 169, 170, 190, 279, 300

stratosphere 10, 11, 280
stratospheric clouds 11, 16, 34–35
stromatoporoids 363
stumps, fossil 324–326, 329, 333
styliolinids 360
Subbotina 96–99, 106, 111
Subbotina eocaenica 99
Subbotina patagonica 98–99
Subbotina triangularis 98–99
Subbotina velascoensis 98–99
subthermocline 79–80, 96, 98, 102–103, 106, 108, 111, 114–115, 122
subtropics 6, 81, 169
succulent forbs 282
sulfate aerosols 10
sulfate reduction 365, 373, 375
summergreen broadleaf trees *see* broadleaf evergreen trees
Supertethys 283
surface temperature (T_s) *see* temperature, mean global
surface temperature gradients *see* sea surface temperature; temperature, latitudinal gradients

Tasmanites 360
temperature
    continental interior 4, 16, 80, 181, 189, 198–199, 228, 277, 285–286, 342, 413, 436
    deep ocean 36, 39, 64–65, 80, 111, 114, 122–123, 264, 290, 436
    estimates from isotopes *see* sea surface temperature
    high-latitude 58, 107, 287, 290–291, 313, 335, 342, 388, 401, 405, 435–436
    latitudinal gradient 3, 4, 6, 12, 16, 53–70, 80, 82, 109–110, 114–116, 133–134, 147, 152, 161–162, 242–243, 247, 254, 265, 267–269, 314, 410, 433, 436, 438
    low-latitude 64, 119, 134, 269, 285, 291–292, 312, 435
    mean annual *see* mean annual temperature
    mean global 5, 12, 16, 31, 34–35, 51, 84, 170, 276, 400, 426, 430, 432–435, 438
    mixed-layer 106–107, 123
    oscillations 8, 60, 215, 222, 228–230
    sea surface *see* sea surface temperature

seasonality 9, 31, 51, 60, 110, 112, 123, 189, 311–312, 322–325, 331–334, 358, 393, 403, 405–408, 429
    stratosphere 11
    variation with altitude *see* lapse rate
    vertical gradients in ocean 6, 61, 90, 96–98, 106–120, 243
terrestrial ecosystems 32, 33, 40, 230, 281, 282, 287, 291, 361
territorial floristic units 299
Tethys Seaway 82, 86–89, 93, 121, 123, 134, 144, 149, 248, 286, 289, 396
thermal expansion coefficient 290
thermal inertia 32, 37, 387, 393, 398, 408
thermohaline circulation 16, 27, 28, 30, 32, 36, 53, 290, 396, 397, 434, 435, 438
Tibet 9
Tibetan Plateau 9, 86, 165, 167, 170, 173
Tiffanian 206, 217, 220
tillite 358, 400
TOC *see* total organic carbon
Todites 303
topography 30, 37, 83, 165, 170–172, 180, 279, 286, 312, 342, 393, 396, 402, 410, 434
total organic carbon 367, 371–376, 397
trace fossils 358, 389
trace gas 34, 35, 280, 429
Transantarctic Mountains 321, 325–326, 328–330, 333, 339–340, 342
transpiration 284, 287, 288, 334, 344, 436
tree architecture 335, 354
tree rings 183, 321–328, 330–331, 333, 335–339, 341–342
Triassic 39, 321–344
trona 183, 188
tropics 4, 5, 9, 12, 14, 51, 54, 55, 61, 63, 69, 85, 147, 167, 242, 312, 313, 432, 435, 436
    fauna 57, 61, 66, 67, 69, 358
    easterlies 167, 169
    rainfall 169, 313
    rainforest 13, 57, 282, 313
    regions 6, 9, 54, 84, 213, 286, 291, 292, 309, 312, 313
    sea surface salinity 63, 265–266
    sea surface temperature 34, 51–55, 57, 58, 65, 66, 68–70, 80, 134, 198, 285, 291, 292, 433
    temperature 34, 64, 68, 134, 262, 265, 266, 268, 269, 391, 435

vegetation 211, 225, 227–228, 277, 334, 335, 341
waters 85, 116, 141, 143, 265
troposphere 10, 12, 32, 279, 280
truncation wave number 24
trunks, fossil 325–327, 331, 333, 336
tuff 184, 185, 203–204, 209, 218
Tully Formation 359
Tully Limestone 356
tundra 33, 83, 170, 278, 283–287, 291, 309, 310

U-shaped valleys 389
UGAMP *see* Universities Global Atmospheric Modelling Programme
Uinta Mountains 181
unconformity 144, 217, 254, 260, 263, 392
    Taghanic Unconformity 359
    Windom–Geneseo Unconformity 373, 375
Union Springs Formation 359, 363, 364
Universities Global Atmospheric Modelling Programme 409
upwelling 57, 61, 89, 137, 142–143, 149, 152, 172, 288, 356, 365, 397
Ural Mountains 177
USA 37, 165, 177, 179, 188, 228, 266, 326 *see also* North America
Utah 181, 182, 184

varved deposits 334
vascular cambium 322, 331, 337–338
vegetation 13–14, 40, 83, 133, 162, 277–278, 298, 299–300, 303–304, 307, 310, 315, 335–336, 436–437
    broadleaf evergreen trees 211, 226, 282, 286–287, 302, 308, 333–336, 342–343
    broadleaf herbaceous 283
    broadleaf trees 83
    effect on climate 26, 32–35, 170–171, 281–288, 291, 343, 394, 411, 436–437
    effects on albedo 13, 170, 190, 394, 411
    forb–fern prairie 283, 285
    forbs 282–283
    grassland 13, 170, 278, 282
    groundcover 13–14, 53, 83, 410
    macrophyllous 302–304, 306, 308
    needleleaf-deciduous 83
    needleleaf-evergreen 83, 282, 283, 286

physiognomy 38, 40, 173, 175, 180, 266, 298, 301, 307, 314, 322, 335
    tundra 83
Vertebraria 339
vertebrates, fossil
    Bunophorus 204–206, 208, 220, 224
    conodonts 418
    Ectocion 208, 224
    Haplomylus 208, 224
    Heptodon 204, 205, 209, 224
    Lambdotherium 204, 209, 224
vesicles 173
Victoria Land, Antarctica 321, 327, 328, 339, 340
volcanism 10, 102, 134, 173, 353, 396, 413
    activity 7, 10
    ash 111, 113, 201–203
    isotopic alteration effects 113
    outgassing 10, 392, 396
Volgian 311, 314

warm saline deep water 81, 264, 288, 291, 435
    see also ocean circulation, halothermal
Wasatchian 199–201, 207–208, 217, 218, 220, 222, 223–230
    Wa0 fauna 199, 207–208, 217–218, 222, 223, 226
Washakie Basin 190
water column stratification 248, 267, 354, 364–366
water stress 211, 302, 323, 327, 342
water vapor 6, 11–12, 32, 33, 34, 35, 55, 85, 102, 105, 114, 117, 280, 281, 283, 286, 288, 291
    transport 105, 117
waves
    stationary 165
    transient 165
    troughs 165
weathering 135–136, 143, 331, 392, 411, 426, 438
    chemical 133, 171, 199, 361, 392
    continental 101, 136
    hydrothermal alteration 102
    submarine 101
WSDW *see* warm saline deep water
West Falls Group 359
white rot fungi 338, 340
Wilkins Peak Member 184–185, 188
Williston Basin 228

Willwood Formation 199–201, 207, 214, 217
Wind River Basin 190, 198
wind stress 28, 79, 83, 85–86, 288
Windom Shale member 369, 373
wood 322–339, 341–342
    development 322–325
    fossil 321, 325–339, 341–342
    wounds 337

Wyoming 181–184, 189–191, 197–199, 215,
    228

xylem 322

Zamites 302
zooplankton 107, 132